KU-166-261

HANDBOOK OF OFFSHORE ENGINEERING

Volume II

Elsevier Internet Homepage – http://www.elsevier.com
Consult the Elsevier homepage for full catalogue information on all books, major reference works, journals, electronic products and services.

Elsevier Titles of Related Interest in this series

Ocean Engineering Series
The Elsevier *Ocean Engineering Book Series* edited by Rameswar Bhattacharyya and Michael McCormick (both at the U.S. Naval Academy) addresses the need for a comprehensive and applied source of literature relevant to both researchers and practitioners alike. For a complete listing of books in this series visit:
http://www.elsevier.com/locate/OEBook

L.K. KOBYLINSKI, S. KASTNER, V.L. BELENKY, N.B. SEVASTIANOV
Stability and Safety of Ships (2 Volume Set)
2003, ISBN: 0-08-044355-9

M. OCHI
Hurricane-Generated Seas
2003, ISBN: 0-08-044312-5

A. PILLAY, J. WANG
Technology & Safety of Marine Systems
2003, ISBN: 0-08-044148-3

J. BROOKE, N. BOSE
Wave Energy Conversion
2003, ISBN: 0-08-044212-9

Related Journals:
Elsevier publishes a wide-ranging portfolio of high quality research journals, encompassing the materials science field. A sample journal issue is available online by visiting the Elsevier web site (details at the top of this page). Leading titles include:

Ocean Engineering
Coastal Engineering
Applied Ocean Research
Marine Structures

All journals are available online via ScienceDirect: www.sciencedirect.com

To Contact the Publisher
Elsevier welcomes enquiries concerning publishing proposals: books, journal special issues, conference proceedings, etc. All formats and media can be considered. Should you have a publishing proposal you wish to discuss, please contact, without obligation, the publisher responsible for Elsevier's Composites and Ceramics programme:

Nick Pinfield
Publisher
Elsevier Ltd
The Boulevard, Langford Lane
Kidlington, Oxford
OX5 1GB, UK

Phone: +44 1865 84 3352
Fax: +44 1865 84 3700
E-mail: nick.pinfield@elsevier.com

General enquiries, including placing orders, should be directed to Elsevier's Regional Sales Offices – please access the Elsevier homepage for full contact details (homepage details at the top of this page).

HANDBOOK OF OFFSHORE ENGINEERING

SUBRATA K. CHAKRABARTI

Offshore Structure Analysis, Inc.
Plainfield, Illinois, USA

Volume II

2005

Amsterdam – Boston – Heidelberg – London – New York – Oxford
Paris – San Diego – San Francisco – Singapore – Sydney – Tokyo

Elsevier
Linacre House, Jordan Hill, Oxford, OX2 8DP, UK
Radarweg 29, PO Box 211, 1000 AE Amsterdam, The Netherlands

First edition 2005
Reprinted 2005, 2006 (twice), 2008

Copyright © 2005 Elsevier Ltd. All rights reserved

No part of this publication may be reproduced, stored in a retrieval system or transmitted in any form or by any means electronic, mechanical, photocopying, recording or otherwise without the prior written permission of the publisher

Permissions may be sought directly from Elsevier's Science & Technology Rights Department in Oxford, UK: phone (+44) (0) 1865 843830; fax (+44) (0) 1865 853333; email: permissions@elsevier.com. Alternatively you can submit your request online by visiting the Elsevier web site at http://elsevier.com/locate/permissions, and selecting *Obtaining permission to use Elsevier material*

Notice
No responsibility is assumed by the publisher for any injury and/or damage to persons or property as a matter of products liability, negligence or otherwise, or from any use or operation of any methods, products, instructions or ideas contained in the material herein. Because of rapid advances in the medical sciences, in particular, independent verification of diagnoses and drug dosages should be made

British Library Cataloguing in Publication Data
A catalogue record for this book is available from the British Library

Library of Congress Cataloging-in-Publication Data
A catalog record for this book is available from the Library of Congress

ISBN: 978-0-08-044568-7 (vol 1)
ISBN: 978-0-08-044569-4 (vol 2)
ISBN: 978-0-08-044381-2 (set comprising of vols 1 & 2)

For information on all Elsevier publications
visit our website at books.elsevier.com

Printed and bound in *Great Britain*

08 09 10 12 11 10 9 8 7 6 5

Working together to grow
libraries in developing countries

www.elsevier.com | www.bookaid.org | www.sabre.org

ELSEVIER BOOK AID International Sabre Foundation

PREFACE

Due to the rapid growth of the offshore field, particularly in the exploration and development of offshore oil and gas fields in deep waters of the oceans, the science and engineering in this area is seeing a phenomenal advancement. This advanced knowledge is not readily available for use by the practitioners in the field in a single reference.

Tremendous strides have been made in the last decades in the advancement of offshore exploration and production of minerals. This has given rise to developments of new concepts and structures and material for application in the deep oceans. This has generated an obvious need of a reference book providing the state-of-the art in offshore engineering.

This handbook is an attempt to fill this gap. It covers the important aspects of offshore structure design, installation and operation. The book covers the basic background material and its application in offshore engineering. Particular emphasis is placed in the application of the theory to practical problems. It includes the practical aspects of the offshore structures with handy design guides, simple description of the various components of the offshore engineering and their functions.

One of the unique strengths of the book is the impressive and encompassing presentation of current functional and operational offshore development for all those involved with offshore structures. It is tailored as a reference book for the practicing engineers, and should serve as a handy reference book for the design engineers and consultant involved with offshore engineering and the design of offshore structures. This book emphasizes the practical aspects rather than the theoretical treatments needed in the research in the field of offshore engineering. In particular, it describes the dos and don'ts of all aspects of offshore structures. Much hands-on experience has been incorporated in the write up and contents of the book. Simple formulas and guidelines are provided throughout the book. Detailed design calculations, discussion of software development, and the background mathematics has been purposely left out. The book is not intended to provide detailed design methods, which should be used in conjunction with the knowledge and guidelines included in the book. This does not mean that they are not necessary for the design of offshore structures. Typically, the advanced formulations are handled by specialized software. The primary purpose of the book is to provide the important practical aspects of offshore engineering without going into the nitty gritty of the actual detailed design. Long derivations or mathematical treatments are avoided. Where necessary, formulas are stated in simple terms for easy calculations. Illustrations are provided in these cases. Information is provided in handy reference tables and design charts. Examples are provided to show how the theory outlined in the book is applied in the design of structures. Many examples are borrowed from the deep-water offshore structures of interest today including their components, and material that completes the system.

Contents of the handbook include the following chapters:

The book is a collective effort of many technical specialists. Each chapter is written by one or more invited world-renowned experts on the basis of their long-time practical experience in the offshore field. The sixteen chapters, contributed by internationally recognized offshore experts provide invaluable insights on the recent advances and present state-of-knowledge on offshore developments. Attempts were made to choose the people, who have been in the trenches, to write these chapters. They know what it takes to get a structure from the drawing board to the site doing its job for which it is designed. They work everyday on these structures with the design engineers, operations engineers and construction people and make sure that the job is done right.

Chapter 1 introduces the historical development of offshore structures in the exploration and production of petroleum reservoirs below the seafloor. It covers both the earlier offshore structures that have been installed in shallow and intermediate water depths as well as those for deep-water development and proposed as ultra-deep water structures. A short description of these structures and their applications are discussed.

Chapter 2 describes novel structures and their process of development to meet certain requirements of an offshore field. Several examples given for these structures are operating in offshore fields today. A few others are concepts in various stages of their developments. The main purpose of this chapter is to lay down a logical step that one should follow in developing a structural concept for a particular need and a set of prescribed requirements.

The ocean environment is the subject of chapter 3. It describes the environment that may be expected in various parts of the world and their properties. Formulas in describing their magnitudes are provided where appropriate so that the effect of these environments on the structure may be evaluated. The magnitudes of environment in various parts of the world are discussed. They should help the designer in choosing the appropriate metocean conditions that should be used for the structure development.

Chapter 4 provides a generic description of how to compute loads on an offshore structure and how the structure responds to these loads. Basic formulas have been stated for easy references whenever specific needs arise throughout this handbook. Therefore, this chapter may be consulted during the review of specific structures covered in the handbook. References are made regarding the design guidelines of various certifying agencies.

Chapter 5 deals with a statistical design approach incorporating the random nature of environment. Three design approaches are described that include the design wave, design storm and long-term design. Several examples have been given to explain these approaches.

The design of fixed offshore structures is described in Chapter 6. The procedure follows a design cycle for the fixed structure and include different types of structure design including tubular joints and fatigue design.

Chapter 7 discusses the design of floating structures, in particular those used in offshore oil drilling and production. Both permanent and mobile platforms have been discussed. The design areas of floaters include weight control and stability and dynamic loads on as well as fatigue for equipment, risers, mooring and the hull itself. The effect of large currents in the deepwater Gulf of Mexico, high seas and strong currents in the North Atlantic, and long period swells in West Africa are considered in the design development. Installation of the platforms, mooring and decks in deep water present new challenges.

Floating offshore vessels have fit-for-purpose mooring systems. The mooring system selection, and design are the subject of Chapter 8. The mooring system consists of freely hanging lines connecting the surface platform to anchors, or piles, on the seabed, positioned some distance from the platform.

Chapter 9 provides a description of the analysis procedures used to support the operation of drilling and production risers in floating vessels. The offshore industry depends on these procedures to assure the integrity of drilling and production risers. The description, selection and design of these risers are described in the chapter.

The specific considerations that should be given in the design of a deck structure is described in Chapter 10. The areas and equipment required for deck and the spacing are discussed. The effect of the environment on the deck design is addressed. The control and safety requirements, including fuel and ignition sources, firewall and fire equipment are given.

The objective of chapter 11 is to guide the offshore pipeline engineer during the design process. The aspects of offshore pipeline design that are discussed include a design basis, route selection, sizing the pipe diameter, and wall thickness, on-bottom pipeline stability, bottom roughness analysis, external corrosion protection, crossing design and construction feasibility.

Chapter 12 is focused on people and their organizations and how to design offshore structures to achieve desirable reliability in these aspects. The objective of this chapter is to provide engineers design-oriented guidelines to help develop success in design of offshore structures. Application of these guidelines are illustrated with a couple of practical examples.

The scale model testing is the subject of Chapter 13. This chapter describes the need, the modeling background and the method of physical testing of offshore structures in a small-scale model. The physical modeling involves design and construction of scale model, generation of environment in an appropriate facility, measuring responses of the model subjected to the scaled environment and scaling up of the measured responses to the design values. These aspects are discussed here.

Installation, foundation, load-out and transportation are covered in Chapter 14. Installation methods of the following sub-structures are covered: Jackets; Jack-ups; Compliant towers and Gravity base structures. Different types of foundations and their unique methods of installation are discussed. The phase of transferring the completed structure onto the deck of a cargo vessel and its journey to the site, referred to as the load-out and transportation operation, and their types are described.

Chapter 15 reviews the important materials for offshore application and their corrosion issues. It discusses the key factors that affect materials selection and design. The chapter includes performance data and specifications for materials commonly used for offshore developments. These materials include carbon steel, corrosion resistant alloys, elastomers and composites. In addition the chapter discusses key design issues such as fracture, fatigue, corrosion control and welding.

Chapter 16 provides an overview of the geophysical and geotechnical techniques and solutions available for investigating the soils and rocks that lay beneath the seabed. A project's successful outcome depends on securing the services of highly competent contractors and technical advisors. What is achievable is governed by a combination of factors, such as geology, water depth, environment and vessel capabilities. The discussions are transcribed without recourse to complex science, mathematics or lengthy descriptions of complicated procedures.

Because of the practical nature of the examples used in the handbook, many of which came from past experiences in different offshore locations of the world, it was not possible to use a consistent set of engineering units. Therefore, the English and metric units are interchangeably used throughout the book. Dual units are included as far as practical, especially in the beginning chapters. A conversion table is included in the handbook for those who are more familiar with and prefer to use one or the other unit system.

This handbook should have wide applications in offshore engineering. People in the following disciplines will be benefited from this book: Offshore Structure designers and fabricators; Offshore Field Engineers; Operators of rigs and offshore structures; Consulting Engineers; Undergraduate & Graduate Students; Faculty Members in Ocean/Offshore Eng. & Naval Architectural Depts.; University libraries; Offshore industry personnel; Design firm personnel.

Subrata Chakrabarti
Technical Editor

ABBREVIATIONS

List of Acronyms

ABS	American Bureau of Shipping
ABL	Above Base Line
API	American Petroleum Institute
BOP	Blowout Preventor
CFD	Computational Fluid Dynamics
CG	Center of Gravity
CRA	Corrosion Resistant Alloys
CVAR	Compliant Vertical Access Riser
DDCV	Deep Draft Caisson Vessel
DNV	Det Norske Veritas
DTU	Dry Tree Unit
EP	Equivalent Pipe
FE	Finite Element
FEA	Finite Element Analysis
FPDSO	Floating Production, Drilling, Storage and Offloading System
FPS	Floating Production System
FPSO	Floating Production Storage and Offloading
FSO	Floating Storage and Offloading
Fr	Froude number
GOM	Gulf of Mexico
HLV	Heavy Lift Vessel
IACS	International Association of Classification Societies
IMO	International Maritime Organization
IRM	Inspection, Repair and Maintenance
ISSC	International Ship Structures Congress
JIP	Joint Industry Project
JONSWAP	Joint North Sea Wave Project
KC	Keulegan–Carpenter Number
ksi	Kips per Square Inch
LF	Low Frequency
LRFD	Load and Resistance Factor Design
MODU	Mobile Offshore Drilling Unit
MPa	MegaPascals (N/mm^2)
NDP	Norwegian Deepwater Programme
NDT	Non-Destructive Testing
PDF	Probability Density Function

PIP	Pipe-In-Pipe
PM	Pierson–Moskowitz
psi	Pounds per Square Inch
QA/QC	Quality Assurance/Quality Control
Re	Reynolds Number
RFC	Rainflow Counting
SCF	Stress Concentration Factor
SCR	Steel Catenary Riser
SLC	Sustained Load Cracking
SPM	Single Point Mooring
SSCV	Semi-Submersible Crane Vessel
St	Strouhal Number
SWL	Still Water Level
TDP	Touch-Down Point
TDZ	Touch-Down Zone
TLP	Tension-Leg Platform
TSJ	Tapered Stress Joint
TTR	Top Tensioned Riser
UKCS	United Kingdom Continental Shelf
UOE	Pipe formed from plate, via a U-shape, then an O-shape, then Expanded
UTG	Upstream Technology Group
VIV	Vortex-Induced Vibration

CONVERSION FACTORS

	English	Metric
Length	1 in	25.4 mm
	1 ft	0.3048 m
	1 lbf	4.448 N
	1 lbm	0.4536 kg
	1 knot	0.5144 m/s
	1 mile	1.609 km
Area	1 ft^2	0.0929 m^2
Volume	1 ft^3	0.0283 m^3
	1 gallon	0.003785 m^3
Velocity	1 ft/s	30.48 cm/s
	1 mile/hr	1.609 km/hr
Acceleration	1 ft/s^2	30.48 cm/s^2
Mass/Force	1 ton (long)	1.016 m. ton
Density	1 lb/ft^3	16.0185 kg/m^3
Pressure	1 psi	6894.76 pascals
Moment	1 ft-lb	1.3558 N-m
Mass Moment of Inertia	1 lbm-ft	20.0421 kg-m^2

LIST OF CONTRIBUTORS

Chapter 1
Historical Development of Offshore Structures
Subrata Chakrabarti, Offshore Structure Analysis, Inc., Plainfield, IL, USA,
Cuneyt Capanoglu, I.D.E.A.S., Inc., San Francisco, CA, USA, and
John Halkyard, Technip Offshore, Inc., Houston, TX, USA

Chapter 2
Novel and Marginal Offshore Structures
Cuneyt Capanoglu, I.D.E.A.S., Inc., San Francisco, CA, USA

Chapter 3
Ocean Environment
Subrata Chakrabarti, Offshore Structure Analysis, Inc., Plainfield, IL, USA

Chapter 4
Loads and Responses
Subrata Chakrabarti, Offshore Structure Analysis, Inc., Plainfield, IL, USA

Chapter 5
Probabilistic Design of Offshore Structure
Arvid Naess and Torgeir Moan, Norwegian University of Science and Technology, Trondheim, NORWAY

Chapter 6
Fixed Offshore Platform Design
Demir Karsan, Paragon Engineering Services Inc., Houston, TX, USA
Vissa Rammohan, Stress Offshore, Inc., Houston, TX, USA (Contributed to the Jackup section)

Chapter 7
Floating Offshore Platform Design
John Halkyard, Technip Offshore, Inc., Houston, TX, USA
John Filson, Consultant, Gig Harbor, WA, USA (Contributed to the Semi, TLP and Hull Structure sections)
(assisted by Krish Thiagarajan, The University of Western Australia, Perth, Australia on Static Stability)

Chapter 8
Mooring, Cables and Anchoring
David T. Brown, BPP Technical Services Ltd., London, UK, and Houston, TX, USA

Chapter 9
Drilling and Production Risers
James Brekke, GlobalSantaFe Corporation, Houston, TX, USA (Drilling section)
Subrata Chakrabarti, Offshore Structure Analysis, Inc., Plainfield, IL, USA (Production section)
John Halkyard, Technip Offshore, Inc., Houston, TX, USA (Contributed to the Top Tension Risers)
Thanos Moros, Howard Cook, BP America, Houston, TX, USA (Contributed to the Steel Catenary Risers) and David Rypien, Technip Offshore, Inc., Houston, TX, USA (Contributed to the Materials Selection)

Chapter 10
Topside Facilities Layout
Kenneth E. Arnold and Demir Karsan, Paragon Engineering Services Inc., Houston, TX, USA
Subrata Chakrabarti, Offshore Structure Analysis, Inc., Plainfield, IL, USA

Chapter 11
Pipeline Design
Andre C. Nogueira and Dave S. McKeehan, INTEC Engineering, Houston, TX, USA

Chapter 12
Design for Reliability: Human and Organizational Factors
Robert G. Bea, University of California, Berkeley, CA, USA

Chapter 13
Physical Modeling of Offshore Structures
Subrata Chakrabarti, Offshore Structure Analysis, Inc., Plainfield, IL, USA

Chapter 14
Offshore Installation
Bader Diab and Naji Tahan, Noble Denton Consultants, Inc., Houston, TX, USA

Chapter 15
Materials for Offshore Applications
Mamdouh M. Salama, ConocoPhillips Inc., Houston, TX, USA

Chapter 16
Geophysical and Geotechnical Design
Jean M.E. Audibert and J. Huang, Fugro-McClelland Marine Geosciences, Inc., Houston, TX, USA

TABLE OF CONTENTS

Handbook of Offshore Engineering
S. Chakrabarti (Ed.)
© 2005 Elsevier Ltd. All rights reserved.

Chapter 8

Mooring Systems

David T. Brown
BPP Technical Services Ltd., London, UK

8.1 Introduction

It is essential that floating offshore vessels have fit-for-purpose mooring systems. The mooring system consists of freely hanging lines connecting the surface platform to anchors, or piles, on the seabed, positioned at some distance from the platform. The mooring lines are laid out, often symmetrically in plan view, around the vessel.

Steel-linked chain and wire rope have conventionally been used for mooring floating platforms. Each of the lines forms a catenary shape, relying on an increase or decrease in line tension as it lifts off or settles on the seabed, to produce a restoring force as the surface platform is displaced by the environment. A spread of mooring lines thus generates a non-linear restoring force to provide the station-keeping function. The force increases with vessel horizontal offset and balances quasi-steady environmental loads on the surface platform. The equivalent restoring stiffness provided by the mooring is generally too small to influence wave frequency motions of the vessel significantly, although excitation by low-frequency drift forces can induce dynamic magnification in the platform horizontal motions and lead to high peak line tensions. The longitudinal and transverse motions of the mooring lines themselves can also influence the vessel response through line dynamics.

With the requirement to operate in increasing water depths, the suspended weight of mooring lines becomes a prohibitive factor. In particular, steel chains become less attractive at great water depths. Recently, advances in taut synthetic fibre rope technology have been achieved offering alternatives for deep-water mooring. Mooring systems using taut fibre ropes have been designed and installed to reduce mooring line length, mean- and low-frequency platform offsets, fairlead tension and thus the total mooring cost. To date however, limited experience has been gained in their extended use offshore when compared to the traditional catenary moorings.

Mooring system design is a trade-off between making the system compliant enough to avoid excessive forces on the platform, and making it stiff enough to avoid difficulties, such as damage to drilling or production risers, caused by excessive offsets. This is relatively easy to achieve for moderate water depths, but becomes more difficult as the water depth increases. There are also difficulties in shallow water. Increasingly integrated mooring/riser system design methods are being used to optimise the system components to ensure lifetime system integrity.

In the past, the majority of moorings for FPS were passive systems. However, more recently, moorings are used for station-keeping in conjunction with the thruster dynamic positioning systems. These help to reduce loads in the mooring by turning the vessel when necessary, or reducing quasi-static offsets.

Monohulls and semi-submersibles have traditionally been moored with spread catenary systems, the vessel connections being at various locations on the hull. This results in the heading of the vessel being essentially fixed. In some situations this can result in large loads on the mooring system caused by excessive offsets caused by the environment. To overcome this disadvantage, single-point moorings (SPM) have been developed in that the lines attach to the vessel at a single connection point on the vessel longitudinal centre line. The vessel is then free to weathervane and hence reduce environmental loading caused by wind, current and waves.

Since the installation of the first SPM in the Arabian Gulf in 1964, a number of these units are now in use. A typical early facility consisted of a buoy that serves as a mooring terminal. It is attached to the sea floor either by catenary lines, taut mooring lines or a rigid column. The vessel is moored to the buoy either by synthetic hawsers or by a rigid A-frame yoke. Turntable and fluid swivels on the buoy allow the vessel to weathervane, reducing the mooring loads.

Although the SPM has a number of good design features, the system involves many complex components and is subjected to a number of limitations. More recently, turret mooring systems for monohull floating production and storage vessels (fig. 8.1) have been developed that are considered to be more economic and reliable than SPMs, and are widely used today. The turret can either be external or internal. An internal turret is generally located in the forepeak structure of the vessel, though a number of turrets have in the past been positioned nearer amidships. Mooring lines connect the turret to the seabed.

In order to further reduce the environmental loading on the mooring system from the surface vessel in extreme conditions, disconnectable turret mooring systems have also been developed. Here the connected system is designed to withstand a less harsh ocean environment, and to be disconnected whenever the sea state becomes too severe such as in typhoon areas.

In this section, the fundamentals of mooring systems are covered, the influence of the relevant combinations of environmental loading is discussed and the mooring system design is considered. Also included is information on mooring hardware, including turrets used on weather-vaning floating production systems, model-testing procedures and in certification issues. There are numerous other sources of information on mooring systems, see for example CMPT (1998).

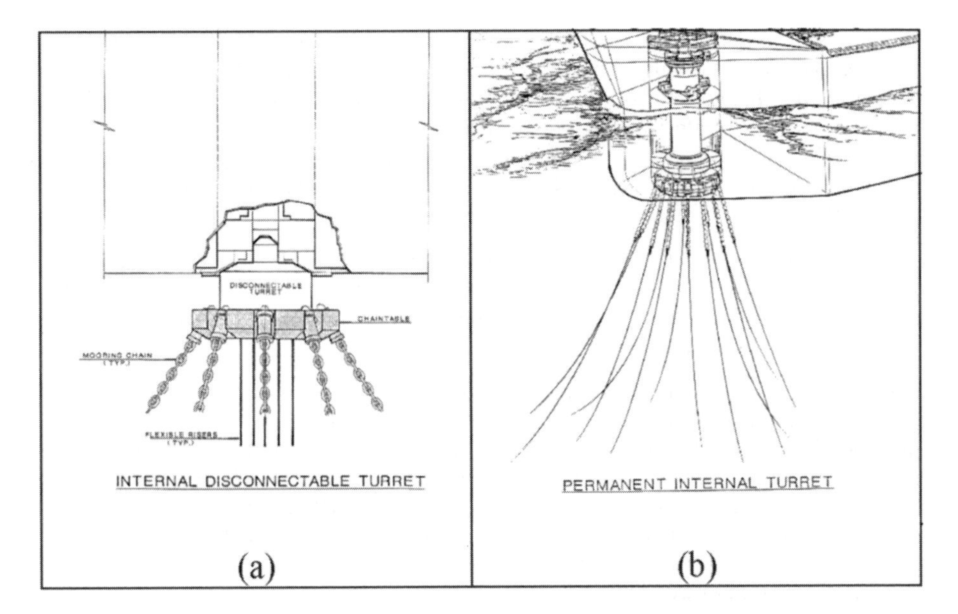

Figure 8.1 Turret moorings. (a) Disconnectible and (b) Permanent

8.2 Requirements

Functional requirements for the mooring system include:

1. offset limitations
2. lifetime before replacement
3. installability
4. positioning ability

These requirements are determined by the function of the floater. MODUs are held to less restrictive standards than "permanent" mooring systems, referring to production platforms. Table 8.1 lists the principal differences in these requirements.

8.3 Fundamentals

It is instructive to review the basic mechanics of a mooring line in order to understand its performance characteristics with respect to station-keeping. The traditional wire or chain catenary lines are considered first, followed by taut moorings of synthetic fibre.

8.3.1 Catenary Lines

Figure 8.2 shows a catenary mooring line deployed from point A on the submerged hull of a floating vessel to an anchor at B on the seabed. Note that part of the line between A and

Table 8.1 Comparison of typical MODU and FPS mooring requirements

MODU	Floating Production
Design for 50-yr return period event Anchors may fail in larger events	Design for 100-yr return period events
Risers disconnected in storm	Risers remain connected in storm
Slack moorings in storm events to reduce line tensions	Moorings are usually not slacked because of risk to risers, and lack of marine operators on board
Components designed for < 10 yr life	Components designed for > 10 yr life
Fatigue analysis not required	Fatigue analysis required
Line dynamics analysis not required	Line dynamics analysis required
Missing line load case not required	Missing line load case required

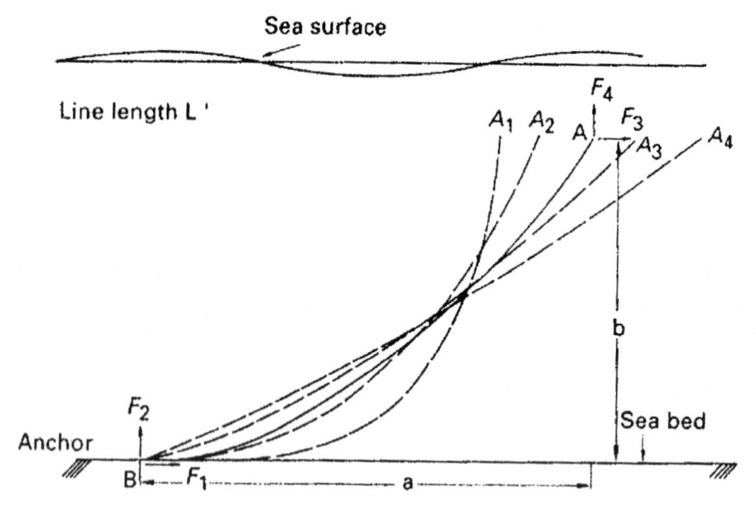

Figure 8.2 Catenary mooring line

B is resting on the seabed and that the horizontal dimension, a, is usually 5–20 times larger than the vertical dimension, b. As the line mounting point on the vessel is shifted horizontally from point A_1, through A_2, A_3, A_4, the catenary line laying on the seabed varies from a significant length at A_1, to none at A_4. From a static point of view, the cable tension in the vicinity of points A is due to the total weight in sea water of the suspended line length. The progressive effect of line lift-off from the seabed due to the horizontal vessel movement from A_1 to A_4 increases line tension in the vicinity of points A. This feature, coupled with the simultaneous decrease in line angle to the horizontal, causes the horizontal restoring force on the vessel to increase with vessel offset in a non-linear manner.

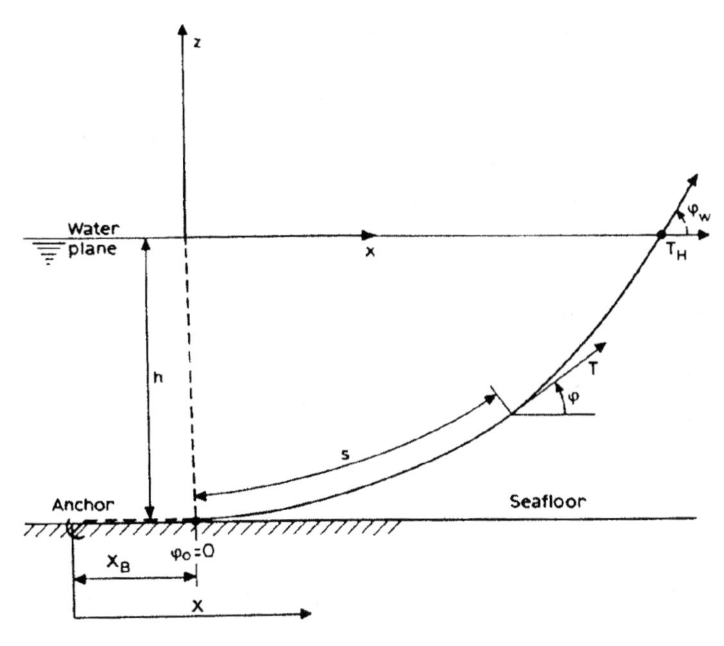

Figure 8.3 Cable line with symbols

This behaviour can be described by the catenary equations that can be used to derive line tensions and shape for any single line of a mooring pattern. The equations are developed using a mooring line as shown in fig. 8.3. In the development that follows, a horizontal seabed is assumed and the bending stiffness effects are ignored. The latter is acceptable for wire with small curvatures and generally a good approximation for chain. It is necessary also to ignore line dynamics at this stage.

A single line element is shown in fig. 8.4. The term w represents the constant submerged line weight per unit length, T is line tension, A the cross-sectional area and E the elastic modulus. The mean hydrodynamic forces on the element are given by D and F per unit length.

Inspecting fig. 8.4 and considering in-line and transverse forcing gives:

$$\mathrm{d}T - \rho g A \mathrm{d}z = \left[w \sin\phi - F\left(\frac{T}{EA}\right) \right] \mathrm{d}s \tag{8.1}$$

$$T\mathrm{d}\phi - \rho g A z \mathrm{d}\phi = \left[w \cos\phi + D\left(1 + \frac{T}{EA}\right) \right] \mathrm{d}s \tag{8.2}$$

Ignoring forces F and D together with elasticity allows simplification of the equations, though it is noted that elastic stretch can be very important and needs to be considered when lines become tight or for a large suspended line weight (large w or deep waters).

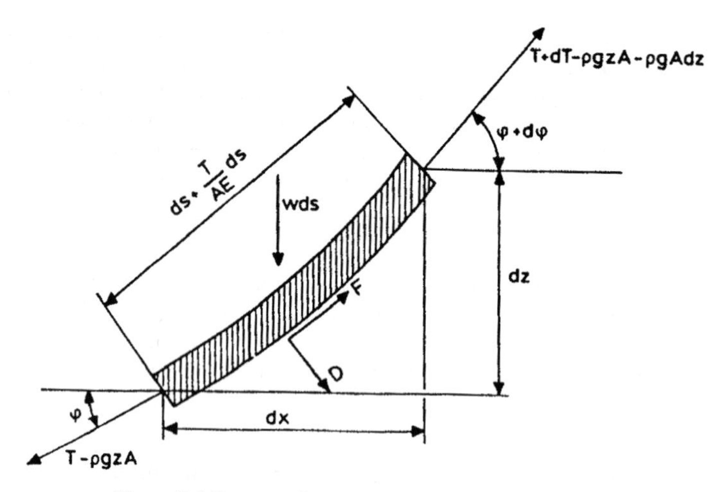

Figure 8.4 Forces acting on an element of an anchor line

With the above assumptions we can obtain the suspended line length s and vertical dimension h as:

$$s = \left(\frac{T_H}{w}\right)\sinh\left(\frac{wx}{T_H}\right) \tag{8.3}$$

$$h = \left(\frac{T_H}{w}\right)\left[\cosh\left(\frac{wx}{T_H}\right) - 1\right] \tag{8.4}$$

giving the tension in the line at the top, written in terms of the catenary length s and depth d as:

$$T = \frac{w(s^2 + d^2)}{2\mathrm{d}} \tag{8.5}$$

The vertical component of line tension at the top end becomes:

$$Tz = ws \tag{8.6}$$

The horizontal component of tension is constant along the line and is given by:

$$T_H = T\cos\phi_w \tag{8.7}$$

It is noted that the above analysis assumes that the line is horizontal at the lower end replicating the case where a gravity anchor with no uplift is used.

A typical mooring analysis requires summation of the effects of up to 16 or more lines with the surface vessel position co-ordinates near the water plane introducing three further variables. The complexity of this calculation makes it suitable for implementing within computer software.

For mooring lines laying partially on the seabed, the analysis is modified using an iteration procedure, so that additional increments of line are progressively laid on the seabed until the suspended line is in equilibrium. Furthermore, in many situations, multi-element lines made up of varying lengths and physical properties are used to increase the line restoring force. Such lines may be analysed in a similar manner, where the analysis is performed on each cable element, and the imbalance in force at the connection points between elements is used to establish displacements through which these points must be moved to obtain equilibrium.

The behaviour of the overall system can be assessed in simple terms by performing a static design of the catenary spread. This is described in Section 8.5.2, but it is noted that this ignores the complicating influence of line dynamics that are described in Section 8.4. The analysis is carried out using the fundamental equations derived above.

8.3.2 Synthetic Lines

For deep-water applications, synthetic fibre lines can have significant advantages over a catenary chain or wire because they are considerably lighter, very flexible and can absorb imposed dynamic motions through extension without causing an excessive dynamic tension. Additional advantages include the fact that there is reduced line length and seabed footprint, as depicted in fig. 8.5, generally reduced mean- and low-frequency platform offsets, lower line tensions at the fairlead and smaller vertical load on the vessel. This reduction in vertical load can be important as it effectively increases the vessel useful payload.

The disadvantages in using synthetics are that their material and mechanical properties are more complex and not as well understood as the traditional rope. This leads to over-conservative designs that strip them of some of their advantages. Furthermore, there is little in-service experience of these lines. In marine applications this has led to synthetic ropes subject to dynamic loads being designed with very large factors of safety.

Section 8.5.5 discusses the mooring system design using synthetic lines in more detail. Detailed mathematical models for synthetic lines are not developed here, but are

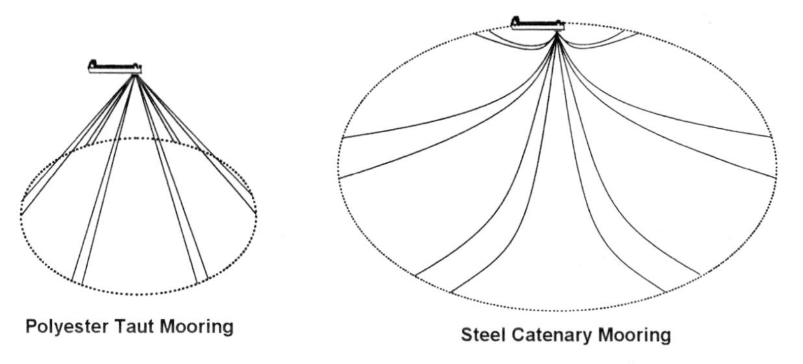

Figure 8.5 Taut and catenary mooring spread

available within the expanding literature on the subject. In particular, these models must deal with:

(i) Stiffness – In a taut mooring system the restoring forces in surge, sway and heave are derived primarily from the line stretch. This mechanism of developing restoring forces differs markedly from the conventional steel catenary systems that develop restoring forces primarily through changes in the line catenary shape. This is made possible by the much lower modulus of elasticity of polyester compared to steel. The stretch characteristics of fibre ropes are such that they can extend from 1.2 to 20 times as much as steel, reducing induced wave and drift frequency forces. The stiffness of synthetic line ropes is not constant but varies with the load range and the mean load. Furthermore the stiffness varies with age, making the analysis of a taut mooring system more cumbersome.

(ii) Hysteresis and heat build up – The energy induced by cyclic loading is dissipated (hysteresis) in the form of heat. In addition, the chaffing of rope components against each other also produces heat. Cases are known in which the rope has become so hot that the polyester fibres have melted. This effect is of greater concern with larger diameters or with certain lay types because dissipation of the heat to the environment becomes more difficult.

(iii) Fatigue – The fatigue behaviour of a rope at its termination is not good. In a termination, the rope is twisted (spliced) or compressed in the radial direction (barrel and spike or resin socket). The main reason for this decreased fatigue life is local axial compression. Although the rope as a whole is under tension, some components may go into compression, resulting in buckling and damage of the fibres. In a slack line this mechanism is more likely to be a problem than in a rope under tension. The phenomenon can appear at any position along the rope.

(iv) Other relevant issues to consider are that the strength of a polyester rope is about half that of a steel wire rope of equal diameter. Additionally the creep behaviour is good but not negligible (about 1.5% elongation over twenty years). Furthermore, synthetic fibre ropes are sensitive to cutting by sharp objects and there have been reports of damage by fish bite. A number of rope types such as high modulus polyethylene (HMPE) are buoyant in sea water; other types weigh up to 10% of a steel wire rope of equal strength. Synthetic fibre lines used within taut moorings require the use of anchors that are designed to allow uplift at the seabed. These include suction anchors, discussed further in Section 8.6.

8.3.3 Single Catenary Line Performance Characteristics

Figures 8.6a and b present the restoring force characteristics of a single catenary line plotted against offset (non-dimensionalised by water depth) for variations respectively in line weight and initial tension. Both figures emphasise the hardening spring characteristics of the mooring with increasing offset as discussed above. While this is a specific example, several observations may be made regarding design of a catenary system from these results.

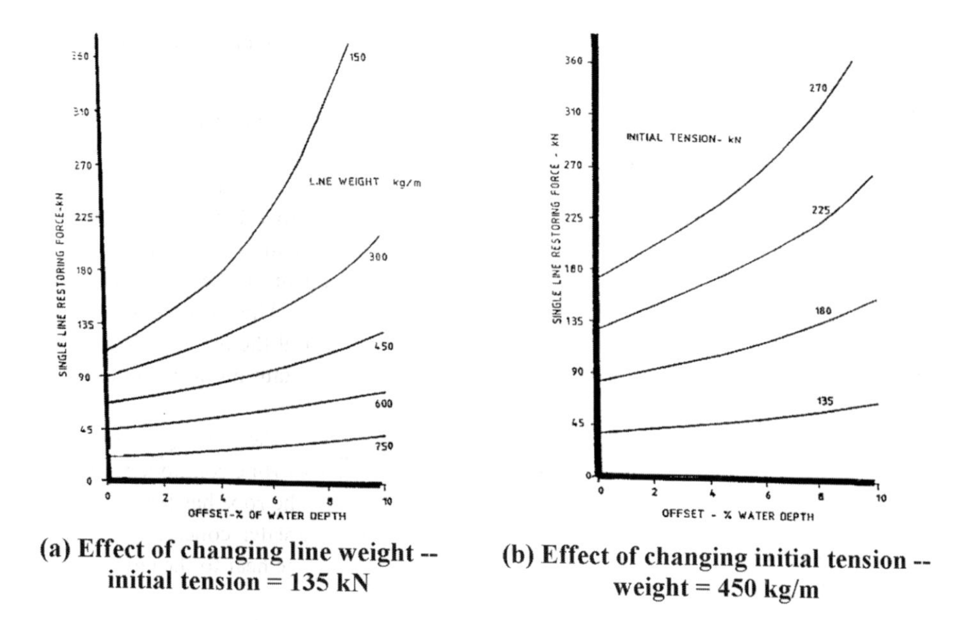

(a) Effect of changing line weight -- initial tension = 135 kN

(b) Effect of changing initial tension -- weight = 450 kg/m

Figure 8.6 Restoring force for a single catenary line (depth = 150 m)

Figure 8.6a shows the effect of line weight for a single line in 150 m of water with 135 kN initial tension. Under these conditions, the mooring would be too hard with lines weighing 150 kg/m. A 300 kg/m system is still too hard, but could be softened by adding chain. Additional calculations would be required to determine the precise quantity. The 450 kg/m line appears acceptable with heavier lines being too soft at this water depth and initial tension.

The softness can be reduced by increasing the initial tension in a given line for the specified water depth. Figure 8.6b shows that latitude exists in this particular system. The choice of initial tension will be determined by the restoring force required. The hardness of a mooring system also decreases with water depth, assuming constant values for other properties.

8.4 Loading Mechanisms

There are various loading mechanisms acting on a moored floating vessel as depicted in fig. 8.7. For a specific weather condition, the excitation forces caused by current are usually assumed temporally constant, with spatial variation depending on the current profile and direction with depth. Wind loading is often taken as constant, at least, in initial design calculations, though gusting can produce slowly varying responses. Wave forces result in time-varying vessel motions in the six rigid body degrees of freedom of surge, sway, heave, roll, pitch and yaw. Wind gust forces can contribute to some of these motions as well.

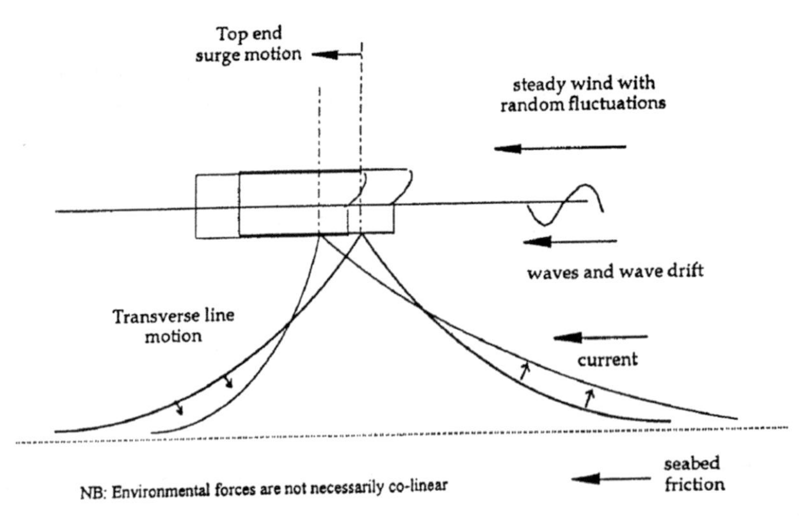

Figure 8.7 Environmental forces acting on a moored vessel in head conditions and transverse motion of catenary mooring lines

Relevant FPS responses are associated with first-order motions at wave frequencies, together with drift motions at low frequencies (wave difference frequencies). In particular, motions in the horizontal plane can cause high mooring line loads. This is because the frequency of the drift forces results in translations that usually correspond to the natural frequency of the vessel restrained by the mooring system. Consequently, it is essential to quantify the level of damping in the system, as this quantity controls the resonant motion amplitude.

Wave period is of great importance and generally the shortest wave period that can occur for a given significant wave height will produce the highest drift forces at that wave height. Furthermore, on ship-shaped bodies, the forces are greatly increased if the vessel is not head on to the waves. This situation will occur if the wind and waves are not in line and the vessel has a single point mooring. For example, on a 120,000 ton DWT vessel the wave drift forces will be doubled for a vessel heading of approximately 20° to the wave direction, when compared to the forces on the vessel heading directly into the waves.

There are a number of contributions to damping forces on a floating vessel and the moorings. These include vessel wind damping caused by the frictional drag between fluid (air) and the vessel, though the effect can be small. This has a steady component allowing linearisation procedures to be used to obtain the damping coefficient. Current in conjunction with the slowly varying motion of the vessel provides a viscous flow damping contribution because of the relative motion between the hull and the fluid. This gives rise to lift and drag forces. Both viscous drag and eddy-making forces contribute. The magnitude of the damping increases with large wave height. Wave drift damping on the vessel hull is associated with changes in drift force magnitude caused by the vessel drift velocity. The current velocity is often regarded as the structure slow drift velocity. It can be shown that when a vessel is moving slowly towards the waves, the mean drift force will be larger than

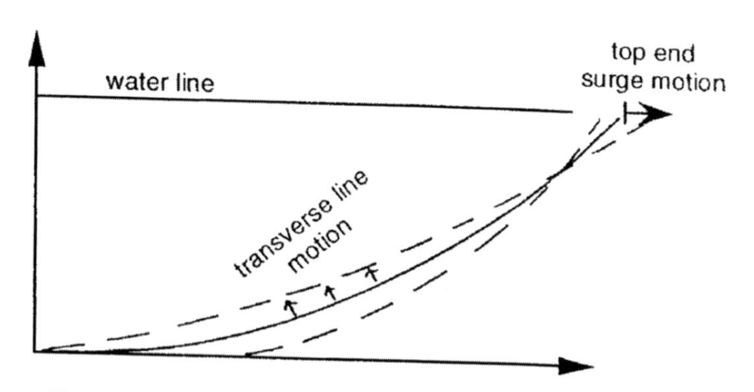

Figure 8.8 Catenary line motions caused by vessel horizontal translation

when it is moving with the waves. The associated energy loss can be thought of as slow drift motion damping.

There are a number of contributions to the overall damping from the mooring system. These are:

(i) Hydrodynamic drag damping – depending on the water depth, line pre-tension, weight and azimuth angle, a relatively small horizontal translation of the vessel can result in transverse motion over the centre section of the line that can be several times larger than the vessel translation itself as indicated in fig. 8.8. The corresponding transverse drag force represents energy dissipation per oscillation cycle and thus can be used to quantify the line damping. Brown and Mavrakos (1999) quantified levels of line damping for variations in line oscillation amplitude and frequency. Webster (1995) provided a comprehensive parametric study quantifying the influence of line pre-tension, oscillation amplitude and frequency and scope (ratio of mooring length to water depth) on the line damping.

(ii) Vortex-induced vibration – vortex formation behind bluff bodies placed in a flow gives rise to unsteady forces at a frequency close to the Strouhal frequency. The forces cause line resonant response in a transverse direction to the flow and the vortex formation can become synchronised along the length resulting in the shedding frequency "locking in" to the line natural frequency [Vandiver, 1988]. This can give a significant increase to the in-line drag forces. It is generally considered that this effect is important for wire lines, whereas for chains it is assumed negligible.

(iii) Line internal damping – material damping caused by frictional forces between individual wires or chain links also contributes to the total damping. Only limited work has been performed in this area.

(iv) Damping caused by seabed interaction – soil friction leads to reduced tension fluctuations in the ground portion of line effectively increasing the line stiffness. Work by Thomas and Hearn (1994) has shown that out-of-plane friction and suction effects are negligible in deep-water mooring situations, whereas in-plane effects can significantly influence the peak tension values.

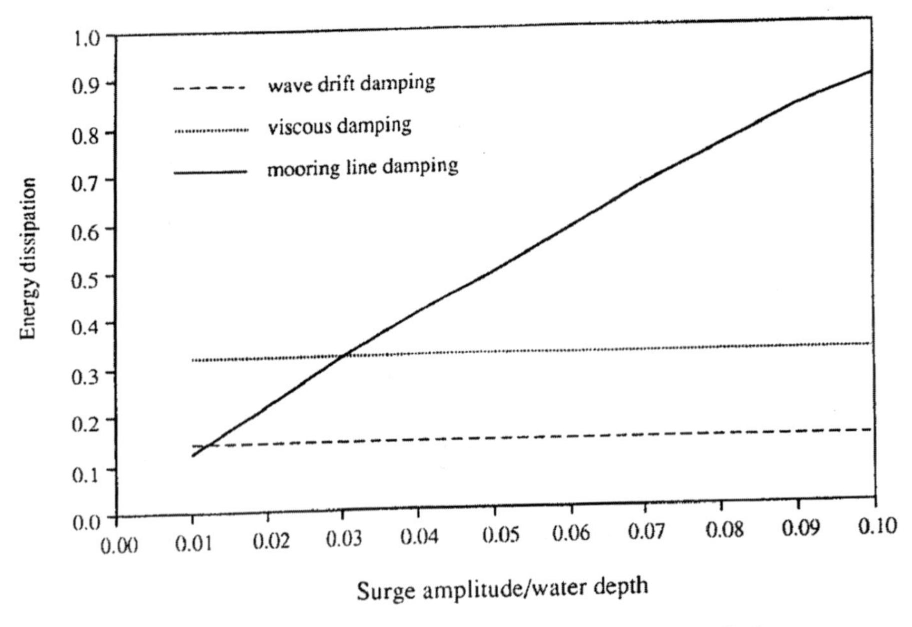

Figure 8.9 Relative energy dissipation caused by surge damping contributions

Table 8.2 Relative % damping contributions for a 120,000 ton DWT tanker in 200 m water

Significant wave height	Peak period	Damping contribution %		
(m)	(s)	Mooring	Waves	Viscous
8.6	12.7	81	15	4
16.3	16.9	84	12	4

The levels of mooring line damping relative to other contributions can, in some situations, be very high. See, e.g. fig. 8.9 showing energy dissipation as a result of wave drift, hull viscous damping and mooring line damping [Matsumoto, 1991] for a catenary mooring spread restraining a model tanker in 200 m water depth. The increased line damping for higher motion amplitudes is caused by the large transverse motion of the catenary lines.

Table 8.2 from Huse and Matsumoto (1989) gives measured results for a similar vessel undergoing combined wave and drift motion. Here, damping from the mooring system provides over 80% of the total with viscous and wave drift giving limited contributions in moderate and high seas. The line damping work is extended in Huse (1991).

8.5 Mooring System Design

In this section the range of available design methods for catenary moorings is considered. Their use with synthetic taut moorings is also outlined. The methods should be read in conjunction with the certification standards outlined in Section 8.7. There then follows some considerations associated with effective water depth, an outline of mooring spreads and a discussion of some uncertainties associated with the design procedures and their input data.

8.5.1 Static Design

This is often carried out at the very initial stages of the mooring system concept design and is described for a catenary system. Load/excursion characteristics for a single line and a mooring spread are established ignoring fluid forces on the lines.

The analysis is carried out by utilising the algorithms described in Section 8.3.1 to calculate the forces exerted on the vessel from each catenary line, given the line end-point coordinates on the surface vessel and seabed together with lengths and elasticity. These forces are then summed for all lines in the mooring spread to yield the resultant horizontal restoring and vertical forces. The restoring force and tension in the most loaded line is then calculated by displacing the vessel through prescribed horizontal distances in each direction from its initial position.

The results of a typical analysis are presented in fig. 8.10. The steady component of environmental force from wind, current and wave drift effects is applied to the vertical axis of this diagram to obtain the resultant static component of vessel offset from the horizontal axis. The slope of the force curve at this offset gives an equivalent linear stiffness C_t of the mooring system in the relevant direction for use in an equation of the form:

$$C_t x = F_x(t) \tag{8.8}$$

where co-ordinate x refers to a horizontal degree of freedom (surge or sway), F_x is force, and the stiffness resulting from the vessel hydrostatics is zero.

The maximum dynamic offset caused by the wave and drift frequency effects is then estimated. Certifying authority standards give guidance on this.

It is necessary to check that line lying on the seabed has no upward component of force at the anchor. If there is insufficient line length, the calculations should be repeated with increased length. The load in the most heavily loaded line is then read off and compared with a pre-set fraction of the breaking strength of the line. If the fraction is too high, it is necessary to adjust the line pre-tension, change material specification for each line, alter the line end co-ordinates or number of lines and repeat the calculations.

Once the intact system has been established, the calculations should be performed for the case where the most loaded line is broken and similar checks carried out.

The method has the disadvantages that conservative assumptions are made in terms of the uni-directional environment and large safety factors need to be applied to account for uncertainties. Furthermore important features of the dynamics are absent from the methodology.

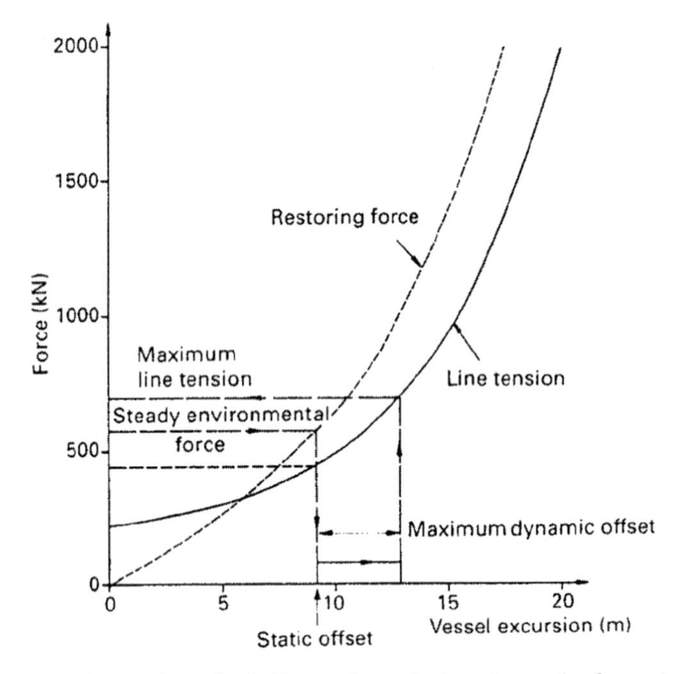

Figure 8.10 Restoring force and most loaded line tension against vessel excursion for a catenary mooring system (static analysis)

8.5.2 Quasi-Static Design

This procedure is the next level of complexity; generally, one of the two types of calculations are carried out:

- A time-domain simulation that allows for the wave-induced vessel forces and responses at wave and drift frequency, while treating wind and current forces as being steady and using the mooring stiffness curve without considering line dynamics.
- A frequency response method where the mooring stiffness curve is treated as linear and low-frequency dynamic responses to both wave drift and wind gust effects are calculated as if for a linear single degree of freedom system.

The basic differences between the static and quasi-static design are that:

- the quasi-static analysis is usually non-linear in that the catenary stiffness at each horizontal offset is used within the equations of motion. Note that a stiff catenary or taut mooring may have essentially linear stiffness characteristics;
- the equations of motion are integrated in the time domain. The influence of, at least, some added mass and damping contributions are included, although these tend to be associated with the vessel rather than accurate values including the influence from the mooring system;

- frequency domain solutions are possible but gross assumptions associated with linearisation of stiffness and damping need to be made.

The analysis solves the equation:

$$(m + A)\ddot{x} + B\dot{x} + B_v\dot{x}|\dot{x}| + C_t x = F_x(t) \tag{8.9}$$

in each degree of freedom to give the motions, x. Coupling between the motions can also be included. The terms m, A, B and B_v refer to vessel mass, added mass, linear and viscous damping respectively with F_x representing the time varying external forcing.

To give reliable answers, the simulation must cover a minimum of 18 h full-scale behaviour in order to provide sufficient statistical data for the low-frequency responses.

8.5.3 Dynamic Design

Full dynamic analysis methods are regularly utilised in design, though there is no universal agreement in the values of mooring line damping. This can influence vessel responses and line loads strongly, particularly in deep water. In outline terms, the methodology is as follows:

Usually a static configuration must first be established with non-linear time domain solutions developed about this initial shape. Often the line is de-composed into a number of straight elements (bars) with linear shape function except for the distributed mass plus added mass that is lumped at the end nodes. Generally, the motions of the platform are calculated independently of the estimates of line dynamics. However for deep-water moorings, the importance of mutual interactions between the mooring lines and the moored platform has been recognised and coupled platform mooring analysis methods need to be used. In this case, the effect of line dynamics on the platform motion is mutually included in a time-domain solution.

Importantly, dynamic methods include the additional loads from the mooring system other than restoring forces, specifically the hydrodynamic damping effects caused by relative motion between the line and fluid. Inertial effects between the line and fluid are also included though the influence is often small.

Simulations use lumped mass finite element or finite difference schemes to model small segments of each line whose shape is altered from the static catenary profile by the water resistance.

Analysis is performed in the time domain and is computationally intensive. Difficulties are:

- time steps must be small so that wave-induced line oscillations are included,
- runs must be long to allow for the vessel drift oscillation period, which in deep water may be of the order of 5 min,
- for a typical floating vessel mooring system design, the weather is multi-directional and a number of test cases must be considered.

Line top-end oscillation must be included, because of vessel motion at combined wave and drift frequencies; otherwise, dynamic tension components may be underestimated, or advantages of line damping contributions neglected. It is noted that line dynamics can, in some cases, result in the doubling of top tension when compared to the static line tension. Furthermore, damping levels vary significantly depending on water depth, line make up, offsets and top-end excitation.

Hybrid methods that work in the time domain but make a number of simplistic assumptions about the instantaneous line shape are currently being investigated. There is some potential here, but further work is needed to provide methods usable in the design.

More efficient frequency domain methods are also being developed that include line dynamics in an approximate manner. At present these do not work well when strong non-linearities, such as those caused by fluid drag forces are present, for example, when large line oscillations occur.

Figures 8.11–8.13 show results from a design study for a turret-moored monohull vessel positioned at a northerly North Sea location. Figure 8.11 depicts the drift force energy spectra for the vessel in head seas with 1 and 100-yr return period weather. The energy spectra are very broad banded, providing excitation over a wide frequency range that includes, as is usually the case, the resonant surge frequency of the vessel on its mooring system.

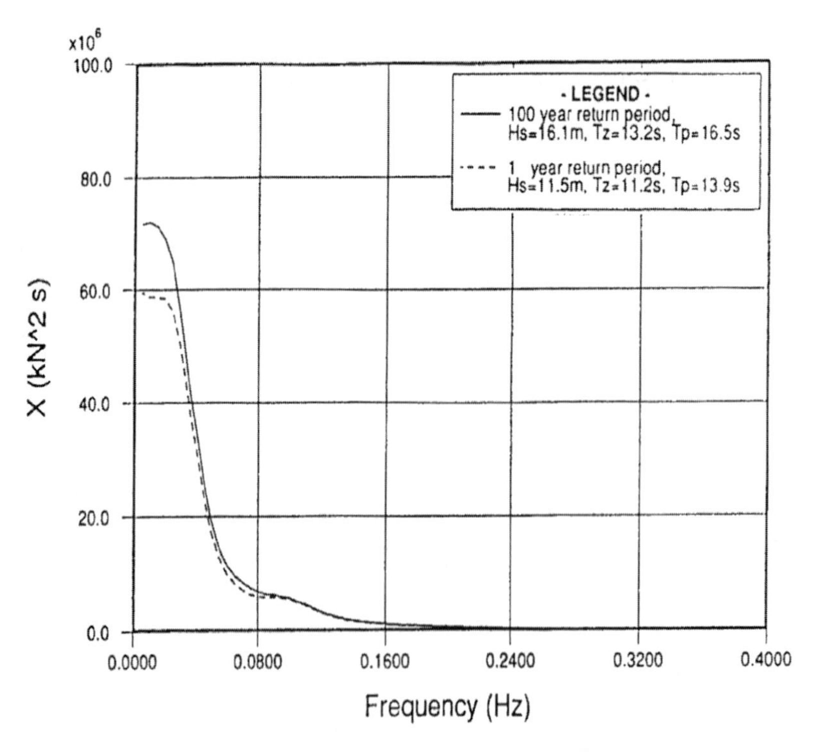

Figure 8.11 Mooring line analysis – head sea drift spectra

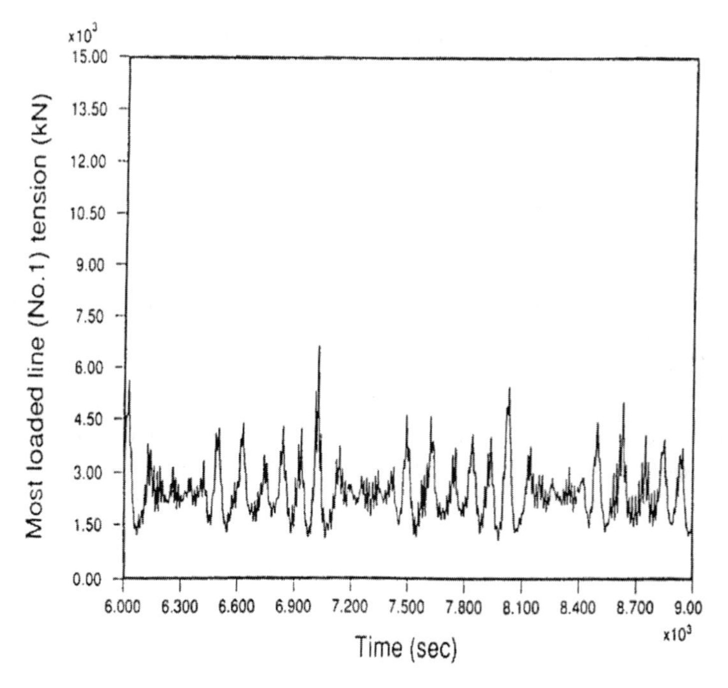

Figure 8.12 Mooring line analysis – Line tension vs. time (intact)

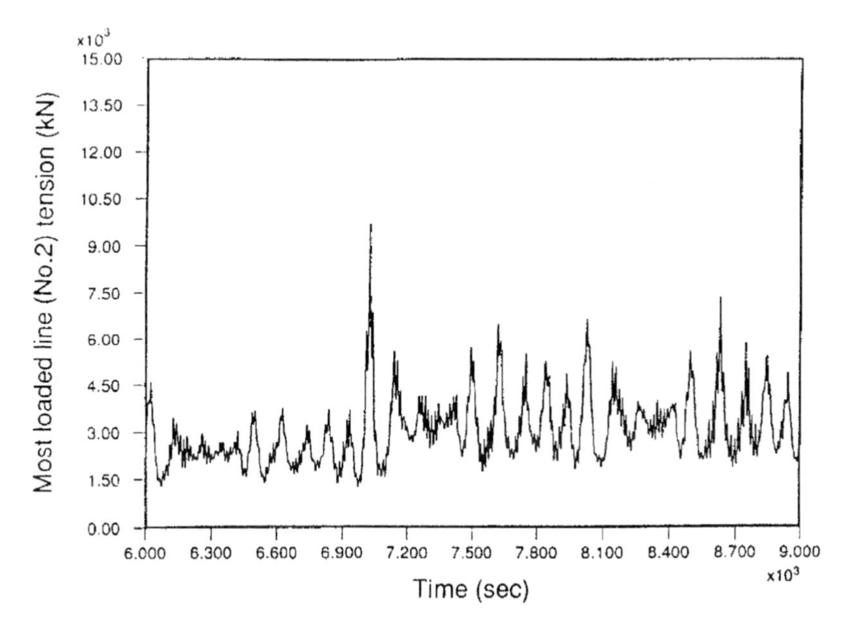

Figure 8.13 Mooring line analysis – Line tension vs. time (transient motion)

Figures 8.12 and 8.13 give the line tension graphs for the intact mooring and transient conditions after line breakage for 1 yr storm conditions. Low amplitude wave and high amplitude drift effects can clearly be seen.

8.5.4 Synthetic Lines

Essentially, the design procedures for taut moorings are similar to those described for catenary systems with the exception that three stiffness values are used in the design calculations:

- Bedding-in stiffness – This is the initial elongation after manufacture and is as a result of fibre extension, which may be partially recovered in some circumstances unless the load is maintained. It is also partly due to a tightening of the rope structure, which is retained unless the rope suffers a major buckling disturbance. The bedding-in elongation becomes negligible after approximately one hundred cycles up to a given load. The response after installation, when the rope has been subjected to a certain load cycling regime, is given by the post-installation stiffness. A minimum estimated value of installation stiffness should be used to calculate offsets in the period after installation.
- Drift stiffness – Cyclic loading under moderate weather conditions, applicable to the mooring during a high proportion of the time, shows a mean variation of tension and elongation which is represented by the drift stiffness. A minimum estimated value of drift stiffness should be used to calculate offsets under normal mooring conditions.
- Storm stiffness – Under more extreme conditions, the mean variation of tension and elongation is represented by the storm stiffness, which is higher than the drift stiffness. A maximum estimated value of storm stiffness should be used to calculate peak load. Creep with time may also occur, and analyses need to consider this, with re-tensioning at site required throughout the installation lifetime.

Calculations must also be performed to assess hysteresis effects inherent in the fibre properties and caused by friction. This will generate heat.

8.5.5 Effective Water Depth

Combinations of tide change plus storm surge, for example, together with alterations in vessel draught, because of ballasting, storage and offloading etc. result in changes in the elevation of the vessel fairleads above the seabed. The example given in fig. 8.14 presents the range of elevation levels for a 120,000 ton dwt floating production unit in a nominal water depth of 136 m. This elevation range is likely to be relatively larger in shallow water. LAT represents lowest astronomical tide. A number of elevations must be considered in the mooring design to establish the resulting influence on line tension.

8.5.6 Mooring Spreads

Although a symmetric spread of mooring lines is the simplest in terms of design, it may not be the optimum in terms of performance. Criteria needing considerations are:

- directionality of the weather; in particular if storms approach from a specific weather window, it may be advantageous to bias the mooring towards balancing these forces,

Site conditions:

water depth at site, to LAT	136m
maximum depth of fairleads below WL (loaded)	16m
minimum depth of fairleads below WL (ballasted)	8m
maximum tide + tidal surge above LAT	2.5m
minimum tide + tidal surge below LAT	0.5m

Maximum vessel fairlead elevation is:

water depth	136m
minimum depth at fairleads (ballasted)	-8m
maximum tide + tidal surge	+2.5m
fairlead elevation	130.5m

Minimum vessel fairlead elevation is:

water depth	136m
maximum depth to fairlead (loaded)	-16m
minimum tide + tidal surge	-0.5m
fairlead elevation	119.5m

Mean elevation is thus 125m.

Figure 8.14 Effective water depth and fairlead position range

- subsea spatial layout; seabed equipment and pipelines may restrict the positioning of lines and anchors in this region,
- riser systems; clashing of risers with mooring lines must be avoided and this may impose limitations on line positions,
- space restrictions in the turret region; it may be beneficial to cluster lines together to gain further space.

Figure 8.15 gives an example of a symmetric spread, while fig. 8.16 depicts an alternative arrangement having wide corridors to accommodate a large number of flexible risers for an extensive offshore development.

8.5.7 Uncertainty in Line Hydrodynamic Coefficients

There are many uncertainties associated with mooring system design. These include the uncertainties in input data, the environment, its loading on the vessel and mooring system together with the response, seabed conditions and line physical properties. Because of the large number of "fast track" projects, research and development work cannot keep pace and consequently, mooring systems are less cost-effective, requiring higher safety factors or, in some cases, lower reliability.

A specific uncertainty is associated with the choice of chain line drag coefficient, required in the design in order to calculate the maximum line tensions including dynamic effects. Furthermore, line drag is the major contribution towards induced mooring damping as discussed earlier.

Figure 8.17 provides drag coefficients plotted against Re for harmonic, sinusoidal oscillations taken from Brown, et al (1997). Various Keulegan–Carpenter (KC) values

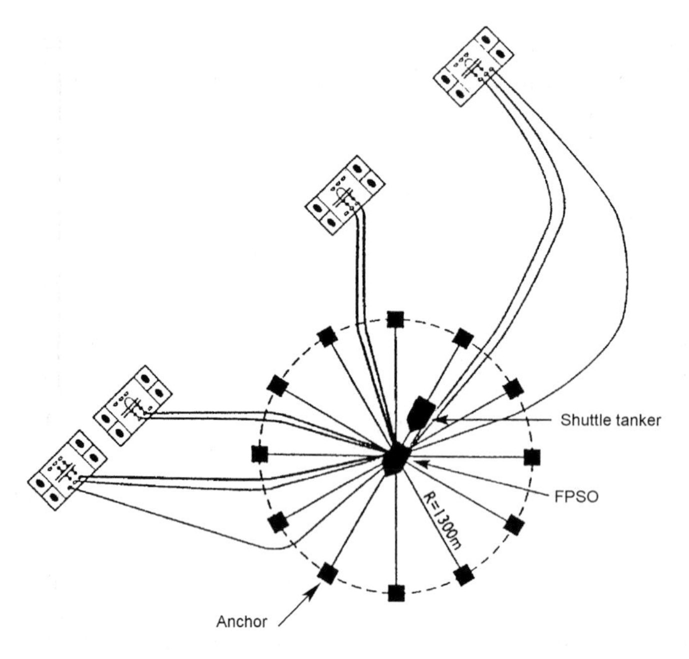

Figure 8.15 Plan view of symmetric spread

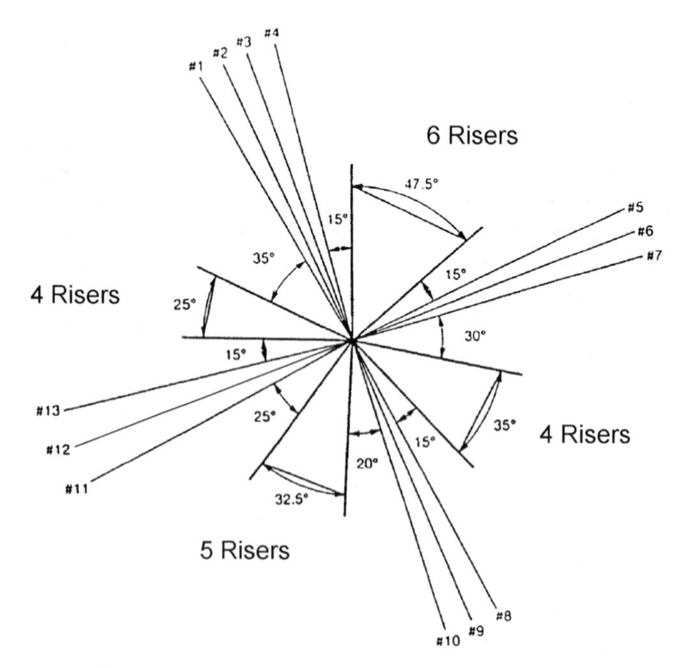

Figure 8.16 Riser corridors between non-symmetric spread

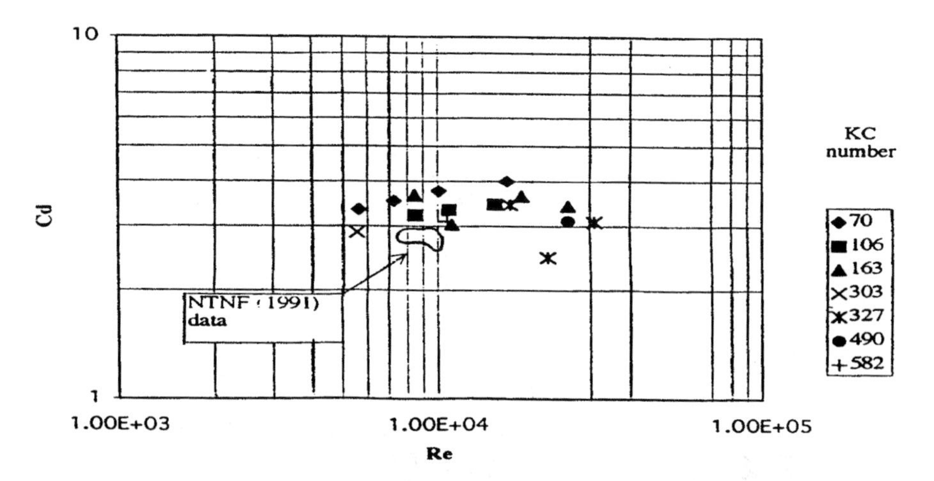

Figure 8.17 Measured drag coefficient for chain in harmonic flow conditions

between 70 and 582 are indicated, and results are for the large-scale stud chain samples. Also plotted are the results from NTNF (1991). These data are based primarily on results from a number of tests with small-scale specimens, cross-flow conditions or harmonic oscillations. It is noted that a drag coefficient of 2.6 for chain without marine growth is commonly used in design, whereas 2.4 is common for studless chain.

Mooring lines undergo bi-harmonic motions caused by the combined wave and drift floater response. It is known, however, that simply superimposing the wave and drift effects gives erroneous results.

The calculation of drag coefficient for harmonically oscillating flow past a body is based on the drag force term of the Morison equation. When there is bi-harmonic flow (i.e. two frequencies of oscillation), the situation is not so simple. In resolving the measured force into drag and inertia components, it is possible to define two drag (and inertia) coefficients, appropriate to either of the two frequencies of oscillation. An additional complication arises as either the wave or drift maximum velocity, or indeed the sum of the two may be used within the Morison formulation. Furthermore, alternative Reynolds numbers and KC values may also be established based on the appropriate oscillation frequency and amplitude.

Figure 8.18 examines the variation of in-line drag coefficient under bi-harmonic oscillation conditions with wave oscillations in various directions to drift motion. C_d is plotted against wave frequency oscillation direction relative to the drift frequency and direction. Drag coefficients are based on the drift frequency of oscillation as the damping contribution to the drift motion of the vessel is of interest. Velocities used to calculate the drag coefficient are based on the combined wave and drift oscillations.

The results show a significant increase in drag for the situation with wave oscillations in the transverse direction to the drift when compared to the in-line wave oscillations. In a sense this can be thought of as a drag amplification effect somewhat similar to that induced by

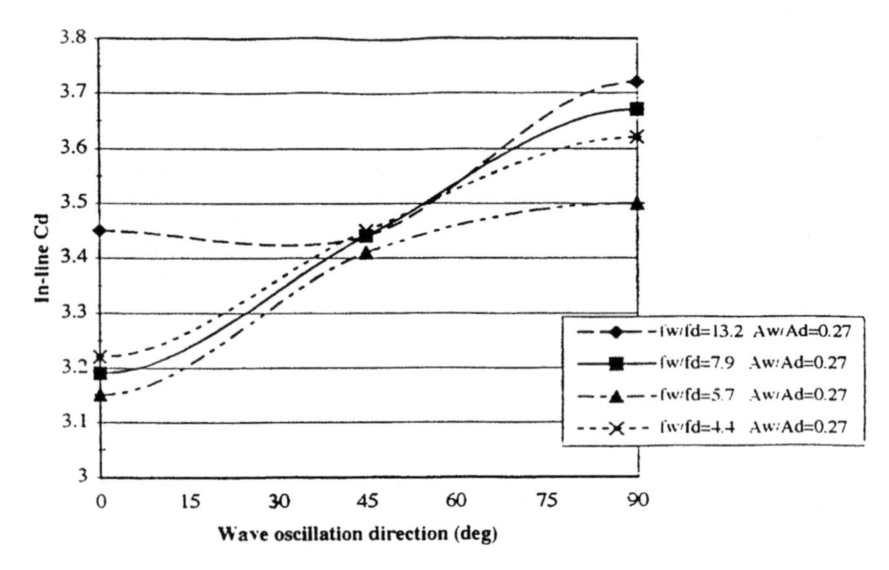

Figure 8.18 Measured in-line drag coefficients for chain in bi-harmonic flow

vortex-induced vibrations, though here the out-of-plane vibrations are caused by top-end motion in the transverse direction, as opposed to flow-induced loading. Curves are plotted for wave to drift motion amplitude ratios (Aw/Ad) of 0.27 and wave to drift motion frequency ratios (fw/fd) from 4.4 to 13.2.

In a realistic sea state, a mooring line will be subjected to motions at wave frequencies both in in-line and transverse directions to the imposed drift motions. Consequently, in order to use the present results in design it is necessary to interpret the vessel surge, sway and yaw motions at wave frequencies to establish the relevant translation angle of the fairlead in the horizontal plane relative to the drift motion. This can then be used in conjunction with the drag coefficient values interpolated from fig. 8.18. It is also necessary to estimate the ratios of wave to drift motion amplitude and wave to drift motion frequency of oscillation. A simple method to establish the latter could be to use the zero-crossing period of the sea state relative to the drift period. Linear and higher-order potential flow analysis methods or model test data can be used to estimate amplitude ratios. In the absence of more refined data, fig. 8.18 provides appropriate results of in-line drag coefficient for use in design.

8.5.8 Uncertainty in Line Damping and Tension Prediction

Work initiated by the International Ship and Offshore Structures Congress (ISSC), Committee I.2 (loads) presents a comparative study on the dynamic analysis of suspended wire and stud chain mooring lines [Brown and Mavrakos, 1999]. A total of 15 contributions to the study were provided giving analysis results based on dynamic time or frequency domain methods for a single chain mooring line suspended in 82.5 m water depth and a wire line in 500 m depth. Bi-harmonic top-end oscillations representing in-line combined wave- and drift-induced excitation were specified.

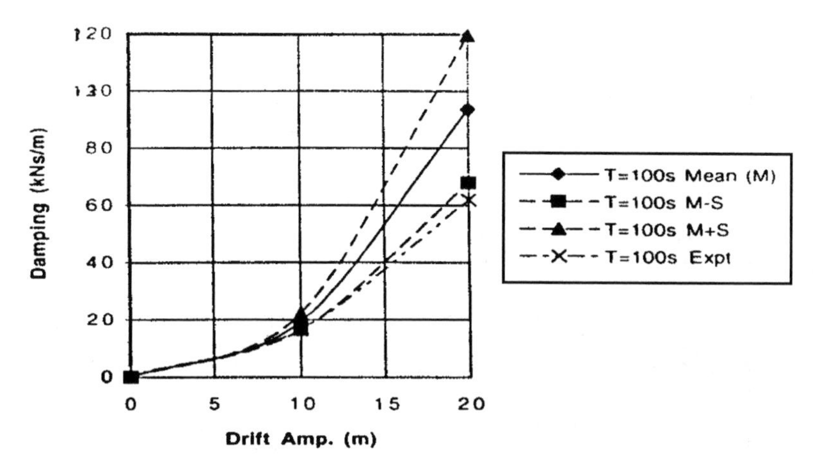

Figure 8.19 Chain line damping vs. drift induced top-end amplitude (drift period = 100 s) – no wave oscillation, water depth = 82.5 m

The mooring line damping results for chain are compared with the limited available experimental data. The results provided by the participants show a fair agreement despite the complexity of the numerical methods. Predictions of dynamic tension based on time-domain methods show scatter, the estimates of damping giving further discrepancies. Some results were based on frequency-domain methods for which there are even more disagreement.

The uncertainty in results is quantified by plotting the mean, mean plus/minus one standard deviation (M+S, M−S) of tension and line damping from the various data provided by contributors. Clear trends in tension and damping with oscillation frequency and amplitude are also revealed.

Calculated line damping values are plotted against drift-induced oscillation amplitude for the chain in 82.5 m water depth in fig. 8.19. Here there is no oscillation at wave frequencies. The results indicate that increasing the drift top-end amplitude from 10 to 20 m causes an increase in damping by a factor of approximately 4.5. It is noted that doubling the oscillation period caused the damping to reduce by 50%. Similar trends with drift-induced amplitude were observed for the wire in 500 m water depth.

Figures 8.20 and 8.21 give dynamic tension components (total tension minus static catenary tension) for the chain (with drift amplitude and period of 10 m and 100 s respectively) and wire (with drift amplitude and period of 30 m and 330 s respectively). It is seen that a number of contributions with the wire results predict total tensions less than the catenary value. A possible reason for this is that the calculation method for catenary tension does not include stretch of the seabed portion and thus may give slightly conservative values. Contributor data may allow stretch of this grounded portion. There is a consistent trend throughout these results in that both the dynamic tension and the mooring line damping increase significantly as the line wave-induced top-end motion increases. There is also large uncertainty in the results; for example, contributor responses given in fig. 8.20 indicate a

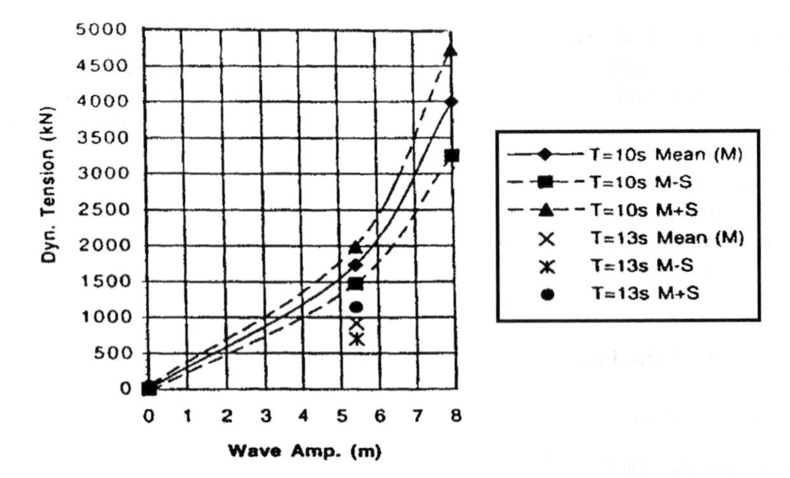

Figure 8.20 Chain maximum dynamic tension vs. wave-induced top-end amplitude – with drift oscillation, water depth = 82.5 m

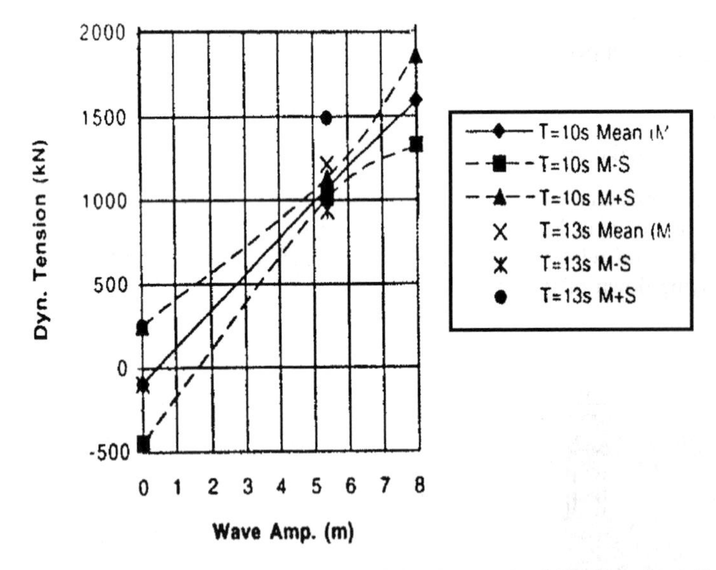

Figure 8.21 Wire maximum dynamic tension vs. wave induced top end amplitude – with drift oscillation, water depth = 500 m

line tension standard deviation at 8 m wave amplitude of over 600 kN about a mean of 4000 kN. The catenary (static) tension not plotted here is approximately 3500 kN.

More recently, a number of studies have developed efficient numerical and analytical solution techniques for the evaluation of mooring line dynamics. Aranha and Pinto (2001b) derived an analytical expression for the dynamic tension variation along the cable's suspended length, whereas Aranha, et al (2001a) followed the same methodology to obtain an analytical expression for the probability density function of the dynamic tension envelope in risers and mooring lines. Gobat and Grosenbaugh (2001a) proposed an empirical model to establish the mooring line dynamic tension caused by its upper end vertical motions. Aranha, et al (2001a) introduced a time integration of the cable dynamics equations. Chatjigeorgiou and Mavrakos (2000) presented results for the numerical prediction of mooring dynamics, utilising a pseudo-spectral technique and an implicit finite difference formulation.

8.6 Mooring Hardware Components

The principle components of a mooring system may consist of

- Chain, wire or rope or their combination
- Anchors or piles
- Fairleads, bending shoes or padeyes
- Winches, chain jacks or windlasses
- Power supplies
- Rigging (e.g. stoppers, blocks, shackles)

8.6.1 Chain

Chain and wire make up the strength members for the mooring system.

There are two primary chain constructions. Stud-link chain (fig. 8.22a) has historically been used for mooring MODUs and FPSOs in relatively shallow water. It has proven strong, reliable and relatively easy to handle. The studs provide stability to the link and facilitate laying down of the chain while handling.

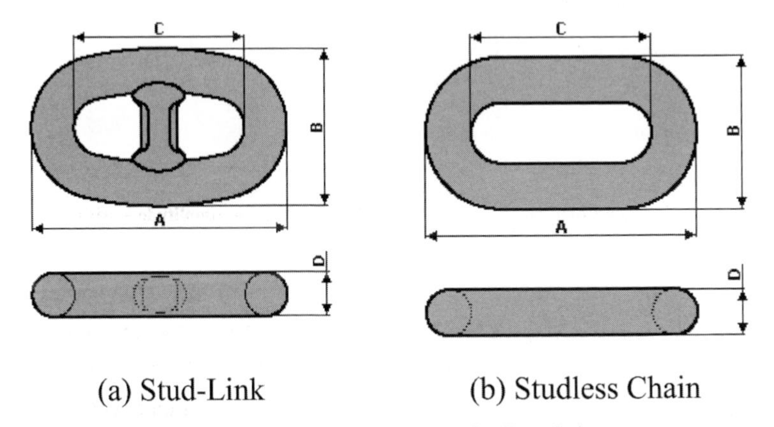

Figure 8.22 (a) Stud-link and (b) Studless chain

Permanent moorings have recently preferred to use open link, or studless chain (fig. 8.22b). Removing the stud reduces the weight per unit of strength and increases the chain fatigue life, at the expense of making the chain less convenient to handle.

Chain size is specified as the nominal diameter of the link, "D" in figs. 8.22a and b.[1] The largest mooring chain manufactured to date is the 6.25 in. (159 mm) studless chain for the Schiehallion FPSO in the North Atlantic (West of Shetlands).

The specification of chain properties is an important function in any mooring system design. The chain is sold in a variety of grades. Grade 4 (K4) is the highest grade chain currently available. Drilling contractors have traditionally used the oil rig quality (ORQ) chain, which has detailed specifications in API Specification 2F.[2] Properties of these chains are presented here.

8.6.2 Wire Rope

Wire rope consists of individual wires wound in a helical pattern to form a "strand". The pitch of the helix determines the flexibility and axial stiffness of the strand.

Wire rope used for mooring can be multi-strand or single-strand construction. The principle types used offshore are shown in fig. 8.23.

Studlink chain and six-strand wire rope are the most common mooring components for MODUs and other "temporary" moorings. Multi-strand ropes are favoured for these applications because of their ease of handling. Six-strand rope is the most common type of multi-strand rope used offshore. Mooring line ropes typically consist of 12, 24, 37 or more wires per strand. The wires have staggered sizes to achieve higher strength. Common "classes" of multi-strand rope include [Myers, 1969]:

- 6 × 7 Class: Seven wires per strand, usually used for standing rigging. Poor flexibility and fatigue life, excellent abrasion resistance. Minimum drum diameter/rope diameter (D/d) = 42.

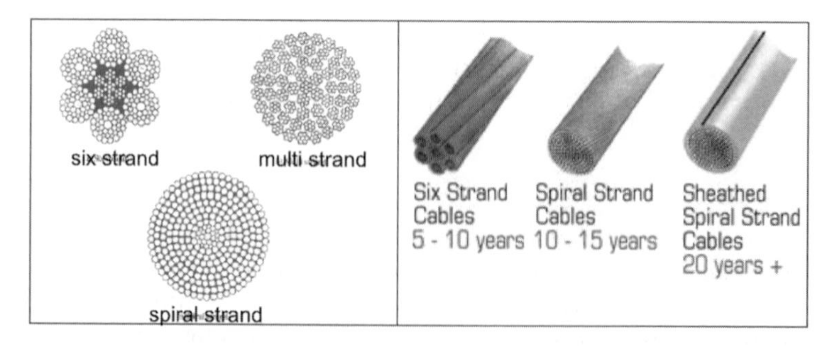

Figure 8.23 Wire rope construction

[1]Note that mooring design guidelines require that the chain be oversized to allow for corrosion.
[2]API, "Specification of Mooring Chain", 2F (latest edition).

- 6 × 19 Class: 16–27 wires per strand. Good flexibility and fatigue life and abrasion resistance. Common in lifting and dredging. Minimum D/d = 26–33.
- 6 × 37 Class: 27–49 wires per strand. Excellent fatigue life and flexibility, but poor abrasion resistance. Minimum D/d = 16–26.

Multi-strand wire ropes may contain either a fibre or a metallic core. The core is important for support of the outer wires, especially on a drum, and in some applications to absorb shock loading. Fibre core (FC) ropes are not generally used for heavy duty marine applications. Metallic core ropes may be one of the two types: independent wire rope core (IWRC) or wire-strand core (WSC). IWRC is the most common core filling for heavy marine applications.

Single-strand ropes are more common in large permanent installations. The wires are wound as a helix with each layer wrapped in a different direction. This provides "torque balancing", preventing the rope from twisting when under load. The spiral strand is more fatigue resistant than the multi-strand rope. Corrosion resistance is enhanced by either sheathing with a polyurethane coating, adding zinc filler wires or using galvanised wires. Sheathing provides the best performance, provided that the handling procedures insure against damage to the sheath.

8.6.3 Properties of Chain and Wire Rope

Tables 8.3 and 8.4 are taken from the Det Norske Veritas OS-E301 and show the mechanical properties of common grades of mooring chain in which *d* is the nominal diameter in mm.

Tables 8.5 and 8.6 show the mechanical properties of the most common types of mooring chain and wire in English units. The quantity "*d*" is the nominal diameter in inches.

The rope and chain properties are constantly being improved. Latest values should be obtained from the manufacturers.

8.6.4 Moorings

Figure 8.24 gives a typical line leg for a catenary moored floating production unit in 140 m water depth. Lower and upper terminations are of chain to avoid seabed wear and excessive bending associated with handling. In a number of moors, one shot (27.5 m) of chain is used at the line top-end and a spiral wound wire over the centre section that does not contact the seabed.

8.6.5 Connectors

Connectors are used to join sections of chain to one another, connecting chain to wire rope, connecting to padeyes on anchors or vessels, etc. The common types of connectors for the stud link chain and studless chain are given in DNV OS-E301. Mooring connectors are designed to take the full breaking strength of the chain or wire rope, but their fatigue properties require special attention. There is very little fatigue data for the standard connectors and their use is therefore not recommended for permanent moorings.

Links used in permanent moorings should be special purpose designs. An example of a triplate is shown in fig. 8.25.

Table 8.3 Mechanical properties of offshore mooring chain (DNV OS-E301)

Grade	Minimum yield strength (N/mm^2)	Minimum tensile strength (N/mm^2)	Minimum elongation (%)	Minimum reduction of area (%)
NV R3	410	690	17	50
NV R3S	490	770	15	50
NV R4	580	860	12	50

Grade	Minimum Charpy V-notch energy (*J*)				
	Temperature[1] (°C)	Average		Single	
		Base	Weld	Base	Weld
NV R3	0	60	50	45	38
	−20	40	30	30	23
NV R3S	0	65	53	49	40
	−20	40	33	34	25
NV R4	0	70	56	53	42
	−20	50	36	38	27

[1]At the option of the purchaser, and when chain segments are intended to be permanently submerged, testing may be carried out at 0°C. Otherwise, testing to be carried out at −20°C.

Table 8.4 Formulas for proof and break test loads (adopted from DNV OS-E301)

Type of chain	Grade	Proof test load, kN	Break test load, kN
Stud Chain Links	NV R3	$0.0156d^2(44\text{-}0.08d)$	$0.0223d^2(44\text{-}0.08d)$
Stud Chain Links	NV R3S	$0.0180d^2(44\text{-}0.08d)$	$0.0249d^2(44\text{-}0.08d)$
Stud Chain Links	NV R4	$0.0216d^2(44\text{-}0.08d)$	$0.0274d^2(44\text{-}0.08d)$
Studless Chain Links	NV R3	$0.0156d^2(44\text{-}0.08d)$	$0.0223d^2(44\text{-}0.08d)$
Studless Chain Links	NV R3S	$0.0174d^2(44\text{-}0.08d)$	$0.0249d^2(44\text{-}0.08d)$
Studless Chain Links	NV R4	$0.0192d^2(44\text{-}0.08d)$	$0.0274d^2(44\text{-}0.08d)$

These links are typically engineered and tested as "fit for purpose" designs for each project.

Cable terminations consist of a socket, which is a cast in-place to achieve a strength equivalent of the wire rope. The connecting socket may be either "closed" or "open", see fig. 8.26.

Table 8.5 Properties of mooring chain and wire rope

	K4 Studless chain	Spiral strand	6 Strand IWRC
Steel area (in.^{2})	$2.64d^2$	$0.58d^2$	$0.54d^2$
Weight in water (lb/ft)	$7.83d^2$	$1.74d^2$	$1.59d^2$
Breaking strength (kip)	$3.977d^2(44\text{-}2.032d)$	$126d^2$	$93.2d^2$
Stiffness (kip)	$10{,}827d^2$	$13{,}340d^2$	$8640d^2$

Table 8.6 Tabulated mooring component data

	K4 studless chain			Spiral strand			IWRC wire rope		
Dia-meter	Weight in water	Breaking strength	EA	Weight in water	Breaking strength	EA	Weight in water	Breaking strength	EA
in.	lb/ft	kip	kip	lb/ft	kip	kip	lb/ft	kip	kip
2.5	48.9	968	67,667	10.9	788	83,375	9.9	583	54,000
2.75	59.2	1155	81,876	13.2	953	100,884	12.0	705	65,340
3	70.5	1357	97,440	15.7	1134	120,060	14.3	839	77,760
3.25	82.7	1571	114,356	18.4	1331	140,904	16.8	984	91,260
3.5	95.9	1797	132,626	21.3	1544	163,415	19.5	1142	105,840
3.75	110.1	2035	152,250	24.5	1772	187,594	22.4	1311	121,500
4	125.3	2283	173,226	27.8	2016	213,440	25.4	1491	138,240
4.25	141.4	2541	195,556	31.4	2276	240,954	28.7	1683	156,060
4.5	158.6	2808	219,239	35.2	2552	270,135	32.2	1887	174,960
4.75	176.7	3083	244,276	39.3	2843	300,984	35.9	2103	194,940
5	195.8	3366	270,666	43.5	3150	333,500	39.8	2330	216,000
5.25	215.8	3655	298,409	48.0	3473	367,684	43.8	2569	238,140
5.5	236.9	3950	327,506	52.6	3812	403,535	48.1	2819	261,360
5.75	258.9	4251	357,956	57.5	4166	441,054	52.6	3081	285,660
6	281.9	4556	389,759	62.6	4536	480,240	57.2	3355	311,040
6.25	305.9	4864	422,916	68.0	4922	521,094	62.1	3641	337,500
6.5	330.8	5176	457,426	73.5	5324	563,615	67.2	938	365,040
6.75	356.8	5490	493,289	79.3	5741	607,804	72.4	4246	393,660
7	383.7	5805	530,505	85.3	6174	653,660	77.9	4567	423,360

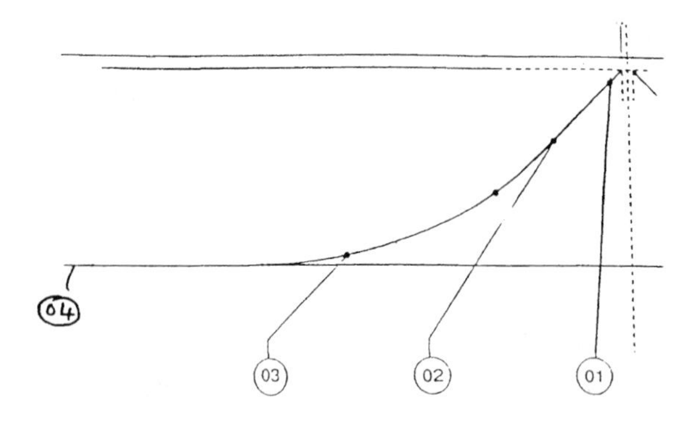

04	5" 1/4 Anchor Chain	L = 320m
03	5" 1/2 Anchor Chain	L = 300m
02	Spiral Strand Rope	L = 100m
01	6" Anchor Chain	L = 27.5m

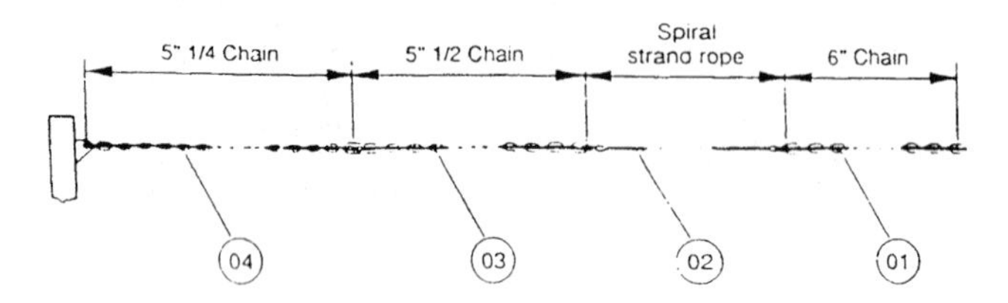

Figure 8.24 Typical mooring line components (shackles not shown)

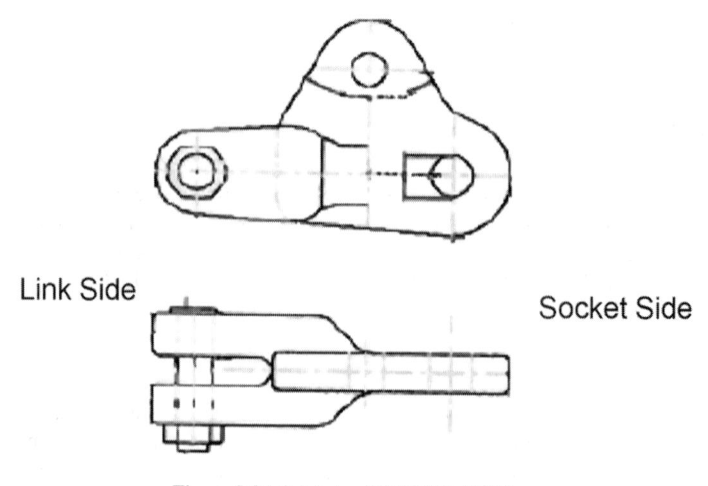

Figure 8.25 Triplates (DNV OS-E301)

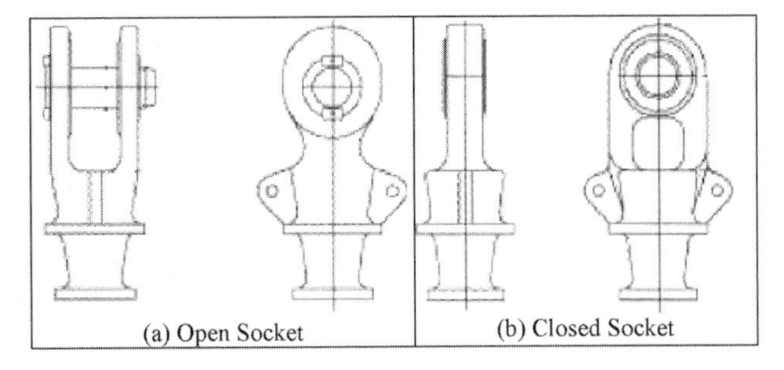

Figure 8.26 Wire rope sockets (DNV OS-E301)

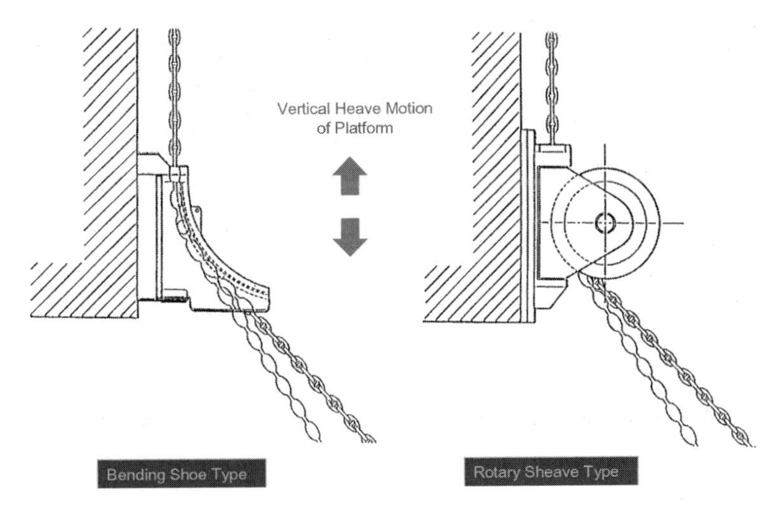

Figure 8.27 Examples of chain fairleads

8.6.6 Shipboard Equipment

Shipboard equipment depends on the type of line (wire rope or chain) connected to the vessel, and whether the mooring is used for positioning or is static. For example, the chain jacking system may be placed on top of a column for a semi-submersible or placed on the platform for an FPSO. A typical fairlead for a chain at the platform end is shown in fig. 8.27. On the left hand-side, a bending shoe-type fairlead is depicted. On the right hand-side the chain is fed through a rotary sheave.

8.6.7 Anchors

Anchors are basically of two types, relying either on self-weight or suction forces. The traditional embedment anchors, as shown in fig. 8.28, are not normally designed for vertical force components. Holding power is related to anchor weight and type of seabed.

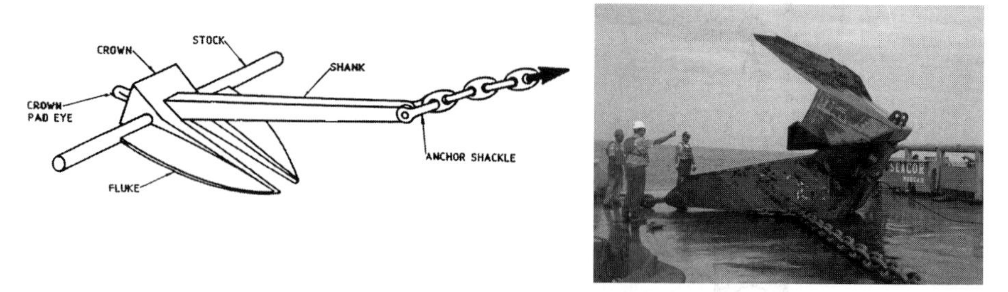

Figure 8.28 Drag anchor

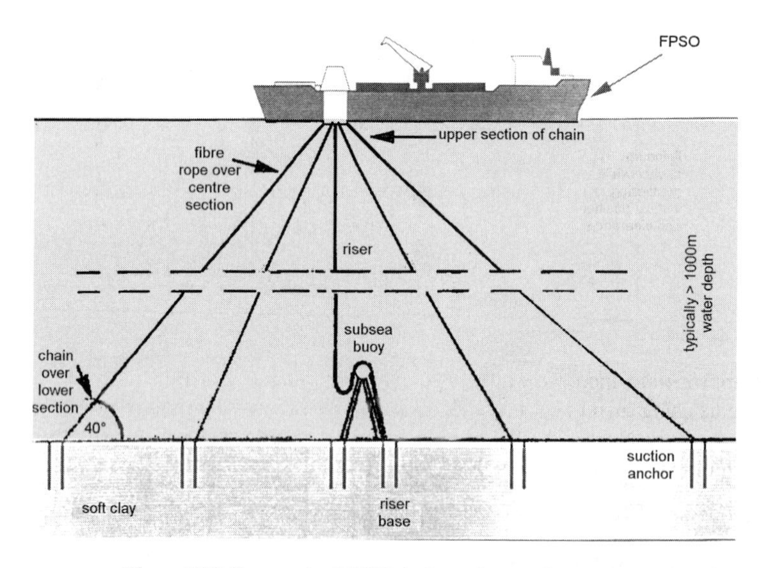

Figure 8.29 Deep water FPSO design using suction anchors

Figure 8.29 depicts a deep water floating production vessel moored with a taut station keeping system of fibre rope using suction anchors. These allow vertical anchor loads. The angle at the line lower-end is noted as being 40° to the horizontal. Figure 8.30 shows a typical suction anchor installation sequence. By reversing the suction process, the anchor can be "pushed" from the seabed using over-pressure. Piles can be used as an alternative to anchors. However, they require a large crane installation vessel with piling capability.

8.6.8 Turrets

The design of monohull turret structures used for single-point moorings in floating production systems must allow for large static and dynamic loading caused by the vessel motions in waves together with forces transmitted by the mooring system. The hull design in the turret region must reflect the fact that the amount of primary steel is reduced here

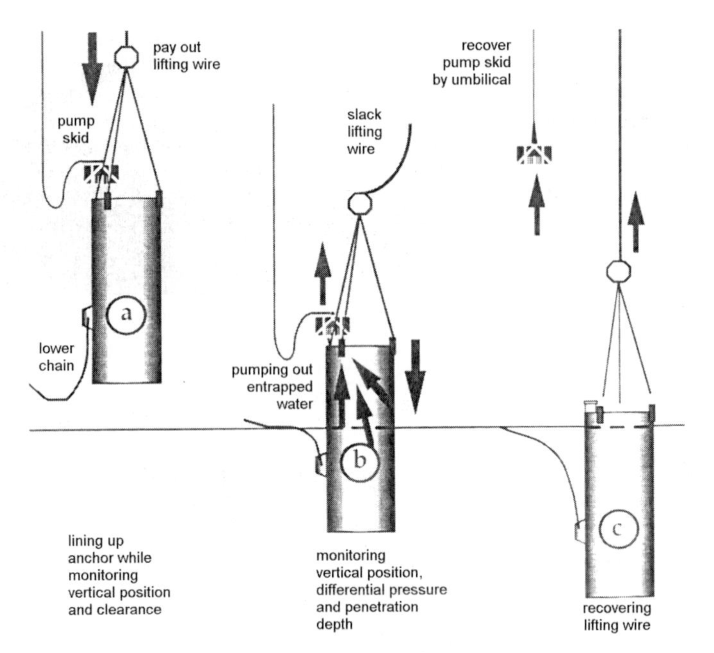

Figure 8.30 Suction anchor installation sequence

with an appropriate increase in the stress concentration. A comparison of the existing developments using turret-moored vessels in use indicates wide variations of turret position. Indeed some early North Sea designs use a turret placed close to the vessel amidships, whereas a number of Far Eastern applications place a disconnectable turret off the bow.

Careful selection of turret position is important because of its influence on:

- Mooring line tension and riser loading – The turret position alters the vessel yaw and hence the surge and sway motions, thus influencing the mooring line tension. This is also affected by the vessel heave and pitch motions. In particular, the pitch contribution to the turret vertical motion is relatively high for the turrets near or off the vessel bow. The combined effects can also result in high loading on the riser system.
- Vessel yaw – The motion response magnitude in yaw is likely to increase significantly if the turret is placed close to amidships, because the yaw restoring moment causing the vessel to head into weather is reduced. Use of azimuthing thrusters, if fitted, can be employed to control the yaw but with an increased capital and operating cost. Increased yaw results in more wear on the turret bearing, together with higher downtime because of inertial loading from the vessel motions. It can also cause yaw instability of the vessel. The low-frequency yaw about the turret also needs to be restricted in order that hydrocarbon off-loading from the vessel stern can be carried out with high operability levels. Figure 8.31 shows the stern horizontal displacements for two vessels [Brown, et al 1998] with turrets positioned at 12 and 36.5% of the hull length from the vessel

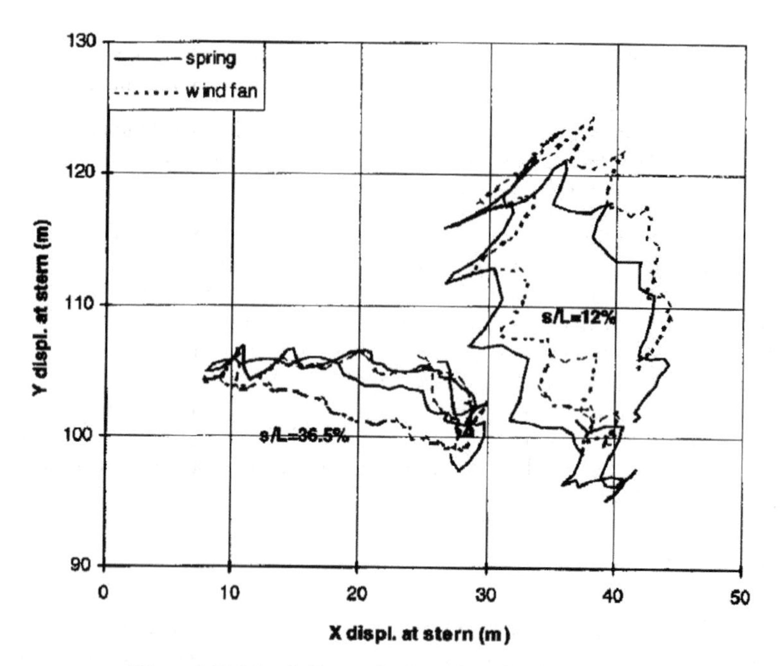

Figure 8.31 Monohull stern horizontal motion in head seas

amidships responding in identical sea states of $H_s = 8.7$ m and $T_p = 11.8$ s wind of 60 kt was also simulated at 60° to the wave direction (using fans and a turret-mounted spring mechanism). The results show large increases in stern transverse (Y) motions when the turret is closer to amidships.

- Rigid body oscillation in the horizontal plane – The natural frequencies and amplitude of oscillation can be affected by the position of the turret. The full low-frequency vibration behaviour of a turret-moored vessels is not well understood.

The turret rotates within the vessel hull using a combination of radial and thrust bearings positioned on roller assemblies at deck and within the hull. Transmission of hydrocarbons from the non-rotating components, such as the turret and risers, to the weather-vaning vessel is carried out using either a stacked swivel or "drag chain" type system. This also permits the continuous transfer of hydraulic and electrical control lines.

8.7 Industry Standards and Classification Rules

The specific requirements for design of mooring systems are defined in Classification Rules and Industry Recommended Practices by API RP 2SK, Det Norske Veritas and Bureau Veritas. Additionally Lloyds, NMD, NPD and IACS provide similar rules and design information. Industry Guidance Notes or Recommended Practices are non-binding recommendations, which are sometimes incorporated into design criteria either in whole or in part. Classification Rules or Offshore Standards are invoked, if the owner of a platform elects to have the platform classed. In this case they become binding rules.

The specific requirements for floating production systems vary among these various references.[3] MODU rules do not explicitly cover mooring and leave the specification of safety factors and other conditions to the owner.

There is a significant difference in the current mooring criteria between European (mainly North Sea) and the U.S. Gulf of Mexico as reflected in API RP 2SK. As an example, the DNV Offshore Standard for Position Mooring specifies different safety factors for design depending on the criticality of the production. The safety factors are also applied differently. The DNV practice applies a separate safety factor to the computed mean load (FOS = 1.4) as opposed to the dynamic load (FOS = 2.1) (for dry tree applications). On the other hand, the API recommendation is for a single safety factor of 1.67 to be applied to the peak load for all types of mooring systems. The European standards also make allowances for application of the quantitative risk assessment methods for the selection of appropriate design loads.

8.7.1 Certification

Representative certification authority rules, such as those issued by DNV (2001) give guidance on relevant issues associated with mooring systems. There is strong emphasis on catenary analysis using chain and wire and, more recently, guidance on taut moorings using fibre ropes. The standards are, in many cases, developed from those for mobile drilling units.

The objectives of the standards are to provide:

- Uniform level of safety to mooring systems,
- Guideline for designers, suppliers and contractors,
- Reference document for contractual considerations between suppliers and contractors.

The standards are typically divided into a number of sections as follows:

- Environmental conditions and loads
- Mooring system analysis
- Thruster-assisted mooring
- Mooring equipment
- Testing

A further description of certification standards is given below for one particular authority. It is necessary to refer to the relevant certification standard for full information.

8.7.2 Environmental Conditions and Loads

Survival environmental criteria for permanent moorings are usually based on a 100-yr return period event. It is common to use two or three environments including the 100-yr

[3] Recommended practices are subject to continual review and updating. These values should not be considered definitive. The latest documentation should be consulted.

wave with associated wind and current, and the 100-yr wind with associated wave and current. In high current environments, such as, the Gulf of Mexico deepwater, North Atlantic and certain areas of West Africa and Southeast Asia, the current may be the controlling event and a 100-yr current plus associated wind and waves is also specified. Specification of "associated wind and waves" is somewhat subjective. A more rigorous method for specifying environment is to perform a "response-based analysis", see for example Standing, et al (2002). This method employs a simplified mathematical model of the platform and mooring responses to various environmental conditions. Hindcast environmental data covering many years, including extreme events, is compiled and used as input to this model. This might involve thousands of cases covering, for example, hindcast conditions every 6 h going back 10–20 yr at the specific site. The statistics of the responses are tabulated to determine a "100-yr response" usually defined as that response having a 0.01 chance of exceedance in any year. The environments which generate this response and responses close to this response are chosen for more refined analysis. "Response-based modelling" is not presently required by any rules or recommended practice, but it may be specified by the owner. The DNV Offshore Standards recommend determining a "100-yr response" for design based on a compilation of wave heights, periods covering a span of 100-yr environments and selecting the combination yielding the worst response.

In order to calculate the mooring line structural response it is necessary to apply appropriate environmental loads for the site under consideration. This usually corresponds to the wave and wind conditions having return periods of 100-yr, together with 10-yr return period current conditions. However, if, for example, current and wind are the dominant features, such as Gulf of Mexico conditions with loop currents and hurricanes, then 10-yr sea conditions combined with 100-yr current and wind should be assessed.

A number of sea states should be selected along a "contour line" representing the joint probability of significant wave height and peak wave period combinations at the mooring location. The contour represents wave height and period pairs for a specified return period, for example, 100-yr. Guidance notes and standards give examples of contour lines. Wind loads should consist of both steady and time-varying components, the latter being specified in both DNV and API documentation.

The weather directions to be considered depend on the vessel mooring arrangement. For vessels that cannot change direction relative to the weather it is necessary to consider waves, wind and current acting from the same directions. These are head, quartering and beam, together along with the mooring line for vessels with the symmetric mooring patterns. For non-symmetric mooring patterns, all directions, with a maximum 45° spacing, should be assessed. For vessels that can weathervane, site data should be used, if available, otherwise collinear weather should be applied at 15° to the vessel bow, together with a non-collinear condition with bow waves, wind and current acting from the same side at respectively 30 and 45° to the bow.

Wind and current loads can be established by model tests and/or calculations, see for example OCIMF (1994). Calculations are based on a drag force formulation, suitable coefficients being established from model tests or computational fluid dynamics. Current forces will increase, if the water depth is typically less than three times the vessel draught, OCIMF providing relevant enhancement factors. Current forces on multiple riser systems should be considered though forces on a system consisting of only a single riser are usually ignored. Current loads on moorings are only considered, if these are dominant, such as at sites with loop currents.

Marine growth on long-term moorings should be included by increasing the line weight and drag coefficient, C_d. A marine growth density of 1325 kg/m^3 is common, and the standards provide equations to calculate the mass of growth depending on the line type and diameter, together with growth thickness and water depth. The line drag coefficient can be assumed to increase linearly with growth thickness. For new lines, the standards indicate the following drag coefficients:

- $C_d = 2.6$ for stud chain,
- $C_d = 2.4$ for studless chain,
- $C_d = 1.8$ for six-strand steel wire rope,
- $C_d = 1.2$ for spiral strand with sheathing,
- $C_d = 1.6$ for spiral strand without sheathing.

Waves provide three loading mechanisms acting on the floating vessel. These result in mean wave drift motions, and responses at wave and low frequency as described in Section 8.3. For catenary moored structures, the restoring stiffness contributions to the wave frequency motions from the mooring and riser system are ignored in deep waters, though must be investigated for water depths below 70 m. For the taut moored structures, the restoring forces from the mooring and riser system must be addressed to establish whether they influence motions at wave frequencies. Shallow water also influences the horizontal motions of the vessel for depths less than 100 m, in that surge and sway motion amplification factors must be included. These can result in a doubling of the deepwater motions for large wave periods in a very shallow water.

Low-frequency motions for semi-submersibles and ships should be calculated in the horizontal directions only, that is, surge, sway and yaw. For deep draft floaters, such as spar platforms, vertical responses also need to be assessed. It is important to establish a stable equilibrium position for the vessel, where the steady forces of current, wind and wave drift balance the restoring forces from the station-keeping system. For systems that are free to yaw, vessel rotation should be included when calculating the mean forces.

The frequency or time-domain methods may be used to establish the vessel low-frequency response about this stable equilibrium position. Alternatively, the model test results may be used. It is important that the model test or simulation is carried out over a suitable length of time to give appropriate statistical quantities. A minimum of 3 h full-scale equivalent time is specified, though usually significantly longer time is beneficial. The model testing has been addressed in detail in Chapter 13.

8.7.3 Mooring System Analysis

Certification standards give guidance on the methods employed to perform the structural design of wire, chain and fibre mooring systems, including their combinations, used on floating vessels, including deep draft floaters, such as spars. The mooring system is assessed in terms of three limit states based on the following criteria:

- Ensuring that individual mooring lines have suitable strength when subjected to forces caused by extreme environmental loads – ultimate limit state (ULS).

- Ensuring that the mooring system has suitable reserve capacity when one mooring line or one thruster has failed – accidental limit state (ALS).
- Ensuring that each mooring line has suitable reserve capacity when subject to cyclic loading – fatigue limit state (FLS).

Guidance on the structural stiffness characteristics of wire, chain and synthetic fibre is given. For wire, this depends on whether the make-up is six strand or spiral strand; for chain, the stiffness depends on chain diameter. For fibre moors, it is necessary to establish the non-linear force-extension behaviour of the rope. If this is not available, then the vessel excursion should be established using the estimated post-installation line stiffness for both the ULS and ALS. Characteristic line tensions for ULS, ALS and FLS can be found using the storm stiffness. Section 8.4.5 describes these stiffness criteria in more detail.

The analysis procedures are divided into those attributable to establishing the platform response, and those associated with calculating the mooring line behaviour. Mooring line analysis must include the influence of line dynamics, if the vessel is to be used for floating production or storage, or if operations in depths greater than 200 m are considered. Additionally, vortex-induced vibration needs to be addressed for platforms of deep draft.

The platform response is, in many situations, strongly influenced by the damping associated with the low-frequency motions. This depends on sea and current conditions, mooring and riser make-up, together with water depth. Model tests can be used to establish damping, though as described in Section 8.4.9, damping levels associated with the mooring are difficult to quantify. Risers can provide restoring, damping and excitation forces making their influence on floater response more complicated.

The mooring analysis should ideally consider line dynamics, i.e. the inertia and drag force contributions acting on the line components, when calculating line loading associated with the platform wave frequency motions. Quasi-static analysis, allowing for submerged weight and elasticity of line, platform motion and seabed reaction/friction forces, is usually appropriate when dealing with platform mean- and low-frequency motions.

In establishing the characteristic line tension for either the ULS or ALS, Gaussian statistical methods are used, recognising the random nature of the platform response and line tensions under realistic environmental conditions. This allows the maximum wave and low-frequency platform excursions to be found, based on the relevant motion standard deviation and the number of oscillations during a specified period, usually taken as 3 h. The above excursions are combined, after including the mean offset, by taking the larger of the sum of the significant and maximum excursions. Finally, if line dynamics are considered, the maximum wave frequency line tension is obtained from its standard deviation. This depends on the excursion about which wave frequency motion occurs and the number of associated platform oscillations. Combining this with the mean and quasi-static tension components gives the characteristic dynamic line tension.

The mooring analysis must also consider the characteristic capacity or strength for the ULS and ALS, recognising that the line strength is likely to be less than the average strength of its components, whether these be chain links or wire fibres. Thus the characteristic capacity includes the influence of the component mean breaking strength and its coefficient of variation. Other connecting links and terminations must be designed with higher strength characteristics than the main line elements, together with improved fatigue lives.

The design equations to be used for ULS and ALS are based on the concept of partial safety factors (see Chapter 5). The design equation is of the form:

$$S_c - T_{c,\mathrm{dyn}}\alpha_{\mathrm{dyn}} - T_{c,\,\mathrm{mean}}\alpha_{\mathrm{mean}} \geq 0 \tag{8.10}$$

where S_c is the line capacity and $T_{c,\mathrm{mean}}$, $T_{c,\mathrm{dyn}}$ are the characteristic mean and dynamic tensions. The partial safety factors, α_{mean} and α_{dyn}, are specified in the standards. These take on values of between 1.1 and 2.5 for the ULS, and 1.0 and 1.35 for the ALS. The values depend on the intended operation of the vessel, in that higher factors are imposed where mooring failure could lead to unacceptable situations such as loss of life, collisions, sinking or hydrocarbon release. The safety factors are also higher, if a quasi-static analysis, as opposed to a more rigorous dynamic analysis is carried out.

In evaluating the vessel excursions and line tensions, care must be taken not to exceed the permissible vessel offset and line length. For example, horizontal offsets will be influenced by gangway connections to another fixed or floating structure. For rigid riser operations, offsets are limited by the maximum allowable riser angle at the BOP flex joint, and must also allow for heave compensation equipment. Manufacturers' limitations must be considered for flexible risers and steel catenary risers. Line lengths are influenced by whether anchors can withstand up-lift loads. No up-lift is allowed for the ULS, but up-lift may be allowed for the ALS, if the vertical loads do not impair the anchor-holding power.

The layout of the subsea architecture must also be considered within the context of mooring system analysis. For the ULS and ALS there must be a minimum vertical clearance between lines and all subsea equipment of respectively 10 and 0 m (no contact).

A further safety factor should also be applied for situations where analysis has been performed at the limiting sea state for normal operations, usually corresponding to mild weather. The safety factor applies to the mean and dynamic tension components, that is the last two terms on the left hand-side of equation (8.10).

For mooring chains designed to be positioned at the same location for greater than four years, the characteristic capacity of the line must be reduced for the effects of corrosion. This corrosion reduction is larger for components at the seabed and in the surface splash zone. If regular inspection schemes are to be carried out, the required corrosion reductions are smaller. For steel wire rope, the lifetime degradation depends on the construction and level of protection applied. Note however that when addressing the FLS, only 50% of the corrosion allowance need be applied.

When considering the mooring FLS, it is necessary to account for the accumulated fatigue damage that occurs from cyclic loading by individual sea states making up the long-term environment. The relevant vessel heading should be allowed for. For each of these sea states, it is necessary to calculate the mooring system response together with the sea state occurrence frequency. In practice, the long-term environment can be discretised into something like 8–12 headings and 10–50 sea states.

In an individual sea state the fatigue damage d_i is given by:

$$d_i = n_i \int_0^{\infty} \frac{f_{Si}(s)}{n_c(s)} ds \tag{8.11}$$

where the number of stress cycles, n_i, is calculated from the product of the mean up-crossing rate of the stress process (in Hz), the probability of occurrence of the sea state, together with the mooring system design lifetime in seconds. The term f_{Si} represents the probability density of peak to trough nominal stress ranges for the individual state. The stress ranges are obtained by dividing the line tension ranges by the nominal cross-sectional area. This is taken as $\pi d^2/4$ for steel wire rope and $2\pi d^2/4$ for chain. The procedure for fibre ropes is described further here. The term n_c in equation (8.11) represents a fatigue property of the line, giving the number of stress ranges of magnitude s that would lead to failure.

For wire and chain, the capacity against fatigue caused by tension is defined in terms of the number of stress range cycles given by:

$$\log(n_c(s)) = \log(a_D) - m \log(s) \tag{8.12}$$

where s is the stress range double amplitude (MPa) and m, and a_D are the slope and intercept on the seawater S–N curves, given in the Standards for various chain and wire rope types.

In practice, the integral given in equation (8.11) can be replaced by discrete terms for each sea state i, in terms of the expected value of the nominal stress range. Additionally, if the stress process has negligible low-frequency content, then narrow-banded assumptions allow the damage to be established in terms of the stress standard deviation. If, however, there are wave and low-frequency contributions to the stress, then rainflow counting will provide the most accurate estimate. For this situation, two alternatives, the combined spectrum or dual narrow-banded approach, described in the standards can be used.

For fibre rope, the capacity against fatigue caused by tension–tension effects is given by:

$$\log(n_c(R)) = \log(a_D) - m \log(R) \tag{8.13}$$

where R is the ratio of tension range to characteristic strength and m, a_D are given in the standards.

The design equation to be used for FLS is similar to that for ULS and ALS, being of the form:

$$1 - d_c \gamma_F \geq 0 \tag{8.14}$$

where d_c is the damage that accumulates as a result of all the individual environmental states over the system design lifetime, and γ_F is a fatigue safety factor. The following guidance is given on safety factors:

- $\gamma_F = 3$, for wire and chain line that can regularly be inspected on-shore.
- $\gamma_F = 5$, for wire and chain line that cannot regularly be inspected on-shore, and is configured so that the ratio of fatigue damage in two adjacent lines is less than 0.8.

- $\gamma_F = 5$–8, for wire and chain line that cannot regularly be inspected on-shore, and is configured so that the ratio of fatigue damage in two adjacent lines is greater than 0.8.
- $\gamma_F = 60$, for polyester rope. Note that this is much larger compared to steel because of the increased variability in fatigue test results.

Fatigue properties of wire and chain are typically defined in terms of T–N relationship derived from tension–tension fatigue tests. Similar to conventional S–N fatigue curves, the design fatigue curve is in the form:

$$N = K \cdot R^{-M} \tag{8.15}$$

where N = number of cycles, R = ratio of tension range (double amplitude) to nominal breaking strength, M = slope of T–N Curve, and K = intercept of the T–N Curve. M and K are given in table 8.7, where Lm = ratio of mean applied load to the breaking strength of wire rope from the catalogue.

The chain fatigue data presented in API RP2SK is for the stud link chain. DNV OS-E301 presents data in the form [API, Chaplin, 1991]:

$$n_c(s) = a_D s^{-m} \tag{8.16}$$

where $n_c(s)$ = number of stress ranges, s = stress range (MPa), a_D = intercept of the S–N curve, m = slope of the S–N curve. Values of a_D and m are given in table 8.8.

Table 8.7 Fatigue curve parameters for wire rope and chain (from API RP 2SK)

Component	M	K
Common Chain Link	3.36	370
Baldt or Kenter Connecting Link	3.36	90
Six/multi strand rope	4.09	$10^{(3.20-2.79\ \mathrm{Lm})}$
Six/multi strand rope, Lm = 0.3	4.09	231
Spiral strand rope	5.05	$10^{(3.25-3.43\ \mathrm{Lm})}$
Spiral strand rope, Lm = 0.3	5.05	166

Table 8.8 Fatigue curve parameters (from DNV OS-E301)

Type	a_D	m
Stud chain	1.2×10^{11}	3.0
Studless chain	6.0×10^{10}	3.0
Six strand wire rope	3.4×10^{14}	4.0
Spiral strand wire rope	1.7×10^{17}	4.8

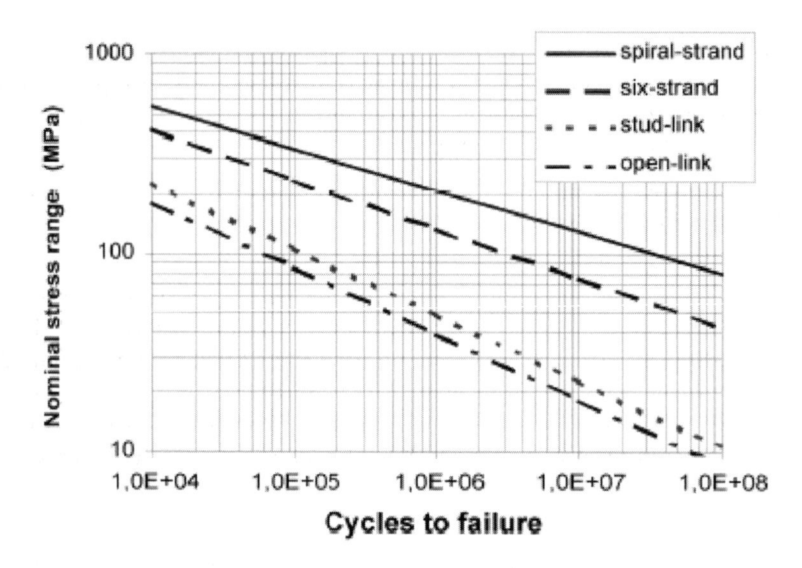

Figure 8.32 Chain wire fatigue curves based on stress (DNV OS-E301)

The DNV curves are shown in fig. 8.32. This relationship is similar to the API curve, but it is based on stress rather than tension. In order to convert from tension to stress the nominal steel areas given in a table in the API RP2SK may be used.

The fatigue of wire rope and chain running over sheaves and fairleads will generally be lower than pure tension–tension fatigue. Additional stress due to bending may be used to account for this effect. For effects other than tension fatigue, for example chain or wire bending and tension–compression for fibre ropes, further consideration, such as experimental testing, is required.

As an alternative to the above procedures, mooring design may be carried out using structural reliability analysis. Standards give guidance on target annual probabilities of failure when performing reliability analysis.

8.7.4 Thruster-Assisted Mooring

This section of the standards gives methods and guidance associated with the design of thruster-assisted moorings. Thrusters can be used to reduce the mooring system loads caused by mean environmental forces, provide damping of the low-frequency motions and assist in heading control.

For manual and automatic remote control systems respectively 70 and 100% of the net thrust can be used when establishing the ULS or ALS. However, if a failure leads to a thruster stop situation during the ALS then this must be considered equivalent to a line failure.

The available (net) thrust can be estimated by calculation at the early design stage based on the propeller thrust at bollard pull. A useful conversion factor is 0.158 kN/kW for nozzle propellers and 0.105 kN/kW for open propellers. These values need correcting to account for in-flow velocity at the propeller, propeller rotation sense and propeller/thrust installation geometry and arrangement – see for example API RP 2SK for further guidance and Ekstrom, et al (2002) for information on the thruster–thruster interaction.

Thrust contributions to station-keeping can be evaluated using the methods of mean load reduction and system dynamic analysis as follows:

- The mean load reduction method involves subtracting the surge and sway components of allowable thrust from the mean environmental loads for spread-moored vessels. For single-point moored vessels, the standards give guidance for methods to establish the contribution to the yaw moment when thrusters are used to influence vessel heading.
- A system dynamic analysis generally consists of a surge, sway and yaw simulator. This can produce mean offset and low-frequency vessel responses corresponding to time-domain records of environmental force. Wave frequency forces are not balanced by the system.

Thrusters can consist of both fixed and rotating configurations and be of variable pitch and speed. The selection is made based on the requirements of the mooring system, but the appropriate configuration must have an automated power management system. There should be a manual or automatic remote thrust control system. Automatic control systems are more sophisticated than manual and can have features such as monitoring of vessel position and line tension alarms, consequence analysis and simulation capabilities, relevant data logging, self-diagnostics and allow system response to major failures. Further details are given in the standards.

8.7.5 Mooring Equipment

Standards provide requirements for all mooring equipment and its installation for temporary and emergency mooring, position mooring and towing. Only a brief overview is given here.

Information on various anchor types is provided including fluke, plate, piled, gravity and suction anchors. Specifications for anchor construction materials are also discussed.

Data on mooring chains and associated connecting links and shackles is also provided. Offshore mooring chain is graded depending on its minimum yield and tensile strength, together with Charpy v-notch energy. For long-term mooring systems, where onshore inspection is not possible, only limited connection elements, such as D shackles or triplates (fig. 8.25), are acceptable. Where mobile offshore units change location frequently, other connections such as Kenter shackles, C links and swivels are allowed in the mooring line make-up. Generally there is a lack of documented fatigue data on these latter connection elements, though API RP 2SK does provide fatigue information on Kenter shackles.

Six-strand wire rope (fig. 8.23) is normally used by mobile offshore vessels for anchor and/or towing lines. This rope is commonly divided into two groups; either 6 by 19, consisting of 6 strands with between 16 and 27 wires in each strand; or 6 by 36, consisting of 6 strands with between 27 and 49 wires in each strand. Long-term floating production vessels use spiral strand steel wire ropes as this has improved fatigue and corrosion behaviour.

Synthetic fibre ropes can be used either as inserts in a catenary mooring layout or as part of a taut leg system. Recognised standards, such as API RP 2SM have been produced that document the use of fibre ropes. The technology is still developing, but fibres being considered for mooring system use include polyester, aramid, high-modulus polyethylene (HMPE) and nylon. Standards specify the relevant load bearing yarn properties and tests to be documented, together with those for the yarn sheathing material. Rope constructions under consideration are parallel strands, parallel yarns and "wire rope constructions". Braided constructions are not considered because of the concerns over their long-term fatigue behaviour. Guidance is also given on stiffness values for polyester, aramid and HMPE for post-installation, drift and storm conditions for deepwater fibre moorings.

Other potential failure modes are also discussed in the standards including:

- hysteresis heating – lubricants and fillers can be included to reduce hotspots,
- creep rupture – in particular this is relevant to HMPE yarns, and the risks need careful evaluation,
- tension – tension fatigue – only limited data exist, indications being that fatigue resistance is higher than for steel wire ropes,
- axial compression fatigue – on leeward lines during storms for example, prevented by maintaining a minimum tension on the rope,
- particle ingress – causes strength loss by abrasion from water-borne material such as sand, prevented by using a suitable sheath and not allowing contact between the rope and seabed.

Fibre rope terminations under consideration included socket and cone, conventional socket and spliced eye, the latter being the only one presently qualified at sizes appropriate to deep-water mooring systems.

The standards give design, material requirements and capacity for additional mooring hardware including windlasses, winches, chain stoppers and fairleads together with end attachments. The necessary structural arrangement for the mooring equipment is also specified, together with arrangements and devices for towing purposes and measurement of line tension. Lee, et al (1999) describe the ABS approach on synthetic ropes, while Stoner, et al (1999) present the contents of an engineer's design guide for fibre moorings, emphasising the limitations in the available test data. Stoner, et al (2002) outline additional work necessary before fibre moorings can be used at harsh weather locations.

8.7.6 Tests

The standards give comprehensive guidance on tests to be carried out on mooring system hardware including the following:

- Fluke anchors for mobile/temporary and long-term moorings,
- Mooring chain and accessories,
- Steel wire rope,
- Windlass and winch assemblies,
- Manual and automatic remote thruster systems,
- Synthetic fibre ropes.

More information can be found in the standards. For example, the UK Health & Safety Executive (2000) gives a comprehensive discussion of model testing techniques for floating production systems and their mooring systems.

References

American Petroleum Institute, "Recommended practice for design and analysis of stationkeeping systems for floating structures", API RP-2SK (latest edition).

American Petroleum Institute (March 2001). "Recommended practice for design, manufacture, installation and maintenance of synthetic fiber ropes for offshore mooring", API RP 2SM, (1st ed.).

Aranha, J. A. P., Pinto, M. O., and Leite, A.J.P. (2001a). "Dynamic tension of cables in random sea: Analytical approximation for the envelope probability density function". *Applied Ocean Research*, Vol. 23, pp. 93–101.

Aranha, J. A. P. and Pinto, M. O. (2001b) "Dynamic tension in risers and mooring lines: An algebraic approximation for harmonic excitation". *Applied Ocean Research*, Vol. 23, pp. 63–81.

Brown, D. T. and Liu, F. (1998). "Use of springs to simulate wind induced moments on turret moored vessels". *Journal of Applied Ocean Research*, Vol. 20, No. 4, pp. 213–224.

Brown, D. T., Lyons, G. J., and Lin, H. M. (1997). "Large scale testing of mooring line hydrodynamic famping contributions at combined wave and frift frequencies", *Proc. Boss 97, 8th Intl. Conf. on Behaviour of Offshore Struct.*, Delft, Holland, ISBN 008 0428320, pp. 397–406.

Brown, D. T. and Mavrakos, S. (1999). "Comparative study of mooring line dynamic loading". *Journal of Marine Struct.*, Vol. 12, No. 3, pp. 131–151.

Chaplin, C. R. (August 1991). Prediction of Wire Rope Endurance for Mooring of Offshore Structures, Working Summary, Joint Industry Project (JIP) Report issued by Noble Denton & Associates, London.

Chatjigeorgiou, I. K. and Mavrakos, S. A. (2000). "Comparative evaluation of numerical schemes for 2-D mooring dynamics". *International Journal of Offshore and Polar Engineering*, Vol. 10(4), pp. 301–309.

CMPT (1998). *Floating Structures: A Guide for Design and Analysis*. Vol. 2, Ed. Barltrop, N. 101/98.

Det Norske Veritas OS-E301 (June 2001). "Position Mooring".

Ekstrom, L. and Brown, D. T. (2002). "Interactions between thrusters attached to a vessel hull", 21st Offshore Mechanics and Arctic Engineering Intl Conf., American Society of Mechanical Engineers, Paper OMAE02-OFT-28617, Oslo, Norway.

Gobat, J. I. and Grosenbaugh, M. A. (2001a). "A simple model for heave-induced dynamic tension in catenary moorings". *Applied Ocean Research*, Vol. 23, pp. 159–174.

Gobat, J. I. and Grosenbaugh, M. A. (2001b). "Application of the generalized-α method to the time integration of the cable dynamics equations". *Computer Methods in Applied Mechanics and Engineering*, Vol. 190, pp. 4817–4829.

Health & Safety Executive, UK. (2000). "Review of model testing requirements for FPSOs", Offshore Technology Report 2000/123, ISBN 0 7176 2046 8.

Huse, E. and Matsumoto, K. (1989). "Mooring line damping owing to first and second order vessel motion", *Proc. OTC*, Paper 6137.

Huse, E. (1991)."New developments in prediction of mooring line damping", *Proc. OTC*, Paper 6593, Houston, USA.

Lee, M., Flory, J., and Yam, R. (1999). "ABS guide for synthetic ropes in offshore mooring applications", *Proc. OTC*, Paper 10910, Houston, Texas.

Matsumoto, K. (1991). "The influence of mooring line damping on the prediction of low frequency vessels at sea", *Proc. OTC*, Paper 6660, Houston, USA.

Myers, J. J., ed. (1969). *Handbook of Ocean and Underwater Engineering*, McGraw-Hill Book Company.

NTNF (1991). "FPS 2000 Research Programme – Mooring Line Damping", Part 1.5, E Huse, Marintek Report.

Oil Companies International Marine Forum (OCIMF) (1994). "Prediction of wind and current loads on VLCCs ", (2nd ed.)

Standing, R. G., Eichaker, R., Lawes, H. D., Campbell, and Corr, R. B. (2002). "Benefits of applying response based analysis methods to deepwater FPSOs", *Proc. OTC*, Paper 14232, Houston, USA.

Stoner, R. W. P., Trickey, J. C., Parsey, M. R., Banfield, S. J., and Hearle, J. W. (1999). "Development of an engineer's guide for deep water fiber moorings", *Proc. OTC*, 10913, Houston, Texas.

Stoner, R. W. P., Ahilan, R. V., and Marthinsen, T. (2002). "Specifying and testing fiber moorings for harsh environment locations", *Proc. OMAE*, 28530.

Thomas, D. O. and Hearn G. E. (1994). "Deep water mooring line dynamics with emphasis on sea-bed interaction effects", *Proc. OTC*, 7488, Houston, USA.

Vandiver, J. K. (1988). "Predicting the response characteristics of long flexible cylinders in ocean currents", *Symposium on Ocean Structures Dynamics*, Corvallis, Oregon.

Webster, W. (1995). "Mooring induced damping". *Ocean Engineering*, Vol. 22, No. 6, pp. 571–591.

Handbook of Offshore Engineering
S. Chakrabarti (Ed.)

© 2005 Elsevier Ltd. All rights reserved.

Chapter 9

Drilling and Production Risers

James Brekke
GlobalSantaFe Corporation, Houston, TX, USA

Subrata Chakrabarti
Offshore Structure Analysis, Inc., Plainfield, IL, USA

John Halkyard
Technip Offshore, Inc., Houston, TX, USA

9.1 Introduction

Risers are used to contain fluids for well control (drilling risers) and to convey hydrocarbons from the seabed to the platform (production risers). Riser systems are a key component for offshore drilling and floating production operations. In this chapter section 9.2 covers drilling risers in floating drilling operations from MODUs and section 9.3 covers production risers (as well as drilling risers) from floating production operations.

A riser is a unique common element to many floating offshore structures. Risers connect the floating drilling/production facility with subsea wells and are critical to safe field operations. For deepwater operation, design of risers is one of the biggest challenges. During use in a floating drilling operation, drilling risers are the conduits for operations from the mobile offshore drilling unit (MODU). While connected much of the time, drilling risers undergo repeated deployment and retrieval operations during their lives and are subject to contingencies for emergency disconnect and hang-off in severe weather. Production risers in application today include top tension production risers (TTRs), flexible pipes steel catenary risers (SCRs), and free-standing production risers. More than 50 different riser concepts are under development today for use in deepwater and ultra-deepwater. A few of the most common riser concepts are shown in fig. 9.1.

According to Clausen and D'Souza (2001), there are more than 1550 production risers and 150 drilling risers in use today, attached to a variety of floating platforms. About 85% of production risers are flexible. Flexible risers are applied in water depths of up to 1800 m, while a top tension riser and a steel catenary riser are used in depths as much as 1460 m. The deepest production riser in combined drilling and early production is in a water depth

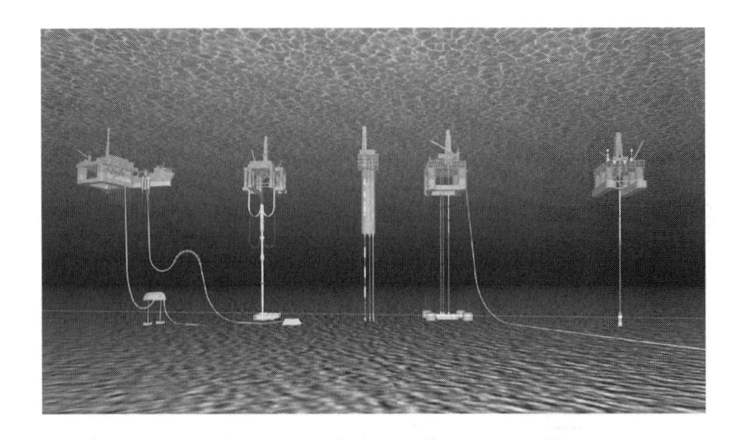

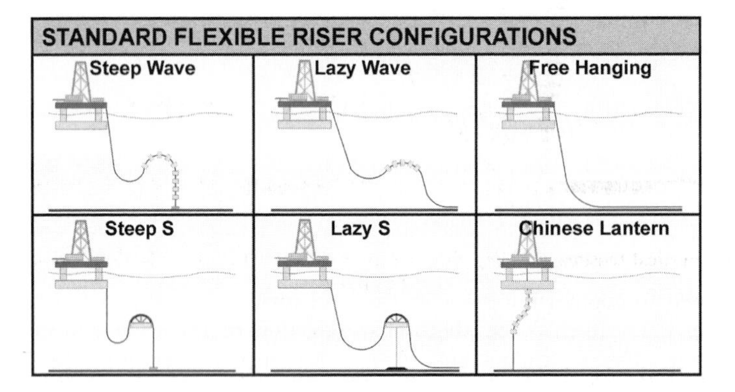

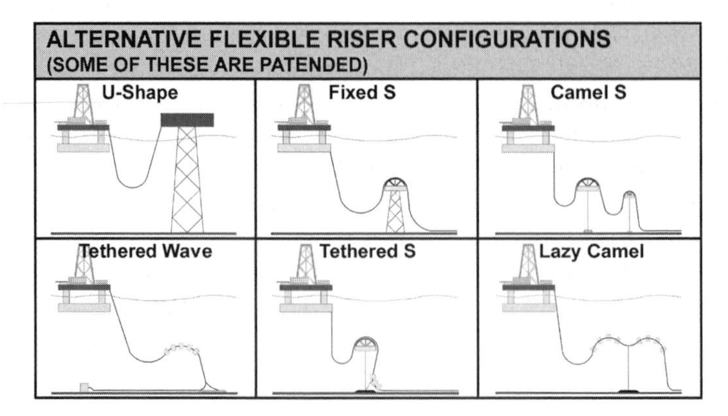

Figure 9.1 Schematic of riser concepts [Courtesy of Clausen and D'Souza, Subsea7/KBR (2001)]

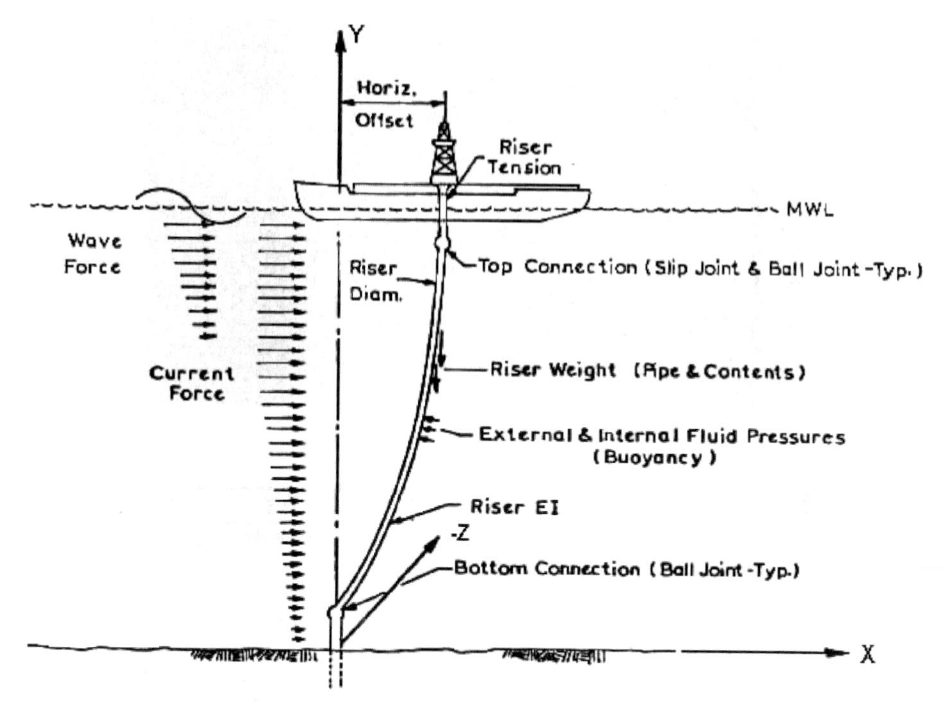

Figure 9.2 Vertical tensioned drilling riser [Note: balljoint (or flex joint) is also located just below drill floor]

of 1853 m in Brazil for the Roncador Seillean FPSO. Drilling risers are in use in greater than 3000 m depth.

A top tensioned riser is a long slender vertical cylindrical pipe placed at or near the sea surface and extending to the ocean floor (see fig. 9.2). These risers are, sometimes, referred to as "rigid risers" or "direct vertical access" risers.

The development of different types of riser with the riser size (diameter in inches) and water depth up to 2000 m is shown in fig. 9.3. The envelopes for the different riser types are given in the figure. The installed SCRs for the floaters are identified in the figure.

The technical challenges and the associated costs of the riser system increase significantly with water depths [Clausen and D'Souza, 2001]. The cost of a riser system for a deepwater drilling and production platform compares with that of the hull and mooring system.

The risers connecting a floating vessel and the seafloor are used to drill or produce individual wells located beneath the floating vessel or for import and export of well stream products. They are connected to a subsea wellhead, which in turn is attached to the supporting sub-mudline casing. The drilling riser is attached via an external tieback connector, while the production riser can be attached via either an external or internal tieback connector [Finn, 1999]. The first joint of the riser above the tieback connector is a

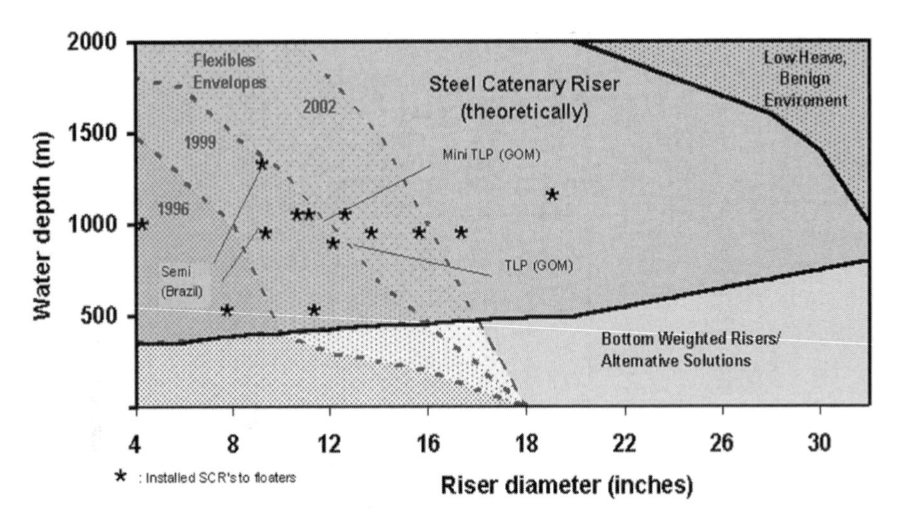

Figure 9.3 Progress of production riser diameters with water depth [Courtesy of Clausen and D'Souza, Subsea7/KBR (2001)]

special segment called the stress joint that is designed to resist the large bending moments, but flexible enough to accommodate the maximum allowable riser angular displacements. Typically these joints are composed of a forged tapered section of pipe that can be made of either steel or titanium. Newer designs call for the stress joints to be composed of a series of pipe segments that are butt-welded or a group of concentric pipes welded to a special terminating flange. In lieu of a stress joint, elastomeric flex or ball joint may be used to accommodate bending at the sea floor.

The top tension risers are initially held in a desired tension which helps in the bending resistance of the riser under the environmental loads. This tension is provided by a mechanical means, as shown in fig. 9.4a for a drilling riser. The tension may also be provided by syntactic foam or buoyancy cans. A top tension riser designed for the application with Spar is shown in fig. 9.4b. The Spar riser uses buoyancy tanks for the top tension. The riser entering the keel of the spar is detailed in the figure. Three different riser pipe configurations are illustrated in fig. 9.4b. In the first case the Neptune Spar uses a single 9-5/8 in. diameter casing, which encompasses the production tubing and other annulus lines. In the second case a dual casing riser is used with internal tubing. In the third configuration the riser tubing strings are separate, requiring fewer riser pipes and less external buoyancy. It is better suited for deeper waters where large riser weight becomes a problem. The selection of the riser configuration is based on a risk/cost benefit analysis.

The general riser dimensions are based on the reservoir information and the anticipated drilling procedures. The size of the tubing is determined from the expected well flow rate. The wall thickness of each riser string is computed from the shut-in pressure and drilling and completion mud weights. The outside dimension of the components that must pass through the pipe, such as subsurface safety value (SSSV), drill bit, or casing connector generally determines the internal diameter of the riser. The hoop stress usually governs the wall

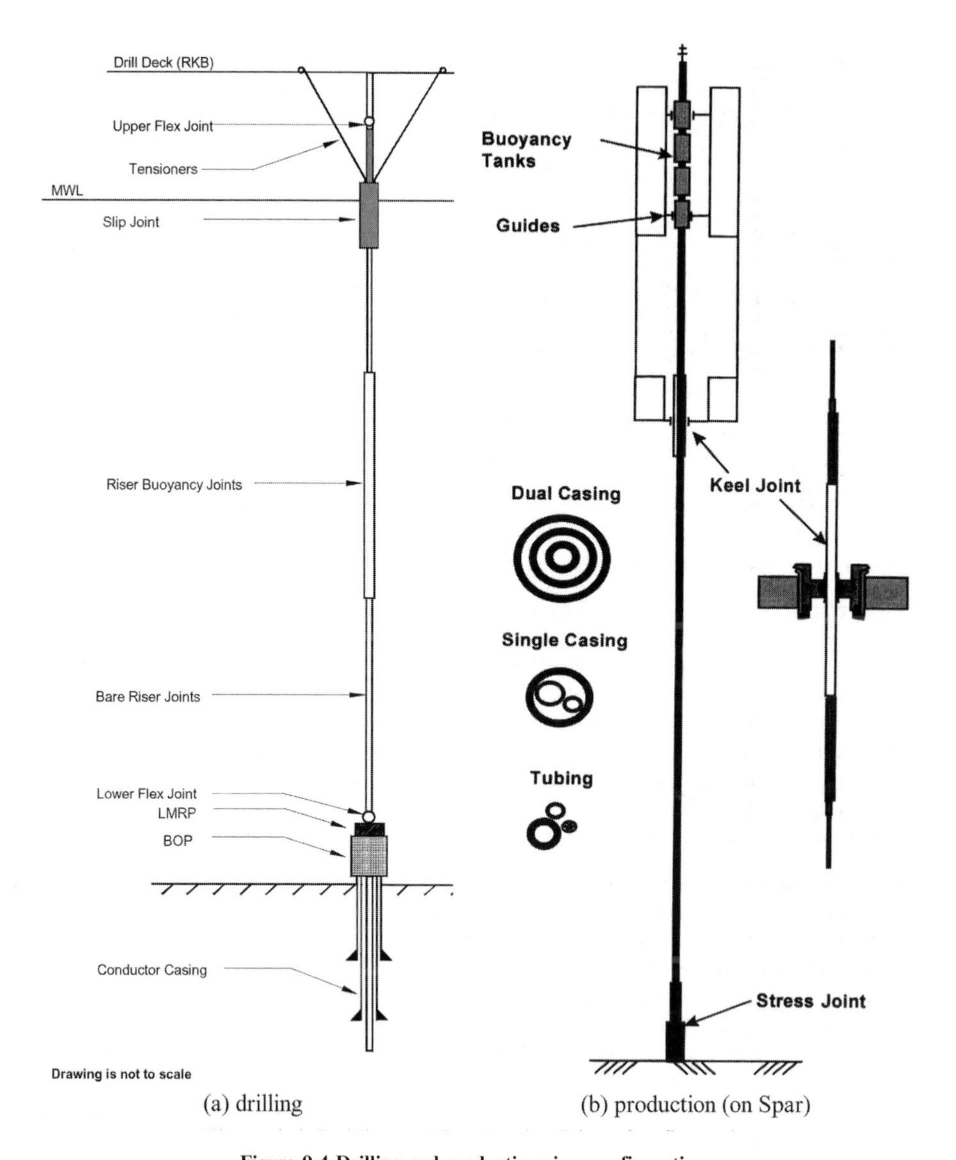

(a) drilling (b) production (on Spar)

Figure 9.4 Drilling and production riser configuration

thickness of the riser pipes. In deeper waters, the wall thickness may depend on the axial stress. The capped-end force generated by the internal pressure should also be considered in computing the axial stress. The bending stress is a determining factor at the upper and lower ball joints of the riser. In these areas thicker riser elements may be required to limit the stresses. The dimensions of the stress joint are more difficult to compute since they must be strong and flexible at the same time. Generally, a finite element program is used that determines the riser bend to the desired maximum angle at the joints. The dimensions are adjusted until the required strength is achieved. The potential for riser interference is also checked during an early determination of the riser component dimensions.

9.2 Drilling Risers

This section provides a description of the analysis procedures used to support the operation of drilling risers in floating drilling. The offshore drilling industry depends on these procedures to assure the integrity of drilling risers, with the goal of conducting drilling operations safely, with no environmental impact, and in a cost-effective manner.

The main emphasis of this section is on drilling risers in deep water (i.e. greater than 900 m or 3000 ft) and some specific coverage is given to drilling from dynamically-positioned drillships in ultradeep water (i.e. greater than 1800 m or 6000 ft of water). Besides analytical procedures, some coverage is given to the operational procedures and the equipment that are peripheral to the drilling riser. However, a comprehensive treatment of drilling riser operations and equipment is outside the scope of this chapter. References to the industry guidelines given below provide additional details.

As the water depths for drilling operations have increased, the importance of the drilling riser has grown in importance. Effective analytical support of the drilling riser and the related operations can substantially reduce the cost and risk of drilling an offshore well. The potential loss of a drilling riser presents high consequences. Currently, the cost of the drilling riser can be tens of millions of dollars; but in addition, the cost of operational downtime for an event involving the loss of a drilling riser can exceed one hundred million dollars. Avoidance of such losses further benefits the entire oil and gas industry through improved safety, reduced environmental impact, and reduced insurance cost.

Some of the guidelines for analysis and operation of drilling risers are contained in API Recommended Practice 16Q (1993). As of this writing, this document, API RP 16Q, is being revised for release by the International Standards Organisation (ISO).

Another related document, API Bulletin 5C3 is referenced for its collapse and burst formulas used in drilling riser design. This document is entitled "Bulletin on Formulas and Calculations for Casing, Tubing, Drill Pipe, and Line Pipe Properties, API Bulletin 5C3, Sixth Edition, October 1, 1994".

This section will cover some of the important aspects in the procedures for drilling riser analysis. This begins with a discussion of metocean conditions, which are a primary driver in determining the operational limitations of a drilling riser at a specific site. This is followed by discussions of the design and configuration of a riser, including the issue of vortex-induced vibration and how the configuration can be modified to help manage it. The remaining sections cover analysis of the drilling riser in various conditions such as disconnected, connected, during emergency disconnect, and as recoil occurs after disconnect.

Sample riser analysis results are reported in this chapter for various water depths as deep as 2700 m or 9000 ft. These results are taken from the analyses done for specific sites for which data are available.

9.2.1 Design Philosophy and Background

To assess whether bending or riser tension dominates, the following non-dimensional number [Moe, 2004]

$$\lambda_{N,\text{tens}} = \frac{T_0 L^2}{\pi^2 n^2 EI} \tag{9.1}$$

may be used. For $\lambda_{N,\text{tens}}$ equal to 1, the stiffness contribution from the bending and tension stiffness will be about the same, while for larger values the tension stiffness will dominate. Here T_0 represents the average tension, L the riser length, EI is the bending stiffness and n the number of half waves. The effects of tension and bending stiffness are both typically included in the riser analysis, and in the water depth of interest, tension dominates the stiffness.

9.2.2 Influence of Metocean Conditions

The selection of accurate metocean conditions for a specific site for use in the analysis of a drilling riser is usually difficult, but it can, sometimes make the difference in whether or not a well can be drilled economically. The drilling riser is analysed based on the collection of wind, waves, and the current profile conditions for a specific well site. These metocean conditions can be based on information for a general region or an area near the well site. Whatever the case, a common understanding of the basis for the metocean conditions between the metocean specialist and the riser analyst is an important part of the process.

The current profile often drives the analytical results used for determining when drilling operations through a riser should be shut down. The steady current loading over the length of the riser influences the riser deflections, and the top and bottom angles that restrict drilling operations. Furthermore, high currents cause vortex-induced vibrations (VIV) of the riser, which lead to increased drag load and metal fatigue. Current profile data at a future well site can be more difficult to collect than data on winds and seastates due to the large amount of data to be gathered throughout the water depth. Furthermore, current features in many regions of the world tend to be more difficult to analyse due to a lesser understanding of what drives them, particularly in the deeper waters.

Winds and waves are important when considering the management of drilling riser operations in storms. Although not as important for determining the shape of the riser, the winds and seastates have a greater bearing on when the drilling riser should be retrieved (pulled) to the surface, i.e. when the mooring system will be unable to keep the vessel within an acceptable distance of the well.

Drilling risers are operated in conditions all over the world. These include large seastates off the east coast of Canada and the North Sea, the combination of high seastates and high currents west of Shetlands, the high currents offshore Brazil and Trinidad, and the cyclonic events combined with high currents in the Gulf of Mexico and offshore northwestern Australia. Typical metocean conditions for the Gulf of Mexico are listed below in table 9.1.

9.2.3 Pipe Cross-Section

The sizing of the pipe is important in order to assure the integrity of the riser for burst and collapse considerations. Collapse is generally checked to ensure the riser can withstand

Table 9.1 Typical design metocean criteria for Gulf of Mexico

		Riser connected/drilling						Riser connected/non-drilling				Riser pulled	
		1-yr winter storm		10-yr eddy		Sudden squall		10-yr winter storm		100-yr eddy		100-yr hurricane	
Winds		m/s	knots	m/s	knots	m/s	knots	m/s	knots	m/s	knots	m/s	knots
Vwind (1 h)		18.0	35.0	15.0	29.2	26.0	50.5	22.0	42.8	15.0	29.2	45.0	87.5
Vwind (1 min)		21.1	41.0	17.7	34.4	31.4	61.0	26.0	50.5	17.7	34.4	53.1	103.2
Seastate		m	ft	m	ft	m	ft	m	ft	m	ft	m	ft
Hs		4.9	16.0	3.5	11.5	1.5	4.9	5.8	19.0	3.5	11.5	12.5	41.0
Tp (s)		10.0		9.0		5.9		10.6		9.0		15.0	
Mean T (s)		7.7		6.9		4.6		8.2		6.9		11.6	
Current													
m	ft	m/s	knots	m/s	knots	m/s	knots	m/s	knots	m/s	knots	m/s	knots
Surface		0.30	0.59	1.40	2.72	0.30	0.59	0.30	0.59	2.00	3.89	1.00	1.94
60	197			1.40	2.72					2.00	3.89	1.00	1.94
76	249	0.30	0.59			0.30	0.59	0.30	0.59				
77	253	0.15	0.30			0.15	0.30	0.15	0.30				
100	328			1.40	2.72					2.00	3.89	0.20	0.39
150	492			1.10	2.14					1.50	2.92		
200	656			0.80	1.56					1.20	2.33		
300	984			0.60	1.17					0.80	1.56		
500	1641			0.30	0.58					0.40	0.78		
Near bottom		0.15	0.30	0.20	0.39	0.15	0.30	0.15	0.30	0.20	0.39	0.20	0.39

NOTES:
– Drilling can be conducted with mud weights of up to 16 ppg mud. Depending on the site-specific current conditions, drilling could be limited for certain mud weights
– Some level of vortex-induced vibration (VIV) could be experienced in the eddy conditions. Depending on the site-specific current conditions, vortex-suppression devices could be warranted

exterior pressure due to a specified voided condition in the riser, while burst is checked to ensure that the riser can withstand the interior pressure from the drilling fluid (mud). The bore of the wellhead housing generally dictates the bore (inside diameter) of the riser pipe, and resistance to collapse and burst pressures generally dictates its wall thickness.

9.2.3.1 Wellhead Housing

The oil and gas industry has generally selected a few standard bore sizes for its subsea wellhead housings. These wellhead bore sizes include 18-3/4-in., 16-3/4-in. and 13-5/8-in. The selection of the bore size determines the size of the casing strings that can be run through the wellhead and hung off in the wellhead housing. The most common of these in use today is the 18-3/4-in. wellhead. With this wellhead size, the drilling riser inner diameter should be greater than 18-3/4-in., so most risers have a 21-in. (or, in some cases, 22-in.) outer diameter, leaving enough margin for the variable riser wall thickness that may be necessary for deeper waters.

9.2.3.2 Burst Check

For the burst check, the water depth, the highest mud weight, the fabrication tolerances and the yield strength of the pipe are used to determine the minimum wall thickness of the riser. API Bulletin 5C3 (1994) is commonly used as the basis for this calculation.

9.2.3.3 Collapse Check

The riser must have sufficient collapse resistance to meet the conditions imposed by the operator. For an ultra deep water well, typical conditions call for collapse resistance sufficient to withstand the riser being void over half its length. This requirement usually covers the case of emergency disconnect in which a column of 17-ppg mud falls out of the bottom of the riser and momentarily becomes balanced with the pressure of seawater after the pressure has been equalised. In shallower water (less than 6000 ft), larger lengths of gas-filled riser may be required based on the risk of other events such as gas in the riser or lost returns. A number of design conditions can be considered when engineering the riser to resist collapse. Among others, these can include the following:

1. A gas bubble from the formation enters the well and expands as it enters the riser. The likelihood of a gas bubble filling the riser in a modern drilling operation is remote. However, it did occur once in 1982 [see Erb, et al 1983]. When this incident occurred, the subsea blowout preventer (BOP) was not shut-in when the flow was detected due to concerns about formation integrity. The surface diverter was being used to direct the flow overboard when it malfunctioned, causing loss of the mud column in the riser. In a modern drilling operation, the likelihood of riser collapse is greatly diminished because the shut-in of the BOP is a standard procedure when dealing with a kick.
2. Returns are lost to the well, leaving a void on the top of the riser. The voiding of a large portion of the riser due to lost returns is a remote possibility. A large amount of lost returns would likely be detected.
3. The contents of the riser (mud) are partially lost during an emergency disconnect of the riser. The u-tube that would occur during an emergency disconnect would typically leave no more than about 50% of the riser tube void after the pressure is equalised, if

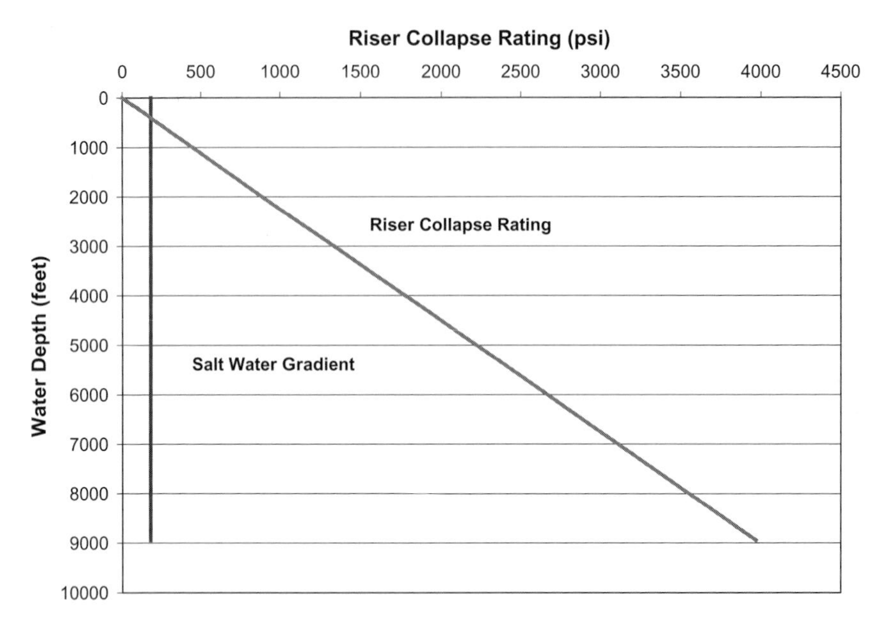

Figure 9.5 Riser collapse profiles (22 in. × 1.125 in. plus 8% machine tolerance)

the mud weight were about 17 lb/gallon (twice that of sea water). The lesser mud weights would void less of the riser.

API Bulletin 5C3 (1994) is commonly used as the basis for selecting the wall thickness to resist collapse. The calculation depends on the voided depth of riser, the yield strength of the pipe (in some cases) and the fabrication tolerances of the pipe.

Collapse calculations using API 5C3 demonstrate that a 22-in. riser with 1–1/8-in. wall thickness resists collapse, if it is completely void in 9000 ft of water. With fabrication tolerances of 8% on wall thickness, the riser resists collapse with the top 8000 ft of riser void. Figure 9.5 shows the external pressure resistance of the riser with an 8% fabrication tolerance vs. depth compared to the applied pressure from the hydrostatic head of seawater. The riser's collapse resistance varies with depth due to a dependence on pipe wall tension.

For various wall thicknesses of 21-in. risers and for various pipe wall tensions, calculations of water depth ratings of a voided riser pipe have been done based on the API 5C3. The results are shown in fig. 9.6. These curves are based on a "no margin" for fabrication tolerances.

9.2.4 Configuration (Stack-Up)

This section covers the issues considered in determining how the drilling riser is configured, or its "stack-up". The key issues in the riser stack-up are to assure the riser is heavy enough to be deployed without excessive angles in the currents expected during deployment and to assure the weight of the riser and Blow-Out Preventor (BOP) is within the hook load capacity of the vessel.

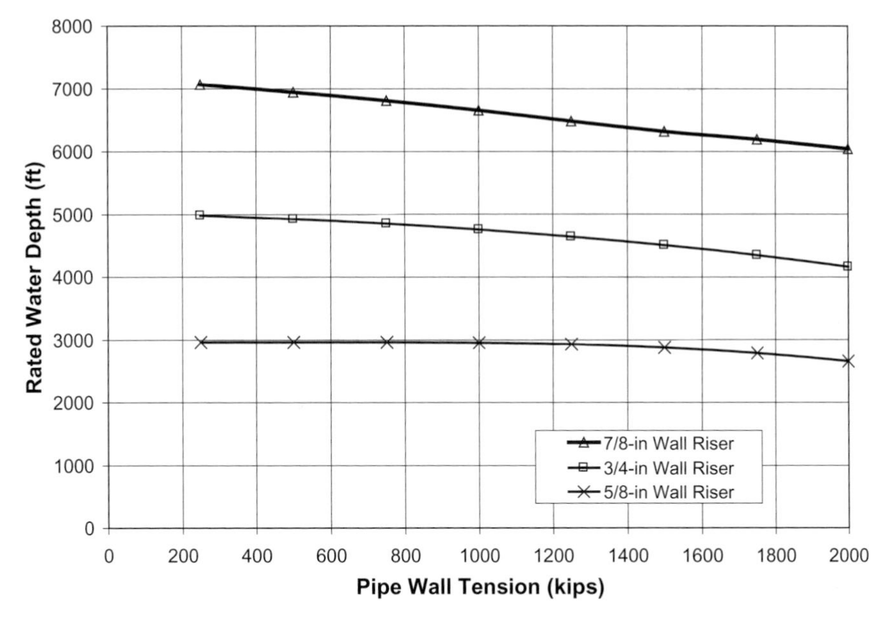

Figure 9.6 Riser collapse ratings (21 in. nominal wall thickness)

9.2.4.1 Vessel Motions and Moonpool Dimensions

The vessel response amplitude operators (RAOs) used in the riser analysis can either be analytical calculations or estimates derived from the model tests. These RAOs are converted into the format required by the riser analysis program. In cases in which the vessel is not in a head seas or beam seas heading, planar riser analysis programs require that the surge and sway motions be combined.

Typical vessel dimensions used for an ultradeep water drillship riser model are as follows:

- Upper Flex Joint Centre above Water Line – 63 ft.
- Drill floor above Water Line – 85 ft.
- Vertical Centre of Gravity (VCG) above Baseline of Vessel (Keel) – 47.55 ft.
- Draft of the Vessel – 29.5 ft.
- Height of the BOP Stack – 63 ft.
- Height of the BOP Stack from Wellhead Connector to Centre of the Bottom Flex Joint – 55 ft.

The terms used above will be illustrated in figures in the upcoming sections.

9.2.4.2 Connection to Vessel

The arrangement of the riser through the moolpool is shown in fig. 9.7. The riser is supported by the vessel through the combination of a tensioned telescopic joint and a top flex joint in an opening in the vessel called the "moonpool". The telescopic joint has an inner

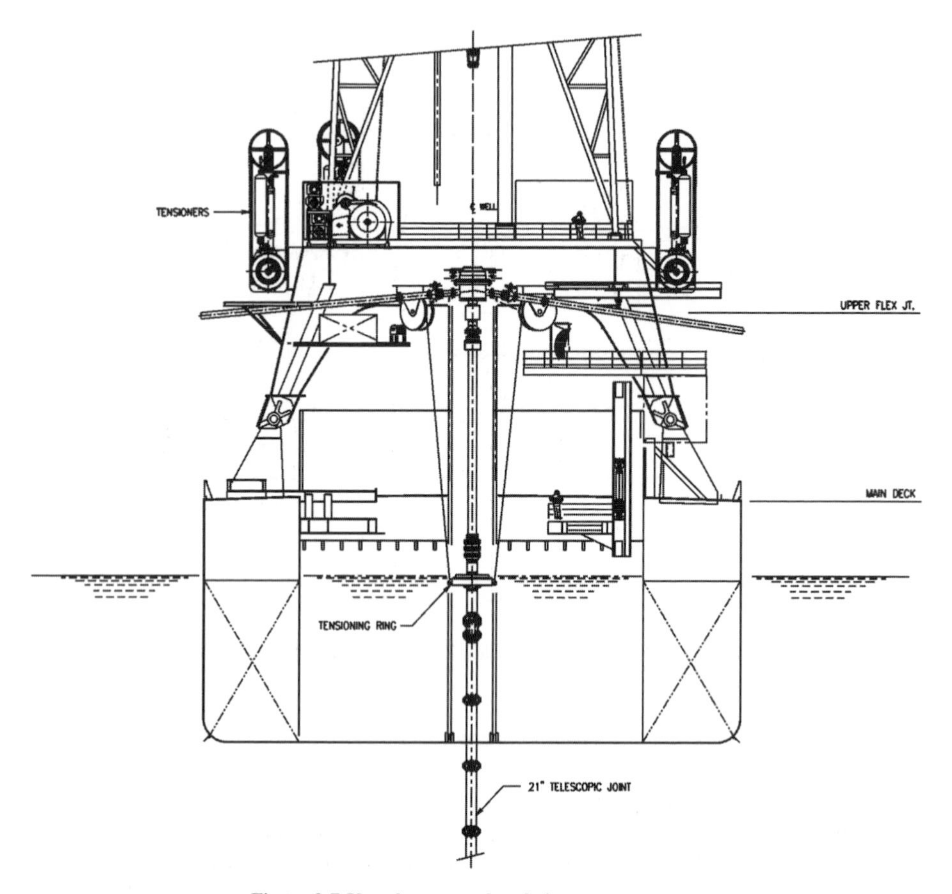

Figure 9.7 Vessel moonpool and riser arrangement

barrel and an outer barrel that allow vertical motion of the vessel while holding the riser with near-constant tension. The tensioning ring at the top of the "outer barrel" of the telescopic joint provides the connection point for riser tensioner lines, which maintain relatively constant tension through their connection to the compensating tensioner units. Top tension variation is minimised through the use of tensioner units that are based on a hydraulic/pneumatic system with air pressure vessels providing the springs. The tensioner lines wrap over "turn-down" sheaves located just under the drill floor. These tensioner lines route back to the tensioner units that are located around the perimeter of the derrick.

The upper flex joint is located above the "inner barrel" of the telescopic joint where it provides lateral restraint and reduces rotation through elastomeric stiffness elements. A diverter located just above the upper flex joint and just below the drill floor allows mud with drill cuttings returning from the well through the riser annulus to be dumped to a mud processing system. A closer view of this arrangement is shown in fig. 9.8.

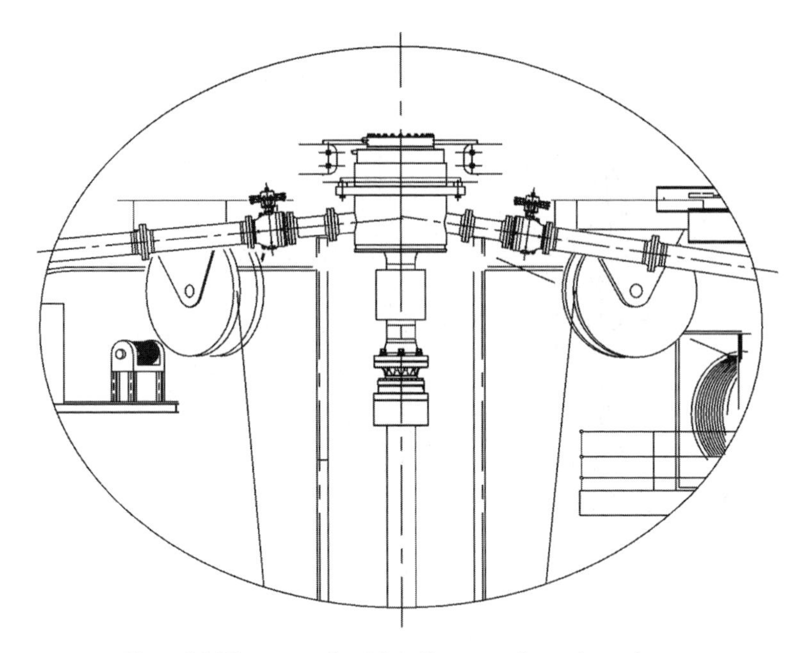

Figure 9.8 Riser upper flex joint, diverter, and turn-down sheaves

9.2.4.3 Riser String

The riser string consists of "joints" (segments) of riser pipe connected at the drill floor and "run" (deployed) into the water. Figure 9.9 shows a typical ultra deepwater riser joint that is 75 ft long and has a continuous steel riser pipe down the middle. As shown, this riser joint has five pairs of buoyancy modules strapped on the outside and flange-type connectors at each end. As discussed below, the riser joints carry auxiliary lines, and thus are made up with bolted flange, dog-type or other non-rotating connections.

The cross-section of a typical riser joint is shown in fig. 9.10. This figure shows auxiliary lines that are clamped to the riser pipe. These lines include choke and kill lines that provide for well control, a riser boost line that can be used to pump mud into the riser annulus just above the BOP stack to improve return of cuttings, a spare line, and a hydraulic line that controls subsea functions. Buoyancy material is shown strapped on the riser and external slots are provided in the buoyancy for attachment of multiplex (MUX) control cables.

Figure 9.9 Typical riser joint

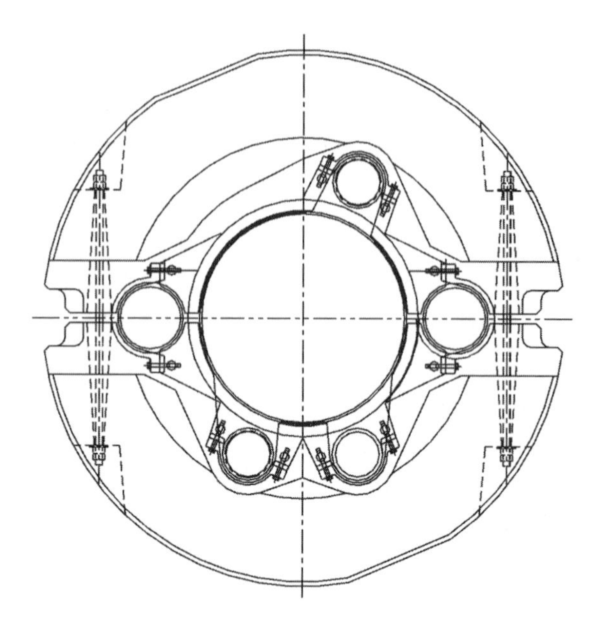

Figure 9.10 Typical riser joint cross-section

9.2.4.3.1 Riser Joint Properties

Riser joint properties include their weights in air, in water, with buoyancy, and without buoyancy. These weights can vary as the joints are deployed in deep water due to compression of the buoyancy and water ingress. Other properties include the joint length and the hydrodynamic properties such as drag diameter, drag coefficient, inertial diameter and inertial coefficient. Typical values for the joint properties used in an ultra deep water riser model are shown below in table 9.2.

9.2.4.3.2 Riser Stack-Up

The riser stack-up consists of joints with lengths typcially ranging from 50 to 75 ft, depending on the drilling rig. Table 9.3 below shows the weight of each component in a riser string for a typical ultra deepwater drilling rig. Each component listed has its submerged weight listed, with the exception of the tensioner ring, which is expected to be above the water line. The total weight of the riser without the LMRP is used for determining the top tension required to support the string. The total weights of the riser with the LMRP and with the full BOP are used to determine the hanging weight of the string.

The considerations in the joint stack-up of a riser string include assuring the riser is heavy enough to be deployed without excessive angles in the currents expected during deployment, and to assure the weight of the riser and BOP is within the hook load capacity of the vessel. This weight is regulated by bare joints or partially-buoyant in the string. The bare joints are often placed at the bottom of the string to get full benefit from the weight as deployment of the string first starts. Due to other considerations, such as VIV due to high currents, the bare joints may be placed in the region of high current often near

Table 9.2 Typical ultra deepwater riser joint properties

Properties	22 in. × 1.125 in. Wall w/55.5 in., 3 k buoyancy	22 in. × 1.125 in. Wall w/56.5 in., 5 k buoyancy	22 in. × 1.125 in. Wall w/59.5 in., 7.5 k buoyancy	22 in. × 1.125 in. Wall w/60 in., 10 k buoyancy	1.125-in. Wall bare joint
In-air weight of bare joint (lbs)[1]	35,644	35,644	35,644	35,644	35,644
Joint length (ft)[1]	75	75	75	75	75
In-air weight/length of bare joint (lb/ft)[2]	475.3	475.3	475.3	475.3	475.3
In-air weight of buoyancy on joint (lbs)[1]	22,080	24,555	31,145	35,000	0
Net lift of buoyancy on joint (lbs)[1]	30,330	30,565	30,335	27,920	0
In-water weight of bare joint (lbs)[3]	30,975	30,975	30,975	30,975	30,975
In-air weight of joint w/buoyancy (lbs)[4]	57,724	60,199	66,789	70,644	35,644
In-water weight of joint w/buoyancy (lbs)[5]	645	410	640	3055	30,975
Buoyancy compensation[6]	97.92%	98.68%	97.93%	90.14%	0.00%
Drag diameter (inches)[1]	55.5	56.5	59.5	60.0	41.3
Drag coefficient[1]	1.00	1.00	1.00	1.00	1.00
Inertial diameter (inches)[1]	55.5	56.5	59.5	60.0	37.5
Inertial coefficient[1]	2.00	2.00	2.00	2.00	2.00

1 – Information provided
2 – In-air weight divided by joint length
3 – In-water weight of bare joint equals 0.869 times in-air weight of bare joint
4 – In-air weight of joint w/buoyancy is in-air weight of buoyancy plus in-air weight of bare joint
5 – In-water weight of joint with buoyancy is in-air weight of a bare joint minus net lift of buoyancy
6 – Buoyancy compensation is (in-water weight of bare joint minus in-water weight of joint with buoyancy) divided by in-water weight of bare joint

Table 9.3 Installed weight of riser string in 9000 ft of water

				In-air		In-water	
Equipment supported by tensioners	Quantity	Unit length	Length	Unit weight	Total weight	Unit weight	Total weight
Tensioner ring*	1	0 ft	0 ft	55,000 lb	55.00 kips	55,000 lb	55.00 kips
Slip joint outer barrel	1	75 ft	75 ft	45,052 lb	45.05 kips	39,150 lb	39.15 kips
Middle flex joint	1	75 ft	75 ft	52,060 lb	52.06 kips	45,240 lb	45.24 kips
10-ft pup joint	1	10 ft	10 ft	11,124 lb	11.12 kips	9667 lb	9.67 kips
20-ft pup joint	1	20 ft	20 ft	15,658 lb	15.66 kips	13,607 lb	13.61 kips
Joint with 3000-ft depth buoyancy	36	75 ft	2700 ft	57,724 lb	2078.06 kips	645 lb	23.21 kips
Joint with 5000-ft depth buoyancy	28	75 ft	2100 ft	60,199 lb	1685.57 kips	410 lb	11.47 kips
Joint with 7500-ft depth buoyancy	33	75 ft	2475 ft	66,789 lb	2204.06 kips	640 lb	21.11 kips
Joint with 10000-ft depth buoyancy	13	75 ft	975 ft	70,644 lb	918.37 kips	3055 lb	39.71 kips
Bare joint with 1.125-in. wall	7	75 ft	525 ft	35,644 lb	249.51 kips	30,975 lb	216.82 kips
LMRP with one annular	1	15 ft	15 ft	225,600 lb	225.60 kips	196,046 lb	196.05 kips
BOP	1	40 ft	40 ft	500,700 lb	500.70 kips	435,108 lb	435.11 kips
		w/BOP	9010 ft	w/BOP	8040.75 kips	w/BOP	1106.14 kips
		w/LMRP	8970 ft	w/LMRP	7540.05 kips	w/LMRP	671.03 kips
		w/o LMRP	8955 ft	w/o LMRP	7314.45 kips	w/o LMRP	474.98 kips
Hang-off ratio of in-water weight to in-air weight of string w/LMRP:							8.90%
Top minus bottom pipe wall tension (all in-water weights except in-air weight for riser tube):							768.62 kips
Bottom pipe wall tension in riser string above LMRP (3000k top):							2231.38 kips

*In-air weights used

the top of the riser string in an alternating, "bare-buoyant" configuration. VIV and methods for mitigating it will be discussed in Sections 9.4 and 9.5.

Another consideration to be discussed later in this chapter is riser recoil. When an emergency disconnect is carried out, the presence of bare joints in the string improves the behaviour of the riser string and thus increases the range of top tensions that allow the riser to meet specified performance criteria. The most important of these criteria are the avoidance of contact between the riser and the rig floor, the avoidance of slacking in the tensioner lines, and the avoidance of subsequent downward movement of the lower marine riser package (LMRP) causing contact with the BOP. These issues will be discussed further in Section 9.2.9.

With a specified length of joints making up the riser, the riser string generally has to include one or two shorter joint lengths to make the string length match up with the water depth. For this purpose, shorter joints are employed just below the telescopic joint. Since the lengths of these pup joints get no shorter than 10 or 5 ft, the telescopic joint is generally not exactly at mid-stroke at a specific location. This inexact match-up becomes a consideration in both Section 9.2.8 on emergency disconnect and Section 9.2.9 on riser recoil.

9.2.4.4 Connection to BOP Stack

At the seabed, the riser connects to the blowout preventer, or "BOP" stack, which provides subsea well control after the well has been drilled to a depth that warrants it. The lowest riser joint connects to a riser adapter on top of the BOP stack. This connects to a lower flex joint located inside the upper portion of the BOP called the lower marine riser package (LMRP). As will be discussed later, the LMRP can be disconnected from the BOP, and this is called a part of the emergency disconnect sequence (EDS). Just above the seabed, the BOP is landed on a wellhead that is connected to the surface casing. The BOP arrangement is shown in fig. 9.11.

9.2.4.4.1 Bottom Flex Joint

At the bottom of the riser, a flex joint provides a connection to the BOP stack. This connection provides lateral restraint and resists rotation through elastomeric stiffness elements. The rotational stiffness improves the performance of the riser by reducing the bottom flex joint angle, thus permitting drilling in more severe conditions.

9.2.4.4.2 BOP Stack

This discussion of drilling riser analysis procedures includes discussion of the BOP stack due its make-up (LMRP plus lower BOP), weight, height, and connection to the seabed. The weight of the LMRP and lower BOP are important when considering deployment and retrieval of the riser as discussed in Section 9.2.6, and riser recoil as discussed in Section 9.2.9. The height of the BOP determines the elevation of the riser's bottom flex joint above the seabed. The connectors in the BOP and the loads passed through to the conductor pipe are an important part of the analysis of wellhead and conductor loading discussed in Section 9.2.7. Furthermore, analysis is often conducted to determine the load expected on each of the BOP connectors under a set of defined loading conditions.

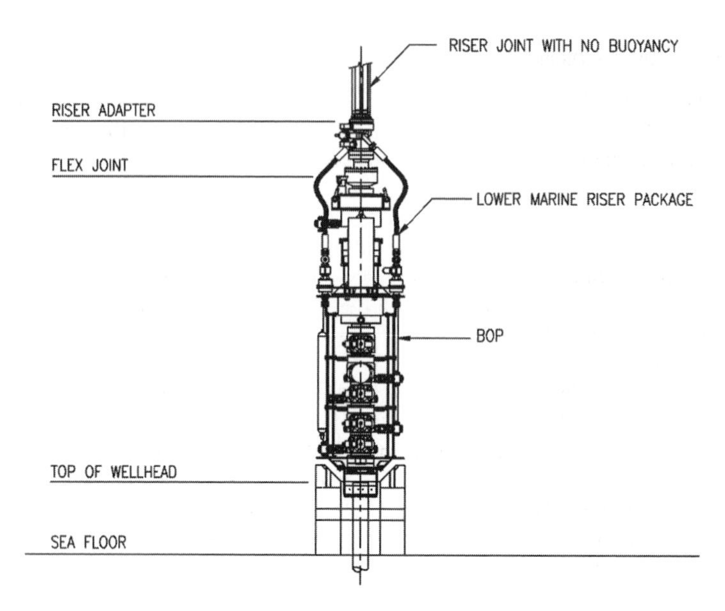

Figure 9.11 BOP arrangement

9.2.4.4.3 Wellhead/Conductor/Soil

The wellhead, conductor, and soil are also part of the drilling riser analysis procedure. Flexibility within these elements alters the behaviour of the riser. For example, soft soils would allow rotation of the BOP stack relative to the mud line. This would reduce the angle of the flex joint (relative angle between the riser and BOP stack) which would permit drilling with larger vessel offsets. The differences could be important, especially in considering limits for drilling or concerns with reaching the limits of the flex joint.

As will be discussed in Section 9.2.8, the wellhead and conductor can become the first to exceed their allowable stresses in a drift off scenario associated with an emergency disconnect. In that case, the loads applied from the riser to the wellhead and on into the conductor pipe are calculated as part of the riser analysis methodology.

The key properties that are included in this analysis are the rated capacity of the wellhead, the cross-sectional properties of the conductor pipe (typically the inner strings are ignored), and the *p–y* curves or the shear strength profiles of the soil.

9.2.5 Vortex-Induced Vibration (VIV)

This section covers the subject of vortex-induced vibration (VIV) as it relates to a drilling riser. The details of riser VIV are covered later (see Section 9.4). Ocean currents can cause VIV of a drilling riser that can lead to costly downtime in a drilling operation and ultimately fatigue failure of the riser as discussed by Gardner and Cole (1982). Such fatigue failure in a drilling riser could result in detrimental effects such as costly inspection and repairs, loss of well control, and compromise of safety. In this text, VIV mitigation measures are considered to be part of configuring the riser.

9.2.5.1 Calculation Methods

Vortex-induced vibration can be calculated using the hand checks, computational fluid dynamics (CFD), and empirical methods. Each of these methods has their place, depending on the current profile being investigated and the level of rigor required.

9.2.5.1.1 Hand Checks

Hand checks for calculating the VIV fatigue damage are most applicable when metocean conditions include currents that are constant with depth. Such conditions can exist in shallow-water locations where the current is driven by tides (e.g. the English Channel) or close to the mouths of rivers. When the current is constant with depth, VIV can be very severe. In these cases, the Strouhal equation can yield a good approximation that can be used to determine the VIV frequency. The amplitude can be estimated as being equal to say, one diameter, or some other value that could be derived from the work of Blevins (1977) or others. Using the mode shape associated with the natural frequency closest to the VIV frequency, the amplitude can be used to determine the curvature of the riser. This curvature can then be used to calculate bending stress which, together with the VIV frequency, can be used to determine a fatigue damage rate and a predicted fatigue life.

9.2.5.1.2 Empirical Methods

High-current conditions in deep waters generally have large amounts of shear (i.e. current velocity that varies with depth). Such sheared currents are most important for the VIV riser analysis for locations in the Gulf of Mexico and offshore Brazil, Trinidad, the UK, and other high-current areas.

Although uniform currents lead to the most severe vortex-induced vibration (VIV), sheared (change of velocity with depth) currents can also lead to VIV. Analysis techniques to predict VIV frequencies and amplitudes are often considered to be a part of a drilling riser analysis procedure. Although research on riser VIV has been ongoing for decades, predictions of VIV amplitudes in real ocean currents still have uncertainties. Empirical techniques for calculating VIV and the resulting fatigue damage have been developed by Vandiver (1998) and Triantafyllou (1999). Related work has been carried out by Fumes, et al (1998).

Current profiles that cause the larger VIV amplitudes are those that have nearly uniform current speed and direction over large portions of the water column. If the current profile has a large amount of shear, the likelihood of VIV is reduced.

9.2.5.1.3 Computational Fluid Dynamics

Computational fluid dynamics (CFD) is another alternative for calculating the vortex-induced vibrations of a riser. This technique simulates the flow of fluid past the riser, models flow vortices, and predicts the riser motions. CFD techniques are under development with the objective to better model the physics, but the method requires large amounts of computer time to simulate VIV of a full length deepwater riser. A simplified analysis using two-dimensional CFD "strips" to represent fluid-structure interaction has been investigated by Schultz and Meling (2004).

9.2.5.2 Detrimental Effects

In VIV induced by high currents, a drilling riser vibrates normal to the flow up to an amplitude of about one diameter, or 50–60 in., since the buoyancy outer diameter must be included. For a drilling riser in high currents, the period of the vibration can be in the range of 2 to 4 seconds, based on the Strouhal equation which shows the frequency (period) linearly dependent on diameter and current speed. The detrimental effects of VIV are two-fold, drag force amplification and fatigue.

9.2.5.2.1 Drag Force Amplification

VIV causes an increase in the drag force on the drilling riser. The effective drag coefficient may be up to twice the value of a riser that is not experiencing VIV.

9.2.5.2.2 Fatigue Due to VIV

Due to the vibration of the riser, alternating bending stresses cause an accumulation of fatigue damage. As a general statement, the most fatigue damage in the riser tends to occur near the bottom or near the top depending on the depth of the current profile. High damage occurs at the top due to the proximity of the current profile; and high damage occurs at the bottom because the effective tension in the drilling riser is low, leading to short bending modes with high curvature. The fatigue of risers due to VIV has been addressed later.

9.2.5.3 VIV Suppression/Management

The metocean criteria (including current profiles) specified by the operator is used to determine if vortex suppression devices such as fairings might be needed to reduce drag force on the riser and suppress VIV. Because of the uncertainties in predicting VIV, this decision is sometimes made using site-specific analysis conducted by the operator and, at times, independent analysis using different methods. Fairings are an expensive option due to the cost of the fairings themselves and the additional rig time required to install them during riser running. Less expensive alternatives include strakes, alternating bare and buoyant joints [Brooks, 1987], and simply increasing the riser tension. The less expensive alternatives are not as effective, but can be adequate in many instances.

9.2.5.3.1 Stack-Up Adjustments

The choice of where fairings are to be installed in the riser stack-up (i.e. the description of joint properties along the string) has a large influence on the cost-effectiveness of well drilling operations. Fairings have been shown to be very effective. They can reduce drag force to as low as one-third of its original value and they suppress VIV almost entirely – provided they cover the portion of the riser where the high currents are predicted to be incident. This estimate of where the current is present in the water column may be highly uncertain. As a further complication, once the fairings are installed, removing or rearranging them would involve pulling (retrieving) the riser – a procedure that could take several days. Furthermore, the notion of placing fairings over the full length of a deepwater riser (i.e. greater than 3000-ft) is cost-prohibitive. Generally, at sites with severe currents, operators have chosen to put fairings over the top portion (500 ft or so) of the riser to cover the most likely high current events.

Strakes are external ribs placed on the riser string, most commonly in a helical shape. When compared to the fairings, these devices are less effective, but are still good at VIV suppression. They allow amplitudes of vibration with 10–30% of a diameter. A disadvantage of strakes is their 30–50% additional drag force when compared to an unsuppressed riser. Typically, strakes can be installed on the riser joints prior to running (installing) the riser, thus minimising the high costs associated with additional rig time.

The concept of using alternating bare and buoyant joints in the riser string (staggered joints) has been documented in Brooks (1987) as a means for reducing the VIV amplitude. This technique also provides a slight reduction in drag force. This is a popular technique because it involves no preparation by rig personnel other than to have bare joints available and sequenced properly. One disadvantage is that bare joints are required, usually near the surface, where their weight cannot be used to full benefit in running the riser.

Additional discussion on this subject may be found in Sections 9.4 and 9.5.

9.2.5.3.2 Operating Tension

Instead of altering the riser stack-up, VIV suppression can be achieved by increasing the operating tension. The concept of this suppression method is to excite lower modes of the riser, which have longer mode lengths. As a result, curvatures and stresses are lower and fatigue damage is reduced. An advantage of this technique is that it helps no matter where the currents are in the water column and it has virtually no effect on the well drilling operation, since the riser does not have to be pulled. However, this technique often has little effectiveness, particularly for a dynamically-positioned vessel requiring emergency disconnect. In these vessels, riser recoil considerations during emergency disconnect usually dictate that maximum riser operating tensions are not significantly higher than the minimum riser operating tensions required to conduct well drilling operations. The margin for increased tension is thus quite small.

Suppression devices may not be necessary if an operator can show that the metocean conditions will not involve high current during the drilling of the well. For example, presently low activity of currents could be used to justify a forecast of low activity for the duration of a well; and this could justify use of an unsuppressed riser. However, loop currents and related or unrelated deep ocean currents are still difficult to predict. Currents that are deep in the water column, whether driven by the loop current or other mechanisms, are particularly difficult to predict (or manage VIV suppression) with any certainty.

A disconnect of the riser due to VIV in high currents is generally avoided, if at all possible. Such a disconnect event in high currents would result in the riser taking on a large angle and possibly contacting the side of the moonpool. If the bathymetry allows, the vessel could be allowed to drift toward deeper water to manage the riser angle and avoid contacting the seabed. If a disconnect does occur in high currents, it will likely be due to an emergency disconnect or a planned disconnect to protect the integrity of the wellhead connector and the conductor pipe.

9.2.5.3.3 On-Board VIV Measurements

The detection of VIV-induced alternating stresses in the riser pipe wall and the associated fatigue damage can be done using a variety of systems. The sensors that are used to measure VIV will not be discussed in this text. The two main categories of systems used to gather information on riser VIV are the so-called "real time" system and the so-called "flight recorder" system. As the name suggests, the real-time system gathers, analyses and displays VIV data virtually immediately after the riser undergoes the response. The flight recorder system gathers and stores the data until the riser is pulled, at which time the stored data can be removed for analysis.

The real-time system provides data so that, if desired, it can be used to base operational decisions on management of the riser. This system generally involves a more complex measurement system, possibly with cables that need to be installed as the riser is being run. The flight-recorder system provides data only after the riser has been pulled, so that the data cannot be used to support operational decisions; it is intended more for the support of inspection decisions or VIV research. This system involves independent canisters mounted at selected locations along the riser.

9.2.6 Disconnected Riser

This section covers the response of the drilling riser when its bottom is in a disconnected condition. This condition can occur during running (deployment or installation) of the riser or during pulling (retrieval) of the riser. Additionally, the riser can be in this condition when the riser has been disconnected for operational reasons. An understanding of the riser's response in this condition is important to avoid damage to the riser and components on or around the riser that could lead to expensive repairs or ultimately loss of the riser or a compromise in safety.

9.2.6.1 Lateral Loading

The lateral force applied to a drilling riser causes it to move into a deflected shape. This shape depends on the distribution of the in-water (submerged) weight of the string, including that of the lower marine riser package (LMRP) or the full blowout preventer (BOP) that are on its bottom. The shape also depends on the current profile being experienced and the lateral velocity of the drilling rig. The effects of weight and drag force plus remedial measures such as "drift running" (to be discussed later in this section) and tilting of vessel determine how well the riser can be deployed in the presence of high lateral loading.

9.2.6.1.1 Lateral Response During Deployment/Retrieval

Lateral response of the disconnected riser string is based on how close to vertical the riser string is at the critical stages of deployment. At the start of deployment, the motion of the BOP and the angle of the riser are important, as the BOP is being deployed into the waves and current. Riser analysis can be used to determine the likelihood of contact between the BOP and the side of the moonpool.

As the riser is lowered further, strong surface currents can cause a large angle of the riser where it passes through the diverter housing (the opening in the drill floor). If the angle becomes large enough, the riser can contact the side of the diverter housing, causing damage to the buoyancy material or causing the riser to become stuck so that it cannot be further deployed or retrieved. As more of the riser becomes deployed, top angles generally reduce, provided the ocean currents are primarily at the surface. However, currents at mid-depth or near the bottom can cause excessive angles leading to problems similar to those noted above. In addition, these currents can cause problems in landing the full BOP or, in particular, the lighter LMRP.

The response discussed above is governed primarily by the drag properties (drag diameter and drag coefficient) of the riser, the riser's distribution of in-water weight, and the bottom weight of the BOP or LMRP. The drag force on the riser can be considered as proportional to the velocity squared according to Morison's Equation [see Krolikowski and Gay, 1980], so that the shape of the riser depends heavily on the current.

Considering a minimal current, a riser that is negatively buoyant above the BOP will tend to take on an approximate catenary shape in the absence of current. This will lead to a bottom angle that is larger than the top angle. By contrast, if the riser is positively buoyancy, it will take on an approximate inverse catenary shape with the top angle larger than the bottom angle. The weight on the bottom, either that of the BOP or the LMRP, determines the straightness and the average angle of the riser.

The same deployment considerations also apply to retrieval. When unlatching a drilling riser at the seabed, ocean currents can cause the riser to take on a top angle that prevents it from being pulled or run back down. In a planned disconnect, this situation can be avoided by using the riser analysis to predict the response. However, in an emergency disconnect that can occur on a dynamically-positioned drillship, no control exists over the metocean conditions in which the disconnect occurs. In this case, the vessel is generally maneuvered to manage retrieval of the riser.

9.2.6.1.2 Deployment/Retrieval Limits

The limits that apply to the deployment and the retrieval process described above depend on the riser and rig equipment. The top angle limits depend on the inner diameter of the diverter housing and the outer diameter of the foam buoyancy on the riser. As a new riser joint is brought in and connected to the top of the string, the weight of the string is transferred to the lifting gear located, say 50–75 ft above the drill floor. As this occurs, the riser deflects about this high pivot point in response to the current. A deflection equal to the undeflected radial gap between the riser and the diverter housing causes contact. Typically, the top angular limit for contact in this configuration is about 0.5°.

As shown in fig. 9.12, when the riser is landed in the spider at the level of the drill floor, the top angular limit of the riser depends on the radial gap between the riser and the diverter housing. In this figure, the riser is shown contacting the top and bottom sides of the diverter housing with an angle of 6.87°. Typically, the riser is centred at the drill floor and the limiting angle for contact at the bottom of the diverter housing is more like 3°. When compared to the configuration with the riser suspended from the lifting gear, the angular limit is larger with the riser landed in the spider because the string pivots about a point that is much lower. The contact again occurs against the side of the diverter housing, which is say 15 ft below the drill floor. The riser can be landed in the spider during high currents,

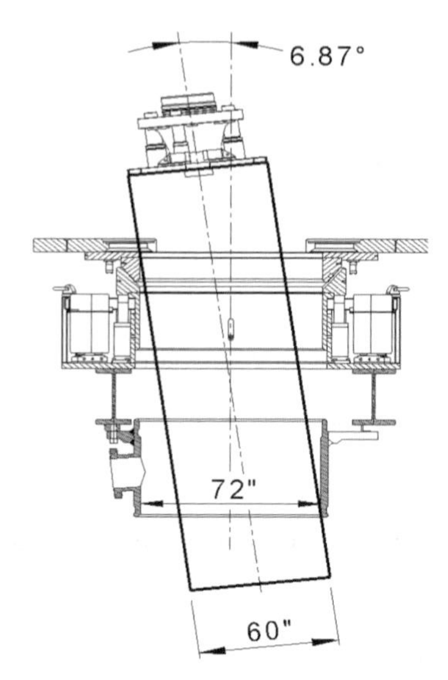

Figure 9.12 Riser clearance in diverter housing

without the need to run or pull. In this case, the limit might be compressive damage to the buoyancy or overstressing of the pipe.

Another limit that applies to the deployment/retrieval process is the geometrical limit associated with the BOP or LMRP contacting the side of the moonpool. Also at the final stage of deployment, the angular limits dictate whether the LMRP can latch up to the lower BOP or whether the BOP can latch up to the wellhead.

9.2.6.1.3 Application of Tensioned-Beam Analysis

A variety of tensioned-beam analysis programs can be used to estimate the response of a riser during deployment or retrieval. Static, frequency-domain or time-domain riser analysis programs can be used, depending on the amount of detail needed.

9.2.6.1.4 "Drift Running" Solution

In various parts of the world such as the Gulf of Mexico, Trinidad, and Brazil, deepwater drilling operations can be interrupted by lateral loading in high currents, particularly while running the riser. To counter this, a "drifting running" procedure is used for running the drilling riser in high currents. In this procedure, a dynamically-positioned vessel drifts towards the well in the direction of the current as the riser is run. This process allows the riser to be run in higher currents than would otherwise be possible and avoids rig downtime while waiting for the current to subside. Running riser without drifting could lead to riser binding in the diverter housing, and could cause excessive stress in the riser pipe and damage to the foam buoyancy. Figure 9.13 shows the deflection of a riser during deployment with the ship stationary.

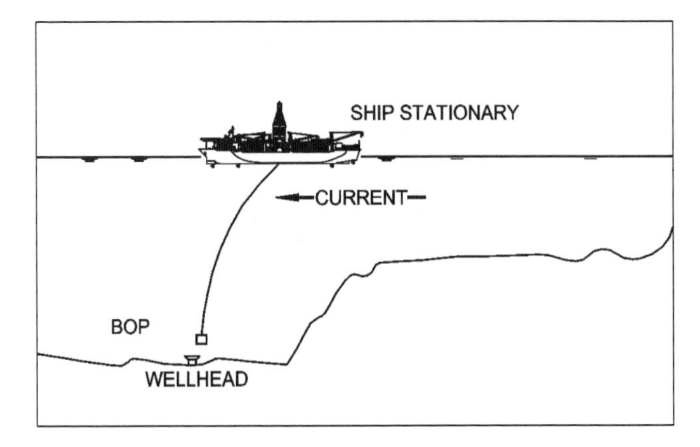

Figure 9.13 Riser deployment in high current – ship stationary

When currents are high during the riser running operations, special equipment or procedures may be warranted to run the riser and land the BOP stack on the wellhead. The terms used in this section will apply only to the riser; however, similar procedures can be used for running casing. A procedure called "drift running" uses controlled down-current motion of the drilling vessel to pass the riser through the rotary and diverter housing. This procedure has been used throughout the industry to successfully land the BOP stack without damaging the riser or the running equipment. Figure 9.14 illustrates the reduced top angle that can be achieved through the use of "drift running".

When the riser string is exposed to high current, it takes on an angle. This angle is a function of the force applied by the current and the weight of the string. If the angle at the top of the string is excessive, the string will see high stresses or bind in the diverter housing,

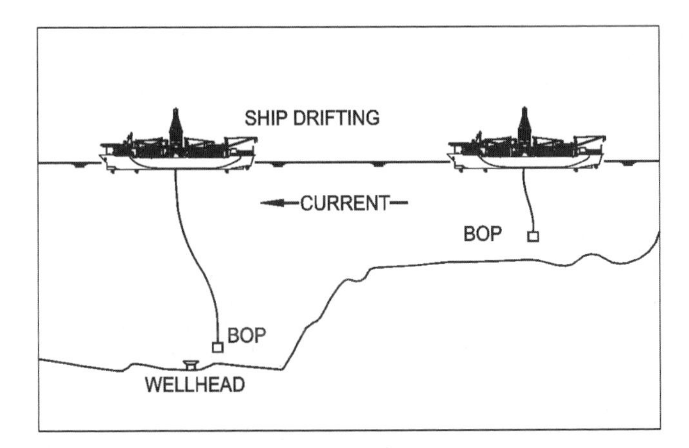

Figure 9.14 Riser deployment in high current – drift running

preventing it from being run. Binding due to excessive side load or high stresses in the riser can occur (1) when the string is hung off in the rotary or (2) when the string is supported by the lifting gear. These two configurations are very different in terms of the forces applied to the riser and the effects of high current.

When a riser is landed out in the rotary, an excessive angle can cause the riser to contact the side of the diverter housing. This can lead to high stresses in the riser and possible damage to the buoyancy material. The angle that causes contact with the diverter housing depends on the inner diameter of the diverter housing and the outer diameter of the riser buoyancy. When the riser is hung off in the rotary table, the consequence is excessive bending stress in the riser or damage to the buoyancy material.

When a riser string is supported by the lifting gear, an excessive angle can cause binding that could prevent running the string. Passing the riser through the rotary table with an excessive angle could damage the buoyancy material by scraping it against the side of the diverter housing. In a more extreme situation, lateral forces can cause binding in the diverter housing as the force against one side of the diverter housing becomes so large that the riser cannot be run. The top angle of the riser that can lead to contact with the diverter housing is generally quite small. In a typical example, the top angle for contact is less than 0.4° considering a 6-in. gap between a centralised riser and the diverter housing just after bringing in a new 75-ft riser joint. As the lateral force associated with this contact increases, binding becomes more likely.

Drift running involves a controlled drift of the vessel down a "track line" in the direction of the current at a speed that minimises the top angle. Ideally, a speed and track are chosen to minimise the top angle of the riser/casing string as it is being run.

In practice, the proper speed can be selected through co-ordination between the captain and the crew on the rig floor. By observing the position of the riser string as it passes through the rotary, the crew on the rig floor can provide information to the captain that can be used to correct the speed and direction of the drift. In this manner, the riser string can be run in whatever current is present, provided VIV concerns have been addressed.

For optimal efficiency in the drift running procedure, the vessel would need to pass over the wellhead just as the riser string has been fully run so that the BOP can be latched up. This requires an informed estimated starting point. The distance and bearing angle of the start-up location with respect to the wellhead can be calculated using an average current profile based on the best information available for current profiles along the track line.

Allowances should be included in this estimated starting point to account for changes in the current profile and bathymetric features. Changes in the current profile can cause overshoot, coming up short, or being off line of the wellhead. In addition, bathymetric features such as escarpments, as shown in fig. 9.14, might require adjustments in the drift running program such as hanging off the riser string in the rotary during certain stages of drifting.

As noted earlier, a relatively small angle (less than 0.4°) could cause contact, just after bringing in a new riser joint when the riser is being supported and run by the lifting gear. Since no contact occurs up to a top angle of, say, 3.3° when the string is hung off in the rotary, this is the configuration in which corrections can be made. When the string is hung off in the rotary, the vessel can slow down or possibly even move up current very slowly without damaging the string, depending on the current conditions. This should be done with care not to overstress the riser pipe or damage the riser buoyancy. This flexibility to slow down or move up current allows the BOP to be latched after corrections are made or when maneuvering over a well near an escarpment.

The captain and the drilling superintendent can carry out the riser running operations and the landing of the BOP stack by estimating the starting point. Such an estimate can be developed with the intent of making the starting point estimate based on an initial measurement of the current profile. It is understood that the current profile will change during the operation, so allowances in the estimate are needed. The intent is to keep the riser in the centre of the diverter housing during running.

Measurements of current can be obtained using current metres such as the acoustic Doppler current profiler (ADCP). ADCPs can be mounted on the ship and on the remotely operated vehicle (ROV) as it is being run, thereby providing current measurements over the full water depth. Measurements of current speeds and directions at the various depths can be used as guidance for the operations.

Joint length and riser running speed (joints per hour) are the other inputs. These quantities are used to calculate the speed of running the riser string and should include testing of the choke and kill lines.

An alternative to the drift running procedure is an equipment solution called the moonpool centering device [Gardner and Cole, 1982]. The centering device is a movable structure that applies a force at several locations on the riser string as it is being run. Rollers on the centering device are used to allow the riser to pass. The centering device is intended to keep the riser string centred and vertical as it passes through the rotary and the diverter housing. The disadvantage of this concept is that the device tends to be a heavy and cumbersome.

9.2.6.1.5 Case History of "Drift Running"

During February of 2001 in Trinidad, the *Glomar Jack Ryan* drillship experienced a block of submerged high current with a peak speed of 2.6 knots and more than 2 knots over a depth interval of 900 ft. This resulted in the riser/BOP running operation, which normally would require 2–3 days to run, requiring nearly 20 days to run. As part of this experience, the following procedure was developed for running riser in such severe conditions.

- Commence running the BOP when set up on DP at a location of 30 miles from the drilling location.
- Continue running the BOP on the DP mode until the drill floor informs the bridge of difficulties due to angle of the riser.
- When this stage is reached, take the vessel off the DP and drift while running joints.
- Make attempts to put the vessel back on the DP while making a riser joint connection and revert to drifting while running it. Anticipate a stage in which the vessel would have to be on continuous drift to run the riser.
- Carry out continuous calculations to ascertain the cut off point for running the riser and using the remaining water depths for recovery.

- During the entire operation the DPOs will log the times for running each riser joint, the drift distance, the current metre data for the depth of the BOP, and the position of the riser/BOP relative to the moon pool.
- Based on the current profiles, estimate the depth at which the BOP will be below the high current and the riser angle will decrease. If this depth cannot be reached by the calculated cut off point, then recovery will begin.

9.2.6.2 Vertical Loading

If the metocean conditions include high seastates while a riser is disconnected and hung off, vessel heave motion could cause dynamic, vertical loading in the riser. Such vessel heave motion could occur if high seastates occur when the riser is hung off in any of the following configurations:

- during deployment or retrieval of the riser
- while the riser is secured in a hang-off configuration

The structural response of a drilling riser that is hung off from a floating drilling vessel is a critical issue for drilling operations in ultra-deep water. A hung-off riser can be exposed to storm conditions prior to its connection to the wellhead or after disconnection. In ultra-deep water, the axial dynamics of the riser are driven by the riser's increased mass and its increased axial flexibility when compared to a shorter riser. With these effects, vessel heave motion and wave and current forces cause riser tension variation, riser motions, and alternating stresses.

If secured in a hang-off configuration, the riser can be put into a "hard" hang-off configuration in which it is rigidly mounted to the vessel or a "soft" hang-off configuration in which the riser is compensated. Brekke, et al (1999) describes the advantages and limitations of the "soft" hang-off configuration when compared to the "hard" hang-off configuration as applied to the *Glomar Explorer* drill ship at a site in 7718 ft of water when subject to winter storms in the Gulf of Mexico. The advantages include:

- peak hang-off loads are minimised;
- compression in the riser is avoided;
- motion of the riser is reduced;
- riser stress variation is minimised.

The limitations of the "soft" hang-off configuration are as follows:

- vessel heave motion does not exceed the stroke limits of the telescopic joint and tensioners
- on-board personnel are available to monitor/adjust the tensioners' set point

During the deployment or retrieval process, the riser is generally in the hard hang-off condition.

9.2.6.2.1 Performance During Hang-off Conditions

Structural analysis of an ultra-deepwater riser will show larger axial (vertical) dynamic response than a shallower water riser due to the influence of the riser's additional mass and increased axial flexibility. Several computer programs are available within the industry for the 3-D time-domain riser analysis required for the combination of axial and lateral dynamic riser analysis.

Brekke, et al (1999) shows that 3-D random wave riser analysis is needed to determine accurate riser response estimates. This analysis discussed the fundamental contributors to tension variation, including (1) mass of the riser string times the vessel's vertical acceleration, (2) resonance at the axial natural period, and (3) lateral motions of the riser leading to additional tension variation. Random analysis is more accurate than regular wave analysis because it models the full spectrum of the seastate and thus avoids artificial response peaks near natural periods.

In random analysis, a realistic random seastate is generated in preparation for the riser analysis. The typical riser simulation is run for 1000 wave cycles, representing about a 3 h storm. In order to determine the results from this analysis, the peak and trough response of each parameter are determined as the maximum and minimum values that occurred during the simulation. If need be, this random analysis approach could be made more accurate by running multiple simulations and averaging the results or using statistical methods to obtain the extreme values.

9.2.6.2.2 Riser Model

The riser computer model is based on the riser joint properties and riser stack-up listed earlier in tables 9.2 and 9.3. For the hard hang-off, the riser is connected directly to the vessel so that it heaves and moves laterally the same amount as the vessel, but it is free to rotate at the top flex joint. For the soft hang-off, the vertical motion of the riser is compensated, but it still moves laterally with the vessel. The riser is connected to the vessel through springs whose total stiffness depends on the stiffness of the tensioner system and the Crown-block Motion Compensator (CMC). The stiffness value also depends on the weight supported by the system (i.e. whether the LMRP or the BOP is suspended) and how the load is shared between the tensioners and the CMC. No damping is typically assumed for the combination tensioner/CMC system because the tensioner recoil valve is assumed to be inactive.

For the hydrodynamic model in the vertical direction, the riser is modelled with a tangential drag coefficient of 0.2 and an inertial coefficient of 0.1 along its length. The BOP/LMRP is modeled according to the dimensions of a horizontal plate consistent with its length and width and a vertical drag coefficient of 1.1.

As noted earlier, vessel RAOs for seas approaching 45° off the bow (135° case) typically give the largest heave and lateral motions for this type of analysis. For the hang-off analysis, heave motions have the most significant influence on the results.

9.2.6.2.3 Metocean Conditions for Hang-Off Analysis

Differing metocean conditions could be rationalised for analysis of the various riser configurations. For deployment or retrieval conditions, a seastate leading to lesser heave such as, 5 ft maximum vessel heave (this peak-to-trough heave (DA) occurs once during a 3 h seastate) may be consistent with the requirement for running riser as stated in a vessel's operating manual. For the storm hang-off configuration, extreme storm conditions (e.g. the 10-yr winter storm) may be required to accommodate the possibility of disconnecting and securing the riser in such conditions.

In the deployment or retrieval configuration, as noted above, only the hard hang-off is generally analysed since the riser is either landed in the spider or supported on the traveling block by the lifting gear. For the storm mode, the riser can be analyzed for both the hard and the soft configurations. Hurricane conditions are not generally analysed, since the riser is expected to be retrieved and secured onboard the vessel during such events.

9.2.6.2.4 Design Limits for Hang-Off Analysis

Design limits used in a typical analysis are as follows:

- Maximum top tension during deployment: 1500 kips (rating of the lifting gear).
- Minimum top tension during deployment: 100 kips (avoid uplift on spider or lifting gear with 100 kips margin).
- Minimum tension along riser during deployment: no explicit limit since momentary compression in the riser does not represent failure. (The consequences of compression are covered by motion/stress limits.)
- Maximum top tension in 10-yr storm: 2000 kips (rating of substructure, diverter, upper flex joint, and other components).
- Minimum riser tension during 10-yr storm: no explicit limit since momentary compression in the riser does not represent failure. (The consequences of compression are covered by the motion/stress limits.)
- Riser Stress: Per limits in API RP 16Q.
- Moonpool Contact: Avoid contact between the riser (intermediate flex joint) and the moonpool with a 10% margin based on the nominal riser position.
- Maintain a sufficiently heavy string to allow deployment and retrieval in a reasonable levels of current without binding in the diverter housing or contacting the moonpool. A heavier string also helps keep the riser from contacting the moonpool after disconnect **during a drift off** and controls riser recoil response during emergency disconnect.

9.2.6.2.5 Interpretation of Analysis Results

Riser analysis for the 10-yr storm conditions can be used to compare riser response in the "soft" and the "hard" hang-off configurations. For a typical ultra deepwater well, the first axial natural period of a hung off riser could be about 5 s. As noted earlier, the soft hang-off configuration with the LMRP is modelled using a spring that connects the top of the riser to the vessel. According to riser eigenvalue analysis, the soft hang-off configuration could have a first axial natural period in the range of 30–50 s.

Riser analysis for the deployment and the storm hang-off conditions was conducted for the *Glomar C. R. Luigs* in 9000 ft of water to estimate peak loads with axial tension variation.

For the riser deployment mode (riser deployment or retrieval), riser analysis is run to determine the tension variation expected with different riser buoyancy configurations.

As noted above, the design limits can be a maximum tension of 1500 kips based on the capacity of the lifting gear and a minimum tension of 100 kips established as a margin above zero tension. This analysis was done for the metocean conditions associated with the 5-ft vessel heave.

For storm hang-off conditions, riser analysis shows that the "soft" (compensated) hang-off configuration has much less riser motion and tension variation than the "hard" (rigid) hang-off. Hard hang-off loads are slightly higher than the 2000-kip capacity of the substructure. The soft hang-off is the preferred option as long as the vessel heave does not exceed slip joint stroke limits and on-board personnel are available to monitor/adjust the tensioner set point. Within these limitations, the risk assumed with a soft hang-off is virtually identical to that assumed when the riser is in its connected configuration.

For the deployment mode (riser deployment or retrieval), riser analysis was run to determine the tension variation expected with different riser buoyancy configurations. As noted above, the design limits are a maximum tension of 1500 kips based on the capacity of the lifting gear and a minimum tension of 100 kips established as the margin above zero tension. This analysis was done for the metocean conditions associated with the 5 ft vessel heave as previously described.

Analyses were run for cases with an LMRP suspended on the bottom of the riser string and for cases with a BOP on the bottom of the riser string. Both of these cases are important because the LMRP case generally gives the lowest minimum tension in the riser and the BOP case generally gives the highest maximum tension in the riser. Riser buoyancy configurations with 2 bare joints, 5 bare joints, and 10 bare joints were run with the LMRP; and buoyancy configurations with 10 bare joints and 15 bare joints were run with the BOP. The results are used to determine the range of configurations that would satisfy the tension limits.

The results of the deployment analysis are summarised in fig. 9.15. This figure shows the variation in riser top tension versus the number of bare joints in the riser, with a minimum, mean (riser string weight in water), and maximum tension curve shown for the LMRP cases on the left side and for the BOP cases on the right side. The minimum and the maximum allowable tensions (100-kip and 1500-kip limits defined earlier) are shown as horizontal dashed lines. Based on this figure, a riser buoyancy configuration with seven or less bare joints would satisfy design limits on maximum tension (with the BOP) and minimum tension (with the LMRP). Based on the considerations of in-water weight noted earlier, a number of bare joints less than seven would result in an in-water to in-air weight percentage less than 9%, so that seven bare joints is the optimal value.

The riser analysis results shown in fig. 9.16 illustrate that riser tension variation during deployment is much higher at the top of the riser than it is near the bottom. This is mainly due to the dominance of inertial loading caused by the mass below each elevation along the length of the riser. Two pairs of curves are shown in fig. 9.16, with each pair made up of a minimum and a maximum tension curve. Each pair represents an extreme case, with the pair on the left representing the LMRP and two bare joints in the riser string, and the pair on the right representing the full BOP and fifteen bare joints. In both cases, the figure

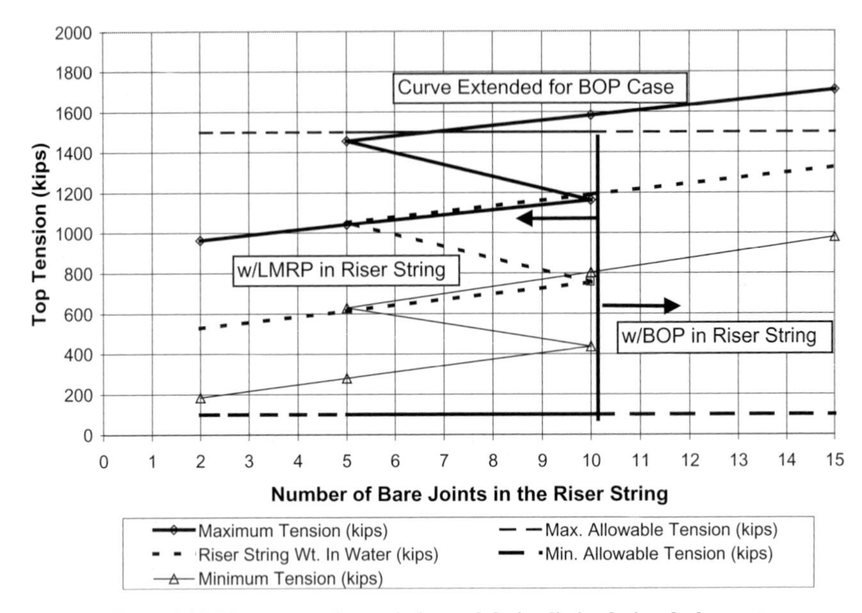

Figure 9.15 Riser top tension variation and design limits during deployment

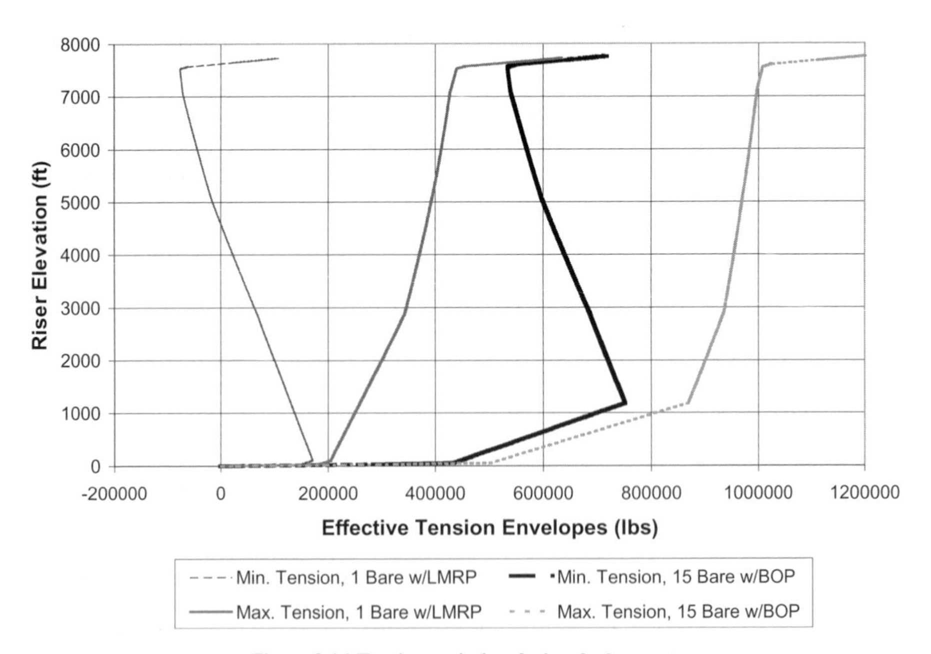

Figure 9.16 Tension variation during deployment

Table 9.4 Hang-off results for deployment configuration

	With LMRP			With BOP	
	2 bare	5 bare	10 bare	10 bare	15 bare
	kips	kips	kips	kips	kips
Mean top tension	529	614	755	1190	1327
Max. top tension	963	1042	1163	1583	1710
Min. top tension	184	279	437	804	980
Min. tension along length	15	110	266	657	836

shows that the top portion of the riser experiences much more tension variation, and stress variation, than the bottom portion. In this 5-ft heave condition, the hung-off riser with the LMRP and two bare joints comes close to compression in its upper portion and the hung-off riser with the BOP and fifteen bare joints experiences a top tension of 1750 kips.

Table 9.4 shows the summary results for the four analysis cases presented for the deployment configuration, including minimum top tensions, maximum top tensions, and minimum tensions along the length.

Riser analysis for the storm configuration was done to compare the riser response in the "soft" and the "hard" hang-off configurations in a 10-yr winter storm. As noted earlier, the design limits on top tension used for the storm configuration are different from those used for the deployment configuration.

In this case, the hard and soft hang-off configurations were analysed with 10 bare joints and the LMRP on the bottom of the riser string. Due to the lighter hanging weight of the LMRP, this configuration is more prone to riser compression than the configuration with the BOP.

The hard hang-off configuration is simply modelled with the top of the riser moving vertically and laterally with the vessel. In this configuration, the first axial natural period of this riser configuration is about 5 s.

As noted earlier, the soft hang-off configuration with the LMRP is modelled using a spring that connects the top of the riser to the vessel. The soft hang-off configuration has a first axial natural period of about 45 s.

Figure 9.17 shows the tension envelopes vs. depth along the riser string, with the LMRP only, for the hard and soft hang-off configurations. The envelopes show the minimum tension on the left side and the maximum tension on the right side. As shown, the envelope for the hard hang-off is much wider than that for the soft hang-off, indicating a large difference in tension variation between them. Additionally, the hard hang-off envelope shows a minimum tension that is below zero (in compression) at the top of the riser and over a large portion of its length. Although this is not considered a failure, it can lead to high bending stresses and lateral deflections. With the LMRP, peak top riser tensions are 750 kips for the soft hang-off and 1620 kips for the hard hang-off.

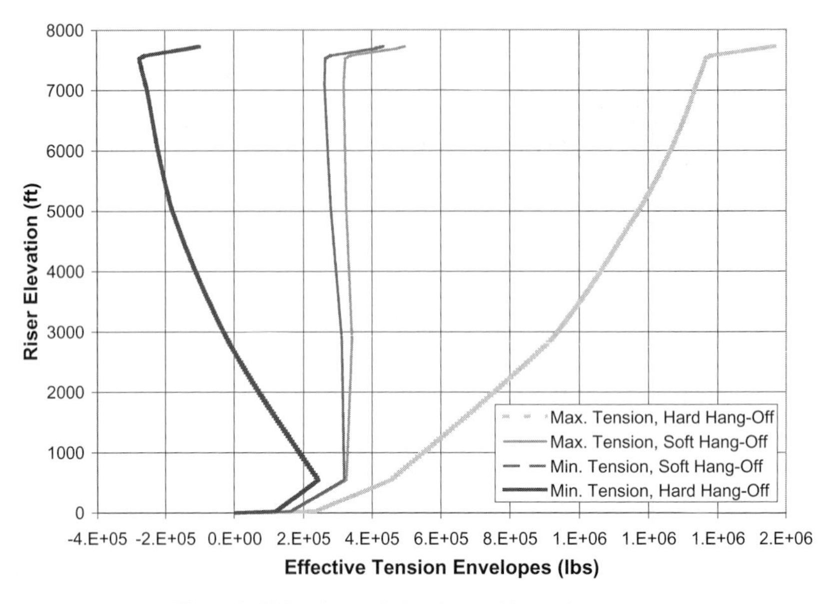

Figure 9.17 Tension variation during 10-yr winter storm

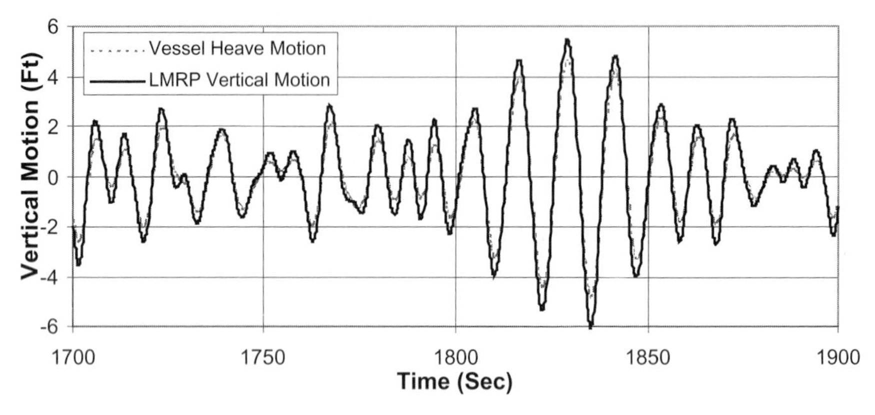

Figure 9.18 Riser vertical motion with hard hang-off, 10-yr winter storm

Figures 9.18 and 9.19 show plots of vertical LMRP motion versus time for a portion of the simulation in which the peak heave motion occurred. For the hard hang-off (fig. 9.18), the peak LMRP motion is 1.23 times the vessel heave motion, which roughly indicates the level of dynamic amplification. For the soft hang-off (fig. 9.19), the LMRP motion is 0.04 times the vessel heave motion.

Table 9.5 gives a typical results summary for the hard and soft configurations in the storm hang-off mode (10-yr winter storm conditions) with the 10 bare joints and the LMRP.

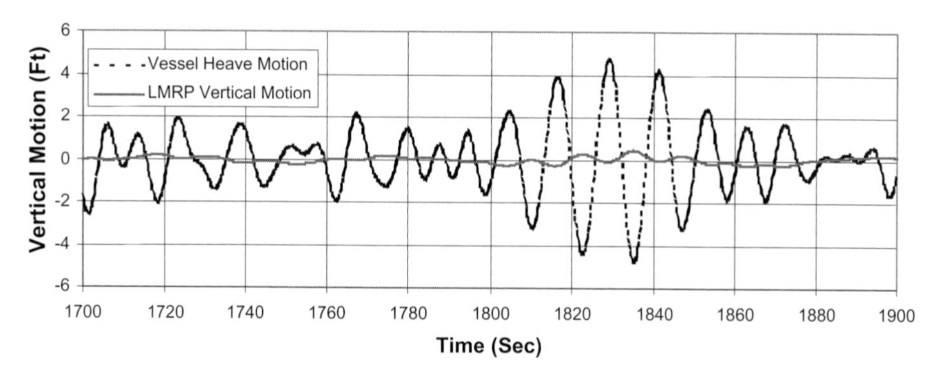

Figure 9.19 Riser vertical motion with soft hang-off, 10-yr winter storm

Table 9.5 Hang-off results for storm configuration, 10 bare joints and LMRP

Storm Configuration. 10 Bare Joints with LMRP		
	Hard hang-off	Soft hang-off
Tensions	kips	kips
Max. top tension	1620	750
Min. top tension	143	732
Min. tension along length	−10	(n/a)
Motions (double amplitude)	ft	ft
Max. heave amplitude	9.2	9.2
Max. LMRP vertical amplitude	11.3	0.4
Stress	ksi	ksi
Max. Von Mises stress	24	13.2

This table shows the maximum top tensions, minimum top tensions, minimum tensions along the length, riser motions, and riser stresses. The peak tensions are consistent with the figures discussed earlier. This shows that for storm hang-off conditions, the soft hang-off configuration has much less riser motion and tension variation than the hard hang-off configuration.

Related work has also been carried out by Miller and Young (1985) studying the effects of a column of mud contained in the riser during hang-off.

9.2.6.2.6 Operational Procedures for Hang-Off

If heavy seas are encountered during riser running or retrieval operations, typical procedures for going into the soft hang-off configuration (load shared between the tensioners and the CMC) are listed below.

1. Make up the telescopic joint in the riser string.
2. Engage the tensioning ring.
3. Make up a landing joint.
4. Lower the riser string until the tensioning lines support about half of the riser string weight and the tensioners are at mid-stroke.
5. Activate the CMC and set it to support the other half of the string weight.
6. Monitor/adjust tensioner stroke and set point.

After an emergency disconnect, assuming the vessel is moved off location per existing procedures, typical procedures for going into the soft hang-off configuration (on tensioners only) are as follows:

1. De-activate the riser recoil valve and open all Air Pressure Vessels (APVs).
2. Reduce pressure on the riser tensioners until they are at approximately mid-stroke.
3. Monitor/adjust the tensioner stroke and set point.

An alternate procedure that uses hard hang-off would call for installing the diverter, lifting the riser string with drill pipe, and locking the slip joint so that the riser is supported under the drillfloor. A second alternate procedure for hard hang-off calls for landing the riser string in the spider; however, this configuration does not provide resistance to uplift (compression at the top of the riser).

9.2.7 Connected Riser

This section discusses the drilling riser in the connected configuration. In this configuration, the riser provides a conduit for drilling operations that guides the drill pipe and casing strings into the well and contains a column of drilling fluid (mud) for well pressure control and circulation of drill cuttings up from the bottom of the well. The assurance of riser structural integrity is provided by an understanding of the riser response in this configuration. Structural integrity is maintained under metocean conditions that include wind, waves, and currents that apply forces to the riser. The associated lateral motions from the vessel are also imposed at the top of the riser. In addition to the external forces and motions, drill string rotation and other operations impose wear and other degradation within the riser.

Analysis of the connected riser configuration is routinely carried out to demonstrate that a rig's top tensioning capacity is sufficient to support the riser at a specific well site or in its design water depth, mud weight, and metocean conditions. In addition, if the metocean conditions include high currents, vortex-induced vibration (VIV) analysis (discussed in Section 9.5) can be carried out to further verify the riser‘s structural integrity.

9.2.7.1 Performance Drivers

The integrity of the connected drilling riser is largely driven by its deflected shape during the various operations that are carried out with it. During drilling operations, greater restrictions are placed on the riser's deflected shape due to the need to rotate drill pipe or strip (run or pull) drill pipe through the drilling riser. When drilling operations are suspended, restrictions on the deflected shape of the riser are reduced significantly.

Due to its length, the stiffness of the drilling riser is derived largely from its tension (similar to a cable), rather than its cross-sectional properties. In the absence of current, the mean deflected shape of the drilling riser is driven by the applied top tension, the mean offset at the top of the riser, the in-water weight of the drilling riser ("effective" tension gradient). A current profile applies force to the riser that further influences the mean shape. The dynamic motion of the riser is driven by the top motion of the vessel coupled with the fluctuating force resulting from the waves and current. Other factors such as end constraints at the top and bottom of the riser also influence the riser's mean shape and the dynamic motion.

9.2.7.1.1 Tensioned Beam Model

Due to its length, the drilling riser is the most accurately modelled as a tensioned beam. The tensioned beam model combines the behaviour of a cable with the local stiffness of a beam. The equation for the tensioned beam is given later.

9.2.7.1.2 Concept of "Effective Tension"

Due to the column of mud inside the drilling riser, differential pressure effects are accounted for in the tensioned beam model for a drilling riser. As discussed by McIver and Olson (1981), differential pressure caused by the mud has a profound effect on the shape of the riser. Instead of using the tension in the wall of the pipe, the "effective tension" includes the internal and external pressures as noted in the equation below.

A simple calculation of the effective tension at any elevation along the riser can be carried out. The effective tension is the top tension minus the "weight" of the riser that is installed above the specified elevation. The "weight" of the riser is the in-air weight of the portion of riser (and contained mud) that is above the water and the in-water weight of the portion of riser (and contained mud) that is below the water.

9.2.7.1.3 Top Motion

Drilling riser analysis includes vessel motions, since the top of the riser is connected to the vessel. The vertical motions, primarily due to heave, roll, and pitch, are not included in drilling riser analysis because of the motion compensation provided by the actions of the slip joint and the marine riser tensioners. However, lateral motions caused primarily by surge, sway, roll, and pitch are accounted for.

The lateral motions imposed on the top of the drilling riser influence the direct wave and current forces applied to the riser by virtue of their phase with the waves. For example, the direct wave and current forces are relatively low if the motion of the vessel is "in-phase" with the water particle motions in the wave. This "in-phase" vessel response generally occurs with surge in large waves. "Out-of-phase" response can occur with smaller, short-period waves and can lead to relatively high direct wave and current forces.

9.2.7.1.4 Hydrodynamic Loading

The direct wave and current forces on the riser are calculated using formulas in Krolikowski and Gay (1980).

A drag coefficient and a drag diameter are characteristics of the riser. Similarly an inertial coefficient and inertial diameter are also characteristics of the riser and are used in the formulas that determine the dynamics of the riser under the action of the current, waves, and top motion.

9.2.7.1.5 Rotational Stiffness — Top and Bottom

Flex joints at the top and bottom of the drilling riser reduce the angle of the riser at its top connection to the vessel and at its bottom connection to the BOP. This local angle reduction provides a moderate reduction in angle that extends the conditions in which drilling operations can be conducted. The flex joint is a passive, elastomeric component, which has become popular for deep water.

Riser flex joints are also used at an intermediate location at the elevation of the keel on dynamically positioned vessels. The purpose of a flex joint at this elevation is to prevent damage in case the riser is disconnected in high currents or while the vessel is drifting after an emergency disconnect. The purpose of the intermediate flex joint is to provide an articulation rather than restrict the angle with its stiffness.

9.2.7.2 Analysis of a Tensioned Beam Model

Mean shape and dynamic motion of a drilling riser are calculated through finite element analysis of a tensioned-beam model. This analysis can be done using static analysis, frequency-domain analysis, or time-domain analysis. Static analysis can be accurate in cases in which no dynamics are expected. For steady-state dynamics, frequency-domain and time-domain solutions are alternatives that depend on solution time requirements, as described below. Time domain analysis is also used to simulate transient processes.

9.2.7.2.1 Time vs. Frequency-Domain Analysis

Time-domain analysis generally provides the more accurate solution than frequency-domain analysis at the expense of more computational time. In time-domain analysis, the equations of motion are solved at each of many small time steps that are used to describe a process such as an extreme storm. Typically, an analysis models an extreme storm with 1000 wave cycles, which roughly corresponds with a 3-h duration.

In the frequency-domain analysis, an extreme storm is described as a spectrum and the equations of motion of the riser are solved at each of many frequencies used to describe the process. The key approximation used in a frequency-domain approach is the technique for linearising any non-linear features in the process. For drilling risers, the most important non-linear feature is the drag force from the waves and current. A commonly used approximation for the drag force is described in Krolikowski and Gay (1980).

9.2.7.2.2 Coupled vs. Uncoupled Analysis

Traditional riser analysis has been performed in an "uncoupled" fashion in which the riser is considered to have no effect on the vessel at its top connection and no effect on the top of the BOP stack at its bottom connection. Usually, these effects are negligible and an uncoupled riser analysis is adequate. However in certain situations, the riser has an effect on the vessel or on the BOP stack that is considered in a "coupled" analysis.

Coupling effect is generally the most important to consider in EDS/drift-off conditions. In these conditions, the riser can take on a large top angle and apply a significant lateral force to the vessel. It can also take on a large bottom angle and thus a significant lateral force to the top of the BOP stack that causes the BOP, wellhead, and conductor pipe to take on an angle.

To accurately analyse the riser under the above conditions, a coupled analysis is required. At the top of the riser, the coupled analysis is carried out in combination with a vessel analysis program. As the vessel moves laterally away from the wellhead, the lateral force from the riser is applied as a restoring force, which reduces the speed of the vessel. This provides a more accurate estimate of the time available to disconnect the riser.

At the top of the BOP stack, the lateral force from the riser causes the BOP, wellhead, and conductor pipe to take on an angle. This angle depends on the soil foundation properties, the conductor dimensions and the elevation of the top of the BOP stack. As the BOP angle increases, coupled analysis considers that the bottom flex joint angle allowable also increases since the "stop" of the flex joint has rotated. In addition, coupled analysis provides an accurate assessment of the loading on the conductor and wellhead. Although, uncoupled analysis generally provides a conservative assessment, the coupled analysis provides an assessment that has many of the unnecessary conservatism removed, particularly in soft soil conditions.

9.2.7.3. Operational Limits

9.2.7.3.1 Minimum and Maximum API Tensions

This section discusses the API guidelines that have been established for minimum and maximum tension. Minimum tension is established to prevent buckling of the riser. Maximum tension is established to prevent top tensions in excess of the installed capacity of the riser.

To prevent buckling of the riser, criteria have been established within the industry to prevent the effective tension in the riser from going below zero. API RP 16Q (1993) provides guidance on this, which provides a margin to account for uncertainties in the weight of the riser steel and the lift of the riser buoyancy. This margin also provides adequate tension in case a tensioner fails. API RP 16Q (1993) also distinguishes the rated capacity of a tensioner and the vertical tension applied at the top of the riser. (The ratio is often in the range of 90–99%.) All of these factors are considered in the calculation of the API minimum tension that is used to prevent buckling.

In practice, the API minimum tension is rarely used as the riser's operating tension. An added margin on tension is warranted to improve the riser performance in high seas or high currents, as will be discussed later in this chapter.

As discussed earlier in this chapter, the lowest effective tension (usually at the bottom of the riser) is calculated as the top vertical tension minus the in-water weight of the riser plus the contained mud. The weight of the riser string and the mud column in the riser must both be supported by the tensioners to avoid riser buckling. To calculate the weight of the mud column, an estimate is made of the capacity (gallons/ft) of the riser pipe and the other lines (choke and kill lines and boost line) that contain mud. Table 9.6 shows how the in-air weight of the mud column above the water line and the submerged weight of the mud column below the water line are added to the string weight to determine the riser string weight with mud.

The API minimum riser tensions are calculated using the installed weight of the riser with mud. The values calculated are vertical tensions at the top of the riser. For this calculation, the following information was used:

- *Tolerances* – 1% on the weight of steel in the riser and 1% on the net lift from the buoyancy material.
- *Tensioners Down* – Positive tension is maintained in the riser if one out of twelve tensioners goes down.
- *Maximum Tension Limit* – API RP 16Q guidance is that top tension should be no more than 90% of the dynamic tensioning limit (same as rated tensioner capacity). This tension multiplied by a reduction factor for fleet angle only (in this example, the tensioner system compensates for mechanical losses, so that the estimate is 0.99) gives the maximum API tension in terms of vertical tension at the top of the riser. A maximum tension limit 90% of the installed capacity prevents the relief valves from popping under most conditions. In practice, a lower maximum tension limit is generally applied.

Table 9.7 shows the calculation of minimum and maximum API tensions for a range of mud weights. Figure 9.20 shows a plot of the results.

Table 9.6 and fig. 9.20 show a slightly higher tension than the API minimum tension at very low mud weights. In this range, a nominal tension (higher than the API minimum tension) is applied to the riser to assure that the riser can have a "planned" disconnect carried out successfully without increasing the tension. This tension is sufficient to support the in-water weight of the riser plus the LMRP (excluding the weight of the mud in the riser).

A significant factor in proper tensioning of an ultra-deep water riser is compaction of the buoyancy material leading to a reduction in the net lift of the buoyancy. API RP 16Q uses the weight of the riser string, the weight of the mud column in the riser and the auxiliary lines, and tolerance values to determine the riser weight installed in seawater. API's specified tolerance values of 5% on steel weight and 4% on buoyancy net lift can be overridden if an accurate weight of the riser is taken during deployment. In one recent example, when comparisons were made to manufacturers' values, weights recorded during deployment of a riser showed that the actual installed weight of the riser string can be matched by using 1% additional steel weight and slightly more than a 3% decrease in net lift due to buoyancy. Although this is within the API tolerance levels, when compared to

Table 9.6 Installed riser weight with mud

Riser string weight with mud in seawater			
	Riser capacity (gal/ft)	18.23	
	Seawater density (ppg)	8.55	
	Length (ft) (FJ to WL)	8940	
	Length (ft) (WL to DH)	50	
	Wt. of riser string (kip)	474.98	
Mudweight	Weight of mud in seawater from flexjoint to waterline[1]	Weight of mud in air from waterline to diverter[2]	Weight of riser string with mud in seawater[3]
ppg	kip	kip	kip
8.55	0.00	7.79	482.77
9	73.34	8.20	556.52
9.5	154.83	8.66	638.47
10	236.32	9.12	720.41
10.5	317.80	9.57	802.36
11	399.29	10.03	884.30
11.5	480.78	10.48	966.24
12	562.27	10.94	1048.19
12.5	643.76	11.39	1130.13
13	725.24	11.85	1212.08
13.5	806.73	12.31	1294.02
14	888.22	12.76	1375.96
14.5	969.71	13.22	1457.91
15	1051.20	13.67	1539.85
15.5	1132.68	14.13	1621.79
16	1214.17	14.58	1703.74

1 – Mudweight in Seawater * riser capacity * length from flexjoint to waterline
2 – Mudweight in air * riser capacity * length from waterline to diverter housing
3 – Riser string weight + weight of mud (FJ to WL) + weight of mud (WL to DH)
FJ = flex joint; WL = water line; DH = diverter husing

Table 9.7 API riser tensions – vertical load at slip ring

API riser tensions – vertical at slip ring – 9000 ft of water									
In-water weight of bare joints				30.97 kips	# of joints				117
Net lift of 3 k buoyant joint				30.33 kips	# of 3 k buoyant joints				36
Net lift of 5 k buoyant joint				30.57 kips	# of 5 k buoyant joints				28
Net lift of 7.5 k buoyant joint				30.34 kips	# of 7.5 k buoyant joints				33
Net lift of 10 k buoyant joint				27.94 kips	# of 10 k buoyant points				13
Remainder of String wt. (excl. LMRP)				162.66 kips					
Mud wt.	Weight of riser string with mud in seawater	Steel weight tolerance[1]	Buoyancy loss/ tolerance[2]	Minimum slip ring tension[3]	1-250 k tensioner loss factor[4]	API min. rec. tension w/1-250 k down (T_{min})[5]	Tension required for LMRP disconnect[6]	Min. rec. tension[7]	Maximum slip ring tension[8]
ppg	kip	kip	kip	kip		kip	kip	kip	kip
8.55	482.77	37.87	33.12	553.76	1.091	604.15	721.03	721	2673
9	556.52	37.87	33.12	627.51	1.091	684.61	721.03	721	2673
9.5	638.47	37.87	33.12	709.45	1.091	774.01	721.03	774	2673
10	720.41	37.87	33.12	791.40	1.091	863.41	721.03	863	2673
10.5	802.36	37.87	33.12	873.34	1.091	952.81	721.03	953	2673
11	884.30	37.87	33.12	955.28	1.091	1042.21	721.03	1042	2673

(Continued)

Table 9.7 Continued

Mud wt.	Weight of riser string with mud in seawater	Steel weight tolerance[1]	Buoyancy loss/ tolerance[2]	Minimum slip ring tension[3]	1-250 k tensioner loss factor[4]	API min. rec. tension w/1-250 k down (T_{min})[5]	Tension required for LMRP disconnect[6]	Min. rec. tension[7]	Maximum slip ring tension[8]
ppg	kip	kip	kip	kip		kip	kip	kip	kip
11.5	966.24	37.87	33.12	1037.23	1.091	1131.62	721.03	1132	2673
12	1048.19	37.87	33.12	1119.17	1.091	1221.02	721.03	1221	2673
12.5	1130.13	37.87	33.12	1201.12	1.091	1310.42	721.03	1310	2673
13	1212.08	37.87	33.12	1283.06	1.091	1399.82	721.03	1400	2673
13.5	1294.02	37.87	33.12	1365.00	1.091	1489.22	721.03	1489	2673
14	1375.96	37.87	33.12	1446.95	1.091	1578.62	721.03	1579	2673
14.5	1457.91	37.87	33.12	1528.89	1.091	1668.02	721.03	1668	2673
15	1539.85	37.87	33.12	1610.83	1.091	1757.42	721.03	1757	2673
15.5	1621.79	37.87	33.12	1692.78	1.091	1846.82	721.03	1847	2673
16	1703.74	37.87	33.12	1774.72	1.091	1936.22	721.03	1936	2673

1 – 1.0% In-water wt. of steel: 0.01*(wt. of bare joints plus remainder of bare string)
2 – 1.0% Net lift of buoyancy: 0.01*(net lift from all buoyancy)
3 – In-water weight plus steel weight tolerance plus buoyancy loss/tolerance
4 – Factor of 1.091 covers loss of one out of twelve 250-k tensioners
5 – Minimum recommended tensions that satisfy API 16Q guidelines for buckling stability:
Min. slip ring tension times tensioner loss factor
6 – In-water string weight with seawater plus in-water LMRP weight plus 50 kips
7 – Maximum value of 5 and 6
8 – 90% of dynamic tensioning limit (rated tensioner capacity) times reduction factor (0.99)

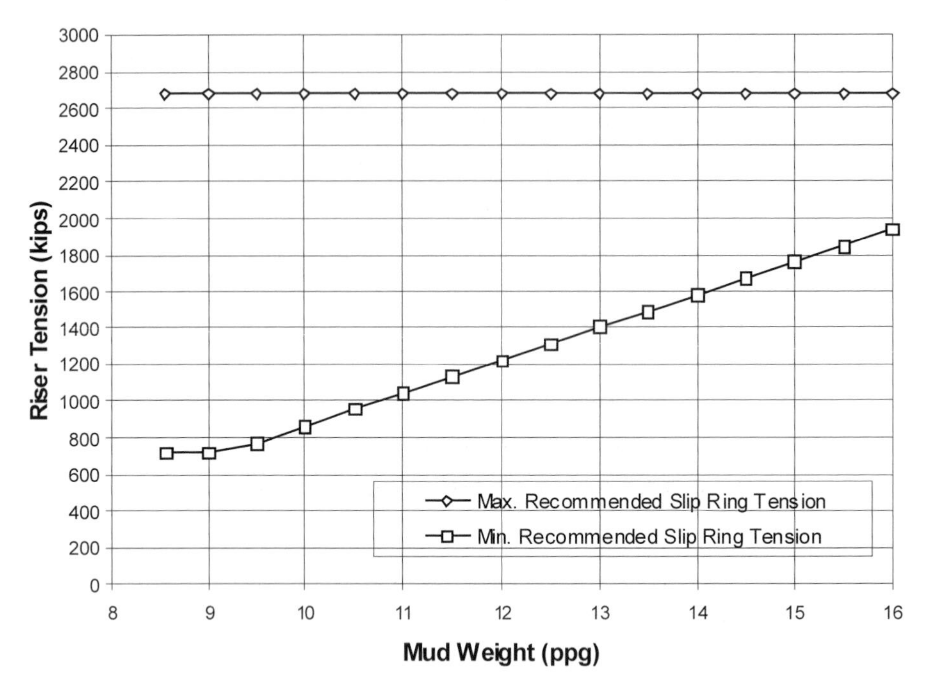

Figure 9.20 API Riser tensions – vertical load at slip ring (Glomar C. R. Luigs – GoM 9000 ft)

the manufacturers' values in 9000 ft of water, this can amount to 150 kips of additional weight for the entire riser string.

9.2.7.3.2 Riser Angle Limits

API RP 16Q has established riser angle limits for drilling and non-drilling operations with the riser connected. The basis for these is generally to minimise wear during rotation of the drill pipe and during tripping of the drill pipe. Angular limits are also necessary in order to conduct certain operations such as landing casing hangers and production equipment. When no drilling operations are being conducted, the limits can be relaxed to simply avoiding bottom-out of the flex joints. Figure 9.21 shows extremely large riser angles on a connected riser during high currents.

Riser wear incidents have continued to occur in drilling operations, with several "keyseating" failures occurring near the bottom flex joint. The key measures for avoiding wear are adequate riser top tension and vessel positioning. The areas susceptible to wear are the inner surfaces of the riser and BOP stack, particularly near the bottom flex joint.

API RP 16Q specifies limits on the bottom flex joint angle and top flex joint angle. During drilling operations, mean top and bottom flex joint angles of 2.0° are specified in API RP 16Q. In ultra-deep water, operations personnel generally use more restrictive targets for top

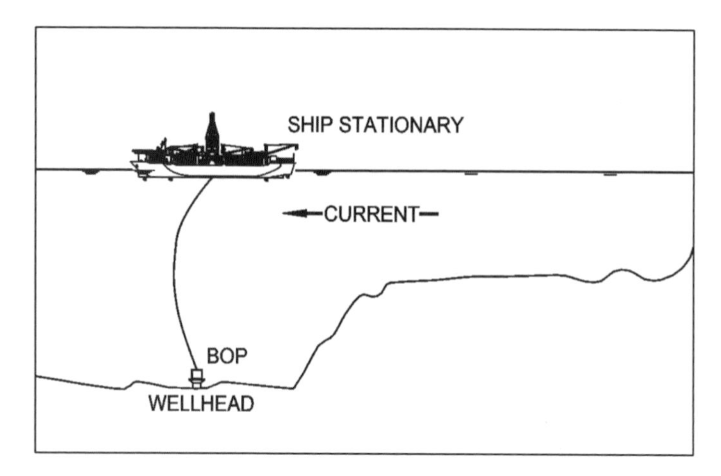

Figure 9.21 Excessive Top and bottom angles on connected riser

and bottom angles, such as 0.5–1.0° due to the cost consequences of tripping the riser. For non-drilling operations, maximum riser angle limitations are generally 9°, based on avoidance of flex joint bottom-out.

9.2.7.3.3 Stress Limits

Riser stresses are checked during the riser analyses. Maximum stresses are generally limited to 67% of yield strength. This limit ensures that the maximum tension applied to the riser is within the capacity of the riser connector. In this check, axial, bending, and hoop stresses are considered.

In addition to maximum stresses, alternating stresses are limited by a recipe given in the API RP 16Q. This recipe is intended to limit the fatigue damage in the connector and the riser pipe. Explicit fatigue analyses are often carried out to provide additional checks of the fatigue damage in a riser under wave loading conditions.

As noted earlier, VIV fatigue analysis is carried out on risers to check the fatigue damage done under high current conditions. The fatigue damage done by VIV is generally considered to be more severe than that done under wave loading.

9.2.7.3.4 Riser Recoil Limits (on DP vessels)

The minimum top tension in a connected riser is often governed by riser recoil considerations whose limits are calculated through analysis. The top tension must be high enough to ensure that the LMRP will unlatch cleanly from the BOP during an emergency disconnect. The limiting value in such a process is the clearance between the LMRP and the BOP after disconnect, if and when the LMRP cycles back downward toward the BOP due to vessel heave motion. A reasonable clearance is chosen to avoid damage based on the physical dimensions of the LMRP and BOP.

The maximum top tension based on riser recoil is limited to no more than the value that could cause excessive slack in the tensioning lines as the riser disconnects and moves upward. The slack could occur soon after disconnect as the riser accelerates upward and the tensioning system cannot keep up. Slack also could occur as the riser is stopped.

Finally, the maximum top tension is limited to no more than the value that the riser recoil system can stop during an emergency disconnect. The riser can be stopped by a combination of the riser recoil system and an arrangement in which the tensioners bottom out before the telescopic joint collapses. This arrangement, sometimes called a deadband, provides for the riser having no force applied to it after the tensioners have bottomed out. This provides some assurance that the riser does not apply force to the rig floor even at relatively high tensions.

These topics will be discussed further under the riser recoil discussion in Section 9.2.9.

9.2.7.3.5 Tensioner Stroke/Telescopic Joint (TJ) Stroke Limits

During EDS/drift-off conditions, the limits on tensioner stroke and telescopic joint stroke become important. The amount of allowable stroke-out depends on how far the telescopic joint is stroked out when it is in its nominal (i.e. calm seas) position at the site. Several factors can cause this nominal position to be "off centre" including the placement of pup joints in the string leading to the outer barrel to be slightly high or low on the inner barrel. As the telescopic joint is stroked out, a margin before complete stroke out of either the tensioners or the telescopic joint must be maintained to allow for wave-frequency variations and other uncertainties. This will be discussed further under the EDS/drift-off discussion in Section 9.2.8.

9.2.7.3.6 BOP, Wellhead, and Conductor Limits

The BOP, wellhead, and conductor pipe are often designed by the loading experienced during EDS/drift-off conditions. The BOP manufacturers provide curves that indicate the rated capacity of the flanges when loaded in tension, bending, and pressure. The wellhead manufacturer provides a similar rated capacity for the wellhead. Finally, the conductor has its connectors and pipe rated for tension and bending. The riser analysis results (including BOP, wellhead, and conductor loading) are compared against these ratings to determine whether the rating of the system is exceeded.

Analysis can be conducted to determine whether riser loading at the bottom flex joint is within the capacity of each of the BOP connectors, the wellhead, and the conductor casing. This analysis is conducted for combinations of vertical load, lateral load, and pressure load conditions specified by the operator. Depending on the component designs, the highest loading occurs during drift-off and the weakest link for bending loads is often either the wellhead connector or the casing connector closest to the wellhead connector.

As noted under operating limits, the assumption of a rigid, vertical BOP is generally a conservative approach for BOP component loads, but a more rigorous approach involves coupled analysis.

After the rig is on site, the misalignment angle of the conductor casing from vertical could be large enough to warrant reanalysis to determine its influence on the component loads.

This could assist in establishing a vessel position that would lead to improved bottom flex joint angles for drilling.

Another topic on operating limits involves torsional loading in special situations in which the vessel rotates and applies torsion to the riser and the wellhead. Depending on the component designs, the weakest link with torsional loading in the system could be the wellhead connector or the casing connector. Operational procedures and limits are set to avoid rotation or damage to these components.

9.2.7.4 Typical Operating Recommendations

9.2.7.4.1 Recommended Top Tension vs. Mean Vessel Offset

Recommended riser top tensions are determined based on the limits defined in Section 9.2.7.3, except riser recoil limits which will be introduced in a later section. These recommended top tensions are discussed in the example below.

Riser analysis for a connected riser configuration was conducted to determine whether the rig's top tensioning capacity is sufficient to support a riser in 9000 ft of water under some representative design metocean conditions. This assessment was done for drilling operations with up to 16-ppg mud in the Gulf of Mexico. The riser stack-up described in Section 9.2.4.3.2 was modelled in a typical riser analysis program. As noted earlier, vessel RAOs from Section 9.2.4.1 are used in the analysis for the riser configuration in 9000 ft of water.

Analysis was carried out for the following conditions, one with extreme waves and the other with extreme current.

- 1-yr Winter Storm – Connected, Drilling
- 10-yr Winter Storm – Connected, Non-Drilling
- High Current – Connected, Drilling
- Extreme Current – Connected, Non-Drilling

As noted earlier, the operational limits that apply for the non-drilling conditions are substantially less restrictive than those that apply for drilling conditions. Also, the high current conditions have a much different influence on the riser than the storm conditions. Besides the high drag loads, vortex-induced vibration of the riser pipe cause increase riser drag coefficients, causing larger riser angles.

For the conditions discussed above, state-of-the-art riser programs are available to calculate the riser's deflected shape, angles, and stresses. As noted earlier, these programs often carry out a solution in the frequency-domain or in the time-domain. Both types of solutions can be used in conducting large parameter studies for determining recommended top tensions with various offsets and mud weights. Frequency-domain programs tend to use less computer time, so they have become more popular. Results from riser analysis programs can be used to assemble parametric results that show plots of top angle vs. top tension, bottom angle vs. top tension, stress vs. top tension, and other relationships for various mean vessel offsets and mud weights. Top tensions that satisfy the operational limits can be derived from these results.

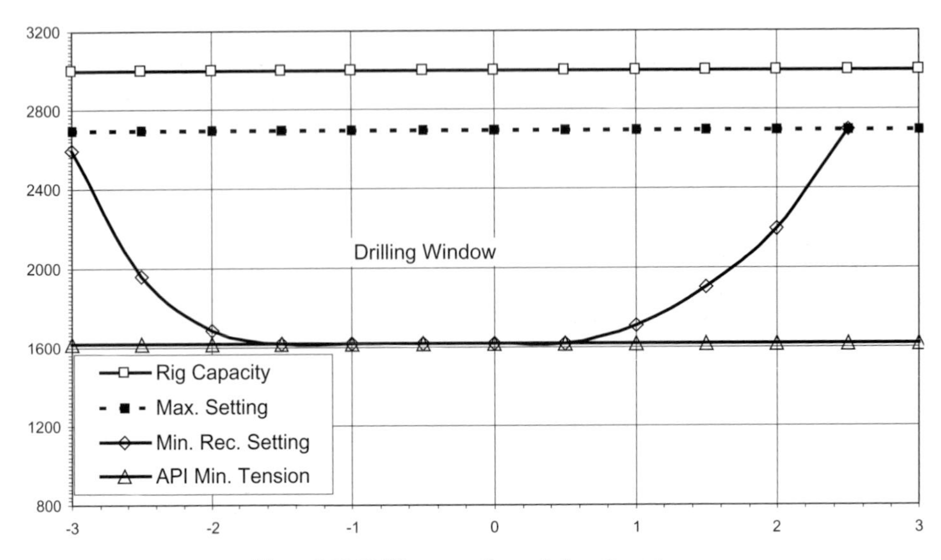

Figure 9.22 Drilling operations window, 1-yr storm

Figure 9.22 shows a curve of tensions that satisfy operational limits for various offsets. As shown, these tensions are within the API maximum tension for a range of offsets. If the vessel can keep station within this range of offsets, the operating tension can be established for that mud weight.

A vessel's mooring system can typically keep the vessel stationed within $\pm 2\%$ of water depth. If the riser angles are too large at these offsets, the vessel can be positioned at a more favorable offset by using "line management". The mooring lines of the vessel can be "managed" by being pulled in or payed out to position the vessel over the well. This requires additional action on the part of the crew and can be restricted under severe metocean conditions.

If a vessel is dynamically positioned, it can typically hold station within an offset circle of 1% of water depth from its set point (not always directly over the well). Given an offset circle of this size, the top tensions needed to satisfy the riser's operating limits vary with metocean conditions. The top tension needed with 16-ppg mud in a one-year winter storm is about 1700 kips, as shown in fig. 9.22. In high currents, (fig. 9.23), the top tension needed to satisfy the same conditions is about 2400 kips.

9.2.7.4.2 Top Tensions for Various Mud Weights

Curves such as those above are generated for various mud weights and are compiled to form a curve of top tension vs. mud weight. This curve is useful because the crew can adjust the top tension as the mud weight is changed, whereas top tension cannot be practically changed as offset varies. A graph of top tension vs. mud weight, specific to each well, is considered a key document on the rig. Figure 9.24 shows a typical curve of top tension

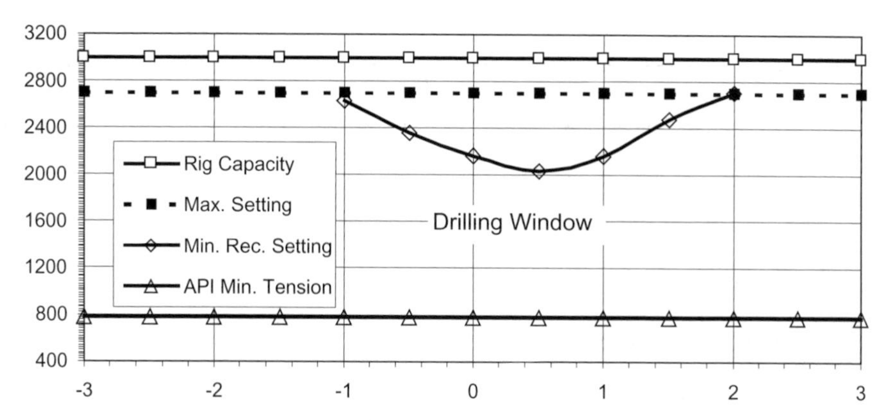

Figure 9.23 Drilling operations window, high current

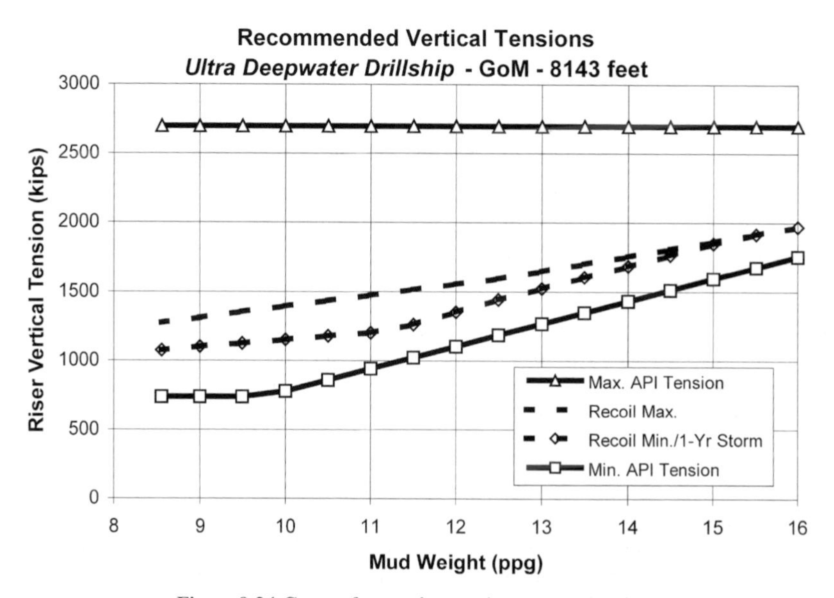

Figure 9.24 Curve of operating tensions vs. mud weight

versus mud weight for a dynamically positioned vessel, including riser recoil limitations that will be discussed in Section 9.2.9.

9.2.8 Emergency Disconnect Sequence (EDS)/Drift-Off Analysis

When a dynamically positioned drilling vessel loses power in ultra deepwater, the resulting motion of the vessel and the response of the riser depends on the intensity of the wind, waves, and current. A "drift-off" begins and the vessel tends to rotate from a heading with the bow into the weather to a heading turned 90° with the weather on the beam. Under the effects of the increasing vessel offset from the wellhead, the riser's deflected shape changes with time and is significantly affected by the vessel's drift-off speed. Through analysis, the riser's deflected shape can be shown to govern the time at which emergency disconnect limits are exceeded. By allowing a specific time for the emergency disconnect sequence to be carried out, yellow and red alerts are established to protect the system. This section discusses the practical application of EDS/drift-off analysis and the techniques that are used.

The response estimates of the vessel and the riser during drift-off conditions are used for setting emergency disconnect limits. The yellow and red alerts are set at vessel offsets and riser limits that will allow an emergency disconnect sequence (EDS) to be carried out while assuring the integrity of the drilling riser and its associated equipment. Potential drift-off scenarios are analysed to establish the yellow and red alert settings.

Results from an EDS/drift-off analysis are generally used to guide the captain in determining DP settings for each well. An analytical simulation of the response of the riser and the vessel is used to determine the time available to disconnect the riser. The resulting prediction of available time are used by the captain to set alert circles for planning the "emergency disconnect sequence" or EDS.

The EDS defines a series of alert circles, each of which has required procedures for the crew to prepare for riser disconnect. For example, a yellow alert circle includes a procedure for discontinuing drilling and hanging the drill pipe off in the BOP stack. A red alert circle signals the captain or the driller to "activate a red button" to start an automatic sequence that causes the drill pipe to be sheared by shear rams in the BOP stack and the riser to be disconnected. The EDS ensures the integrity of the riser and the related equipment, particularly the BOP stack, connectors, and conductor pipe that provide well pressure containment. The disconnect times are governed by exceedance of limits on top riser angle, bottom riser angle, slip joint stroke, wellhead moment, and conductor moment.

The vessel is considered to be in either one of the following two modes when a drift-off occurs: the first mode can be termed "drilling operations" and is associated with metocean conditions that are suitable for drilling; and the second mode can be termed a "state of readiness" and is associated with metocean conditions or other conditions that prohibit normal drilling activities. When comparing drift-offs in the two modes, starting with drilling operations, more time is required to carry out the procedures required to disconnect the riser (say 150 s). When starting from a state of readiness, the captain or driller is ready to activate the red button to start the emergency disconnect sequence so that less time is required (say 60 s). Because of this time difference, drilling operations are discontinued in certain metocean conditions and a state of readiness can be continued into larger metocean conditions. (Please consider the times quoted above as examples only; actual times vary with drilling vessel.)

An example set of metocean conditions used for the state of readiness mode is a 10-yr winter storm with a 1-min wind speed of 50.5 knots, a significant wave height of 19 ft, and a surface current of 0.6 knots. An example set of conditions for drilling operations is a one-minute wind speed of 25 knots, a significant wave height of 7.6 ft, and a surface current of 0.3 knots.

9.2.8.1 Drift-Off During Drilling Operations

For a drift-off that occurs during "drilling operations", the yellow and red alert circles are set using the time history of vessel motion and riser response resulting from the EDS/drift-off analysis. The point in time at which the first riser allowable limit is exceeded is termed the "point of disconnect", or POD. Disconnect at any later time would exceed a system allowable. With the POD as the basis, the vessel motions data time history is used to move backward according to the time required from "activating the red button" to the POD. As noted earlier, an example time allowed for this portion of the sequence is 60 s. This determines the time and offset position associated with the red alert circle. From the red alert circle, the vessel motions data time history is used to move backward again according to the time required to move from ongoing "drilling operations" to a "state of readiness". An example time for this portion of the sequence is 90 s. This determines the time and offset position associated with the yellow alert circle.

As noted above, the point of disconnect (POD), which drives the yellow and red alert circles, is governed by first exceedance of an allowable limit within the system. In this process, allowable limits are set for any component whose integrity could be compromised as the vessel drifts off. The limits are generally set for the top riser angle, the bottom riser angle, stroke-out of the slip joint, stroke-out of the tensioners, loading on the BOP, loading on the wellhead connector, loading on the wellhead, and loading on the conductor pipe. Typical limits for top and bottom angles are 9° (90% of the flex joint stop, per API RP 16Q) and stroke-out values of say 25 ft based on some margin within a 65-ft stroke capacity, for example.

In 4500 ft of water and 9000 ft of water in the Gulf of Mexico, summary results for an EDS/drift-off analysis in a reasonable set of metocean conditions (i.e. the 95% non-exceedance environment) used for drilling operations are as follows:

4500 ft – Red Alert Circle = 225 ft (5% WD); Yellow Alert Circle = 72 ft (1.6% WD)

9000 ft – Red Alert Circle = 360 ft (4% WD); Yellow Alert Circle = 180 ft (2% WD)

In these examples, the results in 4500 ft of water are governed by yield of the conductor pipe; whereas the results in 9000 ft of water are governed by stroke-out of the slip joint.

As shown above, drift-offs tend to be more difficult to manage in the shallower water depths. In 4500 ft of water, the size of yellow alert circle has reduced to a relatively low, but manageable level when compared to the larger yellow circle in 9000 ft.

9.2.8.2 Drift-Off During a State of Readiness

For a drift-off that occurs during a state of readiness, the metocean conditions used are design values in which the riser will remain connected. An example of this is the 10-yr winter storm in the Gulf of Mexico. In the state of readiness mode, only the red alert circles are set and this is done using the time history of vessel motion and riser response resulting from the EDS/drift-off analysis. The point in time at which the first riser allowable limit is exceeded is termed the "point of disconnect", or POD. From the POD, the vessel motions data time history is used to move backward according to the time required from "activating the red button" to the POD. As an example, the time allowed for this portion of the sequence is 60 s. This determines the time and offset position associated with the red alert circle. The allowable limits for the system are the same as they are in the drilling operations mode.

In 4500 ft of water and 9000 ft of water in the Gulf of Mexico, summary results for an EDS/drift-off analysis in a reasonable set of metocean conditions used for a state of readiness are as follows:

4500 ft – Red Alert Circle = 90 ft (2% of WD)

9000 ft – Red Alert Circle = 225 ft (2.5% of WD)

As with drift-offs from a drilling operations mode, drift-offs from a state of readiness tend to be more difficult to manage in a shallower water depths. In 4500 ft of water, the size of red alert circle is reduced to a relatively low, but again manageable level when compared to the larger red circle in 9000 ft. Figure 9.25 shows how much more rapid the drift-off in a 10-yr storm is when compared to the drift-off in a one-year storm. The comparison of results in the 4500-ft and 9000-ft water depth cases is also influenced by the larger riser restoring force in shallow water.

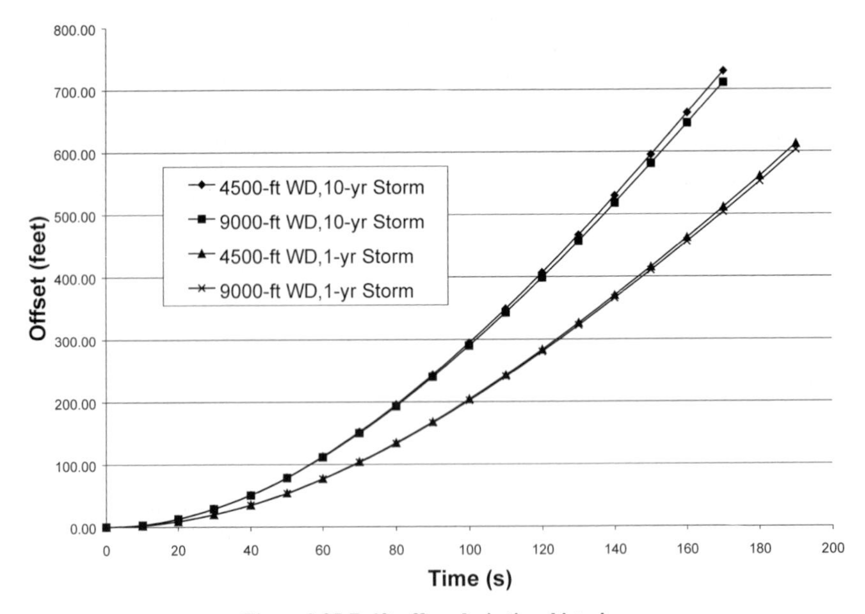

Figure 9.25 Drift-off analysis time histories

9.2.8.3 EDS/Drift-Off Analysis Technique

This section describes a transient coupled analysis technique for calculating drift-off of a dynamically-positioned vessel and the associated effect on the emergency disconnect sequence for a drilling riser. The drift path of the vessel is calculated in the time-domain, taking into account the transient response of the riser and the vessel's change of heading under the influence of current, wind, and waves. The effect of vessel rotation on horizontal motion is important in calculating the yellow and red alert offsets for the EDS. Also, the effect of riser restoring force on the vessel will be shown to be significant.

9.2.8.3.1 Riser Response Analysis

Transient dynamic analysis in the time domain provides a reasonable estimate of riser and vessel response during drift-off. An alternative approach is the quasi-static technique in which inertial forces are approximated and applied as loads distributed along the riser. A third alternative, the static analysis technique, is accurate only for certain combinations of very slow drift speeds or shallow water.

A transient riser analysis can be used to model the inertial effects of the riser and the relative velocity effects between the current and the speed of the riser. Wave-frequency forces are often not a significant factor in these results. The vessel's linear (offset) motion time history is specified at the top of the riser and the analysis is run to generate the riser analysis results including top riser angle, bottom riser angle, slip joint stroke, riser stresses, and wellhead loads.

Figure 9.26 shows the time history of slip joint stroke for conditions associated with a 10-yr storm (non-drilling, state of readiness) and with a reduced storm (drilling operations). Note that the slip joint stroke does not show any appreciable movement until about 50 and 100 s into the drift off, for the 10-yr storm and the reduced storm, respectively. As shown, the rate of increase in the slip joint stroke is much higher for the 10-yr storm. A typical allowable limit for slip joint stroke is between 20 and 30 ft depending on its stroke limits, the water depth, the top tension, and the space-out of the pup joints.

Figure 9.27 shows the time history of bottom flex joint angle for both the 10-yr storm and the reduced storm. Note that the bottom flex joint angle does not show any motion until about 70–80 s into the drift off, regardless of the storm size. After the initial response, the rate of increase in flex joint angle is higher in the 10-yr storm, as expected. A typical allowable limit used against this curve of bottom flex joint angle is 9°.

9.2.8.3.2 Importance of Coupled Riser Analysis

The riser and vessel motions analysis programs are coupled to include the effects of riser restoring force on vessel motion. Depending on the water depth and specific conditions, this can provide a 15–20% reduction of offsets in the time history of vessel motion. A simplistic coupled analysis is illustrated below.

- First, the vessel analysis is done with no riser loads.
- Second, the resulting vessel motions are used in the riser analysis.

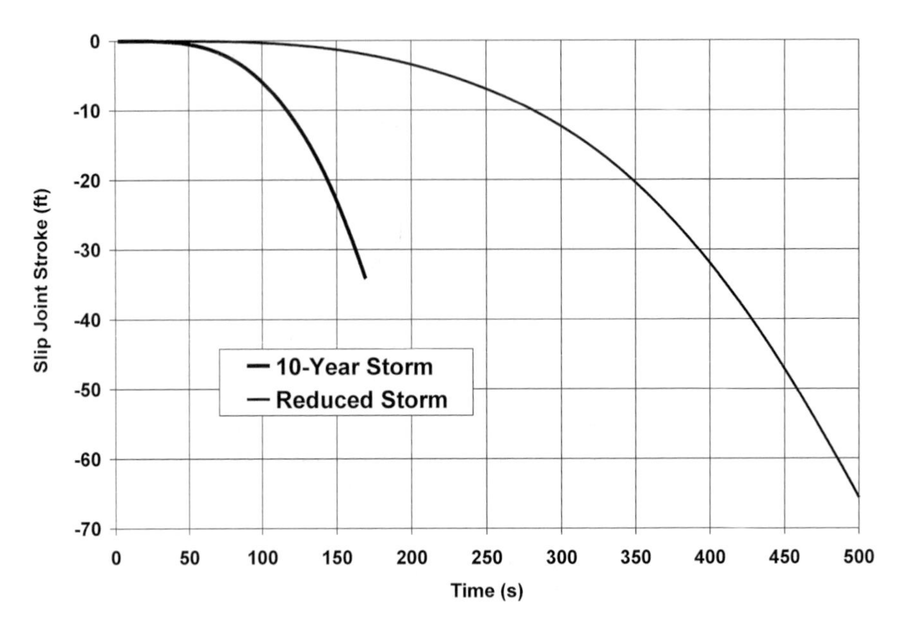

Figure 9.26 Time history of slip joint stroke during drift-off

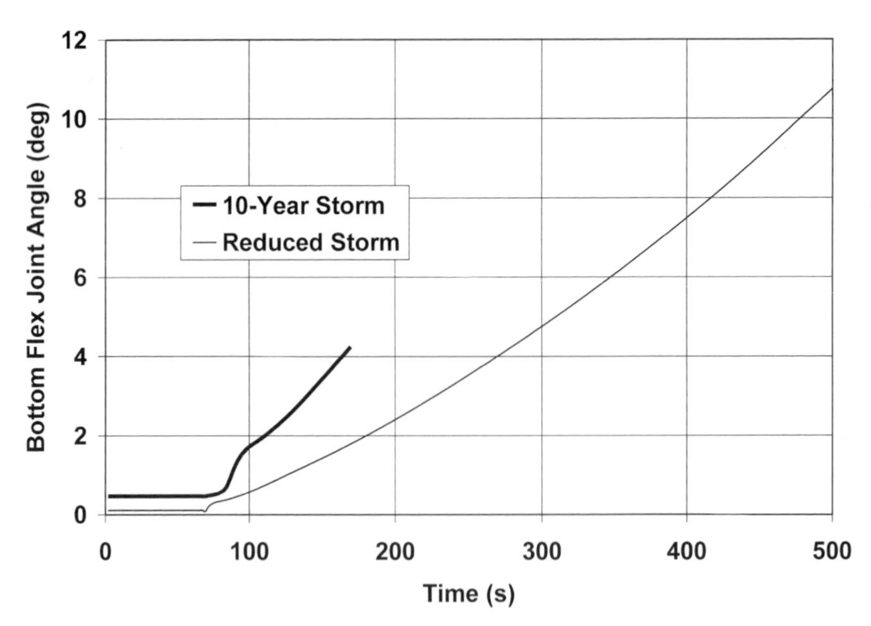

Figure 9.27 Time history of bottom flex joint angle during drift-off

- Third, the vessel analysis is redone with lateral riser loads from the previous riser analysis.
- Fourth, the riser analysis is redone with the updated vessel motions.

A more sophisticated analysis approach would solve for the complete system (vessel and riser) at each time step. This would result in a fully coupled analysis.

9.2.8.3.3 Importance of Vessel Rotation

The results of vessel motions analysis depend heavily on the heading of the vessel with respect to the incident weather (wind, waves, and current). The force on a drillship is much lower when it is headed into the weather than when the weather is on its beam (turned by 90°). To minimise force and vessel motions, the captain generally heads the vessel into the weather. When a vessel loses power, it will tend to rotate such that the weather is on the beam – a stable orientation. The speed at which this rotation takes place can be calculated through the vessel motions analysis. Due to the differing force coefficients in the different headings, the rotational speed has an influence on how quickly the vessel translates away from its set point over the well.

Vessel motions analysis can be carried out simply using the equations of motion for a rigid body based on Newton's 2nd law. The translational and the rotational motions are described by:

$$m\ddot{x} = F(t) \tag{9.2}$$

$$I\ddot{\varphi} = M(t) \tag{9.3}$$

where m represents the mass of the vessel, $\ddot{x}$ represents the translational acceleration of the vessel at the centre of gravity (CG) in the surge and sway modes, I is the vessel mass moment of inertia, $\ddot{\varphi}$ denotes rotational acceleration in the yaw direction, the "dot" represents differentiation with respect to time (t), and F and M represent the exciting force vector and moment vector acting in the horizontal plane.

The applied forces and moments are due to:

- Environmental forces and moments due to wind, current, and mean wave drift;
- Hydrodynamic forces and moments proportional to the vessel acceleration represented by the added mass term and added inertia terms at zero frequency;
- Hydrodynamic drag forces and moments proportional to the vessel velocity; and
- Riser reaction forces in the horizontal plane.

The wind, current, and waves are applied collinearly and concurrently. The initial conditions of the vessel heading and velocity are defined. In a fully coupled analysis, as discussed in 9.8.3.2, the forces (including the riser restoring force) and moments are updated at each time step and the corresponding vessel motion and rotation in the horizontal plane are calculated.

Example vessel characteristics are shown in table 9.8.

The added mass and added mass moment of inertia at zero frequency are calculated using a diffraction program. The current and wind force and moment coefficients can be determined from a wind tunnel model test.

Table 9.8 Vessel principal particulars

Length (perpendiculars)	m	210
Breadth	m	36
Depth	m	17.8
Draft	m	9
Displacement	ton	54,709

9.2.8.4 Trends in Analysis Results with Water Depth

Trends show that EDS/drift-offs are more difficult to manage in shallow water than in deep water because, in deeper water, a specific amount of distance traveled by the vessel results in a lesser percentage offset and a lesser angle. Not all of this advantage can be retained, however, because of the shape of the riser and the different allowable limits involved. In waters shallower than 5000 ft, the wellhead or conductor moment may be the governing limit that establishes the point of disconnect (POD) discussed earlier. The moment values are determined by the soil properties and the dimensions and yield strengths used in these components. Figure 9.28 shows a typical conductor pipe bending moment profile based on the drift-off trajectories for beam sea and the rotating ship conditions.

Figure 9.29 shows a summary of drift-off analysis results for site in 4227 ft of water in the Gulf of Mexico, with a riser top tension of 1371 kips and a mud weight of 10 ppg. The curve represents the horizontal vessel excursion (offset) vs. time. A vertical line is drawn at the time of POD, which is the minimum of the times at which the allowable limits for stroke, angles, wellhead bending moment, and conductor bending stress were reached. In this example, the POD occurs at 254 s and the associated offset is 467 ft. If the time is reduced by 60 s (to 194 s), the red circle radius is established as 290 ft. If the time is reduced by a further 90 s (to 104 s), the yellow circle radius is established as 89 ft. In dynamic-positioning operations, the yellow circle defines the offset at which drilling operations are suspended and the red circle defines the offset at which the EDS sequence is initiated.

9.2.8.5 Operational and Analytical Options

If the yellow or red circles are not large enough to be practical, options may be available by looking at the system as a whole. A first option is usually to find an analytical fix and the second to propose an operational fix. Analytical fixes can include exploring options for reduced top tensions, which if set too high initially, could cause difficulties in either riser recoil or connected riser recommendations. Reduced riser tensions and other such compromises may be needed to reduce the loads on conductors for EDS/drift-off, for example.

In many regions of the world, metocean conditions are so severe that they cause difficulties in managing the possibility of EDS/drift-off. In areas such as the Gulf of Mexico, Trinidad, Brazil, and the Atlantic margin, high currents can cause a vessel to drift off rapidly. If currents exceed conditions associated with a state of readiness mode, steps to provide operational management might be necessary such as positioning up current or simply

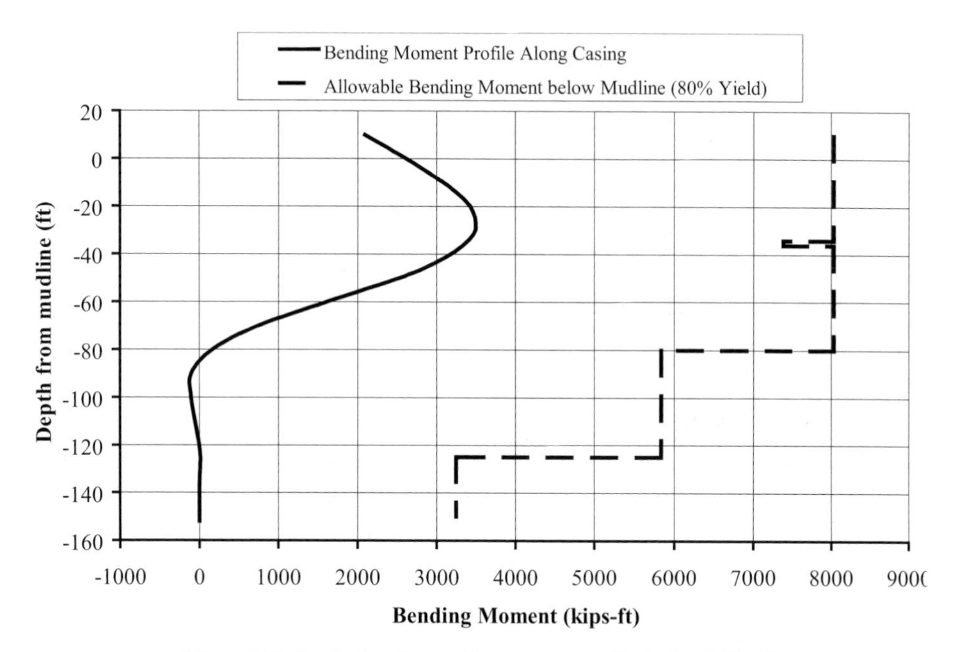

Figure 9.28 Conductor pipe bending moment profile during drift-off

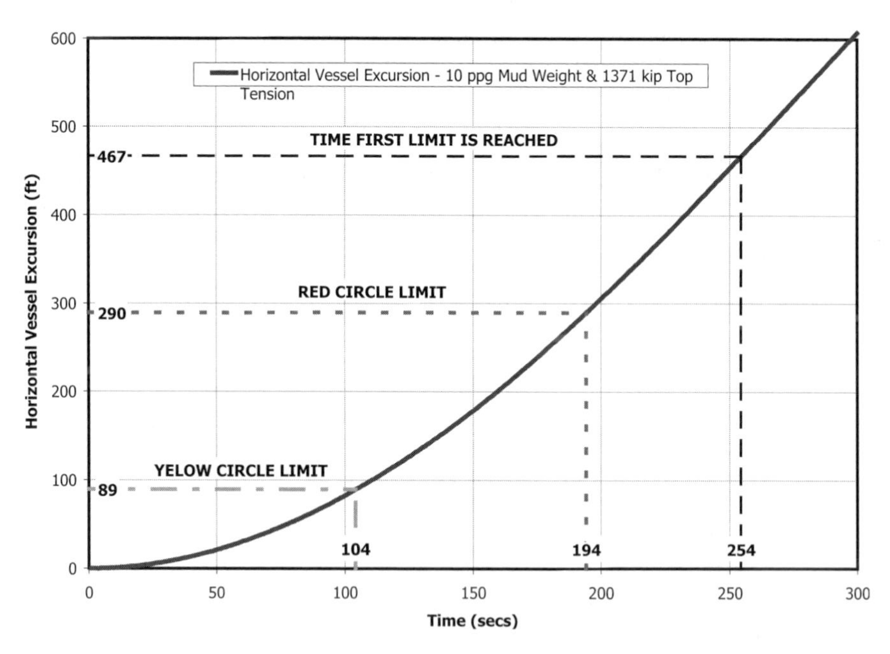

Figure 9.29 Summary of drift-off analysis results

disconnecting the riser in some conditions. However, a disconnected riser in high currents is also difficult to manage due to the large angle that it will take on.

In areas of the world that have high wave conditions that build rapidly, the possibility of an EDS/drift-off event poses another type of riser management issue. If the riser can survive EDS and hang-off in design level wave conditions, the management issue is simply a matter of when to disconnect and ride out the storm. Disconnection of the riser protects the pressure-containment components, i.e. the BOP, wellhead, and conductor. However, when a site has design wave conditions in which EDS and hang-off can jeopardise the free-hanging riser, the riser is pulled before the storm is encountered. Depending on the water depth and the forecasted seastates, the riser pulling operations are begun well in advance.

9.2.9 Riser Recoil after EDS

This section covers the response analysis of the riser as the LMRP is released from the BOP during an emergency disconnect sequence (EDS). An understanding of this process is important in order to maintain safety and avoid damage to the riser and its related components. Additionally, riser recoil considerations often dictate the top tensions that are pulled on the drilling riser.

Riser recoil analysis is conducted to determine the axial response of the riser after an emergency disconnect of the LMRP from the BOP at the seabed. In practice, this analysis is used to optimise riser tensioner system settings and define riser top tensioning bands to prevent excessive response of the riser. Typical allowable limits are aimed at ensuring the system behaves as follows after disconnect: the LMRP connector lifts off the BOP mandrel without reversal that could cause re-contact; the riser stops before impacting the drill floor, and slack in the tensioner lines is limited. To check these limits, some form of riser recoil analysis is generally done for each well site.

This section provides a discussion of the riser recoil process, riser response analysis, allowable limits, results and interpretation of some example cases, and sample operational recommendations. Although some sample guidelines are discussed here, general guidance would be highly dependent on the riser tensioning system and site-specific guidance would depend on the site and the selected operating parameters. The process and its analysis are discussed in more detail in Stahl (2000).

9.2.9.1 Definition of Process

As the riser goes through an emergency disconnect sequence (EDS), it automatically disconnects near the seabed. This disconnect is carried out at the interface between the lower marine riser package (LMRP) and the lower portion of the blowout preventer (BOP). As the riser releases, it responds with upward axial movement that is managed through the tensioners and the associated riser recoil system.

Management of the riser's upward movement is carried out by adjusting the stiffness and/or damping of the tensioner system. This can be done in a variety of ways and the examples below do not cover all of them. In one example system, the EDS includes an automatic command to close air pressure vessels (APVs) normally kept open to maintain small tension variations during operations. This causes a sudden increase in the system's vertical stiffness. Also, a so-called "riser recoil" valve is shut to increase the damping by constricting the orifice for fluid flow. In another example system, the riser's upward movement is managed by changing the orifice size based on tensioner stroke or velocity, with no closure of APVs.

Several properties of the riser also influence the riser's vertical response. First, the in-water weight of the riser string and the LMRP affect the dynamics of the riser. In certain cases, bare joints of riser are included in the riser stackup to help control the upward movement. Secondly, the weight of mud contained in the riser alters the response after disconnect; the frictional effects of the mud stretch the riser downward for some duration after disconnect. Thirdly, in deep water, stretch in the riser can be significant (several feet) and this leads to a rapid upward response (slingshot effect) after disconnect.

9.2.9.2 Riser Response Analysis

Some of the key modelling parameters and analysis cases are considered in a riser recoil analysis. Riser recoil analysis is generally carried out assuming only axial response, with fluid flow through the tensioning system, vessel heave, effects of offset on vertical tension, and mud flow all playing a big part in the response. For this discussion, due to its rig-specific nature, the tensioning system is simply considered a spring-damper device. Heave is an important input parameter, with its selection generally based on a relationship to a vessel in a design storm. Top tension used in the analysis is altered depending on the offset that is of interest. This is due to the build up of tension that can be caused in some systems when the APVs close some time prior to disconnect. Mud flow is typically modelled in the analysis, with higher mud weights give higher frictional loads on the sides of the riser as they fall out, thereby pulling the riser downwards for some duration after disconnect.

9.2.9.3 Allowable Limits

The allowable limits on riser recoil set the following riser top tensions: minimum top tensions to keep the LMRP from damaging the BOP during disconnect; maximum top tensions to avoid slack in the tensioner lines just after disconnect; and maximum top tensions to avoid the riser impacting the drill floor.

Minimum tensions are limited by avoidance of contact between the LMRP and the BOP, as the LMRP cycles back downward toward the BOP after disconnect. Such movement could occur if the disconnect were to occur at the "worst phase" of a vessel's heave cycle. Such phase considerations cannot be controlled because of the duration (about 60 s) of the EDS sequence. Allowable limits on such motion are dependent on the BOP equipment and the tolerance for damage, but leaving a few feet of clearance is generally considered reasonable.

Maximum riser top tensions are limited by avoidance of slack in the tensioner lines during riser recoil. The upward motion associated with this limit could be exacerbated if the disconnect were to occur at the "worst phase" of a vessel's heave cycle. As noted above, such phase considerations cannot be controlled because of the duration (about 60 s) of the EDS sequence. Reasonably small amounts of slack are allowed with certain systems, but no specific limits have been established.

To avoid the riser impacting the rig floor, a "deadband" might be available to provide further protection. This deadband can be defined as an arrangement whereby the tensioners and slip joint stroke ranges are offset. In this arrangement, when the tensioners have pulled their line to their full upward extent, the telescopic joint should still have some travel available (say 5 ft) before it bottoms out. Thus, the tensioners would apply no force to the riser while in this deadband. This arrangement provides a cushion that would help to slow down the riser if it strokes upward further than expected. Slack in the tensioner lines would have to be managed, however. Through means of this deadband arrangement, further limits on maximum top tension can be avoided.

9.2.9.4 Operational Issues

As noted above, some form of riser recoil analysis is generally used for every deepwater well site. Due to the impact of the results on riser top tensions, sensitivity cases are sometimes run to investigate ways to allow a larger band of allowable tensions thus making better use of the rig's installed tensioner capacity. The nature of these sensitivity cases would depend on the rig's tensioner and recoil system. Examples of such cases could include closing varying numbers of APVs, thus altering the stiffness at the time of disconnect; or a larger orifice or a different program for changing the orifice size.

9.3 Production Risers

Four types of production risers were mentioned in the introduction:

1. Top-tensioned (TTR)
2. Free Standing
3. Flexible
4. Steel Catenary (SCR)

Figure 9.1 illustrates the various kinds of risers. All are designed to convey well fluids to the surface. Each type has unique design requirements.

Flexible risers are the most common type of production riser. They may be deployed in a variety of configurations, depending on the water depth and environment.

Flexible pipes, long the standard riser for floating production, have traditionally been limited by diameter and water depth. Deepwater projects in the Gulf of Mexico and Brazil are now employing SCRs for both export and import risers. Figure 9.30 shows the capability of flexible pipes as of this writing. This will undoubtably grow in the future. The choice between a flexible riser and an SCR is not clear cut. The purchase cost of flexible risers for a given diameter is higher per unit length, but they are often less expensive to install and are more tolerant to dynamic loads. Also, where flow assurance is an issue, the flexible risers can be designed with better insulation properties than a single steel riser.

Flexible risers and import SCRs are associated with wet trees. Top tensioned risers are almost exclusively associated with dry trees and hence are not usually competing with flexibles and SCRs except at a very high level: the choice between wet and dry trees.

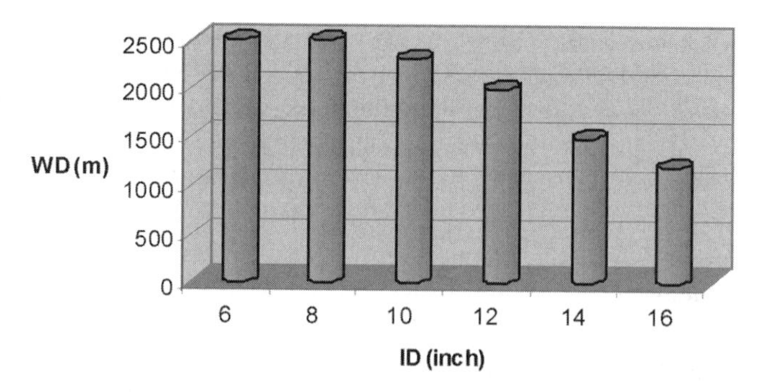

Figure 9.30 Capability of flexible pipe (Technip Offshore)

9.3.1 Design Philosophy and Background

9.3.1.1 Metocean Data

Each location may have critical design conditions; e.g. loop currents in the Gulf of Mexico and highly directional environments in the West of Africa. Vessel motions and offsets have a major influence on riser design and should be paid due attention (see Section 9.3.1.4). Metocean data used in riser analysis are water depth, waves, currents, tide and surge variations and marine growth. For the extreme waves and currents, the 1, 10, 100-yr and higher return periods may be considered. The 95% non-exceedance values may be used as temporary installation design condition. Long-term waves are defined by an H_S–T_P scatter diagram, with directionality if required.

Interfacing between the riser analysts and the metocean specialists at an early stage in the design process is recommended, so that riser-critical environmental conditions do not get overlooked. The importance of both directionality and of joint wave/current behavior varies from one location to another and should always be carefully considered.

Riser response is period sensitive, and analyzing the maximum wave-height case with a single wave period may not result in the worst response of the riser by reference to vessel RAOs, ensuring that important peaks in vessel response are not missed.

It should be recognised that the confidence with which metocean design data is derived varies considerably from one geographical location to another. Currents in the deepwater Gulf of Mexico, for example, are considerably higher than on the shelf. This has a large impact not only on the design of risers and mooring systems but also on the methods used for installation, and this emphasises the need for reliable site specific data.

It is recommended that currents specified for the riser design include an allowance for uncertainties in the derivation of data. No general rule for this is laid down here; such decisions should be taken in consultation with metocean specialists.

9.3.1.2 Materials Selection (This section contributed by David Rypien, Technip Offshore, Inc., Houston, TX)

Materials for riser pipe and components are selected based on design criteria, environmental conditions, and economics. In most cases, the governing criterion is the economics determined by trade-offs for the type of material, e.g. using carbon steel vs. titanium. Titanium was selected for stress joints (Oryx Neptune Spar, Placid Green Canyon 29), and in one case for an entire drilling riser (Heidrun). However, it is generally uneconomic for normal applications. Composite material has also been proposed for risers, but until now has been considered too expensive or immature. A composite string is currently being tested on the Magnolia TLP in the Gulf of Mexico.

Once the material type is selected, a material specification is developed that considers the operating environment; lowest anticipated service temperature, sour service, and/or cathodic protection. The key material properties include:

1. hardness,
2. strength,
3. toughness.

Weldability considerations generally limit use of steel to yield strengths of 80 ksi or less. Higher strength steels may be used with threaded and coupled joints; however, these joints have higher stress concentrations, lower fatigue resistance than is typically required for floating production systems. Finally, inspection, testing (including fatigue testing) and packaging requirements need to be specified.

Common standards and specifications used for carbon steel riser pipe and components are listed below:

API RP 2 RD	Design of Risers for Floating Production Systems and Tension Leg Platforms
API 5L	Specification for Line Pipe
API RP 2Z	Recommend Practice for Preproduction Qualification for Steel Plates and Offshore Structures
ASTM A370	Methods and Definitions for Mechanical Testing of Steel Products
BS 7448	Fracture Mechanics Toughness Tests. Methods for determination of fracture resistance curves and initiation values for stable crack extension in metallic materials
DNV-OS-F101	Offshore Standard – Submarine Pipeline Systems
DNV-OS-F201	Standard for Dynamic Risers
NACE MR-01-75	Sulphide Stress Cracking Resistant Metallic Materials for Oilfield Equipment

Line pipe material specifications are often combined with casing pipe sizes to be compatible with well systems.

API Spec 5L specifies two classification levels: PSL 1 and PSL 2 to define, generally, lower and higher strength steels. Most riser applications call for PSL 2 classification, typically X52, X60 or X80.

Table 9.9 Strength range for API 5L pipe

Grade	Yield strength, ksi	Ultimate strength, ksi
X52	52–77	66–110
X60	60–82	75–110
X65	65–87	77–110
X70	70–90	82–110
X80	80–100	90–120

Specifications of chemistry and heat treatment that will achieve the required material strength, hardness, and toughness need to be developed with the assistance of the pipe manufacturer.

9.3.1.2.1 Strength

Tensile strength is defined in terms of yield, σ_y, and ultimate, σ_u. Yield strength is defined as the tensile stress required to produce a given percentage of strain, e.g. API 5L determines σ_y, corresponding to the value is 0.5% ε (strain). If a tensile test continues past the point of yield, the material elongates and, in a ductile material, the area is reduced. The stress, based on the original area, is the ultimate tensile strength.

API 5L specifies a minimum range of strength levels for the various steel grades as shown in table 9.9. X65 or X80 are the most common steel grades for top tensioned production risers.

The amount of elongation before failure is a measure of ductility. API 5L specifies a minimum elongation, e, in 2 in. length as

$$e = 625{,}000\frac{A^{0.2}}{U^{0.9}} \tag{9.4}$$

where e = Minimum elongation in 2 in. to the nearest percent, A = Specimen area, in^2, U = Minimum ultimate tensile strength, psi. For example, the elongation of a round bar specimen with $A = 0.2$ in^2, and $U = 100$ ksi would be 14%. API 5L also requires that the ratio of σ_y/σ_u shall be less than 0.93 to insure a level of ductility.

9.3.1.2.2 Hardness

The following discussion is taken from www.tpub.com/doematerialsci/.

"Hardness is the property of a material that enables it to resist plastic deformation, penetration, indentation, and scratching. Therefore, hardness is important from an engineering standpoint because resistance to wear by either friction or erosion by steam, oil, and water generally increases with hardness.

Hardness tests serve an important need in industry even though they do not measure a unique quality that can be termed hardness. The tests are empirical, based on experiments and observation, rather than fundamental theory. Its chief value is as an inspection device

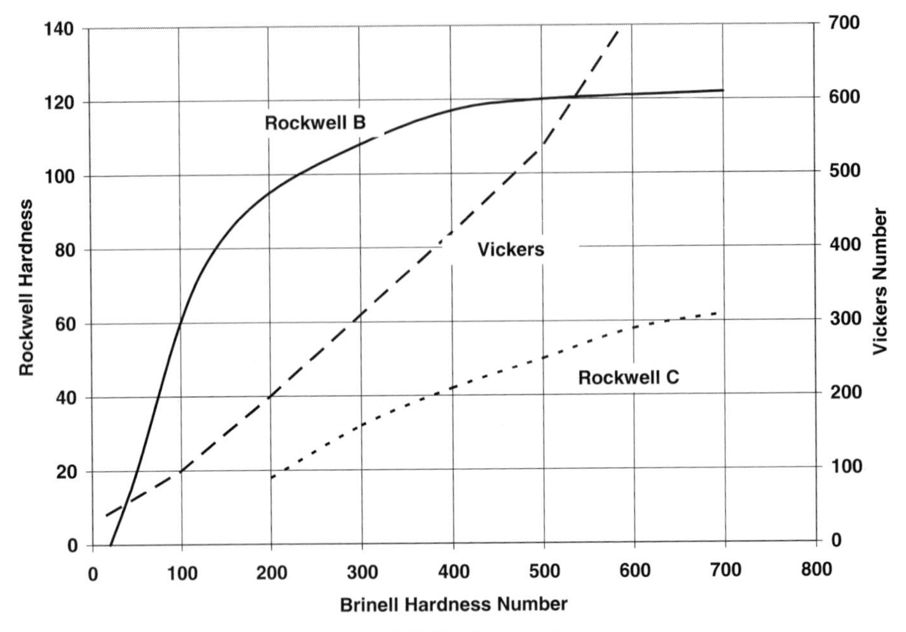

Figure 9.31 Hardness number

able to detect certain differences in material, when they arise, even though these differences may be undefinable. For example, two lots of material that have the same hardness may or may not be alike, but if their hardness is different, the materials certainly are not alike.

Several methods have been developed for hardness testing. Those most often used are Brinell, Rockwell, Vickers, Tukon, Sclerscope, and the files test. The first four are based on indentation tests and the fifth on the rebound height of a diamond-tipped metallic hammer. The file test establishes the characteristics of how well a file takes a bite on the material".

The indentation tests are most commonly used in material qualification. Each method uses a different indentation ball size and results in a different value. Figure 9.31 shows the relationship of Rockwell and Vickers hardness numbers to the Brinnel Hardness.

Hardness is directly correlated with strength, and inversely correlated with ductility. This is shown in fig. 9.32. Although hardness is normally used for testing purposes and not as an independent design criteria, a maximum hardness of 22 Rockwell C (275 HV 10 maximum at cap pass) is specified for risers and pipelines in sour service.

High strength is desirable for weight reduction in deepwater. High hardness, however, increases the risk of brittle fracture. This is a critical concern for tensile members like risers where it is generally desirable to be ductile against failures, allowing time for detection and corrective action (e.g. a through wall crack should not cause fracture of the pipe). Also, high strength and hardness typically require an increase in the carbon content. Figure 9.33 shows the maximum attainable hardness for quenched steel as a function of carbon

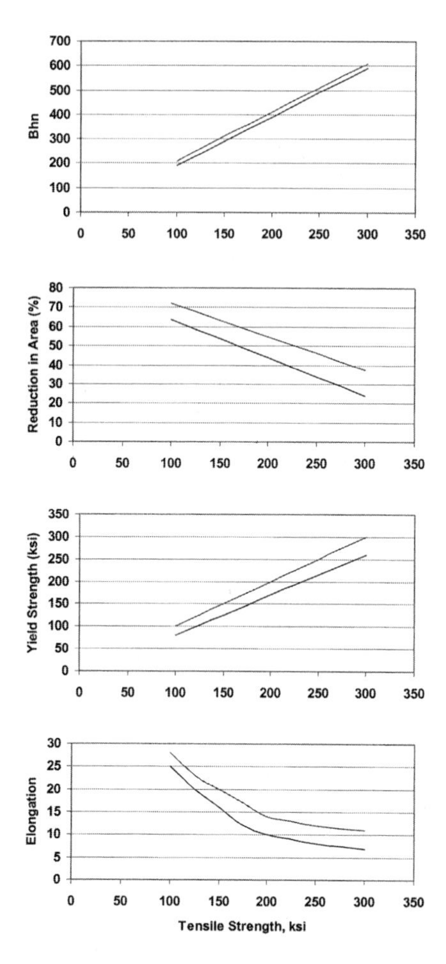

Figure 9.32 Tensile strength (Rothbart, 1964)

content. API 5L specifies a maximum carbon equivalent for use in line pipe to be less than 0.43%, where carbon equivalent, CE is defined as

$$CE = C + \frac{Mn}{6} + \frac{(Cr + Mo + V)}{5} + \frac{(Ni + Cu)}{15} \tag{9.5}$$

The carbon equivalent provides a guideline for determining welding preheat to minimise hardenability issues and reduce the cooling rate.

9.3.1.2.3 Toughness

"The quality known as toughness describes the way a material reacts under sudden impacts. It is defined as the work required to deform one cubic inch of metal until it fractures. Toughness is measured by the Charpy test or the Izod test.

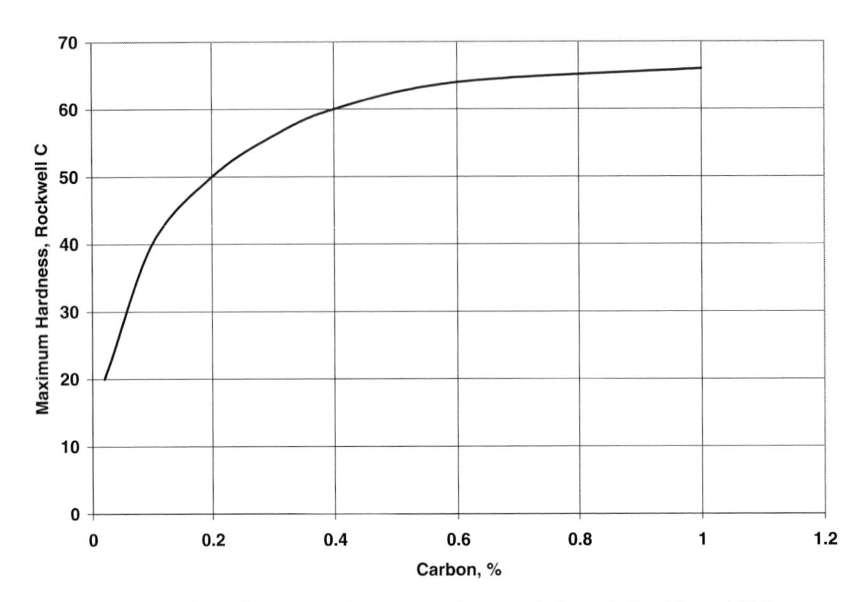

Figure 9.33 Hardness vs. carbon content for quenched steels (Rothbart, 1964)

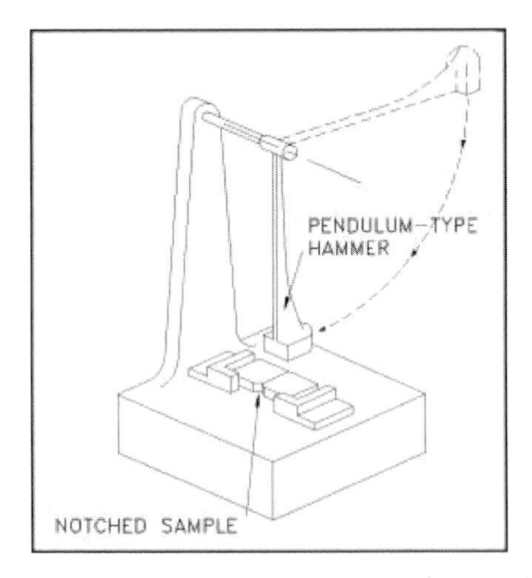

Figure 9.34 Charpy impact test (www.tpub.com/doematerialsci/)

Both of these tests use a notched sample. The location and shape of the notch are standard. The points of support of the sample, as well as the impact of the hammer, must bear a constant relationship to the location of the notch.

The tests are conducted by mounting the samples as shown in fig. 9.34 and allowing a pendulum of a known weight to fall from a set height. The maximum energy developed by the hammer is 120 ft-lb in the Izod test and 240 ft-lb in the Charpy test. By properly calibrating the machine, the energy absorbed by the specimen may be measured from the upward swing of the pendulum after it has fractured the material specimen. The greater the amount of energy absorbed by the specimen, the smaller the upward swing of the pendulum will be and the tougher the material is".

A history of impact tests is given by Siewert, et al (1999). Charpy test results (CVN) are reported as absorbed energy for a standard test specimen. Results are presented in units of ft-lbs, or Joules for SI units (1 ft-lb = 1.35582 J). Charpy Impact tests are not required for PSL 1 pipe. PSL 2 pipe must meet minimum requirements for the absorbed energy as spelled out in API 5L.

Charpy tests are a fast and low cost method for measuring the toughness of steel plate. More elaborate CTOD testing (BS 7448) is sometimes used for measuring toughness of weld heat affected zones. API 5L specifies a Weld Ductility Test, which requires that a pipe be flattened with the weld at 90° to the point of application of the forces. In this test, no cracks or breaks of greater than 1/8" are allowed until the pipe is flattened to a prescribed distance.

Increased demands on strength while maintaining an acceptable hardness for sour service (e.g. Vickers Hardness <275 HV 10), and toughness performance in deep water operations are currently on the edge of formulating a material chemistry that will meet these requirements. This dilemma has promoted the use of corrosion resistant alloys (CRA's) and the use of cladding to try and meet service requirement while trying to keep the material costs down.

9.3.1.2.4 Manufacturing Capability

Another issue is the actual ability to manufacture riser pipe with the specified wall thickness and diameter for riser applications in deep water. Most pipe manufacturers cannot produce or handle these sizes. There are only a few, to date, that have material handling capacity for thick-walled, large diameter seamless pipe. Two manufacturers, which are currently capable of supplying pipe in these sizes, include:

SUMITOMO PIPE & TUBE CO., LTD.
23-1 Sugano 3-Chome
Ichikawa 272-8528, Chiba 272-8528
JAPAN
http://www.sumitomokokan.co.jp/
+81 47 322 3322
+81 47 322 2448

and

Tenaris Pipeline Services
Carretera Mexico-Veracruz
Via Xalapa, km 433.7
(91697) Veracruz, Ver. Mexico
www.tenaris.com
(51) 2 989 1255
(52) 2 989 1600

Many heat treat, tempering, and quench facilities are not capable to produce pipe with uniform material properties along the length and through thickness of the pipe. For heavy wall pipe, a pre-manufacturing test using the material specification should be conducted to verify the capability of potential riser pipe manufacturers. The pipe produced during these tests can be used as test pieces for weldability and fatigue testing, keeping some of the material testing costs down.

9.3.1.2.5 Field Welding

It used to be assumed that once riser pipe is produced with acceptable mechanical properties that we are ready to conduct welding tests using the welding procedure specification called out for installation. This philosophy is changing. Prior to conducting installation weld procedure qualification testing, a weldability test of the material is conducted as outlined in API RP 2Z. The test enables the project to understand the material response to welding conducted at a low, medium, and high range window of heat inputs, e.g. ranging from 15 to 75 kJ/in, and representing heat inputs for manual to automatic weld processes to be used. The weldability of the riser material is verified prior to installation welding as a result of these tests.

After the riser material is supplied to the installation welders, formal weld testing is conducted using the actual installation welding procedures and conditions. The next step is to verify the welding procedure specification with appropriate welding procedure qualification records and testing. A review of these results will determine if the weld procedure is adequate to meet the material specification and produce high quality welds. One area of concern is to minimise the use of pre-heat during welding in order to facilitate installation of the riser. This is often difficult to do because of the carbon equivalent or chemistry of the material, i.e. keeping in mind the purpose of pre-heat to reduce the cooling rate and hardenability of the material.

Inspection of riser welds in the US by automated ultrasonic testing (AUT) has replaced manual and radiographic testing. Prior to conducting AUT, it is recommended that the project review the AUT procedure and the acceptance criteria. A demonstration test by the AUT contractor should be conducted, and a follow-up verification of indications by manual ultrasonic testing should be performed to insure a reliable test. Follow-up audits should be conducted to verify continued weld quality during fabrication of riser sections.

9.3.1.3 Analysis Tools

Riser analysis tools may be classed as frequency or time domain. Most tools for riser response to waves and vessel motions require vessel motions input in the form of Response Amplitude Operators (including phase angles), which permits appropriate marriage of vessel motions with forces from the wave kinematics.

Analysis of riser VIV is widely carried out using the program SHEAR7 developed at MIT under a joint industry research study, and with the more recently developed program VIVA (2001). The programs enable prediction of riser VIV response under uniform and sheared current flows.

Whilst time-domain analysis remains the preferred option in some cases (e.g. confirmatory extreme storm response analysis) the most commonly used VIV software (including

SHEAR7, VIVA and VIVANA) are frequency-domain programs. Reasonable accuracy may well be provided by such programs under many conditions, since VIV motions are typically small, as are the associated structural non-linearities. Furthermore, the reasonable allowance can often be made for some non-linearities by suitable post-processing of results where fatigue prediction is the main concern. Programs such as Flexcom-3D and Orcaflex are used for analysis to determine bending and deflection of the productioin riser systems.

9.3.1.4 Vessel Motion Characteristics

Characteristic vessel motions and their applicability to different design checks are discussed in table 9.10. Vessel RAOs are used throughout the whole design process and it is important for them to be well-defined. Spacing of periods in the RAO curve must be sufficiently close – especially near peaks – to maintain good accuracy. A useful reference on this subject is Garrett, et al (1995).

Noting that the riser attachment location can have a significant influence on both the riser extreme and fatigue response, as may vessel orientation relative to waves and current, it is important to be able to correctly and efficiently manipulate and transform the RAO data.

9.3.1.5 Coupled Analysis

Vessel, risers and mooring lines make up a global system, which has a complex response to environmental loading. The interaction of these components creates a coupled response, which may be significantly different to that predicted by treatment of each component on its own. Fully coupled analysis may be conducted as part of the final riser verification. However, it may be worth considering a coupled analysis at an earlier stage in the design process so that problems with the riser, vessel or mooring line design are highlighted and possible cost savings identified.

The design of offshore structures operating in hostile environment and in water depth more than 5000 ft requires the development of integrated tool which are accurate, robust and efficient. A hull/mooring/riser fully coupled time domain analysis may meet such requirements. For some systems, the coupling effects may magnify the extreme hull responses. Whereas, for most platforms in deep waters, the coupling effects more likely lead to smaller extreme responses due to additional damping from slender members, which results in less expensive mooring/riser system.

Not accounting for the riser stiffness, drag or damping when calculating vessel offsets may result in a conservative estimate of extreme vessel offset which may or may not be acceptable for storm analysis. Conversely, increased vessel offset may indicate that riser fatigue damage from first order effects is spread over a greater length of riser than is truly the case, resulting in an underestimate of riser fatigue damage.

The effects of current and damping are interlinked. Current loads on risers can significantly affect vessel offset (e.g. current loading on risers accounts for 40% of total loading on one FPSO known to the authors) and may increase due to drag amplification if the risers are subject to VIV. On the other hand, the riser hydrodynamic damping is related to the riser drag and will tend to reduce the amplitude of riser first-order response to wave loading. In general, simplifications cannot be assumed to be either conservative or unconservative in

Table 9.10 Characteristic vessel motion summary table

Characteristic vessel motion	Relevant design cases	Discussion
First order/ wave frequency (RAOs)	Extreme, clashing, fatigue	RAOs describe vessel response to wave-frequency excitation. They are typically determined by diffraction analysis and are used in all stages of riser design.
Extreme offsets	Extreme, clashing, fatigue	Extreme offsets represent expected extreme positions at the riser's point of attachment. To avoid undue conservatism, any first order contribution should be removed prior to riser dynamic analysis. Horizontal offsets are usually given, but TLP set-down and spar pitch can also be important. Flooded compartment conditions can give rise to appreciable set-down.
Low frequency/ second order motions	Fatigue	Drift data should be used in detailed riser fatigue analysis. The typical format is mean offset + one standard deviation with period for a range of sea-states/bins in the scatter diagram. The offset data is typically given for surge/sway but may include other degrees of freedom, such as, pitch for a spar. Current, wind and wave forces should be considered as contributors to these motions.
Vessel springing, ringing (for vertically tethered vessels; e.g. TLPs)	Fatigue	The amplitude of vessel springing may be relatively small but could cause high levels of fatigue, especially at the TDP of an SCR, if it occurs a large proportion of the time.
Vessel VIV	Fatigue	Vessel VIV is theoretically possible with any floating vessel subjected to current loading that has cylindrical sections with aspect ratios (L/D) greater than three. The frequency of excitation will be equal to the vessel's natural frequency, which is typically 200–400 s depending on the mooring system. The implication for riser design is high levels of fatigue damage.
Coupled motions	All	More important in deepwater.

their overall effect. However, larger errors can be expected as water depth and the number of risers increase.

A global coupled analysis may be conducted using riser analysis software, though there may be limitations in representing the vessel. Alternatively some seakeeping codes could include risers and moorings, though it may be necessary to simplify these in order to limit computer time.

The issues of coupled analysis have been addressed in the Integrated Mooring and Riser Design JIP using a range of example vessel/mooring/riser systems. The results of this work are available in Technical Bulletin (1999) describing the analysis methodology for preliminary and detailed analysis of integrated mooring/riser systems and outlining the relative importance of the various parameters and integration issues involved.

9.3.2 Top Tension Risers

Top Tensioned Risers (TTRs) are long flexible circular cylinders used to link the seabed to a floating platform. These risers are subject to steady current with varying intensity and oscillatory wave flows. The risers are provided with tension at the top to maintain the angles at the top and bottom under the environmental loading. The tensions needed for the production risers are generally lower than those for the drilling risers. The risers often appear in a group arranged in a rectangular (or circular) array.

9.3.2.1 Top Tension Riser Types

Top tensioned risers are used for drilling and production. Figure 9.35 shows the various types. Conventional exploration drilling risers use a low pressure riser with a subsea BOP. The subsea BOP was also used on the Auger TLP [Dupal, 1991], however most floating production platforms with drilling may now use a surface BOP. The Hutton TLP was the first floating production, dry tree unit, to use a surface BOP [Goldsmith, 1980]. The Split BOP has recently been used for exploration drilling in relatively benign environments [Shanks, et al 2002; Brander, et al 2003].

9.3.2.2 Dry Tree Production Risers

The earliest use of top tensioned risers was for offshore drilling in the 1950s. The first tensioners consisted of heavy weights attached to cables. These cables ran over pulleys to support the riser. These "deadweight" tensioners were replaced by pneumatic tensioners as shown in fig. 9.36. These tensioners used hydraulic cylinders to control the stroke of a block and tackle system. The riser was suspended by cables as in the deadweight system. This method has been replaced by direct acting hydraulic cylinders (fig. 9.37), which has been used on TLPs, and on the Genesis spar drilling riser. The hydraulic rods are in tension

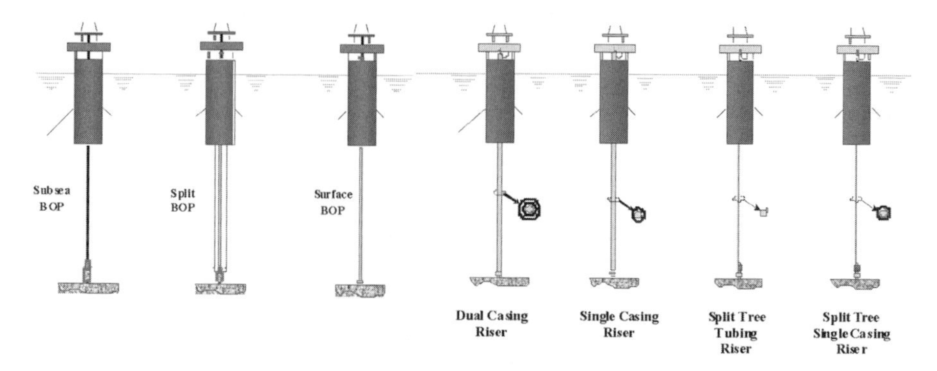

Figure 9.35 Types of drilling and production top tensioned riser systems

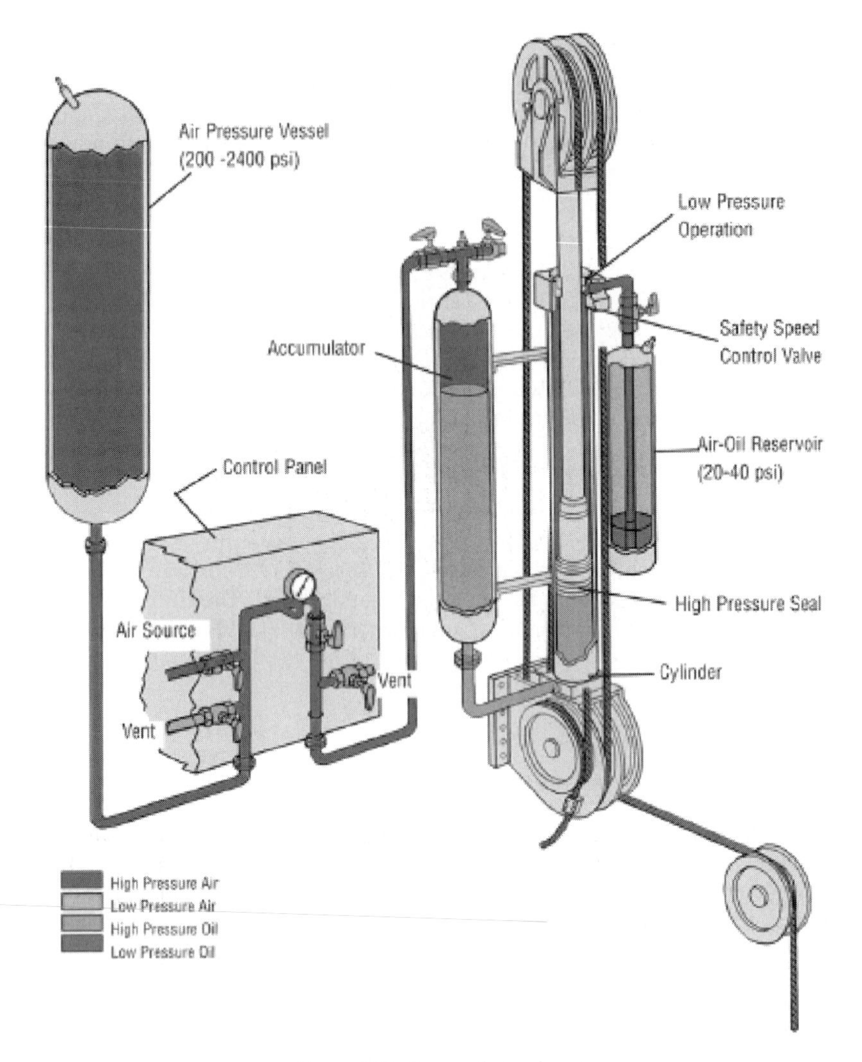

Figure 9.36 Pneumatic tensioner

for the direct acting tensioners. The Diana drilling riser was supported by the first ram style tensioner, fig. 9.38. This more compact arrangement of the tensioner was possible on the spar, because the riser does not take on an angle at the deck. The angle is taken at the keel of the spar and bending is accommodated by intermediate guides in the centrewell.

The first top tensioned production riser was used on the Argyll field in the North Sea [National Supply Company, 1975]. This was actually a tubing riser connected to a wet tree.

Figure 9.37 Direct acting tensioner

9.3.2.3 Single vs. Dual Casing

Figure 9.39 shows typical cross-sections for top tensioned risers, production and drilling. Top tension production risers utilised to date do not have separate insulation. However, the annulus between the tubing and the inner casing is, sometimes, filled with nitrogen to provide thermal insulation.

The choice of single vs. dual casings is a trade-off between the capital cost and the potential risk of loss of well control. There is almost no risk of loss of well control during normal operation because of the sub-surface safety valves [SCSSV, see, e.g. Deaton, 2000] and the dual barrier effect of the tubing and riser (remember that a single casing represents dual barriers during normal operations). The risk of blow out occurs during workover. In this phase the tubing and SCSSVs are pulled and mud is introduced into the riser to provide overpressure in the well relative to the formation pressure. The amount of overpressure offered by the mud is termed the "riser margin" (or "riser loss"). Goldsmith, et al (1999) describe a methodology for analysing the risk cost (RISKEX) and capital cost (CAPEX) trade-offs for a single and dual casing risers. Figure 9.40 shows riser loss as a function of mud weight and water depth. Typical riser margins are 300–400 psi as indicated.

Existing dry tree units, spars and TLPs, use both single and dual casing risers in about equal numbers [Ronalds, 2001].

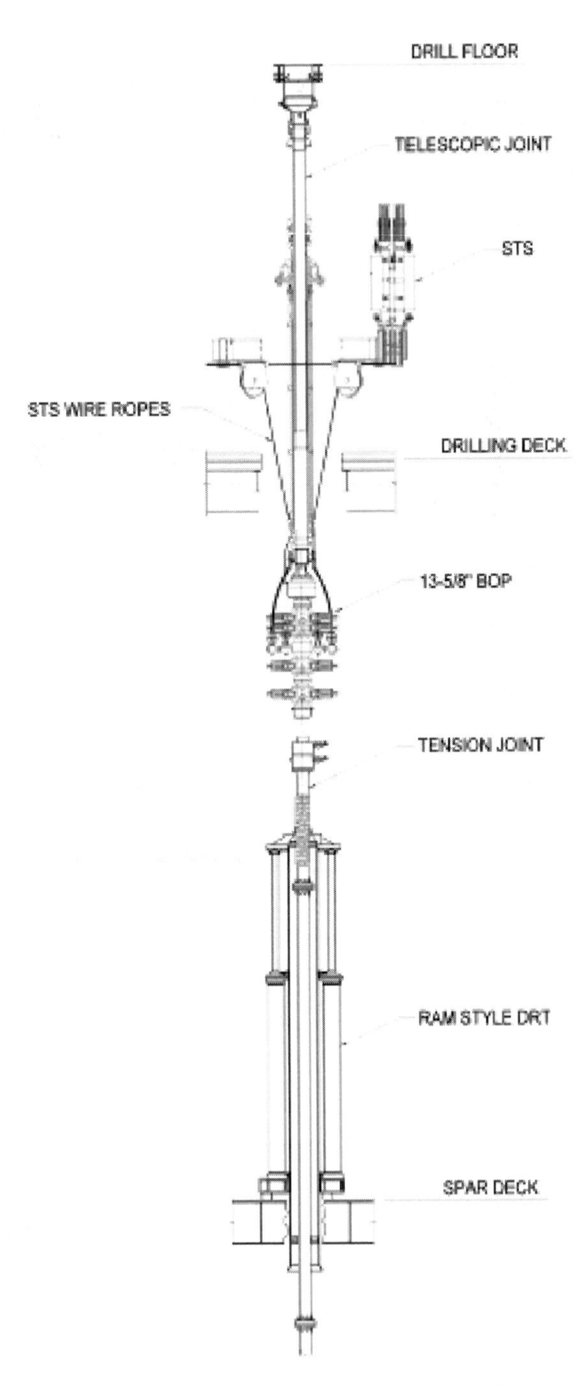

Figure 9.38 Ram style tensioners [Bates, et al 2001]

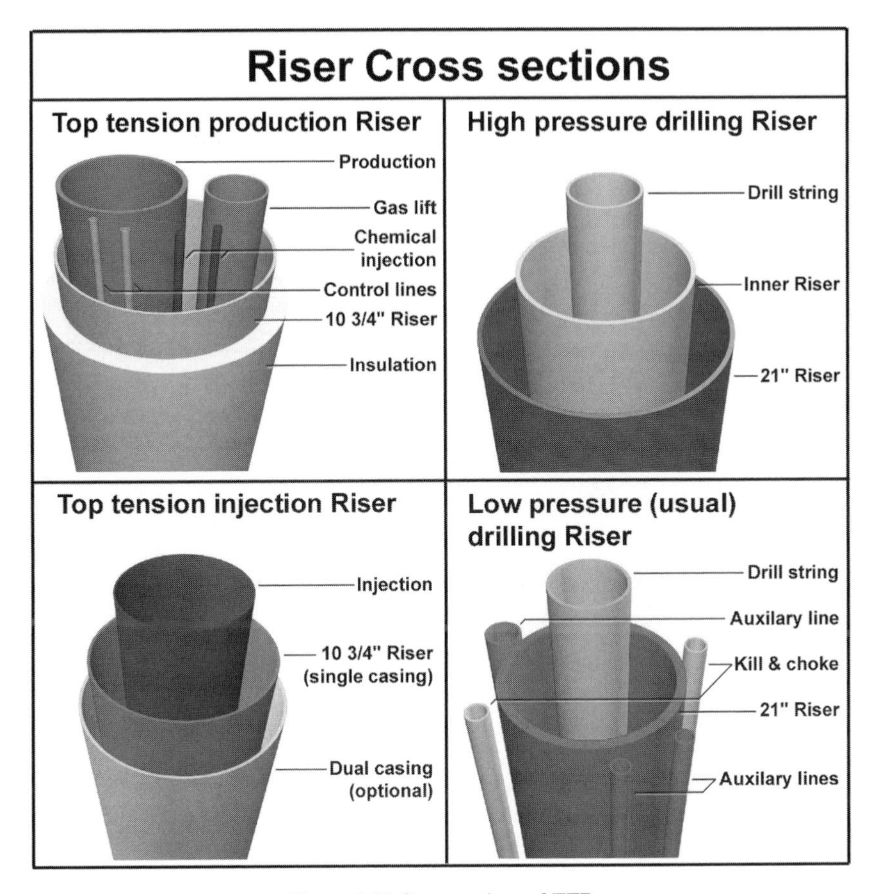

Figure 9.39 Cross sections of TTR

9.3.2.4 Codes and Standards

A valuable source for tracking industry codes and standards for risers and all sorts of oilfield systems may be found at http://www.rigcheck.com/codespecs.html.

The primary industry recommended practices for production of riser design are API RP2RD and DNV OS-F201. These apply to all tensioned risers from floating production systems. Flexible risers are covered in API RP17B and Bulletin 17J. DNV has separate rules for flexible pipe, and recommended practices for titanium (RP F201) and composite (RP F202) risers. MODU completion/workover risers are covered in API RP 17G. Subsea tiebacks are covered in API RP 1111.

The API Recommended Practice is based on a Working Stress Design (WSD) method. This is a prescriptive approach using a single utilisation parameter to account for all the failure

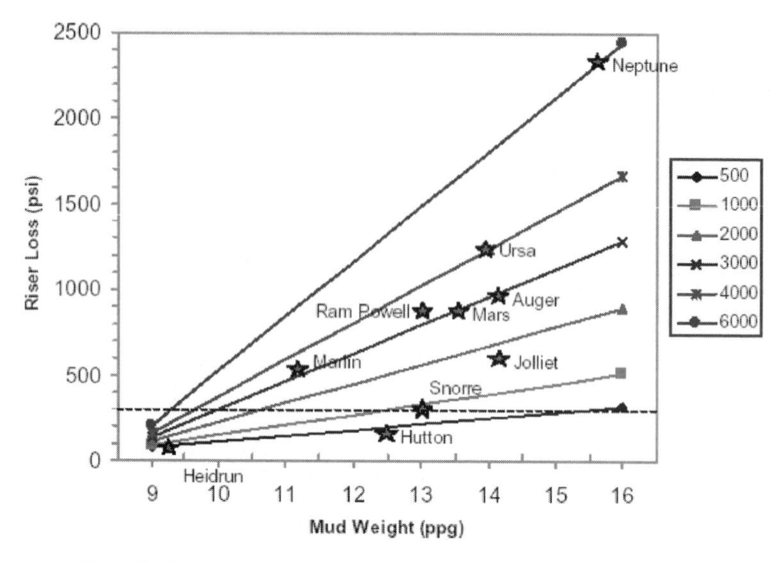

Figure 9.40 Riser loss vs. mud weight and depth [Goldsmith, et al 1999]

mechanisms. The DNV standard allows either the WSD or a Load and Resistance Factor Design method (LRFD). LRFD is aimed at achieving a particular target safety level by utilising partial safety factors for each failure mode, or limit state. The intent of the DNV code is to achieve a certain reliability level by applying probabilistic analysis to the various failure mechanisms. Table 9.11 shows the target failure probabilities for different limit states and safety classes. The limit states are defined as:

Serviceability Limit State (SLS) – Acceptable limitations to normal operations
Ultimate Limit State (ULS) – Structural failure
Fatigue Limit State (FLS) – Cyclic loading
Accidental Limit State (ALS) – Infrequent loading

Table 9.11 Target safety levels (OS-F201)

Limit state	Probability bases	Safety classes		
		Low	Normal	High
SLS	Annual per riser	10^{-1}	10^{-1}–10^{-2}	10^{-2}–10^{-3}
ULS	Annual per riser	10^{-3}	10^{-4}	10^{-5}
FLS	Annual per riser			
ALS	Annual per riser			

Table 9.12 Classification of safety cases (OS-F201)

Riser status (phase)	Riser content					
	Fluid category 1,3		Fluid category 2		Fluid category 4,5	
	Location class		Location class		Location class	
	1	2	1	2	1	2
Testing	Low	Low	Low	Low	NA	NA
Temporary with no pipeline/well access	Low	Low	Low	Low	Low	Normal
In-service with pipeline/well access	Low	Normal	Normal	Normal	Normal	High

The classification of safety cases is defined in table 9.12. Fluid categories are defined by the International Standards Organisation (ISO) as:

1. Water based fluids
2. Oil
3. Nitrogen, argon and air
4. Methane
5. Gas

LRFD, while more complex to apply, allows for optimisation that is not achievable using WSD methods.

Figure 9.41 shows an example plot of utilisation vs. water depth for a top tensioned TLP oil production riser under combined extreme North Sea conditions and external overpressure. In this example, API RP2RD is seen to be the most conservative approach. LRFD would allow optimisation of the riser.

9.3.2.5 Riser Components

The conventional TLP production riser is made up of the following components (fig. 9.42):

- Tieback connector at the bottom
- The bottom tapered joints or flex joints
- The riser joints and connectors
- The tensioner spool pieces
- The tensioner load rings
- The guide rollers at platform deck
- The surface tree
- The tubing strings inside
- Flowline connectors at deck level to trees or valves

The Spar production riser is made up of the following components (fig. 9.43):

- Tieback connector at the bottom

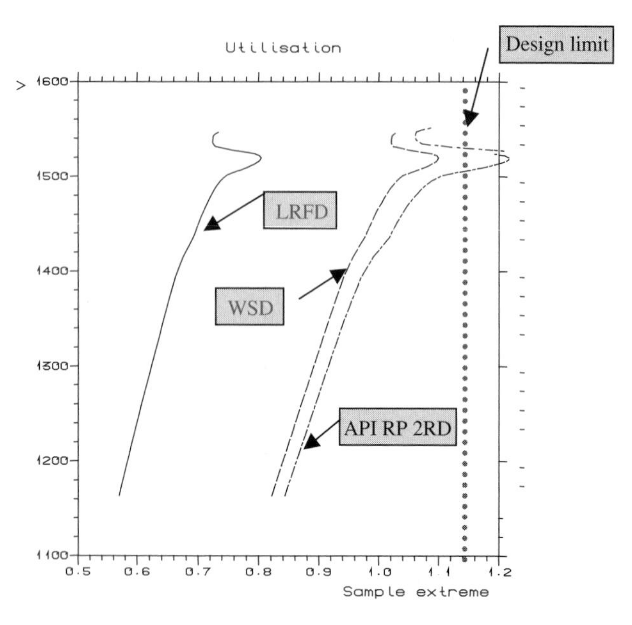

Figure 9.41 Example of the application of LRFD to riser design (Courtesy of DNV)

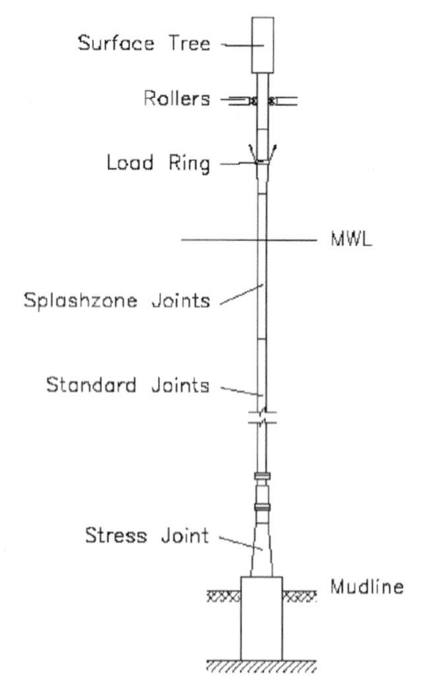

Figure 9.42 TLP top tensioned riser

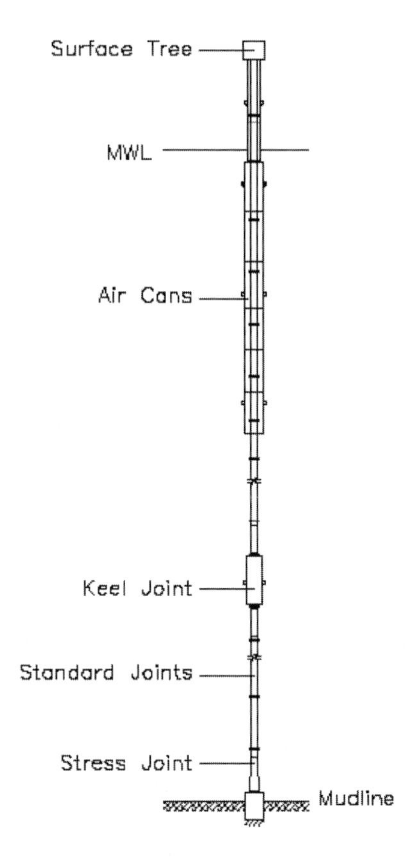

Figure 9.43 Spar top tensioned riser

- The bottom tapered joints or flex joints
- The riser joints and connectors
- The keel joint or lower stem
- The air cans
- The upper stem
- The surface tree
- The tubing strings inside
- Flowline connectors at deck level to trees or valves

TLP and Spar riser systems are virtually identical below the floater. The differences are in the manner of supporting and tensioning the riser at the top. TLP risers are supported by the hull buoyancy. The tension is provided at the load ring, which is supported by tensioners (see fig. 9.42). Rollers at the TLP deck centralise the riser and accommodate the angle between the riser and the hull.

Spar risers are supported by air cans (sometimes called buoyancy cans), not by the spar hull itself. Buoyancy cans may be either "Integral" or "Non-Integral" type, fig. 9.44. The

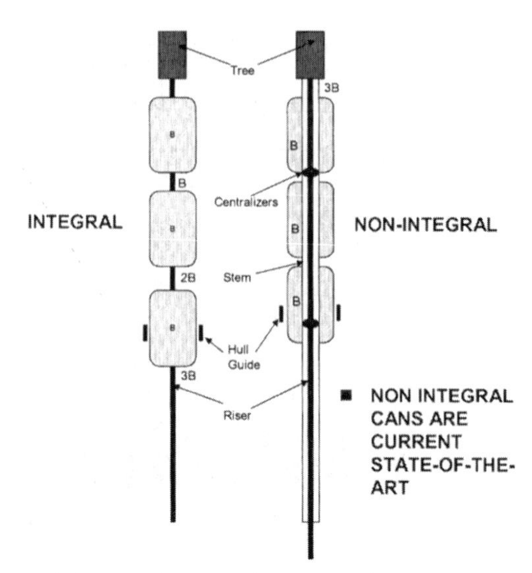

Figure 9.44 Integral and non-integral spar buoyancy cans

Integral air can consists of an air can attached to a riser joint. The air can is installed along with all the rest of the riser joints. This type of air can has only been used once on the Genesis spar. "Non-integral cans", fig. 9.44b, consists of an inner pipe called a "stem" which supports the air cans. The non-integral cans are deployed separately. The riser is run through the stem and landed on a shoulder at the top of an extension of this stem called the "upper stem". The upper stem carried the entire tension of the riser and the weight of the surface tree.

The angle between the spar and the riser is accommodated by a keel joint. Early keel joints on classic spars consisted of a single joint made up of a large diameter pipe on the outside connected to the riser pipe with flexible connections at the ends [Berner, et al 1997; Bates, et al 2002]. The outer riser contacts the spar hull at a keel guide; both items include a sufficient wear allowance to accommodate the loss of material caused by the relative motion over the lifetime of the project. This type of keel joint is not used on truss spars. Instead, the lower stem of the buoyancy cans is extended through the keel. The riser is centralised near the bottom of this stem with a ball joint, fig. 9.45.

Mini-TLPs using top tensioned risers have also used keel guides. These have a similar function to those on spars, but typically require less stroke. One design is described by Jordan, et al (2004), fig. 9.46.

The riser pipe typically follows standard casing dimensions and materials (API Bulletin 5C3), however they may be designed to line pipe specifications as well (API RP 5L). The choice of riser size and materials is discussed below.

Riser pipe may be joined by threaded or bolted connections. Bolted connections are heavy and expensive. They are used primarily where superior fatigue performance is

Figure 9.45 Keel centraliser for the Matterhorn TLP [Jordan, et al 2004]

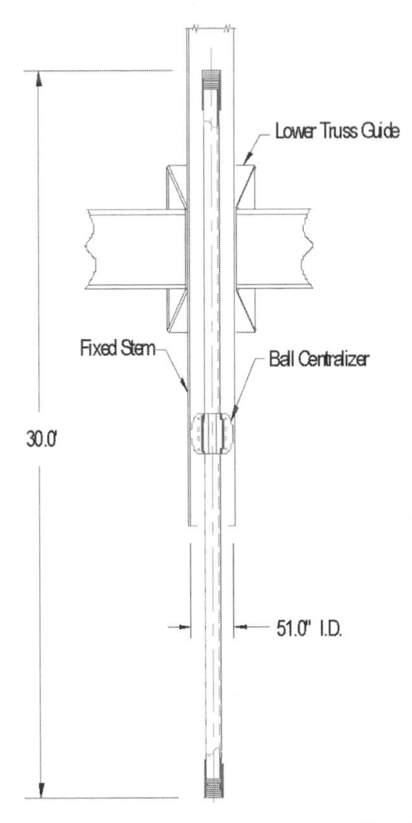

Figure 9.46 Lower stem and ball joint for truss spar [Wald, et al 2002].

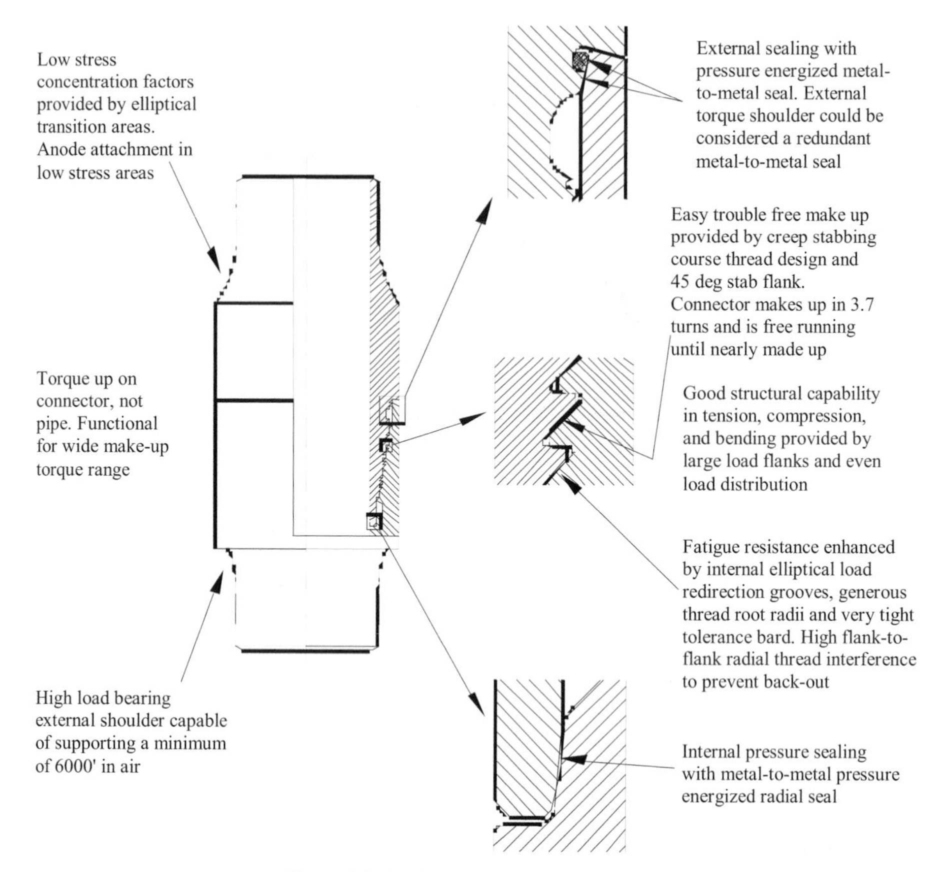

Figure 9.47 Typical weld-on upset connector

required, e.g. for stress joints and keel joints. Threaded connections may be either weld-on (fig. 9.47) or couplings with threads machined into the pipe (fig. 9.48). Casing couplings are cheaper but until recently were normally not suitable for fatigue sensitive applications [Cargagno, et al 2004]. For example, stress concentration factors (SCFs) for the thread root of the casing connectors are typically five or greater. SCFs are usually stated relative to the nominal pipe wall stress. Upset weld-on connecters, on the other hand, can achieve SCFs as low as about 1.2. Both type of connectors can achieve 100% of the strength of the pipe. Weld-on connectors, however, are not suitable for high strength steel, greater than 95 ksi yield strength, because of insufficient data for weld performance in riser applications. Thus the choice of connector is tied to the overall performance requirements of the riser, which requires a significant amount of analysis.

9.3.2.6 Riser Sizing

Riser sizing requires consideration of a number of load cases. The size may be dictated by pressure, collapse, tension, bending or a combination of these factors depending on the

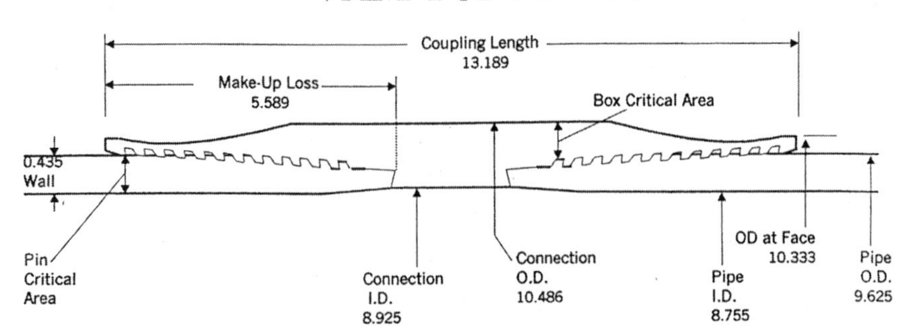

Figure 9.48 Typical casing connector

system. It is important to develop as early as possible a design and functional specification which spells out the various load cases, and that this document become a primary reference document throughout the course of the project. Changes in functional requirements can have "ripple" effects and should be communicated to designers and analysts as soon as they occur.

It is also important to keep a close communication between the riser designers and others involved in the riser interfaces throughout the process:

- Oceanographers
- Vessel and mooring designers
- Global response analysts
- Process engineers

Riser design, especially in deep water, is an iterative procedure. Initial assumptions about topsides weight, vessel size and response and even well characteristics might change several times in the course of a project. A rapid and accurate model for riser sizing is important for keeping up with these inevitable changes.

Riser definition starts with specification of:

1. Number of tubulars: single casing or dual casing (see discussion above)
2. Concentric or non-concentric tubulars
3. Well layout and layout on the seafloor

The inner riser size will be dictated by the size of tubulars, umbilicals, subsurface valves and connectors that have to fit within the internal diameter. In some deepwater applications, the production riser has been used for drilling [Craik, et al 2003]. If this is the case, the outer riser size will be dictated by the drilling program.

Figure 9.49 shows the layout for concentric and non-concentric tubulars. The diameter selected should allow sufficient clearance for the connectors of the inner tubing, i.e. the space should allow for the drift diameter of the inner tubulars.

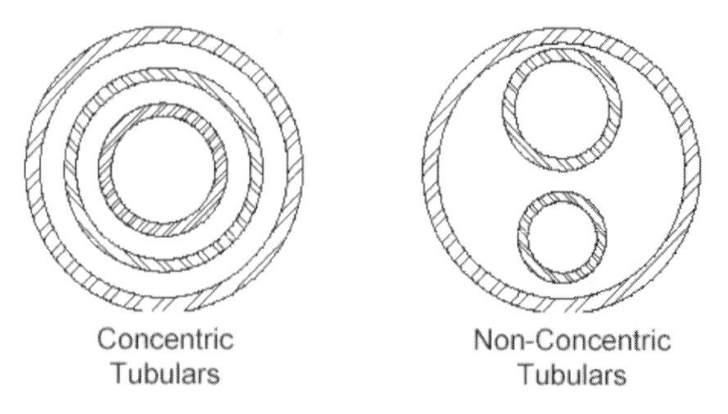

Figure 9.49 Concentric and non-concentric tubulars

For the minimum performance properties for common casing sizes used for risers, consult API RP 2RD. The collapse resistance (in psi) and pipe body yield values (in lbs.) are listed for different riser outside diameters in six tables. There is no requirement that riser sizes fit the standard casing dimensions. However, special sizes normally increase the cost and schedule. Drift diameters in these tables do not reflect weld-on, fatigue resistant connectors.

The wellbay layout and seafloor spacing have primary impact on the size of the vessel and the method of running risers. While initial sizing of the risers may be performed independent of the wellbay layout, e.g. by reference to pressures and operating conditions alone, important parameters like riser stroke, local bending at the seafloor and the keel (in the case of a spar), process deck height, etc. will depend on these parameters. Also, the vessel cannot be sized until the wellbay is determined (see Chapter 7). This means that the vessel motions can not be finalised, and hence the final dynamic stresses cannot be determined. The importance of early consideration of the wellbay layout on the whole design of a floating production system cannot be overemphasised.

Once the basic configuration of the number of tubulars, their makeup and a minimum ID for the inner riser are determined, analysis of a set of load cases is required to determine the controlling environment. At this point a selection of the governing design guideline is required, e.g. API RP2RD, OS-17201 or other. As was shown above, an LRFD code allows room for optimisation; however, the selection will usually be a function of the certifying agency and country and their familiarity with various codes. The following discussion assumes that API RP 2RD is the governing design code.

Table 9.13 shows the recommended minimal design matrix in API RP2RD. Table 9.14 shows an example Load Case Matrix for the Matterhorn TLP. The factor C_f, or the allowable load stress increase, indicates the increase in the allowable load from the nominal value given in Section 5 of API RD 2RD. The basic allowable stress is

$$\sigma_a = C_a \sigma_y \tag{9.6}$$

Table 9.13 Design matrix for rigid risers (API RP2RD)

Design case	Load category	Environmental condition	Pressure	Reduced tensioner capacity or one mooring line broken	$C_{f^{a,b}}$
1	Operating	Maximum operating	Design	No	1.0
2	Extreme	Extreme	Design	No	1.2
3	Extreme	Maximum operating	Extreme	No	1.2
4	Extreme	Maximum operating	Design	Yes	1.2
5	Temporary	Temporary	Associated	No	1.2
6	Test[d]	Maximum operating	Test[d]	No	1.35
7	Survival	Survival	Associated	No	1.5
8	Survival	Extreme	Associated	Yes	1.5
9	Fatigue	Fatigue	Operating	No	Note[c]

Notes:
Anisotropic materials may require special consideration
[a]Use of C_f is described in Section 5: strength issues are discussed is 5.2, deflections in 5.3, collapse issues in 5.4 and 5.5, fatigue in 5.6
[b]Pipeline codes may require lower C_f for risers that are part of a pipeline
[c]Not applicable
[d]Plant testing for rigid risers should be agreed between user and manufacturer

where $C_a = 2/3$, and σ_y is the material yield stress, defined in API RP2RD, for steel and titanium, as the stress "required to produce an elongation of 0.5% of the test specimen gage length".

API RP 2RD defines three stresses: primary membrane, primary bending and secondary.

A "Primary" stress is "any normal or shear stress that is necessary to have static equilibrium of the imposed forces and moments. A primary stress is not self-limiting. Thus, if a primary stress substantially exceeds the yield strength, either failure or gross structural yielding will occur".

A "Secondary" stress is "... any normal or shear stress that develops as a result of material restraint. This type of stress is self-limiting, which means that local yielding can relieve the conditions that cause the stress, and a single application of load will not cause failure".

A primary membrane stress is the average value of the stress across a solid cross section, excluding effects of discontinuities and stress concentrations. For a pipe in pure tension this would include the total tension divided by the cross section of the pipe. For a pipe in tension and global bending, the membrane stress would include the global bending effect as well. The primary bending stress is the portion of primary stress proportional to the distance from the centroid of the solid section, excluding stresses due to discontinuities and stress concentrations.

Table 9.14 Design load cases, Matterhorn TLP [Jordan, et al (2004)]

Case ID	Riser condition description	Pressure & contents(1)	Design environment	Allowable stress increase factor
P-N1	Normal production and shut-in with surface tree and completion tubulars supported from the top of riser	PNO, PSI	1-yr. winter storm	1
P-N2		PNO	100-yr.winter storm	1.2
P-N3		PNO	100-yr. loop current	1.2
CCE		PNO	Cold core eddy	1.2
P-N4		PSI	100-yr.hurricanc	1.2
P-N5		PSI	1000-yr. hurricane	1.5
P-K1	Well killed with the surface tree and completion tubulars supported from the top of the riser.	PK	1-yr. winter storm	1
P-K2		PK	100-yr. loop current	1.2
P-K3		PK	100-yr. hurricane	1.2
P-K4		PK	1000-yr. hurricane	1.5
P-L1	Shut-in with the surface tree and completion tubulars supported at the top of the riser	PSLNO	95% non-exceedance	1.2
P-L2(2)		PSL	1-yr. winter storm	1.2
P-L3(3)		PSL	100-yr. loop current	1.5
P-L4(4)		PSL	100-yr. hurricane	1.5
P-C1	Well killed with BOP stack and completion tubulars supported from the sap of the riser	PK	10-yr. winter storm	1
P-C2		PK	10-yr. loop current	1.2
P-C3		PK	10-yr. hurricane	1.2
P-C4		PK	100-yr. hurricane	1.5
P-TD	Tensioner damage(4)	PSI	100-yr. hurricane	1.5

Notes:

1. The following arc definitions of the "Pressure & Contents" column abbreviations:
   - PNO Normal surface operating pressure under normal flowing conditions
   - PSI Shut-in tubing pressure with no tubing leak
   - PK Well killed with heavy liquid in the tubing and the annulus
   - PSL Surface pressure shut-in tubing pressure with a tubing leak
   - PSLNO Normal operating surface shut-in tubing pressure with a tubing leak. To overcome this situation and replace the leaking tubing, a"Bullhead Pressure" must be imposed at the surface wellhead that is greater than the shut-in tubing. "Bullhead Pressures" will not be present during a hurricane when the platform is abandoned
2. Add a 15 kip snubbing unit BOP, a 41.5 kips snubbing unit, and 25 kips of work string to the top of tree
3. Operator can remove everything but the 15 kip snubbing unit BOP off the tree before a hurricane or a severe loop current situation
4. Tensioner damage is defined as the loss of one tensioner element without any adjustment to the remaining elements

Stresses due to discontinuities and stress concentrations fall into the category of secondary stresses. Primary stress components are combined using an equivalent von Mises stress.

$$\sigma_e = \frac{1}{\sqrt{2}}\sqrt{(\sigma_1 - \sigma_2)^2 + (\sigma_2 - \sigma_3)^2 + (\sigma_3 - \sigma_1)^2} \quad (9.7)$$

where σ_e = von Mises equivalent stress, σ_1, σ_2, σ_3 = principal stress. A common combination of stresses includes the hoop and axial tensile stresses, both of which are primary stresses.

$$(\sigma_p)_e < C_f\sigma_a \quad (9.8)$$

$$(\sigma_p + \sigma_b)_e < 1.5C_f\sigma_a \quad (9.9)$$

$$(\sigma_p + \sigma_b + \sigma_q)_e < 3.0C_f\sigma_a \quad (9.10)$$

where σ_p = Primary membrane stress, σ_b = Primary bending stress, σ_q = Secondary stress. Table 9.15 summarises the allowable stress criteria for API RP2RD.

Primary bending stress and secondary stresses are typically associated with changes in riser section near connectors and transitions. Their evaluation requires an assessment of the through thickness stress profile, and separation graphically or mathematically of the average, linear (bending) and non-linear components of the stress distribution. The example in Annex C of RP 2RD should be consulted for application of these criteria.

Table 9.15 Summary of allowable stress criteria

Stress category	Stress allowable	Allowable stresses (ksi) based on 80 ksi yield material		
		Normal operating	Extreme event	Survival event
Cf factor		1.0	1.2	1.6
Primary membrane	Cf (Sm)	53.3	64.0	80.0
Primary membrane plus bending	1.5 Cf (Sm)	80.0	96.0	120.0
Primary membrane plus bending plus secondary	3.0 Sm	160.0	160.0	160.0
Range of primary membrane plus bending plus secondary plus peak	Based on fatigue curve	Based on fatigue curve	Based on fatigue curve	Based on fatigue curve
Average bearing stress	0.9 Sy	72.0	72.0	72.0
Average primary shear stress	0.6 Sm	32.0	32.0	32.0

Sm = 2/3 Sy

Initial riser sizing, excluding stress joints, typically considers only the primary membrane stresses. The steps include the following:

Select nominal tubular sizes (OD) as described above.
Make a trial selection of IDs

For each load case,

Compute the top tension required to achieve a zero effective tension at the mudline (this is the weight in water of the submerged portion of the riser and its contents, plus the dry weight of the riser and contents above the waterline to the tensioner ring).

Apply a nominal "tension factor" (TF) to insure positive bottom effective tension under dynamic loadings. The tension factor for floating production systems is typically in the range of 1.3 to 1.6. However, the tension factor could change upon further analysis of riser interference or Vortex Induced Motions. This is an iterative process. The top tension is the tension required to yield a zero effective tension at the mudline times the tension factor.

Considering the top tension and pressure in the riser, compute the combined primary stresses and a utilization factor

$$U = \sigma/(C_f \sigma_a) \tag{9.11}$$

Vary wall thickness and repeat the above procedure until $U < 1.0$.

Selection of the material yield strength is required at this point. As was mentioned above, weld-on connectors are presently limited to strengths below 95 ksi. For very high-pressure wells where dynamic stresses are likely to be low, higher yield strength may be selected provisionally to reduce riser tension. However, the fatigue of the riser couplings will need to be checked before this decision is validated. Ductility and toughness are also critical concerns for dynamic risers to avoid the possibility of brittle fracture. The majority of deepwater risers are designed for 80 ksi yield strength.

Another important consideration in deepwater is the minimum effective tension at the seafloor. In deepwater the riser may have a negative effective tension without failing. This is because the bending that occurs is limited by the displacement of the top of the riser, i.e. it is a secondary stress rather than a primary stress. Detailed analysis may indeed indicate that suitable criteria may be met with reduced tension factors. This is part of an optimisation process.

For low motion platforms such as spars, initial sizing for the main body of the riser might ignore dynamic effects. Experience may indicate that higher safety factors applied to this "static" riser sizing approach will lead to good results with a minimum of iteration.

In any event, the next step is to perform dynamic analysis of the risers. This analysis requires

1. Definition of the seastates corresponding to the load cases defined above,
2. Vessel motions

The analysis may be frequency domain or time domain, coupled or uncoupled. Often, global vessel motions will be computed by one group and riser dynamics by another. This paradigm is very risky. For example, vessel motion programs often use different coordinate systems than riser programs. Translating motions from the origin of a vessel to the riser hangoff point using a different coordinate system can and often does lead to errors. It is best to perform a quality check of the procedure by analysing a simple case, e.g. a monochromatic sine wave, and performing some hand calculations prior to doing the bulk of the analysis with random wave input.

Another issue with frequency domain analysis is that it can neglect important non-linear effects such as slowly varying motions and damping. Coupled analysis is the simultaneous solution of the vessel and riser motions. In deep water the riser loads may actually reduce the vessel responses [see, e.g. Prislin, et al (1999) Halkyard, et al (2004) for comparison of full-scale data with calculations].

The above discussion focuses on strength design. The riser may also fail from fatigue, hydrostatic collapse, buckling and thermal effects.

Fatigue analysis needs to consider fatigue for vessel motions, wave loadings on the riser and vortex induced motions. The analysis is similar to that for the drilling riser discussed above; however, the difference is that production risers are designed to remain in place for the life of the field, whereas drilling risers are routinely retrieved and inspected. API RP2RD requires the fatigue life of the riser to be:

- Three times the service life (usually the life of the field) for areas accessible for inspection (or, where safety and pollution risk are low), or,
- Ten times the service life for areas not accessible for inspection (or, where safety or pollution risk are high).

In practice, deep water risers are invariably designed for ten times the service life.

Figure 9.50 shows the DNV fatigue curves used for the analysis of riser components. These are identical to the DOE curves discussed in Chapter 7 [DNV CN 30.2, HSE, 1995]. Riser connectors without welds use the "B" curve with an appropriate stress concentration factor. Preliminary fatigue analysis is often performed using the "B" curve to determine an allowable SCF to achieve the required design life. If this allowable SCF is five or greater, it is likely that threaded couplings may be used for the connectors. New couplings with lower SCFs are becoming available as was mentioned above.

The difference between the C, D and E curves, which apply to welded connections, depends on the quality of the weld. Special considerations need to be given to the presence of H_2S and corrosion. These fatigue curves assume cathodic protection.

The reader is encouraged to review the literature on riser sizing and analysis for deepwater floaters [e.g. Wald, et al 2002; O' Sullivan, et al 2002; Jordan, et al 2004; and Bates, et al 2002].

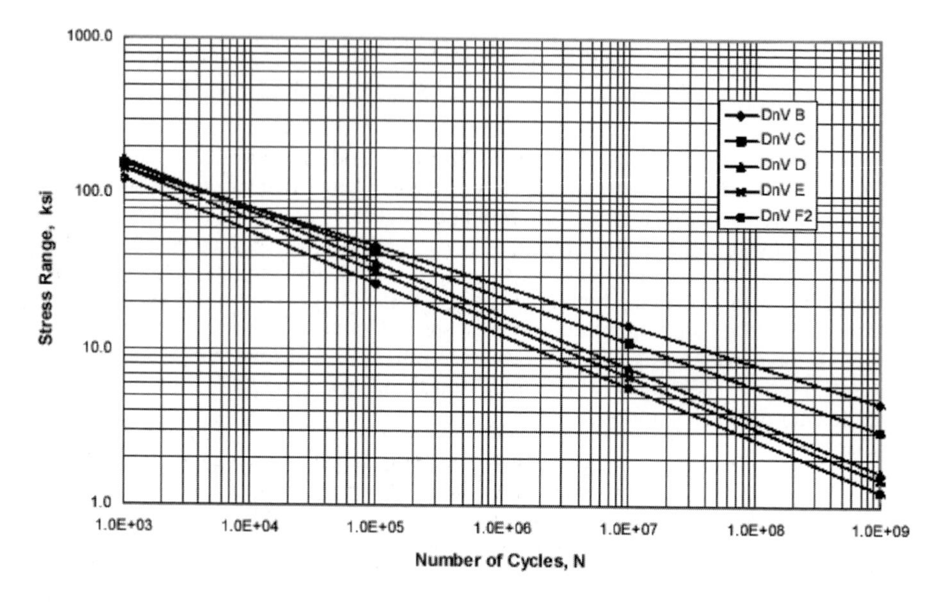

Figure 9.50 DNV fatigue curves

9.3.2.7 TTR Analysis Procedures

The outer geometry of the riser is not uniform because of various elements attached to it. The equation of motion for a riser and its different components is given on the assumption that the riser represents a bent tubular member in one plane and only one plane of motion is considered. Similar equations may be applied to the orthogonal plane, and the two motions may then be combined with the coupling between them, coming from the external forces. The equation of motion is explicitly written as

$$\begin{aligned} &\frac{d^2}{dy^2}\left(\frac{EI(y)d^2x}{dy^2}\right) \text{(flexural rigidity)} - \frac{d}{dy}\left[F_A(y)\frac{dx}{dy}\right] \text{(axial tension force)} \\ &-\frac{d}{dy}\left[\{A_o(y)p_o(y) - A_i(y)p_i(y)\}\frac{dx}{dy}\right] \text{(external \& internal fluid pressure)} \\ &+ m(y)\ddot{x} \text{ (riser inertial resistance)} = f_{xs}(x, y, t) \text{ (external horizontal force)} \end{aligned} \tag{9.12}$$

Additional constraints are needed to solve this equation, which are specified at the top and bottom joints as end restraints. The restraints could be fixed, pinned, free or a specified top offset from the vessel displacement.

This horizontal equation may be solved for both the static and dynamic analysis of the riser. For the static analysis, the fluid inertia is absent and the external loading is due to the current load. In this case, the equation becomes,

$$\frac{d^2}{dy^2}\left(EI(y)\frac{d^2x}{dy^2}\right)-\frac{d}{dy}\left[F_e(y)\frac{dx}{dy}\right]+m(y)\ddot{x}=\frac{1}{2}\rho C_D(y)D(y)|U(y)|U(y) \tag{9.13}$$

where the right hand side is the current force. $F_e(y)$ is called the effective tension due to axial and pressure force, $U(y)$ is the current velocity as a function of the vertical coordinate y, and $C_D(y)$ is the corresponding drag coefficient for the riser.

In order to solve the dynamic problem due to oscillatory excitation, the right hand side of equation (9.12) should represent the dynamic load, e.g. from wave and vessel motion. The two solutions are combined into one, when the static and dynamic external loads, on the right hand side are combined. Either finite difference or finite element methods are used to solve for the deflected riser mode shapes and structural properties under static or dynamic loads. Because of its versatility, the finite element method (FEM) becomes an obvious candidate for the numerical tool. Indeed, most of the general-purpose riser analysis packages are based on the FEM, and the reader is referred to the vast literature that exists on the FEM for details of these analyses [see, for example, Meirovich (1997), and Moe, et al (2004)].

Example Problem: Transverse envelope

The maximum and minimum transverse displacements of a top tensioned riser are computed for several current speeds. The following input are considered as example:

- water depth = 100 m,
- riser length = 120 m,
- outside diameter = 0.25 m, inside diameter = 0.2116 m,
- top tension = 200 kN (~1.5 times the riser weight),
- modulus of elasticity of riser pipe = 2.10 × 108 kN. m^{-2},
- specific weight of the fluid outside = 1025 kg. m^{-3},
- specific weight of the fluid inside = 800 kg. m^{-3},
- specific weight of the riser wall material = 7700 kg. m^{-3},
- mass, $m^* = 2.7$
- damping parameters, $m\zeta^* = 0.054$,
- riser model elements = 80 below, 20 above still water,
- riser ends = fixed but free to rotate,
- uniform flow velocities = 0.16 to 0.93 m/s,
- Reynolds numbers = $4.0 \times 10^4 \leq Re \leq 2.3 \times 10^5$.

The results for the first four mode shapes are shown in fig. 9.51. The current speeds for these modes and the corresponding reduced velocities are included in the figure.

The typical drag coefficients used for a production riser for a SPAR of draft of 2000 ft are summarised in table 9.16.

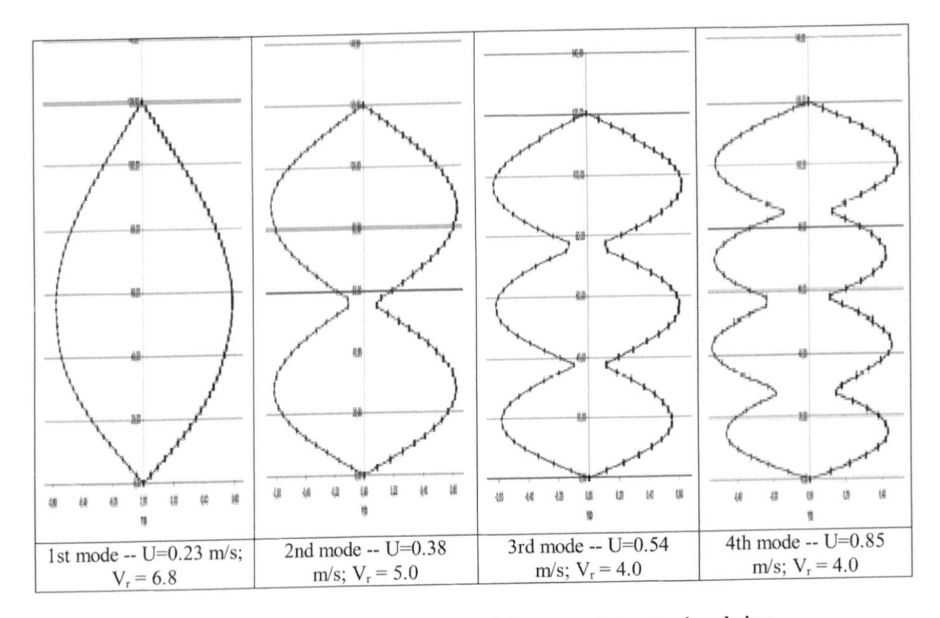

Figure 9.51 First four mode shapes of the example top tensioned riser

Empirical formula [API, 1992] for the blockage based on the relative spacing of the risers with respect to the diameter is given by:

$$C_{BF} = \begin{cases} 0.25S/D & \text{for } 0 < S/D < 4.0 \\ 1.0 & \text{for } S/D = 4.0 \end{cases} \tag{9.14}$$

where S = centre to centre distance of risers of diameter D. The value of the current blockage factor for a row of cylinders is given in Table 9.17.

9.3.2.7.1 Effective Tension

The effective tension, physically a very meaningful quantity, represents a composite tension, which incorporates the effects of internal and external fluid pressures. It is defined for a single-walled riser as follows:

$$T_{\text{eff}} = T_{\text{wall}} - p_i A_j + p_e A_e \tag{9.15}$$

where T_{eff} is effective tension, T_{wall} is tension in the riser wall, p and A denote pressure and enclosed section area respectively and subscripts i and e mean internal and external respectively. In general, all of these quantities vary along the riser length. Broadly, T_{eff} is used in force calculations and T_{wall} is used in stress calculations. However, engineers are advised to familiarise themselves fully with these terms and associated interpretations, particularly with regard to multi-pipe risers, and are referred to Sparks (1983) for a detailed explanation.

Table 9.16 Typical drag coefficients for a production riser

Riser section	C_D	Length (ft)	Remarks
Upper section	0.9	100	Inside SPAR
Straked section	1.4	450	Below SPAR keel
Bare section	1.2	1450	Down to sea bed

Table 9.17 Current blockage factor for cylinder group

No. of cylinders	Current heading	Blockage factor
3	All	0.90
4	End-on	0.80
	Diagonal	0.85
	Broadside	0.80
6	End-on	0.75
	Diagonal	0.85
	Broadside	0.80
8	End-on	0.70
	Diagonal	0.85
	Broadside	0.80

Care is required in communicating tensions to others, since many mistakes have been made. Where there is room for doubt (except discursively) it should always be made clear which tension quantity is meant. Special care is required for riser terminations, where load-paths are diverted and for riser connectors, where component manufacturers may assume a different terminology from analysts.

9.3.2.7.2 Soil Riser Modelling

The ability to predict the behaviour of laterally loaded conductor casing embedded in the seabed is an important consideration in the design of conductor casing systems and in the prediction of lower flex-joint angle and wellhead bending moments. If the soil immediately below the mudline has low strength, as is frequently the case, little resistance is provided against lateral deflection in this region and the area of highest bending of the structural casing can occur some distance below the mudline. For this reason, the characterisation of lateral resistance of the soil near the mudline is an important input to a reliable structural model of a coupled casing-riser system. Under lateral loading, soils typically behave as a non-linear material, which makes it necessary to relate soil resistance to conductor casing lateral deformation. This is achieved by constructing lateral soil resistance-deflection (p–y)

curves, with the ordinate of these curves being soil resistance per unit length, p and the abscissa being lateral deflection, y. The analysis of such a problem may be accomplished by structural analysis of the casing structure using non-linear springs to model the p–y behaviour of the soil and by the solution of the following equilibrium equation:

$$EI\frac{d^4y}{dx^4} = p \tag{9.16}$$

where y = casing deflection, x = length along casing, EI = equivalent bending stiffness of casing system, p = soil resistance per unit length. This equation is solved applying the casing geometry and soil stiffness boundary conditions, typically in terms of a family of p–y curves developed for the soil. These p–y curves, which represent the increasing non-linear soil stiffness with depth, are typically based on empirical formulations proposed for soft clay, stiff clay and sand respectively. The draft API Technical Report (API 16TR1) provides guidance on the derivation of these curves [Kavanagh, et al 2004].

9.3.3 Steel Catenary Risers (Portions contributed by Thanos Moros & Howard Cook, BP America, Houston, TX)

The steel catenary riser is a cost-effective alternative for oil and gas export and for water injection lines on deepwater fields, where the large diameter flexible risers present technical and economic limitations. Catenary riser is a free-hanging riser with no intermediate buoys or floating devices. Flexible riser is a free-hanging riser with intermediate buoys or floating devices. See fig. 9.52 below. A typical profile of a SCR is shown in fig. 9.53.

In 1998, a 10-in. steel catenary riser (SCR) was installed in P-18 platform, a semi-submersible production vessel moored in Marlim Field, at 910 m water depth. This was the first SCR installed at a semi-submersible platform.

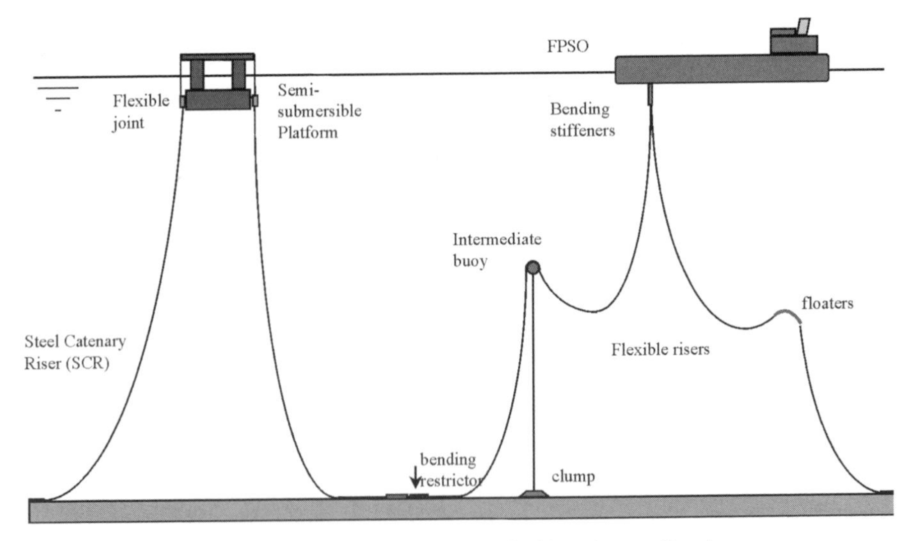

Figure 9.52 Free hanging SCRs with and without intermediate buoys

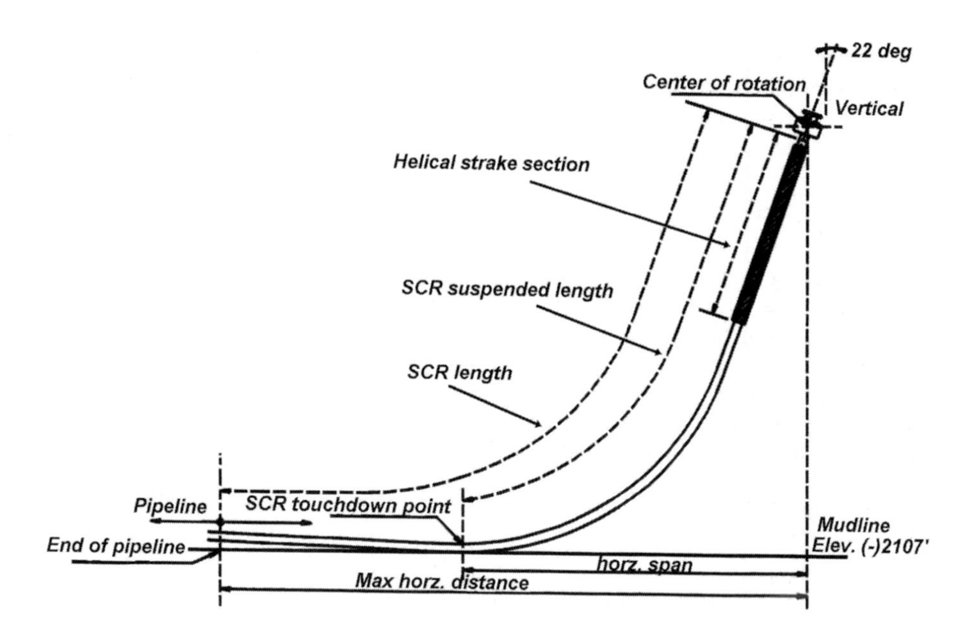

Figure 9.53 Typical profile of steel catenary riser

Figure 9.54 shows the worldwide population of floating production systems (FPS) with steel catenary risers of 12 in. or greater in diameter. The FPSs are ranked by criticality of the SCRs in terms of water depth divided by the diameter of the largest SCR on the platform. The smaller this ratio, the more critical the touchdown point fatigue is likely to be. As can be seen from the chart, Typhoon's 18 in. gas export riser in approximately 2100 ft of water has the smallest ratio of any SCR installed to date.

This section provides guidelines for the design of simple and lazy wave steel catenary risers (SCRs). Such risers are being considered or built for use in many deepwater fields. There are currently two dedicated riser design codes relevant to SCR design, API RP 2RD (1998) and DNV-OS-F201 (2001), and their scope is similar. They provide recommendations on structural analysis procedures, design guidelines, materials, fabrication, testing and operation of riser systems.

The steel catenary risers (SCRs) are designed by analysis in accordance with the API codes (API RP 1111 and API RP 2RD) or the DNV codes. The analysis generally follows the following steps:

- Size the SCRs for pressure and environmental loads;
- Select the minimum top angle required for resisting environmental loads and providing adequate fatigue performance;
- Generate design parameters (angles and loads) for the flex joints and their attachments to the floater;

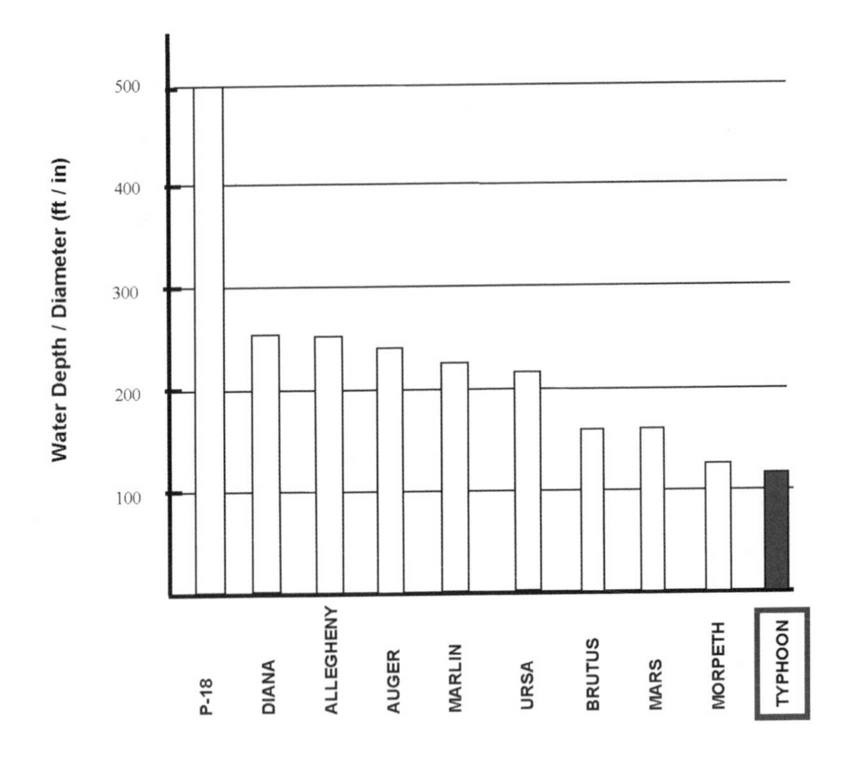

Figure 9.54 Large diameter SCRs for FPS

- Compute fatigue life of the SCRs based on "suitable" design fatigue curves for the proposed welds and assess the criticality of the welds;
- Compute cyclic load histograms for use in fracture mechanics analyses for defining weld acceptance criteria;
- Assess procedures for abandoning (laying on bottom), retrieval (lifting off bottom) and installation of SCRs; and
- Perform Vortex Induced Vibration (VIV) analyses to determine if vortex suppression devices were required and if so in what quantity.

The principal difference between the codes is in the approach to structural design. API is based on a working stress design approach. DNV provides a limit state approach, which is less conservative, although a simplification to a working stress approach is allowed for in the document.

There are several factors that influence the sizing of the riser diameter and its wall thickness. Some of them are the following:

- Metocean conditions,
- Host vessel offsets and motions, and

- Structural limitations – burst, collapse, buckling, post-buckling,
- Construction issues – manufacturability, tolerances, weld procedures, inspection,
- Installation method – tensioning capacity of available vessels,
- Operating philosophy – transportation strategy, pigging, corrosion, inspection,
- Well characteristics – pressure, temperature, flowrate, heat loss, slugging.

The producing well characteristics determine variations in line contents and properties over time, which should be defined for operation in normal and abnormal/shutdown conditions. The designer should take into account the full range of contents for all stages of installation, commissioning and operation.

9.3.3.1 Influence of Construction/Installation Method

The designer should take account of the effects of construction and installation operations, which may impose permanent deformation and residual loads/torques on the riser system whilst consuming a proportion of the fatigue life. In-service requirements determine weld quality, acceptable levels of mismatch between pipe ends and out-of-roundness, whilst NDT requirements are determined from fatigue life and fracture analysis assessments.

The following, in particular, should be quantified:

- In collapse design, the effects of the sag bend strain levels during installation as well as extreme loading, shut down/depressurised and minimum wall thickness cases.
- Residual torque resulting from curves in the pipeline, installation vessel tensioner crabbing or plastic deformation during laying operations, as regards components such as flex-joints.
- SCFs from geometric discontinuities, regarding pre-weld fit up (hi-lo) limits resulting from out of roundness (UOE pipe), non-uniform wall thickness (seamless pipe) and tolerances of weld preps.
- Stress concentrations induced by plastic deformation during installation (reeling, S-Lay).
- Residual ovality induced by plastic deformation during installation (reeling, S-Lay).
- Installation load cases.
- Weld procedure and tolerances, NDT methods and thresholds, which should be related to the required fatigue resistance.

Connelly and Zettlemoyer (1993) and Buitrago and Zettlemoyer (1998) may be found useful in the determination of SCFs for girth welds.

Annealing after seam welding may reduce residual stresses with consequent improvement in hydrostatic collapse resistance.

Mechanical properties of protective coating or thermal insulation systems should be able to accommodate all construction activities. For example, where thermally insulated risers are to be installed from a reel barge, environmental conditions at the spool base may differ considerably from those in the field, particularly if reeling is done in winter in northern Europe or the northern USA. External pipe insulation systems are often made up of several layers of material – with field joints having a different make up. How the system will behave, when reeled and unreeled, can only be reliably assessed by carrying out bending trials under the worst conditions (usually the coldest).

9.3.3.2 Geotechnical Data

As an SCR comes in contact with the seafloor at the touch down point (TDP), an interaction (force-reaction) takes place between the riser and the seafloor. This interaction is usually characterised through the use of three sets of perpendicular "springs", which represent the axial (or longitudinal), horizontal (or transverse lateral), and vertical (or transverse vertical) soil restraints against the riser motions.

The soils at or close to the seabed in deep water are generally very soft, to soft clays, although the presence of sand layers cannot be discounted.

The interaction of the SCR with the seabed will depend on the riser motions and soil conditions. The riser may cut a trench several riser diameters wide and may load or severely disturb soil up to 5 or more riser diameters below the mudline. It is therefore important that any geotechnical data that are obtained from the site are representative of the conditions within the riser zone of influence.

Arguably the most significant soil parameters for modelling the interaction of the riser with a clay seabed are the undisturbed and remolded undrained shear strengths. However, other soil properties such as plasticity, particle size and permeability are important for characterising soil suction and dynamic response, including viscous damping effects. Soil chemistry may be important in some cases in designing for external corrosion.

For sands the most important mechanical properties for assessing riser interaction are the relative density and permeability, as characterised by the angle of internal friction and the particle size distribution or grading.

The definition and units of spring stiffness used in structural codes are not necessarily consistent, which may lead to misinterpretation and misuse. In order to reduce the risk of analytical ambiguities and errors, the units commonly used to describe soil springs are discussed below. Guidelines for selecting soil spring stiffness are also given.

9.3.3.2.1 Soil Springs for Modelling Riser–Soil Interaction

One of the simplest and most popular ways of modelling the support or restraint provided by soil surrounding a pipeline or riser pipe is by using discrete uncoupled soil springs.

Many structural codes can handle non-linear soil springs such as those frequently used to model interactions with offshore piles and conductors, usually called *p–y* curves (lateral springs) and *t–z* curves (axial springs). Others may be limited to equivalent linear elastic soil springs, often referred to as Winkler Springs.

Experience indicates that the definition and units of spring stiffness used in structural codes are not necessarily consistent. This may result in some misinterpretation and misuse, particularly if the spring stiffness is obtained from independent sources.

The aim of this section is to summarise the units commonly used for soil stiffness to reduce the risk of analytical ambiguities and errors. For this purpose it is helpful to assume the soil as elastic and to consider the classical problem of a flexible strip or beam on an elastic foundation.

9.3.3.2.2 Soil Stiffness

The modulus of elasticity of an elastic material E is defined as:

$$E = \sigma/\varepsilon \tag{9.17}$$

The units of E are Stress or Force/length squared, e.g. kN/m^2. The deflection of a vertically loaded area supported on a semi-infinite elastic half space is related to E by the following expression:

$$\delta = Ip\, q\, B(1 - \nu^2)/E \tag{9.18}$$

where Ip = an "elastic" influence factor, q = the average stress applied over the loaded area, B = the width of the loaded area, ν = Poisson's Ratio. The deflection at the centre of a uniformly loaded flexible strip or beam, such as a riser pipe, on a quasi-elastic seabed is given by the following specific solution:

$$\delta = 2\, q\, B(1 - \nu^2)/E \tag{9.19}$$

The traditional way of defining soil stiffness for a beam on an elastic foundation is by the Modulus of Subgrade Reaction, Ksu, which can be obtained by re-arranging equation (9.19):

$$\mathrm{Ksu} = q/\delta = \frac{E}{2B(1 - \nu^2)} \tag{9.20}$$

The units of Ksu are Force/length cubed, e.g. kN/m^3. An alternative way of describing the same soil stiffness is by Ku, defined as:

$$\mathrm{Ku} = Q/\delta \tag{9.21}$$

The quantity Q is the total load on the strip or beam so

$$Q = B\, L\, q \tag{9.22}$$

where: B and L are the width and the length of the loaded area, respectively. Substituting for Q in equation (9.21)

$$\mathrm{Ku} = B\, L\, q/\delta \tag{9.23}$$

or

$$\mathrm{Ku} = B\, L\, \mathrm{Ksu} \tag{9.24}$$

From equation (9.20)

$$\mathrm{Ku} = \frac{EL}{2(1 - \nu^2)} \tag{9.25}$$

The units of Ku are force/unit length, e.g. kN/m and it is independent of the strip or beam width, B. The main potential source of confusion with units arises from a variant of equation (9.25) obtained by considering a unit length of the strip or beam, i.e. by assuming L is 1.0. In this case equation (9.25) reduces to:

$$\mathrm{Ku}^* = \frac{E}{2(1-\nu^2)} \tag{9.26}$$

The dimensions of Ku* are force/length squared (stress), but the actual units are force/length/length. In the case of a riser pipe the units are force per unit deflection per unit length of pipe, e.g. kN/m/m length of pipe. Note Ku* is also independent of width.

The use of stress units for Ku* can be and has led to misinterpretation. Therefore, when expressing soil stiffness in this form it is important to use units of force/unit deflection/unit length of pipe.

Preliminary indications from recent research work are:

(i) Soil stiffness under vertical compressive loading is important for wave-related riser fatigue. An increase in soil stiffness reduces fatigue life.
(ii) Suction effects due to riser embedment appear to be less important for riser design, but may in some circumstances need to be accounted for.

Interaction between the seabed and the riser is dependent on many geotechnical factors, including non-linear stress–strain behaviour of soil, remolding, consolidation, backfilling, gapping and trenching, hysteresis, strain rate and suction effects.

Riser analysis codes presently in use have limited seabed/riser interaction modelling capabilities, but typically allow the use of soil springs to model load-deflection response and the product of submerged weight and friction coefficients or soil-bearing capacity theory to calculate maximum resistance force, as follows:

- Friction coefficient for lateral movement across the seabed
- Friction coefficient for axial movement along the seabed
- Seabed resistance or stiffness to bearing loads.

Guidance on seabed friction coefficients can be obtained from BS 8010 (1973), which gives ranges for lateral and axial coefficients based on experience in shallower waters. However, as stated in the standard, these coefficients are an empirical simplification of actual pipe/soil interaction, particularly for clays.

Soil models that capture some of the key features of riser–clay interaction much better are currently being developed in recent industry research programs, such as "STRIDE" and "CARISIMA". These models may include refinements such as soil non-linearity, hysteresis, plastic failure, suction and viscous damping. Meanwhile, a simplified modelling approach combined with sensitivity checks that can bound the problem and identify key parameters can be used.

An analysis method with a two-step approach is:

1. Conduct global riser analysis using simplified soil/riser interaction model – for example, linear elastic soil springs with maximum soil resistance based on sliding

friction or bearing capacity. In lieu of other data, a rigid or very stiff seabed is recommended for fatigue analysis, as this provides a conservative estimate of damage.

2. Conduct analyses for critical storm load cases and fatigue sea-states, using a detailed soil/riser model, such as that being developed by STRIDE or CARISIMA. If a detailed model is not available, conduct sufficient analysis to bound the seabed interaction problems.

9.3.3.3 Buoyancy Attachments and Other Appurtenances

Lazy-wave risers are similar to simple catenary risers except that they have an additional suspended length that is supported by a buoyant section. This provides a compliant arch near the TDZ on the seabed.

Analysis is required to optimise the arrangement and to define the required arch size and buoyancy distribution. The arch height and riser response can be sensitive to variations in the density of fluid contents. In addition, there may be a loss of buoyancy with time through water intake and degradation of the buoyant material. Analysis should be conducted to confirm that the riser has an acceptable response for the complete range of fluid contents and buoyancy.

Buoyancy modules affect normal and tangential drag, mass and buoyancy upthrust. When modelling auxiliary buoyancy, consideration should be given to hydrodynamic loading at the bare pipe/buoyancy interfaces. Buoyancy is typically supplied in modules that provide a discontinuous distribution of buoyancy and hydrodynamic properties. Analysis may be conducted assuming a continuous distribution. But it is recommended that sensitivity analyses be conducted to confirm that this is an acceptable assumption both in terms of storm and fatigue response. Care should be exercised in modelling to ensure accurate representation of (distributed) buoyancy, mass, added mass/inertia and drag.

9.3.3.4 Line-end Attachments

SCR attachment to the floating vessel may be achieved using a flex-joint or a stress joint:

Flex-joints. Correct understanding of the flex-joint stiffness is important in determining maximum stresses and fatigue in the flex-joint region. Flex-joint stiffness for the large rotations which typically occur in severe storms is much less than for the small amplitudes occurring in fatigue analysis. Temperature variation can also result in significant changes in flex-joint stiffness. In addition, it should be verified that the flex-joint can withstand any residual torque that may be in the riser following installation or released gradually from the seabed section of the line. Steps may be taken to relieve torque prior to attachment.

Another design consideration for flex-joints, especially in high-pressure gas applications, is explosive decompression. Under pressure, gas may permeate into any exposed rubber faces of the flex-element. When de-pressurised, the gas expands and can migrate outwards with time. However, if the reduction in pressure is rapid the expanding gas can cause breakaway of the rubber covering the steel/rubber laminates. With repeated, rapid de-pressurisation, the steel laminates become exposed, the edges of the rubber laminates become damaged and functionality of the flex-joint is impaired. Explosive decompression risk is increased in gas applications and at high pressures (say 3000 psi) may cause structural damage to the flex-joint. Suppliers may apply proprietary methods to avoid these problems.

Stress joints – may be used in place of flex-joints, but they usually impart larger bending loads to the vessel. They are simple, inspectable, solid metal structures, and particularly able to cope with high pressure and temperature. They may be either steel or titanium, the latter material having the advantage of good resistance to attack from sour and acidic well-flows and, of course, gas permeation. Titanium gives lower vessel loads than steel and typically has better fatigue performance than steel.

When conducting analysis with either flex-joints or stress joints as the attachment method, sufficient load cases should be considered to define the extremes of response. Angle change across the component is a key input for both types of termination, as are tension, pressure and temperature. An assessment of long-term degradation is also important from both a technical and an economic standpoint.

9.3.3.5 Pipe-in-Pipe (PIP) SCRs

Thermal insulation is required for some production risers to avoid problems with hydrate, wax or paraffin accumulation. The use of external insulation may in some cases impair a riser's dynamic performance by increasing drag and reducing weight-in-water. However, PIP thermal insulation technology can often be used to satisfy stringent insulation requirements (lower U-values) whilst maintaining an acceptable global dynamic response with the penalty of a heavier and perhaps more costly structure.

Inner and outer pipes of a PIP system may be connected via bulkheads at regular intervals. Bulkheads limit relative expansion and can separate the annulus into individual compartments, if required. Use of bulkheads can be a good solution for pipelines, but for dynamic SCRs one must consider the effects of high stress concentrations, local fatigue damage and local increase in heat loss. Alternatively, regular spacers (centralisers) may be used, which allow the inner and outer pipes to slide relative to each other whilst maintaining concentricity.

A detailed discussion of all analysis issues is beyond the present scope, but a checklist follows. The items listed are in the most cases additional to those for single-barrier SCRs and it is not claimed that the list is exhaustive. Also, according to engineering judgement, some of these effects may be omitted in the early phases of design, though justification for doing that should be given wherever possible.

(i) Residual curvature (which may change along the SCR) following installation
(ii) Residual stresses due to large curvature history
(iii) Residual axial forces between the two pipes
(iv) Connection between the inner and outer pipes, including length and play of centralisers
(v) Boundary conditions and initial conditions at riser terminations
(vi) Fatigue life consumed during installation
(vii) Pre-loading of inner and outer pipes
(viii) Axial forces and relative motions during operation, due to thermal expansion and internal pressure

(ix) Poisson's ratio effect on axial strains
(x) Local stresses in inner and outer pipes due to centraliser contact, including chattering effects
(xi) Frictional effects between inner and outer pipes
(xii) Thermal stresses and thermal cycling effects
(xiii) Buckling checks (including helical buckling) due to thermal and general dynamic loading
(xiv) Soil forces on outer pipe
(xv) Internal and external pressures having different effects on stress in inner and outer pipes
(xvi) Effect of packing material in reversal of lay direction on a reel should be assessed and cross-section distortion minimised; the pipe yields as it is reeled and it is very soft at the reel contact point
(xvii) Effects of PIP centralisers on pipe geometry during reeling
(xviii) Wear of centralisers
(xix) Validity of VIV calculations (e.g. as regards damping)
(xx) Possible effect of any electrical heating on corrosion rates
(xxi) Effect of damage (e.g. due to dropped objects striking outer pipe) on thermal and structural performance

The capabilities of software intended for performing PIP analysis should be carefully considered, since commercially available programs vary widely in this respect.

A PIP SCR can be modelled as a single equivalent pipe (EP), although it should be recognised that the technology is new, and careful attention must be paid to several aspects of the analysis. Development of stress amplification factors, to estimate loads and stresses in individual pipes following global EP analysis, is acceptable in the early stages of design. However, it is important to appreciate the conditions under which such factors become inaccurate, which will vary from case to case. Ultimately, full PIP analysis is required for verification.

Two useful references on PIP SCR analysis are Gopalkrishnan, et al (1998), and Bell and Daly (2001). The first of these illustrates the large disparities, which can arise between the simplified EP approach and full PIP analysis, especially regarding the static stress.

A JIP on deepwater PIP (including tests) named RIPIPE has been conducted by Boreas and TWI in the UK, and results will become public domain in due course.

9.3.3.6 Analysis Procedures

The SCR is a 3-D structure, which in terms of design planning implies that directionality of loading (wave and current) must be included in the engineering analysis.

Analysis methods for flexible risers include a complex finite-element structural method coupled with more simplistic hydrodynamic models (e.g. Morrison equation or potential theory). Empirically derived hydrodynamic coefficient databases are combined with the structural dynamic models. CFD method for computing excitation is combined with finite-element dynamic response analysis.

The analysis process typically falls into two phases. A preliminary analysis is performed in which the global behaviour of the riser is examined to confirm that the proposed configuration is practical and to provide preliminary data relating to key components in the system. A detailed analysis refines the definition of components and further examines all aspects of riser operations.

In the preliminary analysis, the riser behaviour is generally examined in the normal operating mode using extreme loading conditions, and design changes are made accordingly. Several combinations of riser configuration and loading conditions may be required to complete this initial assessment and to determine preliminary design load data for specific components. Initial VIV and fatigue life assessments should also be included.

A flow chart showing the interaction between all aspects of the riser design and analysis is given in fig. 9.55.

9.3.3.7 FEA Codes and Modelling Methods

A number of commercial finite element or finite difference codes are available that may be used for SCR analysis, mostly time-domain. Frequency-domain analysis uses minimal computational effort, but does not account for non-linearities in riser response. Time-domain analysis accounts for non-linearities in riser response and the computational time and effort, whilst much greater than the frequency-domain analysis, can be acceptably low.

When modelling SCRs the element mesh should be refined at locations of high curvature and dynamic response; e.g. directly below the interface with the vessel and in the TDZ. Guidance is given in API RP 2RD on calculating the required element mesh. Appropriate convergence checks should be conducted in any case at a suitable stage in the analysis.

Riser boundary conditions are the connection to the vessel and the termination and contact at the seabed, and care should be taken to model these accurately. Flex-joints can be modelled as articulation elements, and the designer should be aware of the sensitivities of flex-joint stiffness to both temperature and dynamic loading. Stress joints with a continuous taper may be modelled as a series of stepped sections, again paying due regard to convergence as well as accuracy. The orientation of the vessel attachment can have a large effect on end loading and termination sizing and should be optimised.

9.3.3.8 Analysis Tools

The software that are used in the design of risers are listed later. In the following, discussions are included in brief in order to illustrate general procedures for the analysis. For details of their capabilities the readers should consult the manuals of the specific software.

Static configuration and mode-shapes should be calculated using an FE model. Alternatively, for quick evaluation studies only, an analytical solution to the catenary equation may be used. In-plane and out-of-plane mode-shapes should be calculated. Such externally generated mode-shapes can account for soil–riser interaction, property changes along the riser, lateral constraints.

The FE model should properly model boundary conditions at the top of the riser. If a flex-joint is used, a suitable rotational stiffness should be implemented (stiffness depends on

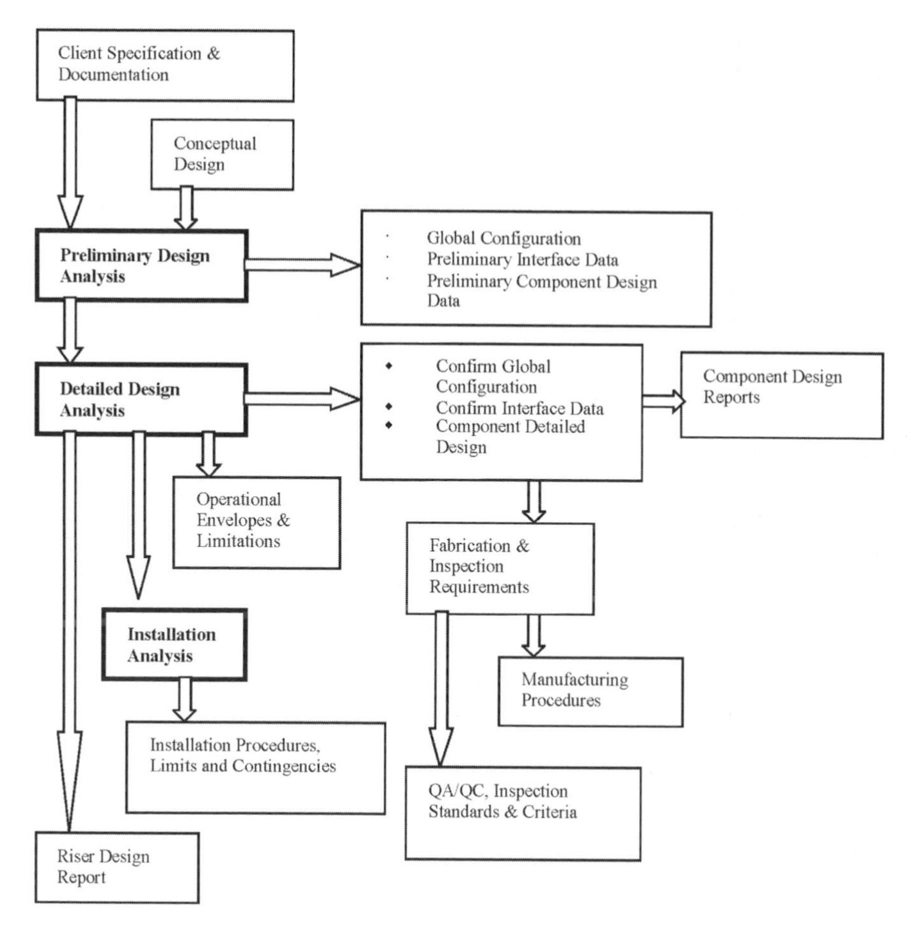

Figure 9.55 Flow chart on riser design

response for flex-joints). The most appropriate bottom boundary condition may vary from case to case. If the modelled riser is terminated at TDP, the use of a fixed (built-in) end is likely to produce a lower fatigue life than use of a pinned end. However, it should not be assumed that a fixed end will produce the lowest of all possible fatigue lives; sometimes an intermediate stiffness case may be worse.

Even for calculations with 2D currents, both in-plane and out-of-plane loading should be considered, which should yield reasonable accuracy, although it is not necessarily conservative compared to a true 3D behaviour. Out-of-plane loading (in-plane response) is often the most critical. For out-of-plane loading the current needs no modification. For in-plane loading the component of current normal to the riser axis should be used:

$$V_N = V \sin \alpha \tag{9.27}$$

where V = horizontal current velocity, V_N = velocity normal to riser, α = angle between riser and horizontal.

Suppression devices are discussed later. The way in which they can be included in VIV modelling programs, such as SHEAR7, is evolving. Calculated damage for each profile should be factored by its probability of occurrence, then added to obtain the overall damage (taking account of the location on the riser periphery where damage accumulates). Sensitivity to profile shape and current intensity should also be evaluated. The following values of structural damping and stress concentration factor may be often suitable:

Parameters	Value
Stress concentration factor	1.3
Structural damping	0.003 (i.e. 0.3% of critical)*

*Based on bare steel welded pipe. PU/foam coating may increase this value and experimental work may be required to obtain representative values

9.3.3.9 Hydrodynamic Parameter Selection

Typical hydrodynamic coefficients for flow normal to the riser axis are given in table 9.18. Two exceptions to these general guidelines are:

(i) For first-order fatigue analysis of non-VIV-suppressed riser sections, a $C_D = 0.7$ may be appropriate.
(ii) For straked risers or parts of risers where strakes have been applied, especially where Keulegan–Carpenter Number, KC, is low, an increased C_D may be appropriate and application-specific data should be sought.

Further data on hydrodynamic coefficients for single risers and riser in arrays, showing dependence on KC, roughness, turbulence, spacing and strakes are also available (See Chapter 4). Effects, which can further influence the drag coefficient, are pipe roughness (due to marine growth, for example), VIV due to current or vessel heave, and interference from adjacent risers and structures. Reynolds number, in this regard, is defined in terms of the relative velocity between riser and water particle.

The tangential drag of a riser is typically small as the structure is slender and the outer profile is even. Buoyancy elements, other appurtenances or marine growth can result in

Table 9.18 Hydrodynamic coefficients for flow normal to riser axis

Flow regime	C_D[1]	C_a[2]
Subcritical, $Re < 10^5$	1.2	1.0
Critical, $10^5 < Re < 10^6$	0.6–1.2	1.0
Post-critical, $Re > 10^6$	0.7	1.0

Note 1: Reference area is area projected normal to riser axis
Note 2: Reference volume is displaced volume of riser per unit length

Table 9.19 Hydrodynamic coefficients for flow tangential to riser axis

Component	C_D	C_a
Riser	0.01[3]	0.0
Riser/buoyancy interface	0.7[4]	Note 5

Note 3: Reference area is surface area of riser
Note 4: Reference area is exposed annular area
Note 5: Reference volume and C_a to be agreed for each case

local increase of tangential drag coefficient. Some typical values for modelling the buoyancy modules are given in table 9.19. Care should be taken to ensure that the reference area/volume associated with the hydrodynamic coefficients is correct for the software being used.

Further guidance is provided in DNV Classification Notes 30.5, whilst the FPS 2000 JIP Handbook (2000) produced a wide-ranging survey of hydrodynamic data applicable to riser design and analysis. As a general rule, if doubt remains about the selection of C_D, the value used should tend towards the conservative side. This means use of a higher value, when and where drag acts as an excitation and use of a lower value, if it acts to produce damping.

An increased added mass coefficient, typically $C_a = 2.0$ (i.e. twice the value given in table 9.18) should be used for straked risers.

9.3.3.10 Sensitivities

The sensitivity of riser response to changes in design and analysis assumptions should be evaluated. Parameters that should be considered include the following:

- Riser length – including installation tolerances, thermal expansion effects, tide and surge
- Weight – including corrosion, fluid density variations and slugging
- Flex-joint stiffness, including sensitivity to deflection, rate of deflection and temperature
- Seabed stiffness and soil/riser interaction effects
- Current directionality
- Drag coefficients
- Marine growth
- Vessel motion (draught and mass distribution dependence).

Expected extremes of the parameters identified above should be incorporated into the riser model. This will allow the effects of parameter changes to be quantified.

9.3.3.11 Installation Analysis

Limiting installation seastate or current, hand-over limitations and expected loads and stresses at each phase of the installation process should be established and the effect of installation methods and operations on fatigue life should be determined.

The installation analysis should establish functional requirements for installation equipment, identify operational sensitivities and establish limiting conditions and key hold points in the procedure. In addition, the analysis should identify contingency procedures/escape routes to be implemented in the event that safe operational limits are exceeded.

Venkataraman (2001) discusses a number of issues relevant to installation of steel risers by reeling, including reeling strain, buckling due to bending on the reel, strain amplification, elastic-plastic fracture mechanics, fatigue and hydrogen cracking.

9.3.3.12 Extreme Storm Analysis

Riser response is period and direction sensitive and highly dependent on vessel motions. Analysis using the maximum wave height with associated wave period may not result in the greatest response.

Extreme storm analysis can be conducted using either regular or random waves. Regular wave analysis is a good preliminary design tool, as required design changes can be quickly evaluated. Regular wave analysis may be validated using random wave analysis, as the latter is able to provide a more realistic representation of the environment. However, if the wave period range is adequately covered, regular wave analysis is sufficient for early feasibility checks. Due to the period sensitivity of dynamic catenary riser systems it is recommended that a range of periods be analysed to confirm riser extreme storm response.

A typical random wave analysis is:

1. Establish riser structural model.
2. Select spectrum type and parameters based on available data. Associated current and directional data should also be established. Representation of wave spreading is not usually required.
3. Apply extreme vessel offset corresponding to assumed environmental conditions.
4. Simulate random sea and calculate resulting vessel motion and riser structural response. In selecting the random sea, consideration should be given to its duration and its statistics. Where practical, these should be reported.
5. Postprocess sample response statistics to estimate extremes (see below).
6. Repeat for other cases, ensuring that period content is suitably represented.

Compatible low-frequency motions may be included, depending on the software used.

9.3.3.12.1 Short-Term Extreme Responses

Short-term extreme responses are those occurring in storms of relatively short duration, typically three hours. There is sometimes a need to post-process sample random dynamic analysis, results (or measurements) in order to establish an extreme response prediction. Alternatively, the sample extreme may be considered a good enough estimate.

Unless it can be demonstrated that a simpler method, e.g. the Rayleigh method gives sufficient accuracy for the particular response under investigation (taking due account of the stage to which a project or study has progressed) a (three-parameter) Weibull method is more appropriate.

Analyses should be long enough to get satisfactory convergence of response statistics, noting that variability between different realisations of the same sea-state is reflected in the response, and extreme order statistics converge more slowly than lower order statistics.

Convergence may best be achieved (and observed) by performing several different three-hour simulations for the same sea-state; i.e. using different seed numbers to produce different realisations of the same wave spectrum.

9.3.3.12.2 Long-Term Extreme Responses

In principle, full prediction of long-term (e.g. lifetime) extreme responses requires probability-weighting and addition of all short-term extreme distributions, including those corresponding to the low and moderate wave height. However, this is usually not possible because not all of the short-term extreme response distributions will have been developed. In fact, very few may be available – perhaps some corresponding to an extreme wave envelope, perhaps just one or two. Some judgement is needed regarding the extent to which limited short-term data can be adapted and extrapolated to provide a suitable long-term extreme prediction. For a detailed coverage of this see Chapter 5.

9.3.4 Diameter and Wall Thickness

The first parameter that should be determined for the design of an SCR is the wall thickness. The minimum wall thickness is calculated on the basis of external and internal pressure and buckling requirements. However, for SCRs the dynamic and fatigue life are, in most cases, the determining factor for the wall thickness [Chaudhury, 1999]. This wall thickness should include for corrosion allowance.

Initial wall thickness estimates are made using assumed riser loads. Further increases in riser wall thickness or changes of material grade may be required for a satisfactory response, based on later and more detailed analysis. Refer to Section 9.3.4.1 for further details.

Wall thickness must account for all potential modes of failure as follows:

- Burst at maximum internal pressure,
- Burst under hydrotest,
- In-service collapse,
- Collapse during installation,
- Propagation buckling in-service and during installation[1],

[1] Propagation buckling checks may be performed, but need not be acted on for the dynamic part of the SCR unless required by regulatory bodies. The emphasis should be on preventing a buckle from initiating, although buckle arrestors in the static flowline, beyond the TDZ, may be advisable to prevent a buckle propagating between pipeline and riser. The primary function of buckle arrestors in pipelines is to restrict damage to limited lengths, which can then be replaced, whereas a buckle in a riser would require replacement of the whole riser, even if only a short length were damaged.

- Combined modes; e.g. external pressure with bending and tension.

Calculations should allow for reduced wall thickness due to manufacturing tolerances, corrosion and wear, although corrosion may be neglected for installation and hydrotest conditions. Increased wall thickness may be required, perhaps only locally, to comply with dynamic response criteria.

More generally, optimisation of wall thickness over the full riser length may help reduce cost and vessel interface loads. However, such an exercise is likely to be more beneficial for designs governed by collapse, and may yield no benefit at all for high-pressure cases governed by burst.

9.3.4.1 Nominal Wall Thickness

The nominal wall thickness of pipeline is the specified wall thickness taking into account manufacturing tolerance.

Corrosion Allowance

The external surface of submarine pipelines is generally protected from corrosion with a combination of external coating and a cathodic protection system. The internal surface, depending upon the service, may be subject to corrosion. This is accounted for by the addition of corrosion inhibitors or applying a corrosion allowance to the pipeline wall thickness. The corrosion allowance is calculated from the anticipated corrosion rate and the design life of the pipeline system.

Manufacturing Tolerance

Manufacturing or mill tolerances are the specified acceptance limits for the linepipe wall thickness during manufacture. The tolerance will depend upon the size of pipe and manufacturing process involved. A negative wall thickness tolerance should be taken into account when calculating wall thickness required for hoop stress criteria. The specified nominal wall thickness is calculated from the minimum required wall thickness as follows:

$$t_{nom} = t_{min} \times \frac{100}{100 - t_{tol}} \tag{9.20}$$

where t_{tol} = negative manufacturing tolerance as specified by codes DNV, IP6 etc. Typical values for the wall thickness tolerance for seamless and welded pipe are ±12.5% and ± 5%, respectively.

Consideration should be given to the nature and consequences of post-buckling behaviour. Under combined loading a pipe may buckle only locally in shallow water, but fail completely under the action of continuing hydrostatic pressure in deeper water.

Guidance on wall thickness sizing against collapse and burst criteria is given in the following. This is drawn from several sources on standard pipeline practice and is

Table 9.20 Design codes

IP6	Institute of Petroleum Pipeline Safety Code (UK).
DNV (1976, 1981)	Det Norske Veritas, Hϕvik, Norway, 1981 and 1976.
ASME 31.8	American Society of Mechanical Engineers, Liquid Transportation Systems for Hydrocarbons, Liquid Petroleum Gas, Anhydrous Ammonia, and Alcohols.
ASME 31.4	American Society of Mechanical Engineers, Gas Transmission and Distribution Piping Systems.
BS 8010 (1973)	British Standard 8010, Code of practice for pipelines. Part 3. Pipelines subsea: design, construction and installation.
DNV (2000)	Det Norske Veritas, Offshore Standard OS-F101, "Submarine Pipeline Systems", 2000.

suitable for initial sizing. However, project-specific requirements or guidance developed, more specifically for risers, may take precedence where there is justification. Propagation buckling, maximum D/t ratio, corrosion allowances, manufacturing tolerance, hydrotest pressure and API standard wall thickness are also discussed.

The design criteria for the wall thickness calculations are as follows:

- Limiting hoop stress due to internal pressure.
- Hydrostatic collapse due to external pressure.
- Buckle propagation due to external pressure.

Minimum wall thickness shall be the larger wall thickness determined from the above design criteria. The design codes in table 9.20 are used for wall thickness design:

These codes are briefly covered below, except DNV (2000), which is relatively new and applies a Limit State approach. Its relevant section is referred to for each of the above design criteria. Section 5, fig. 5.3 of DNV (2000) gives an overview of the required design checks.

9.3.4.2 Maximum Diameter to Thickness Ratio

The pipeline and riser wall thickness may be specified independently of the static design criteria due to installation stress limits. If the pipeline or riser is to be installed by the reel method, a maximum diameter to thickness ratio is recommended to avoid excessive out of roundness of the line during reeling. The ratio will depend upon the line size, reel diameter and total length of line to be reeled. As a general guideline, a diameter to thickness ratio of less than twenty three (23) is used for reel barge installation.

The American Petroleum Institute (API) specification for line pipe is based upon a range of standard diameters and wall thickness. These values are different for imperial and metric sizes. Pipe mill tooling and production is set up around this specification.

Non-standard line sizes are, sometimes, used for risers, where a constant internal diameter is specified, or for alloy steels, which are manufactured to special order and sized to meet production requirements.

API RP 2RD (1998) also gives recommendations on collapse pressure and collapse propagation.

9.3.4.3 Resistance to Internal Pressure (Hoop Stress Criteria)

Two load cases, namely, maximum design pressure and hydrotest pressure, need to be considered with respect to resistance to internal pressure. Design codes and standards stipulate that the maximum hoop stress in a pipeline shall be limited to a specified fraction of yield stress.

The design pressure used in the analysis is based upon the maximum pressure occurring at any point in the pipeline and riser system. The maximum operating pressure will be limited by pump capacity or reservoir pressure and determined during a hydraulic analysis of the system. Design pressure may also take into account the transient surge pressure effects due to valve closure or shutting down of the transfer pump.

The minimum or nominal wall thickness required to resist internal pressure may be calculated from any of the formulas given in table 9.21 below. Alternatively DNV-OS-F101

Table 9.21 Formulas for nominal wall thickness

Code	Formula	Comments
IP6	$t_{\min} = \frac{P_i D_{\text{nom}}}{2 n_h \sigma_y} + t_{\text{cor}}$	
DNV	$t_{\text{nom}} = \frac{(P_i - P_o)}{2 n_h \sigma_y} D_{\text{nom}} + t_{\text{cor}}$	
ASME 31.4/ ASME 31.8	$t_{\text{nom}} = \frac{(P_i - P_o)}{2 n_h \sigma_y} D_{\text{nom}} + t_{\text{cor}}$	
BS 8010 (1973)	$t_{\min} = \frac{(P_i - P_o)}{2 n_h \sigma_y} D_{\text{nom}} + t_{\text{cor}}$	Ratio $D_{\text{nom}}/t_{\min}$ greater than 20.
BS 8010 (1973)	$t^2 - t + D_{\text{nom}} + \frac{D_{\text{nom}}^2}{2\left[1 + \frac{n_h \sigma_y}{P_i - P_o}\right]} = 0$ $t_{\min} = t + t_{\text{cor}}$	Ratio $D_{\text{nom}}/t_{\min}$ less than 20. Positive root of quadratic equation is used.

Table 9.22 Usage factors for internal pressure

Design code	Usage factor (n_h)	
	Riser	Linepipe
Det Norske Veritas, DNV 1981	0.5*	0.72*
Institute of Petroleum, IP6	0.6	0.72
ANSI/ASME 31.4 & 49 CFR195	0.6	0.72
ANSI/ASME 31.8 & 49 CFR192	0.5#	0.72#
British Standard 8010, Part 3	0.6	0.72

* DNV (1981) specifies the riser (zone 2) as the part of the pipeline less than 500 m from any platform or building and the pipeline (zone 1) as the part of the pipeline greater than 500 m from any platform building

#ANSI/ASME 31.8 specifies the riser zone as the part of the pipeline, which is less than 5 pipe outside diameters from the platform and the pipeline zone as the part of the pipeline, which is more than 5 pipe diameters from the platform

can be used. It must be observed that DNV and ASME codes specifically refer to a nominal wall thickness, while IP6 and BS 8010 (1973) refer to a minimum wall thickness. If a minimum wall thickness is specified, the nominal wall thickness may then be calculated using a corrosion allowance and a manufacturer's tolerance (see Section 9.3.4.1).

In table 9.21 t_{cor} = corrosion allowance, t_{min} = minimum wall thickness, t_{nom} = nominal wall thickness, D_{nom} = nominal outside diameter, σ_y = specified minimum yield stress (SMYS), n_h = usage factor or fraction of yield stress, P_i = internal design pressure, P_o = external design pressure.

The usage factor n_h, which is to be applied in the hoop stress formulae, is specified by the applicable design code and the zone or classification of the pipeline. For submarine pipelines and risers the code requirements governing design usage factors are summarised in table 9.22.

Temperature de-rating shall be taken into consideration for the risers and pipelines operating at high temperatures (typical $>120°C$). The pressure containment check (bursting) should be performed according to DNV 2000 DNV-OS-F101 Section 5 D400 [equation (5.14)].

9.3.4.4 Resistance To External Pressure (Collapse)

Two load cases need to be considered with respect to resistance to external pressure:

- In-service collapse
- Collapse during installation

Failure due to external pressure or hydrostatic collapse is caused by elastic instability of the pipe wall. For wall thickness determination the external pressure is calculated from the hydrostatic head at extreme survival conditions. The maximum water depth taking into account the maximum design wave height and storm surge should be used. The minimum wall thickness required to prevent hydrostatic collapse is determined from Timoshenko and Gere (1961) for DNV, API 5L and IP6 and from BS 8010 (1973).

In the first case, the Timoshenko and Gere (1961) formula is used to calculate the minimum wall thickness as follows:

$$P_o^2 - \left[2\sigma_y\left(\frac{t_{\min}}{D_{\text{nom}}}\right) + \left(1 + 0.03e\frac{D_{\text{nom}}}{t_{\min}}\right)P_c\right]P_o + 2\sigma_y\left(\frac{t_{\min}}{D_{\text{nom}}}\right)P_c = 0 \tag{9.29}$$

where P_o = external hydrostatic pressure, P_c = critical collapse pressure for perfectly circular pipe given by:

$$P_c = \frac{2E}{(1-\nu^2)}\left(\frac{t_{\min}}{D_{\text{nom}}}\right)^3 \tag{9.30}$$

ν = Poisson's ratio, E = Young's modulus of elasticity, E = eccentricity of the pipe (%) (see below), σ_y = specified minimum yield stress (SMYS).

BS 8010 advocates use of the formula described in Murphy and Langner (1985) and this is described as follows:

$$\left\{\left(\frac{P_o}{P_c}\right) - 1\right\}\left\{\left(\frac{P_o}{P_y}\right)^2 - 1\right\} = 2f_o\left(\frac{P_o}{P_y}\right)\left(\frac{D_{\text{nom}}}{t_{\text{nom}}}\right) \tag{9.31}$$

where the notation is as above and in addition,

$$P_y = 2\sigma_y\left(\frac{t_{\text{nom}}}{D_{\text{nom}}}\right) \tag{9.32}$$

f_o = the initial ovalisation (see below).

9.3.4.5 Pipe Eccentricity, Out of Roundness and Initial Ovalisation

Pipe eccentricity is a measure of pipe out of roundness. This is generally a specified manufacturing tolerance, which is measured in a different way depending on which code or standard is used. The various ways of measuring it and permitted values are given by various codes. The out of roundness definitions and tolerances with reference to API, DNV and BS 8010 are given in table 9.23.

Summarising:

IP6% eccentricity = API% out of roundness
2 × API% out of roundness = DNV% out of roundness
BS 8010% initial ovalisation = API% out of roundness

Table 9.23 Out of roundness formulas

Code or standard	Out of roundness or initial ovalisation	Tolerance
IP6	$\frac{D_{max} - D_{nom}}{D_{nom}} \times 100$ and $\frac{D_{nom} - D_{min}}{D_{nom}} \times 100$	1.0%
API 5L	$\frac{D_{max} - D_{nom}}{D_{nom}} \times 100$ and $\frac{D_{nom} - D_{min}}{D_{nom}} \times 100$	1.0%
DNV	$2\left[\frac{D_{max} - D_{min}}{D_{max} + D_{min}}\right] \times 100$	1.0%
BS 8010 (1973)	$\frac{D_{max} - D_{min}}{D_{max} + D_{min}} \times 100$	2.5%

The combined loading criterion is to be performed according to DNV 2000 DNV-OS-F101 Section 5 D500 [equations (5.22)–(5.26)]. If the riser is in compression between the supports the global buckling shall be checked according to Section 5 D600.

9.3.4.6 Resistance to Propagation Buckling

Two load cases need to be considered with respect to resistance to internal pressure: in-service and during installation. The required buckle propagation wall thickness is the wall thickness below which a buckle, if initiated, will propagate along the pipeline or riser, until a larger wall thickness or a reduced external pressure is reached. The wall thickness required to resist buckle propagation can be calculated from the formulas in table 9.24.

Generally, the DNV formula is marginally more conservative than the Shell Development Corporation formula and they are both considerably more conservative than the Battelle formula. The degree of conservatism required depends on the installation technique in terms of risk, the length and cost of the line and the water depth in terms of how easily a repair can be made. In practical terms, changing of any criteria will change the required wall thickness. However, since designers are normally limited to selecting from API pipe sizes, there is quite often no actual change in the pipe specified.

If during design, a pipeline is found to be governed by the buckle propagation criteria, then there are two options open to the designer; the first option is to make the wall thickness of the pipe sufficient so that a buckle once initiated will not propagate. The second option is to make the wall thickness of the pipe sufficient to only withstand external pressure (hydrostatic collapse) and to use buckle arresters. Buckle arresters consist of thick sections of pipe or welded fittings, which a buckle cannot propagate through. If buckle arresters are fitted, the damage will be limited to length of the pipeline between arresters, should a buckle initiate.

Table 9.24 Wall thickness to resist buckle

Code or standard	Formula	Remarks
DNV 1981 and DNV 1976	$t_{\text{nom}} = \dfrac{kD_{\text{nom}}}{1+k} \quad k = \sqrt{\dfrac{P_o}{1.15\pi\sigma_y}}$	Conservative
Battelle Columbus Laboratories [Johns, et al 1976]	$t_{\text{nom}} = \dfrac{D_{\text{nom}}}{2}\left(\dfrac{P_o}{6\sigma_y}\right)^{0.4}$	
Shell [Langner, 1975]	$t_{\text{nom}} = \dfrac{D_{\text{nom}}}{2}\sqrt{\dfrac{P_o}{\sigma_y}}$	1.0%
BS 8010 Part 3 (1973)	$t_{\text{nom}} = D_{\text{nom}}\left(\dfrac{P_o}{10.7\sigma_y}\right)^{4/9}$	2.5%

Note: The propagating buckling check should be performed according to DNV 2000 DNV-OS-F101 Section 5 D500 [equation (5.27)]

This risk, however, is considerably reduced after installation. The choice between the two is determined by considering the potential cost saving in wall thickness and possibly installation benefits due to the reduced submerged weights. This is paid off against the risk of having to replace a relatively large section of pipe or riser possibly in deep water.

9.3.4.7 Hydrotest Pressure

The BS 8010 (1973) design code gives criteria to calculate the minimum hydrostatic test pressure for a pipeline. The test pressure required to qualify for a MAOP (Maximum Allowable Operating Pressure) equal to the specified design pressure is either the lower of 150% of the internal design pressure, or the pressure that will result in a hoop stress (based on specified minimum wall thickness) equal to 90% of the specified minimum yield stress. The test pressure should be referenced to the Lowest Astronomical Tide (LAT) and due allowance made for the elevation of the pressure measurement point and parts of the system above LAT.

For definition of design pressure see DNV 2000 DNV-OS-F101 Section 1. Hydrotest criteria are discussed in Section 5 B200.

9.3.5 SCR Maturity and Feasibility

Three views of SCR maturity and feasibility are given in figs. 9.56 and 9.57. Figure 9.56 shows the existing SCRs against diameter and water depth. The choice of diameter and depth as axes is largely motivated by collapse considerations, although installation capabilities are also relevant. Figure 9.3 shown earlier (similar to fig. 9.56) puts more emphasis on water depth and diameter than on waves and vessel-types. It also shows recent

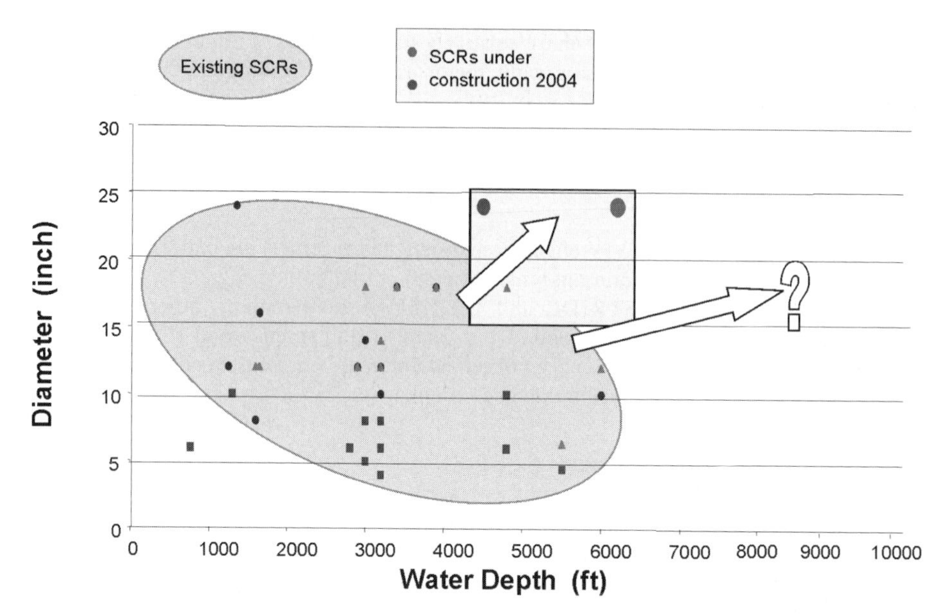

Figure 9.56 Existing SCRs and technology stretch

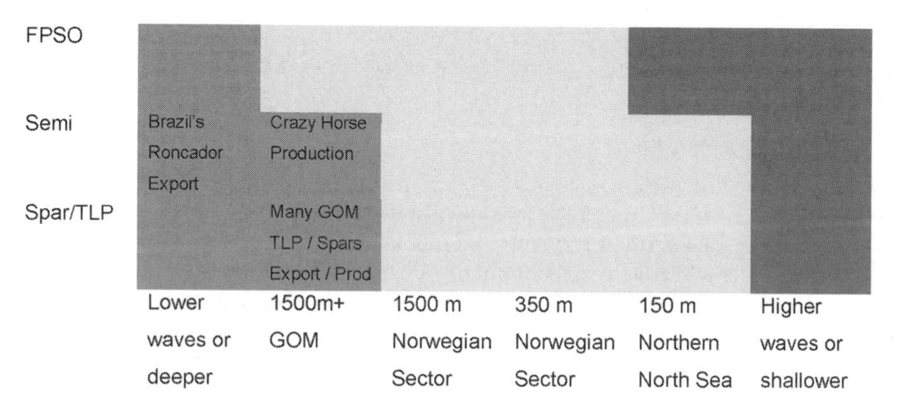

Figure 9.57 Estimated feasibility [from Spolton and Trim (2000)] and maturity of SCRs [Note: feasibility colour scheme developed for 10-in. HPHT oil production SCRs (left shading = steel, middle light shading = steel-titanium, right = unproven)]

chronological development of flexible-pipe riser regimes, indicating that feasibility for typical production sizes is now approaching 2000 m water depth.

In fig. 9.57 [Spolton and Trim, 2000], titanium parts were progressively substituted for steel parts as environments and vessel motions considered became worse. Thus, in the left and the lower light shading parts of the chart, relatively small amounts of titanium are used; e.g. in TSJs and TDZs, whereas the top right region represents SCRs composed mostly or completely of titanium.

These figures are guidelines only but give a reasonable first impression of what has been and what can be achieved, and what "technology stretch" is required for harsher conditions.

Soil-structure interaction and VIV effects (separately and together) are major uncertainties remaining in the SCR design and analysis, although considerable progress has been made in both areas through the STRIDE and CARISIMA JIPs (among others). Findings continue to be extended and assimilated into mainstream practice and software codes. Trenching of SCRs in the TDZ has been observed and may represent an additional critical case for careful examination during detailed design.

9.3.6 In-Service Load Combinations

The in-service design cases of table 9.25 should be assessed for the most severe loading direction, which may vary according to the response quantity of interest. Allowable stresses are given in table 9.26. Latest versions of the API and the DNV riser codes, such as API RP 2RD, and DNV-OS-F201 are commonly used.

Von Mises stress is defined in accordance with API RP 2RD (1998) as:

$$\text{von Mises stress} = \sqrt{\frac{1}{2}\left[(S-h)^2+(S-r)^2+(h-r)^2\right]} \tag{9.33}$$

Table 9.25 In-service design cases

Design case	Wave	Current	Wind	Contents	Other*
Normal operating	To be agreed (typically 1 yr.)	Associated	Associated	Product	Design pressure
Extreme	100 yr.	Associated	Associated	Product	Design pressure
	Associated	100 yr	Associated	Product	Design pressure
	Associated	Associated	100 yr.	Product	Design pressure
Survival	100 yr.	Associated	Associated	Product	Failed mooring
	100 yr.	Associated	Associated	Variable	Accidental**
	1000 yr.	Associated	Associated	Product	–
Pressure test	1 yr.	Associated	Associated	Water	Test pressure

*Use associated pressure for survival. Maximum pressures are given; i.e. lower pressures should also satisfy checks
**Accidental conditions are discussed further in Section 9.3.7

Table 9.26 Allowable stresses

Design case	Von Mises/yield	Design case factor C_f[1]	
		Upper section	Sag bend[3]
Normal operating	0.67	1.0	1.0
Extreme	0.8	1.2	1.5
Survival	1.0	1.5	1.5–1.8[3]
Pressure test	0.9	1.35[2]	1.35

[1]API RP 2RD (1998) definition. Plain pipe allowable stress is 2/3 yield × C_f

[2]At the riser top the distinction between load- and curvature-controlled stress may not be clear. If so, stress should be considered load-controlled and C_f reduced to 1.2

[3]Where primary membrane stress exceeds yield (corresponding to $C_f = 1.5$) a strain-based formulation should be used in which the strain at yield is substituted for the yield stress. Non-linear strain analysis is then required in order to demonstrate compliance. Also, for any case where yielding is predicted, further consideration and consultation should take place, and the higher value of $C_f = 1.8$ for Survival may be acceptable if the exceedance is isolated. In general, it should not be assumed that increased sag bend factors can always be used; the effects of the various forces and motions applied to the riser should first be carefully considered

where S = stress due to equivalent tension and bending stress, r = radial stress, h = hoop stress:

$$\begin{aligned} S &= \frac{T}{A} \pm \frac{(d_o - t)M}{2I} \\ r &= -\frac{(P_o d_o + P_i d_i)}{d_o + d_i} \\ h &= \frac{(P_i - P_o)d_o}{2t_{\min}} - P_i \end{aligned} \tag{9.34}$$

Most sag bends are predominantly curvature-controlled, not load-controlled, and higher bending stresses are then allowable, since yielding does not of itself constitute failure. Increased values of API RP 2RD (1998) Design Case Factor C_f for curvature-controlled sag bends are shown in table 5.2.

In accordance with the API RP 2RD (1998), tangential shear and torsional stresses are not included and can be treated as secondary stresses, which are self-limiting. Torque, however, can influence the integrity of flex-joints (see Section 9.3.3.4).

The term "associated" in table 9.25 is defined in API RP 2RD (1998) as "to be determined by considering joint wind, wave and current probabilities". Often a 10-yr return period is assumed, unless there is a very strong correlation (positive or negative) between these items, or, project-specific requirements dictate otherwise. Associated Pressure is the greatest pressure reasonably expected to occur simultaneously with survival environmental conditions.

Other design considerations include flex-joint rotational limits, interface loads, compression in the TDZ, tension on flowlines and clearance from vessels, mooring lines, umbilicals and other risers.

9.3.7 Accidental and Temporary Design Cases

A failed mooring line with a 100-yr wave condition is an accidental design case typically used in SCR design, table 9.25. However, one failed mooring line is not the only potential failure mechanism that will have an effect on riser integrity. Other accidental design cases applicable to SCRs are listed as follows:

- Two or more failed mooring lines
- Breached hull compartments
- Failed tethers
- Internal pressure surge

The likelihood of each accidental design case needs to be addressed on an individual basis. For example, two failed mooring lines combined with a 100-yr wave condition may be highly unlikely, especially if a failure is fatigue and not strength related. In this case an increased design allowable or less severe environmental condition may be considered. The likelihood of each accidental design case may be defined with a quantitative risk assessment.

For guidance on analysis and criteria for temporary conditions; e.g. transportation and installation, see Section 4.3.3 and tables 1 and 2 of API RP 2RD (1998). When calculating fatigue in towed risers, due allowance should be made for variability of environmental conditions and uncertainties in forecasting weather windows. Additional damage may be justified if there is a realistic risk that changing weather conditions will force an altered course or return to port.

9.4 Vortex Induced Vibration of Risers

9.4.1 VIV Parameters

Important hydrodynamic quantities that influence VIV are:

- Shedding frequencies and their interactions,
- Added mass (or mass ratio) and damping,
- Reynolds number,
- Lift coefficient, and
- Correlation of force components.

For the hydrodynamic design a few important non-dimensional numbers in fluid-induced vibration are given in table 9.27.

If $V_R < 10$, there is strong interaction between the structure and its near wake. If $V_R < 1$, VIV is usually not critical

Table 9.27 Basic non-dimensional VIV parameters

Structural aspect ratio	$a^* = \frac{l}{D}$
Vibration amplitude ratio	$A^* = \frac{A}{D}$
Mass ratio (total mass includes hydrodynamic added mass)	$m^* = \frac{\text{total mass}}{\text{structure length}} = \frac{m}{\rho D^2}$
Damping ratio	$\zeta = \frac{c}{c_c}$
Reynolds number	$\mathrm{Re} = \frac{UD}{\nu}$
Strouhal number – related to the fluid	$\mathrm{St} = \frac{f_s D}{U}$
Reduced velocity – related to the structure	$V_R = \frac{\text{path length per cycle}}{\text{structure width}} = \frac{U}{fD}$

Vortex shedding regions may be checked on the basis of fig. 9.58. The figure suggests that for a fixed cylinder, the vortex shedding frequency is proportional to the fluid velocity. For a cylinder at the intermediate Reynolds number of $1.18 \times 10^5 < \mathrm{Re} < 1.91 \times 10^5$ and $\mathrm{St} = 0.2$, i.e. the vortex shedding frequency is unaltered. At the transition region for Re of 10^5–5×10^6 the shedding frequency has a scatter and is broad banded. Note that drag coefficient also dips in this range (drag crisis).

For large amplitude motion of cylinder, the shedding is correlated along the span and vortices become two-dimensional.

9.4.2 Simplified VIV Analysis

The VIV of riser may be investigated by a simple Wake Oscillator Model (fig. 9.59). For fixed cylinder or small amplitude motion, the vortex shedding along the cylinder span is uncorrelated (no fixed-phase relationship).

Equation of motion for the above model is written as

$$\ddot{y} + 2\zeta_T \omega_y \dot{y} + \omega_y^2 y = a_3 \ddot{w} + a_4 \dot{w} U/D \tag{9.35}$$

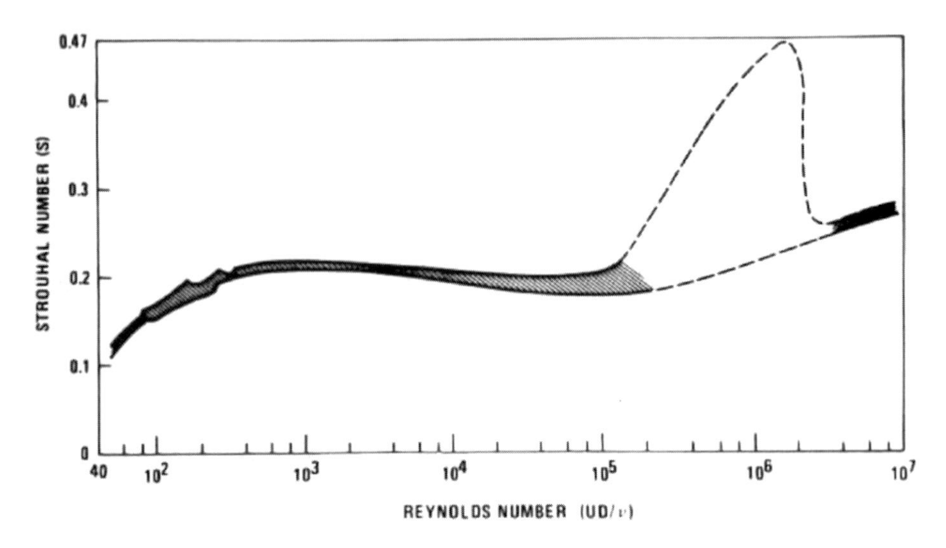

Figure 9.58 Reynolds vs. Strouhal number for a fixed circular cylinder

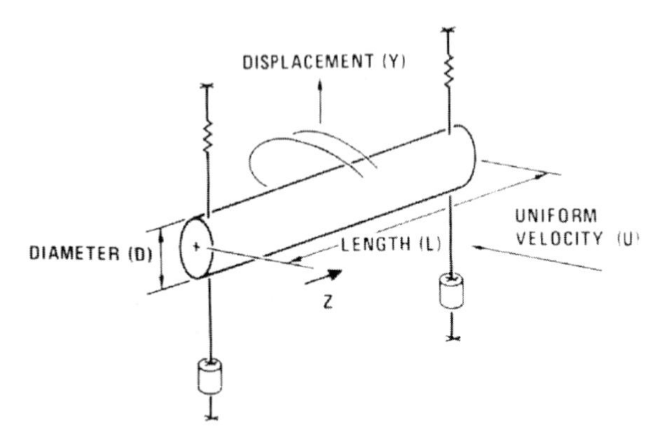

Figure 9.59 Wake oscillator model

ω_y = cylinder natural frequency, a_3, a_4 = dimensionless constant, $\dot{w}$ = transverse fluid flow velocity in the wake, $\ddot{w}$ = transverse fluid flow acceleration in the wake. The quantities $\dot{w}$ and $\ddot{w}$ are functions of the shedding frequency, which depends on U/D. Note that the fluid force on the right hand side is inter-dependent on the cylinder motion. As the natural frequency of fluid oscillation approaches the natural frequency of the cylinder, resonance occurs. (See Blevins, Flow-induced Vibration, pp. 25–32 for details.)

Solving equation (9.35), the amplitude ratio is given by

$$\frac{A_y}{D} = \frac{0.07\gamma}{(\delta_r + 1.9)\mathrm{St}^2}\left[0.03 + \frac{0.72}{(\delta_r + 1.9)\mathrm{St}}\right]^{1/2} \tag{9.36}$$

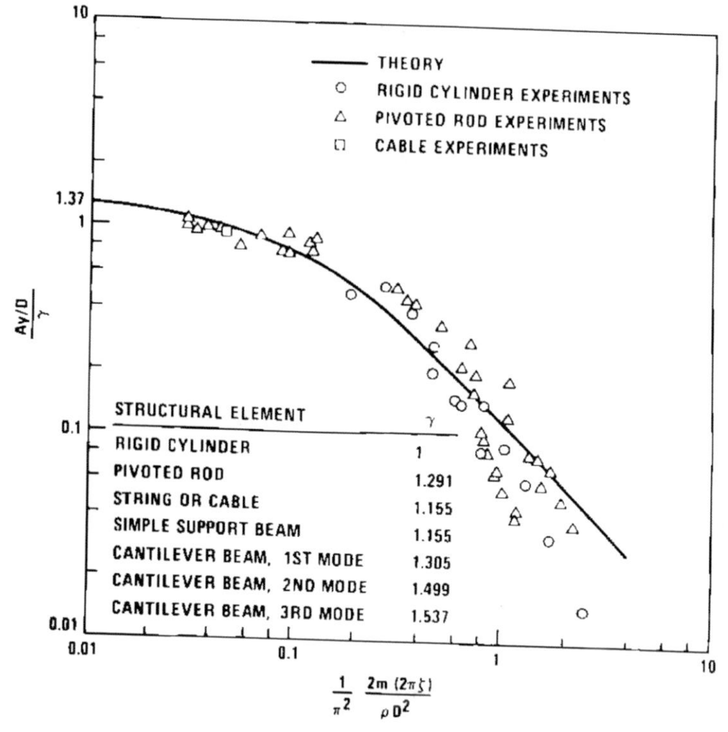

Figure 9.60 Amplitude ratio vs. reduced damping (applicable 200 < Re < 2×10^5)

where γ = shape factor (function of mode shape) and δ_r = reduced damping,

$$\delta_r = \frac{2m(2\pi\zeta)}{\rho D^2} = 4\pi m^* \zeta \tag{9.37}$$

The results are shown in fig. 9.60. The analysis shows that the amplitude ratio decreases with increasing mass ratio and increasing damping.

An FE analysis problem was run with a TTR in shallow water, in which the transverse envelope, maximum and minimum transverse displacements were computed for the riser subjected to uniform current. The results of the anaylsis are shouwn in table 9.28. It shows the predominant modes of vibration for various current speeds, the corresponding reduced velocity and frequency of vibration.

For the discussion purposes assume that the value of St = 0.2. Then, the Strouhal number realtionship, $St = f_s D/U$ where D is the riser diameter and U is the current velocity provides the vortex-shedding frequency f_s. For example,

$D = 0.25\,\text{m},\ U = 0.23\,\text{m/s}$ gives $f_s = 0.18$ vs. $f_r = 0.15$

$D = 0.25\,\text{m},\ U = 0.40\,\text{m/s}$ gives $f_s = 0.32$ vs. $f_r = 0.20$

Table 9.28 Current velocity, vibration mode, and reduced velocity and frequency

U (m/s)	Mode	V_r	f_r	U (m/s)	Mode	V_r	f_r
0.16	1st	4.51	0.22	0.54	3rd	4.00	0.25
0.23	1st	6.76	0.15	0.62	3rd	4.57	0.22
0.31	2nd	4.00	0.25	0.70	3rd	5.15	0.19
0.39	2nd	5.00	0.20	0.78	3rd	5.72	0.17
0.40	2nd	5.14	0.20	0.86	4th	4.03	0.25
0.47	2nd	6.00	0.17	0.93	4th	4.40	0.23

Therefore, in the first case, we have a VIV lock-in, while the second case shows that lock-in is avoided.

9.4.3 Examples of VIV Analysis

A typical example of VIV analysis is illustrated below.

An analysis by VIVANA for the deflected shape of an SCR is shown in fig. 9.61. The analysis results are compared with model towing test for a towing speed of 0.13 m/s. The Strouhal number was calculated to be 0.24 for this example.

9.4.4 Available Codes

There are many design codes available for the analysis of risers. A few of those are listed in table 9.29. The details of the capabilities of these codes may be obtained from their websites.

9.5 VIV Suppression Devices

Several types of vibration spoilers are used in the offshore industry. To prevent the VIV and lock-in, vortex suppression devices interrupt the regularity of the shedding and stop vortex streets from forming. In a test program at the US Navy facility with cylinders in steady flow, a fiberglass cylinder model was built with a super smooth ground surface. The tests in supercritical Reynolds number demonstrated the absence of VIV.

VIV of risers can cause high levels of fatigue damage but can be reduced using suppression devices such as:

- Strakes
- Fairings
- Shrouds

The typical cross-section of a streamline fairings, such as rudders, fins, etc. (taper ratio $>=6$ to 1) shown in fig. 9.62 is effective for VIV suppression of a marine riser. The slender

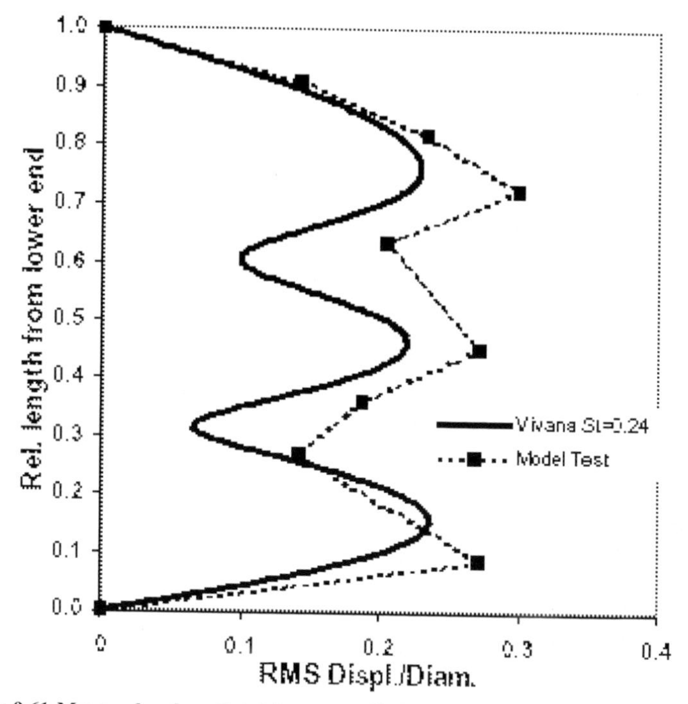

Figure 9.61 Measured and predicted transverse displacement of an SCR [Lie, et al (2001)]

Glasgow (UK) Science tower was designed in the shape of an aerodynamic foil and allowed to rotate 360° with the mean wind direction with the help of a turntable at its base.

One of the most common types of VIV suppression devices for the production risers is helical strakes (fig. 9.63). The width of the strake is typically about 10% of the cylinder diameter. Strakes generally increases the overall drag force as well as the hydrodynamic damping of the riser, which are counteracting for the motion of the riser.

In developing/designing a riser the questions to ask on VIV are:

- Is VIV a problem for the riser under the given environment at the site?
- If VIV is a problem, will an alternative riser design avoid the problem?
- If suppression is necessary, what is the best practical method available?

Analysis should account for effects of suppression devices on riser behaviour, via changes in weight and hydrodynamic coefficients.

VIV-suppression strakes are an incorporated design element in all SCRs (fig. 9.64). Various manufacturers offer these strakes. A contracting philosophy needs to be prepared before ordering these elements. Nominated strake manufacturers should have wet tank test results of their product design in a similar diameter application, which demonstrates their efficiency.

Table 9.29 Some available software for riser analysis

Software	Source	Description	Website
DeepVIV	IFP	VIVs as well as fluid-structure interactions in riser bundles	http://www.ifp.fr/IFP/en/researchindustry/
Skaas	Global maritime	Riser and station keeping advisory systems	http://www.globalmaritime.com/software/
NOBSystem		Analysis of floater motions and mooring-riser system response	http://www.name.ac.uk/research/off_eng/
ABAQUS			http://www.hks.com/
ORCAFLEX	Orcina Ltd.	Analysis of floater motions and mooring-riser system response	http://www.orcina.com/OrcaFlex/
Flexcom3	MCS International		http://www.mcs.com
	JP KENNY	Design and code checks; J-pull – J-tube analysis; riser calculation	http://portal.woodgroup.com/pls/portal30/url/page/external_jpkenny_home/techsoft
CAVIAR	DHI	Deepwater riser design product	http://www.dhi.dk/
DERP	Stress engineering services, Inc.	Frequency domain riser analysis program	http://www.stress.com/oilgas/riser_tech.htm
DEEPLINES	PRINCIPIA	Global analysis of risers, moorings and flowlines	http://www.principia.fr/principia-deeplines-eng.html
Mentor subsea risers	J. Ray McDermott	Induced vibration, soil structure interaction and buckling	http://www.jraymcdermott.com/mentor/mentor_risers.htm
	DHI	Numerical dynamic pipeline-seabed interaction	http://www.dhi.dk/consulting/offshore/pipelinesrisers/
Riser analysis	BPP	Prediction of riser displacements and stress status	http://www.bpp-tech.com/vertical.htm
Seaflex	DNV		http://www2.dnv.com/elnib/
FLEXRISER	ZENTECH	Static and dynamic analysis of flexible risers	http://www.zentech.co.uk/flexrise.htm
Shear 7	MIT	Riser analysis	email: kimv@mit.edu

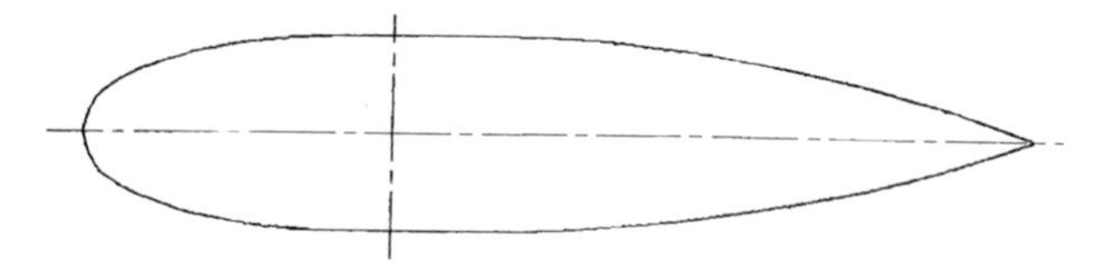

Figure 9.62 Streamline geometry

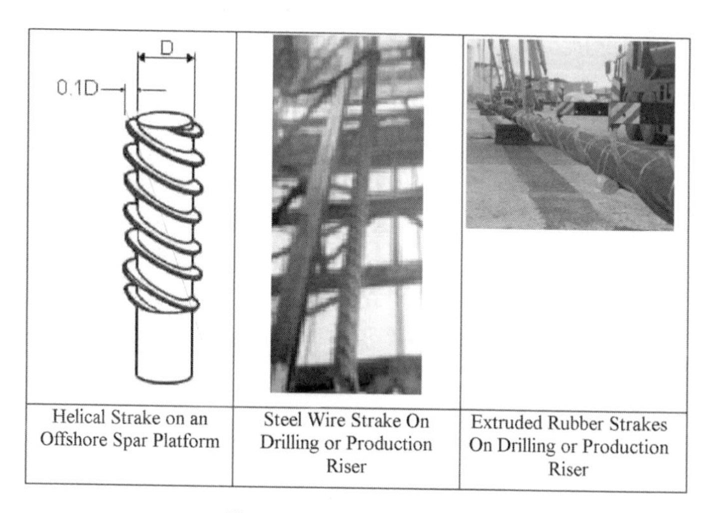

Helical Strake on an Offshore Spar Platform	Steel Wire Strake On Drilling or Production Riser	Extruded Rubber Strakes On Drilling or Production Riser

Figure 9.63 Strakes on risers

Figure 9.64 The Prince SCR during hang-off installation showing the pre-installed VIV strakes [Gore and Mekha, 2002]

Strake suppression efficiency (percentage reduction of motion amplitude compared to bare pipe) depends on pitch (P) and height (H). Common values are $P/D = 17.5$ and $H/D = 0.25$ (where D is hydrodynamic diameter, including insulation and strake shell, and H is height above this level). For these values a suppression efficiency of 80% may typically be assumed, in addition to an increased drag coefficient compared to the underlying bare pipe.

Strakes near the water surface may need to be treated against marine growth and strakes near the seabed may need to consider abrasion performance. The designer should consider various factors when planning to use strakes, including the following:

- Strake drag and lift coefficients
- Required coverage
- Alternative P/D and H/D values
- Strake and fairing suppression efficiency (including any Reynolds number effects)
- Performance of strakes (or fairings) in tandem

The strake suppliers and some consultants and operators now have performance data from model testing to address the above.

The performance of fairings is in some respects better (e.g. lower drag) but can present increased challenge in other areas, e.g. installability. However the field use of fairings as an alternative to strakes does appear to be increasing.

Both strakes and fairings can reduce the VIV induced motion, can reduce fatigue damage due to VIV by over 80%, will, however, introduce handling difficulties. Strakes increase riser drag, whereas fairings reduce drag loading. Fairings need to rotate with current direction and add to design complexity.

9.6 Riser Clashing

Riser deflections may need to be controlled to avoid collision with adjacent risers, umbilicals, moorings or the host vessel. Often a target minimum clearance is specified e.g. five times the outside diameter of the riser. If this criterion cannot be met the designer may elect to demonstrate that the probability of collision during field life is of an acceptably low probability (e.g. less than 10^{-4}) or demonstrate that collision can be resisted without compromising riser integrity. This logic may also apply to installation operations.

9.6.1 Clearance, Interference and Layout Considerations

Analysis should be conducted to confirm that interference with other parts of the production system does not occur. Interaction may occur between the following:

- Riser and vessel;
- Riser and riser;
- Riser and mooring lines;
- Riser and umbilicals.

The results of a clearance analysis can have an effect on the layout of the risers, umbilicals, mooring and orientation of the flowlines. The layout of the risers should also take into account the overall field layout, the requirement for discrete flowline corridors, anchor system prohibited areas, crane locations, supply boat loading positions and the trajectory of dropped objects. The designer should usually avoid collision among risers. But, if the layout is such that this ideal cannot be achieved, the cumulative probability of risers contacting other risers, umbilicals, mooring legs, the hull or any other obstruction during field life including installation may be assessed and compared to some target value (e.g. 10^{-4}) as well as resistance to consequence damage.

A model test of risers in a deep-water fjord was performed to investigate riser collision [Huse, 1996]. The test site was chosen at Skarnsund, 100 km north of Trondheim. The sound has a water depth of 190 m, and tidal currents well above two knots. An existing bridge spanning the sound was used as the work platform. A set of riser models were suspended from a surface catamaran with a weight attached to their bottom end, and supported by a pulley system to introduce the desired tension in the risers.

The riser group consisted of an array of risers in a 3×4 rectangular arrangement (fig. 9.65) with equal spacing. One riser in the middle of the array represented a drilling riser, while the others were smaller diameter production risers. The array represented a riser system for a Tension Leg Platform. The spacing at the top and bottom end among the risers were maintained at equal distances in the inline and transverse directions.

The drilling riser had a pretension of about 1205 kN, while that of the production risers was varied from 412 to 862 kN for two test conditions. Several tests were performed at different current velocities and shear type profiles encountered at the site. At low current velocities, no collision of risers was observed. As the current velocity increased, the collision between neighbouring risers was initiated and the frequency of collision increased with the increase in the magnitude of current velocity.

Vortex induced vibration increases the mean inline drag force, causing large static deflection in the middle of the risers. This, in turn, induced collisions between the

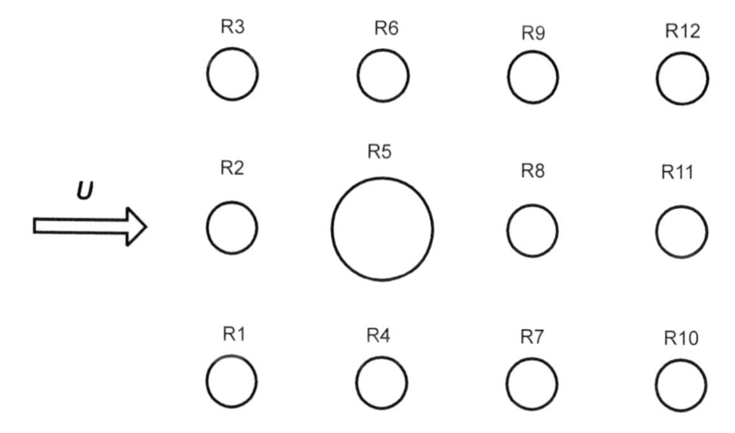

Figure 9.65 Setup of riser array in the Fjord

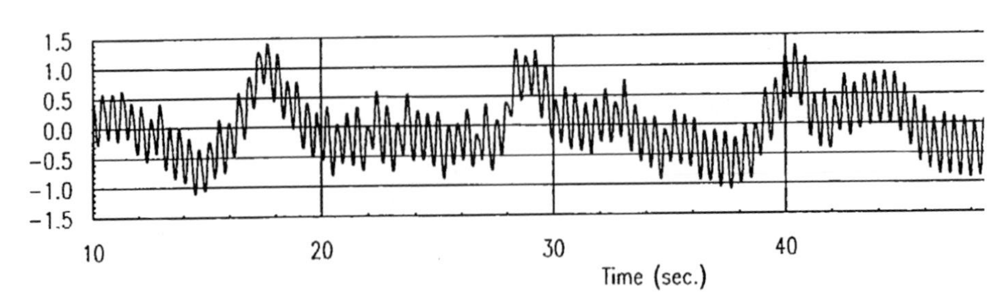

Figure 9.66 Displacement time history of drilling riser [Huse, 1996]

neighbouring risers. The collision generated a loud audible noise indicating a collision between the risers.

The displacement time history shown in fig. 9.66 shows that the drilling riser experienced a clear evidence of lock-in vibration at its natural frequency. The VIV amplitude was about half the riser diameter. Additionally, the risers experienced an irregular low frequency inline oscillation of large magnitude, almost of a chaotic nature. The peak-to-peak amplitudes of these motions were as much as 30–40 diameters. Typically, the far upstream risers remained stationary. The next riser collided with the upstream riser and then moved far downstream in a slow motion before returning upstream and colliding again with the upstream riser. This situation arose at or above the collision velocity of current. In a practical design, of course, it is undesirable to have collisions and they should be avoided in a design. Thus, the low frequency oscillation of the intermediate risers, while of interest, should not arise in a properly designed spacing of a riser system.

9.7 Fatigue Analysis

Fatigue damage verification is an important issue in riser design, demanding a high number of loading cases to be analysed. The random time domain non-linear analysis is considered an attractive and reliable tool for fatigue analysis, as non-linearities are properly modelled and the random behaviour of environmental loading is considered. However, time domain analysis consumes large computer time. The frequency domain analysis is considered an efficient alternative tool for the initial phases of riser design used mainly for a fatigue damage verification.

Riser fatigue analysis is conducted using a stress–cycle (S–N) approach. The equation used to determine fatigue life of steel components is:

$$N = K \cdot S^{-m} \tag{9.38}$$

where S = stress range (MPa), including the effects of stress concentration due to misalignment, but excluding that due to the weld itself, N = the allowable number of cycles for the stress range and K and m = parameters depending on the class of weld/constructional detail.

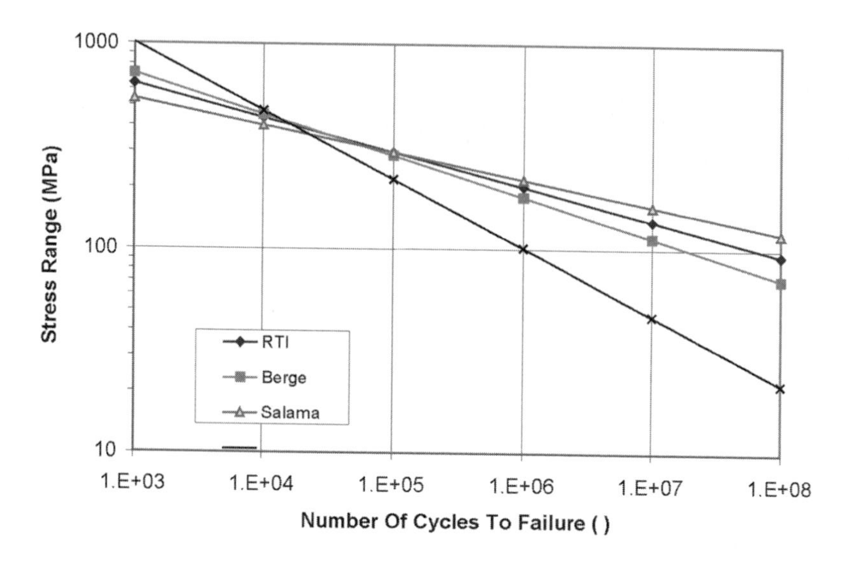

Figure 9.67 Titanium S–N curves

Table 9.30 Basic parameters defining fatigue curves for steel in air

Class	K	m	Reference
X	1.15×10^{15}	4.38	API RP 2A-LRFD (1993)
X′	2.50×10^{13}	3.74	”
B	5.73×10^{12}	3.00	HSE: 1995 Offshore Guidance Notes
C	3.46×10^{12}	3.00	”
E	1.04×10^{12}	3.00	”
F	6.30×10^{11}	3.00	”
F2	4.27×10^{11}	3.00	”

For titanium alloys such as Grades 23 and 29, the following S–N curve [Baxter, et al 1997] is widely applicable for good quality girth welds,

$$N = 6.8 \times 10^{19} \cdot S^{-6} \tag{9.39}$$

The S–N curves for titanium are shown in fig. 9.67. The choice of fatigue design curve will depend on many factors specific to a particular design, construction detail, materials, and the level of conservatism desired. UK HSE Guidance (1995) is given in table 9.30 below for steel in air.

Adjustments may be required to fatigue curves such as those above to account for the endurance-limit effect at low stress/high cycles in air, cathodically protected joints in sea-water: and freely corroding joints in sea-water.

Other parameters that may affect riser fatigue are thickness, mean stress correction for unwelded or stress-relieved components, stress concentration factors (SCFs), and temperature. Based on published codes and standards it is recommended that for thickness T greater than 25 mm the DNV (2000) correction of $(25/T)^{0.2}$ should be applied to the design (or allowable) stress-range obtained from S–N curves. A value of 1.3 may be assumed in the absence of more detailed information, although SCFs as low as 1.1 have been achieved for some risers.

Fatigue damage in risers comes from three main sources:

First-order wave loading and associated vessel motions
Second-order/low frequency platform motions
Riser VIV due to current or vessel heave (see Section 9.3.1.5 for comments on the latter)

Stresses due to 1 and 2 may in some cases be combined prior to calculating fatigue. At the present time it is not considered necessary to combine a riser VIV stresses with these, although that is possible in principle and would be the most accurate approach.

Second-order effects are sometimes larger than first-order effects. Also, it is pointed out in Campbell (1999) that introducing second-order effects does not necessarily increase or necessarily decrease fatigue life. An example shows a reduced life (compared to the case where only first-order fatigue is calculated) for a spar-mounted SCR but an increased life for a semisubmersible-mounted SCR.

Additional fatigue may accumulate from vessel VIV, slugging, pressure pulses and installation. The fatigue calculation methods use the above stress-cycle (S–N) approach. Fracture mechanics analysis may also be applied.

The following methods are possible (among others) for obtaining the combined fatigue effects of 1 and 2:

1. Rainflow Counting (RFC) of stress from a combined (wave-frequency + low-frequency) analysis. The most accurate method for any stress time-history, such as output from most riser analyses, requires specialist software and uses more computer time than alternative methods, but is nowadays fairly widely used. Simpler methods may be better for rapid turn around of results; e.g. early feasibility checks or parameter studies.
2. Assume a Rayleigh distribution for the stress peaks from a combined analysis. This overestimates fatigue damage significantly unless stress is highly narrow-banded. There are potential ambiguities in counting the cycles as the response becomes more wide-banded.
3. Use a bimodal method. This still overestimates damage but less so than the Rayleigh assumption, and it is quicker than RFC. A method by Jiao and Moan (1990) is valid when bimodal peaks of the stress spectrum are distinct and well separated. The method

can be used under the right circumstances, but is cumbersome and requires continual checking of the spectrum.

4. Separate wave-frequency and LF stresses. The damage for each frequency region can then be calculated assuming a Rayleigh distribution, and these are summed to get the total damage. This method usually underestimates damage, sometimes significantly.
5. As in 4, but factor the result by $(\Sigma S_i)^m/\Sigma S_i^m$, where S_i are individual stress process rms's. Theoretically, this is a somewhat crude correction, but in practice it often works fairly well. However, no attempt is made to correct for the different upcrossing-rates of the different stress processes, which can lead to serious error.
6. A number of investigators have developed correction factors to the Rayleigh approach; e.g. Wirsching and Light (1980), Ortiz and Chen (1987), Lutes and Larsen (1990, 1991). The most accurate and most easily applied of these methods is the single moment method of Lutes and Larsen (note, however, that the spectrum of stress is required, which may require specialist post-processing software, depending on the riser analysis program which has been used).

One view on the order of preference is

(i) RFC,
(ii) Rayleigh or other method with similar or better accuracy,
(iii) Lutes–Larsen.

In a single moment method of Lutes and Larsen (1990, 1991) the fatigue damage expression given involves one moment of the spectral density function and can be written as follows:

$$\overline{D} = \frac{T}{2\pi K}\cdot\left(2\sqrt{2}\right)^m\cdot\Gamma\left(\frac{m}{2}+1\right)\cdot\left(\lambda_{2/m}\right)^{m/2} \tag{9.40}$$

where T is duration, and K and m are the parameters of the S–N curve defined by equation (9.40). The single moment in the fatigue damage equation is

$$\lambda_{2/m} = \int_0^\infty \omega^{2/m}\cdot G(\omega)\cdot d\omega \tag{9.41}$$

where $G(\omega)$ is the spectral density function of stress-range and ω is frequency in rad/s. This method requires no more effort than the Rayleigh method, but the results are generally more accurate, approaching the accuracy of direct RFC for practical purposes.

It is recognised that many factors influence the selection of a method, including the domain and format of riser stress data, available software, available time, the relative importance of different terms and the required accuracy at a particular stage in a particular project. However, as a design moves in to final detailed design, there will be a strong expectation that RFC will be used, unless comfortable margins of safety are demonstrated.

The use of combined stresses; i.e. LF and wave-frequency components calculated in the same dynamic analysis, is preferred, and the level of accuracy should be commented on in all cases. Other methods are possible. For example, regular wave analysis may be sufficient in some cases, especially where fatigue is not a governing criterion; it may also enable more rapid design evolution.

Similarly, although time-domain analysis is generally regarded as essential for extreme and confirmatory assessment of riser fatigue, enhanced frequency domain analysis may have a part to play in feasibility studies, parameter studies and fatigue estimation. This is especially true for deepwater risers, where large regions are not subject to grossly non-linear structural response and where accurate random time-domain analysis can be time-consuming. In these cases RFC is not applicable and the Lutes–Larsen method may see greater use.

For fatigue analysis it is usually assumed that the riser is installed and operating. Fatigue life is influenced by many factors, and the designer has many techniques at his disposal, for example:

- Use of thick-end forging (increased thickness and better S–N curve)
- Use of project-specific S–N curves, generated by a dedicated test program
- Refinement of current profiles through further analysis or site measurements
- Inclusion of wave spreading
- Increased wall thickness in TDZ
- Use of auxiliary buoyancy in TDZ
- Optimisation of hang-off angle
- Use of lazy-wave rather than free-hanging configuration
- Review and refinement of inertia coefficient (e.g. if straked pipe is used)
- Review of structural damping coefficient used in analysis

The relative importance of the parameters varied depends on numerous factors, including geographical location and vessel type.

9.7.1 First- and Second-Order Fatigue

There are a number of methods available for conducting fatigue analysis of SCRs and the more reliable methods require more computational time and effort. The most important considerations are to include all the relevant sources of fatigue loading and to account correctly for the interaction of first- and second-order contributions. Two example methods for dealing with first- and second-order fatigues are discussed below. The second approach, rainflow counting applied to the combined response, is probably the most accurate.

Selecting which method to use depends on a number of factors, such as the required level of detail, design stage, type of vessel, and whether or not a wave scatter diagram is available. Other approaches and variations are possible, including cruder but quicker regular wave analysis.

The earlier discussion on preferred methods for estimating the statistics in specific sea-states provides input to the example methods below.

Methodology 1: Add Separately Calculated First and Second Order Random Fatigue Damages

First-Order Fatigue

- Discretise wave scatter diagram into linearisation windows, as in fig. 9.68.
- Select sea-state from each window, to give equal or greater damage than for original sea-state.
- Use selected sea-states in non-linear time-domain analysis, with associated mean offset.
- Combine tension and bending to obtain total stress.
- Fourier analysis to get stress RAOs around circumference for each window, as in fig. 9.68.
- Apply statistics (e.g. Rayleigh distribution) to obtain damage due to each sea-state in window.
- Multiply damage by probability of occurrence and sum for all sea-states in window.
- Repeat for each window.
- Repeat for other loading directions and the sum for total damage.

Second-Order Fatigue

- Discretise scatter diagram into windows or analyse every sea-state, depending on required level of detail.
- Conduct quasi-static riser analysis using second-order vessel motions.
- Determine RMS stress response in each case.
- Apply statistics (e.g. Rayleigh distribution) to obtain damage due to each sea-state.
- Multiply damage by probability of occurrence and sum damage for all sea-states.
- Repeat for required number of loading directions and sum for total damage.

Combining First- and Second-Order Fatigue

- Sum the first- and second-order damages at each point on circumference and along the riser length.

A variation on this approach, which allows greater flexibility to use the methods already discussed is to calculate the total (first- plus second-order) damage in each sea-state before applying the probabilities. When the preferred approach (RFC) is used in conjunction with this variation, the analysis is essentially the same as the second example methodology, given below.

Methodology 2: Apply Rainflow Counting to a Combined First- and Second-Order Random Response

- If necessary, condense the scatter diagram to manageable number of "bins" (say, 10–20).
- For each bin, apply mean offset and conduct non-linear time-domain analysis with vessel second-order motions included.

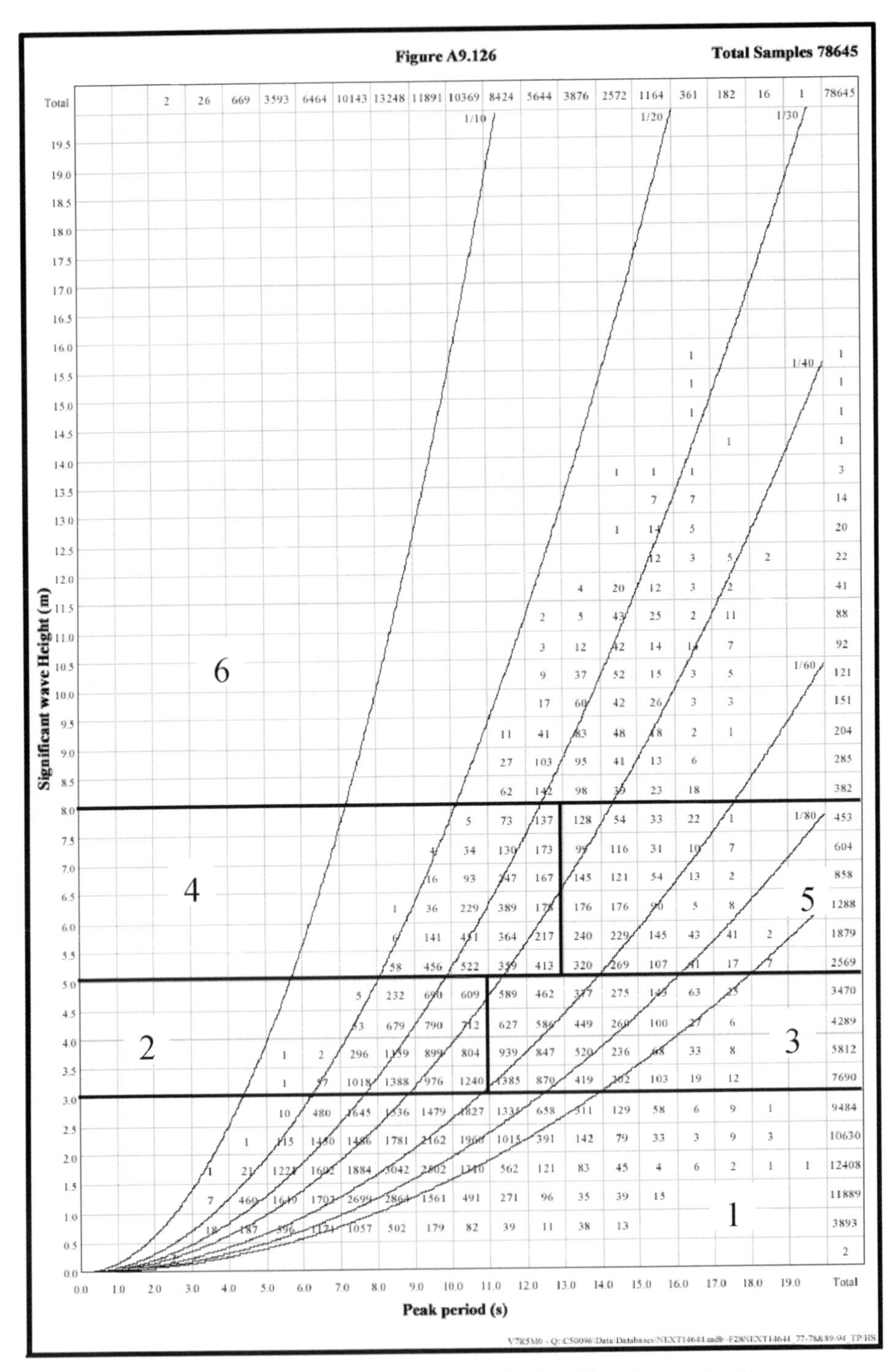

Figure 9.68 Example windowing and sea-state selection of long-term scatter diagram

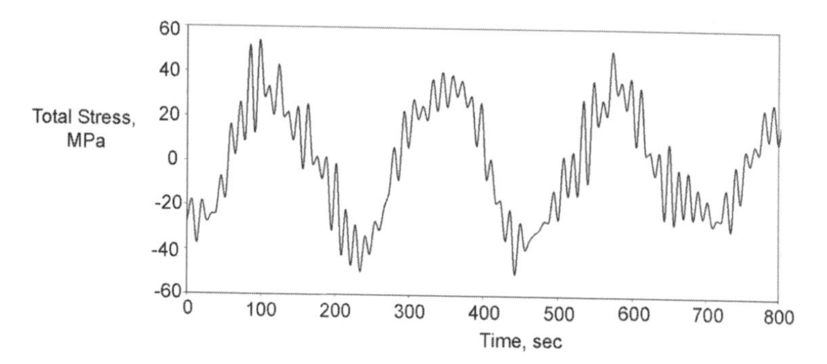

Figure 9.69 Combination of HF and LF narrow-banded Gaussian processes

- Combine tension and bending to obtain total stress (fig. 9.69).
- Rainflow count total stress time traces to get fatigue damage due to each bin at points around the circumference and along the riser length.
- Multiply damage by probability of occurrence of bin and sum over bins.
- Repeat for required number of loading directions and sum for total damage.

As for all random analyses, convergence of statistics needs to be understood and checked. In this method, use of a minimum of ten low-frequency cycles to achieve meaningful results is one rule of thumb, though this should not be taken as a substitute for checking.

9.7.2 Fatigue Due to Riser VIV

To estimate long-term riser VIV fatigue damage:

- Establish current data. Normally, at least ten profiles are required, and directional variations should be included. If available, concurrent data; i.e. actual profiles, are preferable to exceedance profiles.
- Conduct VIV analysis using a suitable VIV analysis tool. The nominal (or neutral) riser configuration may be used, but this is not essential.
- Factor calculated fatigue damage in each current according to the probability of current and sum of all such damages to obtain total damage and hence predict fatigue life.

Sensitivity analyses may be conducted in which currents and riser configuration are varied. Justification and a methodology for spreading (or smearing) fatigue damage in the TDZ, based on the fact that the TDP and riser system properties will vary over time, is given in Section 9.2.5.2.2.

VIV fatigue in risers is commonly assessed using dedicated software such as SHEAR7 or VIVA. It should be noted that there are other prediction tools available, such as

VIVANA and Orcaflex. The tools chosen for discussion in this section should not be taken as any form of recommendation, rather as typical examples.

Most VIV programs allow input of only 2D current, although advances are expected in this area. As a general rule, for a SCR, resolution of velocity on to planes parallel with and perpendicular to the plane is acceptable. It is assumed that an initial VIV fatigue analysis is performed (e.g. a modal analysis using SHEAR7) where the vessel is in the neutral position. Apart from the VIV, no dynamic forces or motions are accounted for in this initial analysis. Under these assumptions it is found that the predicted fatigue damage in the TDZ peaks sharply at anti-nodes of the calculated mode-shapes, where curvature and bending stress peak. This results in large fluctuations in overall predicted fatigue life between anti-nodes, the extent of this effect depending on which modes, and how many, are mobilised.

In reality, riser system properties and boundary conditions will vary continuously. The TDP will move under the influences of vessel motion and direct hydrodynamic loading on the riser, and the riser mass will change for various reasons over various time-scales. This means that mode-shapes will also be continuously changing, and the locations on the riser of the modal anti-nodes may move around significantly. The effect of this will be to tend to even out peaks and troughs in the calculated damage curve. If this region governs the fatigue, then the true life of the riser will be greater than that predicted by the "constant riser system" assumed in the initial VIV analysis.

Reasons for variation of riser (e.g. SCR) system properties and boundary conditions are numerous, and include both short-term and long-term effects over the design life; e.g.

1. Wind, second-order wave loads and varying current introduce low-frequency vessel motion and affect mean vessel offset, causing the location of the TDP to change.
2. Variation of current force applied directly to the riser will also move the TDP.
3. Vessel draught and tidal variations will move the TDP.
4. Depending on the field development plan, vessel drilling offsets may be applied over a substantial period, and risers phased in at a later stage may impose incremental offsets.
5. Density of riser contents may vary. Short-term density variations in production risers may arise from variable well fluids and conditions. There may also be long-term variations as a reservoir becomes depleted and the composition of both produced and exported fluids changes. Even if these variations are small, they may be sufficient to shift natural modes and frequencies enough to have an important effect on fatigue peak locations.
6. Riser mass may depend on several long-term effects; e.g. corrosion and water absorption in auxiliary buoyancy.

It is emphasised that this list is not exhaustive and that not all of these effects need to be considered in every analysis.

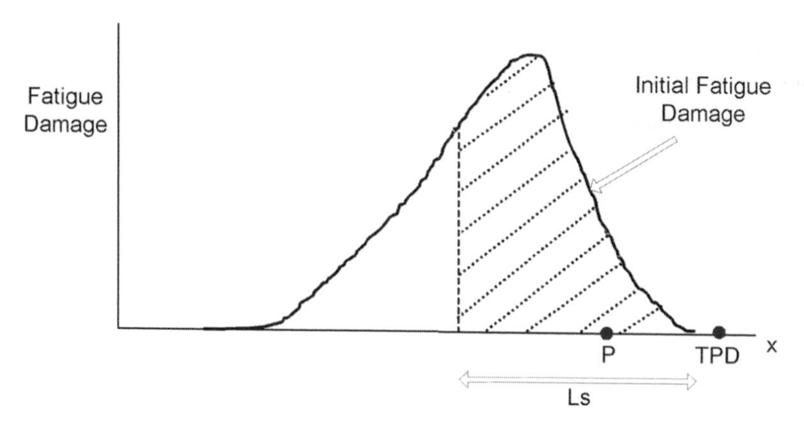

Figure 9.70 Approximate fatigue calculation

Effects on SCR TDP boundary conditions and response may also arise from trenching, suction and other soil-related phenomena; and the way the TDP is modelled in the VIV analysis can be crucial. However, whilst important, these are considered to be the aspects of detailed structural modelling which should be addressed elsewhere.

Comprehensive statistical treatment of all influences on the fatigue damage distribution is possible but will normally be unnecessary. The excess conservatism of an assumed constant riser system should be avoided, however, although it is possible to make reasonable allowances without performing an unduly complex analysis. The preferred approach depends on specific risers, field development plans, available software and individual company practice. In some cases it may be considered necessary to perform a separate VIV fatigue analysis for numerous variations from the neutral configuration, to cover all scenarios in 1–6, above.

In general, however, the depth of analysis required to get the right balance of accuracy, conservatism and economy will vary. One simple approximation, which may either be useful as a preliminary check or give sufficient confidence in itself, is:

(a) Determine a characteristic movement along the riser of the anti-node nearest to the TDP, allowing for all effects, such as those stated above. This is denoted L_S.
(b) For each point P in the region of the TDP, take the fatigue damage as being that calculated from initial VIV analysis, averaged over a distance L_S, centred on P. This may be described as a "moving average" calculation. It applies to all points around the circumference, although the averaging is performed in the lengthwise sense, only. The essentials of this calculation are illustrated in fig. 9.70.

L_S can be determined from,

$$L_S^2 = \sum L_{S,i}^2 \qquad (9.42)$$

where $L_{S,i}$ is a characteristic movement of the anti-node due to the *i*th effect acting in isolation, and it is assumed that all effects are uncorrelated. There is some freedom in the choice of the $L_{S,i}$, each of which is some representative value of a random variable. But it is suggested that a value of two standard deviations of the amplitude of movement will ensure that benefits are realised, whilst a degree of conservatism is maintained. Correlation between the various effects and use of more realistic distributions can be incorporated into the analysis, if the information required to do this is available. However, this may complicate the analysis considerably without yielding a great improvement in results. One relatively simple adjustment which could be reasonable in some cases is to assume Gaussian behaviour and weight the initial fatigue damage distribution accordingly (i.e. instead of using the uniform distribution implied by step (b) above) but this approach is not assumed here.

It is possible that only a single value of L_S will be required, applicable across all initial VIV fatigue analyses. However, if currents from different directions make significant contributions to fatigue damage, it may be necessary to use more than one value of L_S – each in conjunction with results for the corresponding current direction and associated probability. Also, a situation may arise in which the initial VIV analysis is not performed for a single neutral position but for, say, two configurations, near and far. It is not possible to anticipate all such scenarios, and judgement and adjustment must be made on a case-by-case basis. Ultimately it is the responsibility of the contractor to identify key influences and account for them appropriately.

In any event, it is recommended that sensitivity checks be performed to determine how much the anti-nodes of typically excited modes move under the influence of effects like 1–6 above.

In addition to first- and second-order fatigues and riser VIV, other possible sources of fatigue damage are vessel VIV, vessel springing, and internal fluid effects, such as slugging and pressure surges. For example, vessels with cylindrical sections subjected to current loading may oscillate due to vortex shedding; e.g. spars (usually straked to reduce this effect) and other deep draft floaters.

Fatigue also depends on riser/seabed interaction. Trenching, suction and seabed consolidation will also have an effect on fatigue. This topic has been the subject of several recent industry research initiatives.

9.7.3 Fatigue Acceptance Criteria

It is necessary to determine overall fatigue resistance, accounting for each relevant effect, which may include:

- First- and second-order loads and motions
- Riser VIV
- Vessel VIV
- Other effects such as slugging, pressure surges

Issues to be addressed when combining fatigue damage are correlation, stress amplification, and interaction. Correlation refers to the fact that (for example) riser fatigue is due to wind and wave effects may not be related to current induced fatigue, such as riser VIV. Fatigue due to slugging may occur at any time. Stress amplification refers to the effect of two or more loading regimes occurring in combination, for example, first-order wave loading and riser VIV. The resulting fatigue damage is greater than that calculated from treating the two separately and adding the damages. This effect is most significant when damage rates are of a similar order of magnitude. Interaction between loading mechanisms may reduce the effect of stress amplification; e.g. wave-induced riser response may disrupt riser VIV.

With due consideration to these and other uncertainties inherent in riser fatigue prediction, the designer should select a safety factor to apply to fatigue life predictions. The choice of safety factor will depend on many factors. Typical ranges applied are from 3, for non-critical applications where in-service inspection is planned, to 20 to applications with increased uncertainty (e.g. VIV) where inspection is not possible. The choice of safety factor(s) should be made in conjunction with the end-user.

The fatigue damage components predicted from all effects are accumulated to arrive at the total damage at each location on the riser, which must satisfy:

$$1 \Big/ \sum S_i D_i > \text{Design Life} \tag{9.43}$$

where S_i = safety factor and D_i = annual fatigue damage for the ith effect. The sum should include damage arising from all effects; e.g. first- and second-order, various types of VIV, installation and pressure surges. In calculating the D_i, allowance should be made for the duration of each effect throughout the year.

9.8 Fracture Mechanics Assessment

Fracture mechanics (FM) analyses may be used to develop flaw acceptance criteria. The FM analysis is very useful not only in controlling fatigue limiting cracks, but also provides guidance for selecting appropriate weld inspection techniques, as well as reducing the number of welds needing to be cut-out and replaced.

The fracture mechanics analysis usually consists of three steps, which are discussed below:

1. Engineering Critical Assessment (ECA) of the riser body
2. Paris Law fatigue analysis
3. Acceptance criteria development

Development of stress histograms for input to FM analysis depends on data available from riser dynamic analysis, and may use a recognised cycle counting scheme [e.g. as in ASTM E1049-85 (1997)] or conservative distribution (e.g. Rayleigh curve, based on combined LF and HF dynamic analysis). This is analogous to determination of stress distributions for use in S–N fatigue analysis. For VIV fatigue a Rayleigh stress-range distribution is often considered suitable regardless of the number of modes responding.

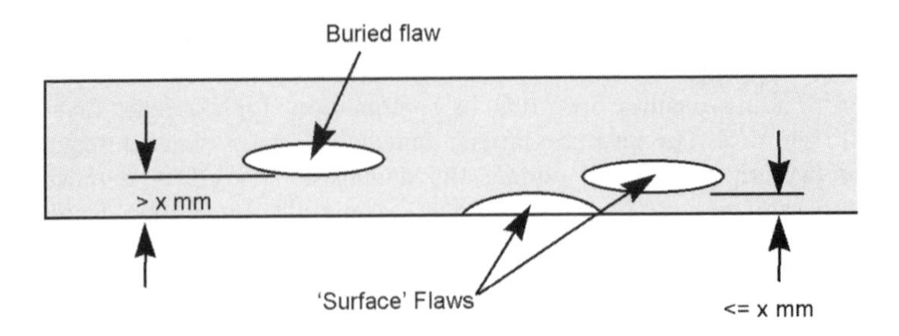

Figure 9.71 Flaw characterisation

9.8.1 Engineering Critical Assessment

The ECA is typically performed using industry accepted practices such as EPRI, CTOD method, or more rigorous analyses such as the R-6 method. SCRs to date have typically been assessed using PD-6493 (1991) or BS7910 (1999) methods. These methods allow for material behaviour ranging from brittle fracture to plastic collapse of the cross section. However, most modern materials with good ductility are often best characterised by nonlinear fracture mechanics, which is well treated using the Failure Assessment Diagram (FAD) approach.

The analyses should result in an envelope of limiting crack sizes which cause failure under the expected extreme event (e.g. 100-yr return period hurricane) for a particular system (e.g. TLP, SPAR, etc.) and worst loading condition.

Material and weld specific CTOD, measured at −10°C or lower, should be used, if available. Codified default values may be assumed. However, the designer should realise that these values might be far from representative depending on the weld process and inspection techniques employed. Material yield and tensile strength should be measured for the parent and weld metal, as well as, for the heat-affected zone. Conservative values should be used properly to account for the weld/parent metal mismatch. The BS7910:1999 Level 2 FAD is appropriate for an initial riser ECA. If material specific ductile tearing data is available, then the Level 3 approach (J_R) may be used. Care should be taken with the Level 3 approach since very large limiting flaws may result.

Cracks are usually assumed to be elliptical for analysis purposes. Surface breaking, buried, and interacting flaws should be considered. An idealisation of the elliptical surface and buried flaws is shown in fig. 9.71. Note that in some cases, the uncracked ligament of a buried flaw may be so close that it is re-characterised as a surface flaw. Refer to PD 6493 (1991) for guidance on values for "x" in fig. 9.71.

Stress intensity factors must be chosen so that the analytical solution accurately mimics the cracked pipe. In many cases, flat plate solutions provide sufficiently accurate results. However, for cases where the crack length and depth are not small with respect to the pipe circumference and wall thickness, the far-field uniform stress plate solutions may be inaccurate. Moreover, thin shells with outer to inner radii greater than 0.8 need curvature correction factors [refer to PD6493 (1991) for guidance].

9.8.2 Paris Law Fatigue Analysis

The so-called Paris Law for fatigue is described using

$$da/dN = A(\Delta K)^m \tag{9.44}$$

where da/dN = crack growth rate of crack of depth a vs. the number of applied stress cycles N, ΔK = stress intensity factor range, while A and m are material specific constants. BS7910 (1999) provides recommended values for the Paris Law, which should be suitable for the fatigue analysis. Material specific data obtained from tests are relatively inexpensive and may be used in-lieu of codified data.

If idealised stress intensity factor solutions are utilised (e.g. smooth plate solutions) in lieu of the finite element fracture mechanics analysis of the actual geometry, then relevant stress concentration factors should be applied to the stress range bins to account for increased applied stress due to local weld geometry, pipe mismatch, etc.

9.8.3 Acceptance Criteria

The industry has typically followed an approach similar to the schematic in fig. 9.72. The approach has been to develop curves showing an envelope of elliptical cracks (edge or embedded), which may grow to the limiting flaw size (see above) in a specified time. The "specified time" is usually established as a safety factor multiplied by the design life. Deciding the safety factor is subjective, but must take into account the type of inspection used during weld fabrication. Additionally, the safety factor should reflect uncertainties in predicted loads (see, also, Section 9.7 on safety factors).

9.8.4 Other Factors To Consider

Some of the other factors are listed below:

- Internal Contents: crack growth may be accelerated in H_2S or other corrosive conditions
- Cathodic Protection: crack growth is dependent on the level of corrosion potential protection
- Hydrogen embrittlement from welding
- Plastic straining (for reeled risers)
- Internal pressure effects on crack growth

9.9 Reliability-Based Design

Reliability-based design is becoming more common in pipeline engineering and other areas of the offshore industry. Its application to risers is limited at this time. Particularly,

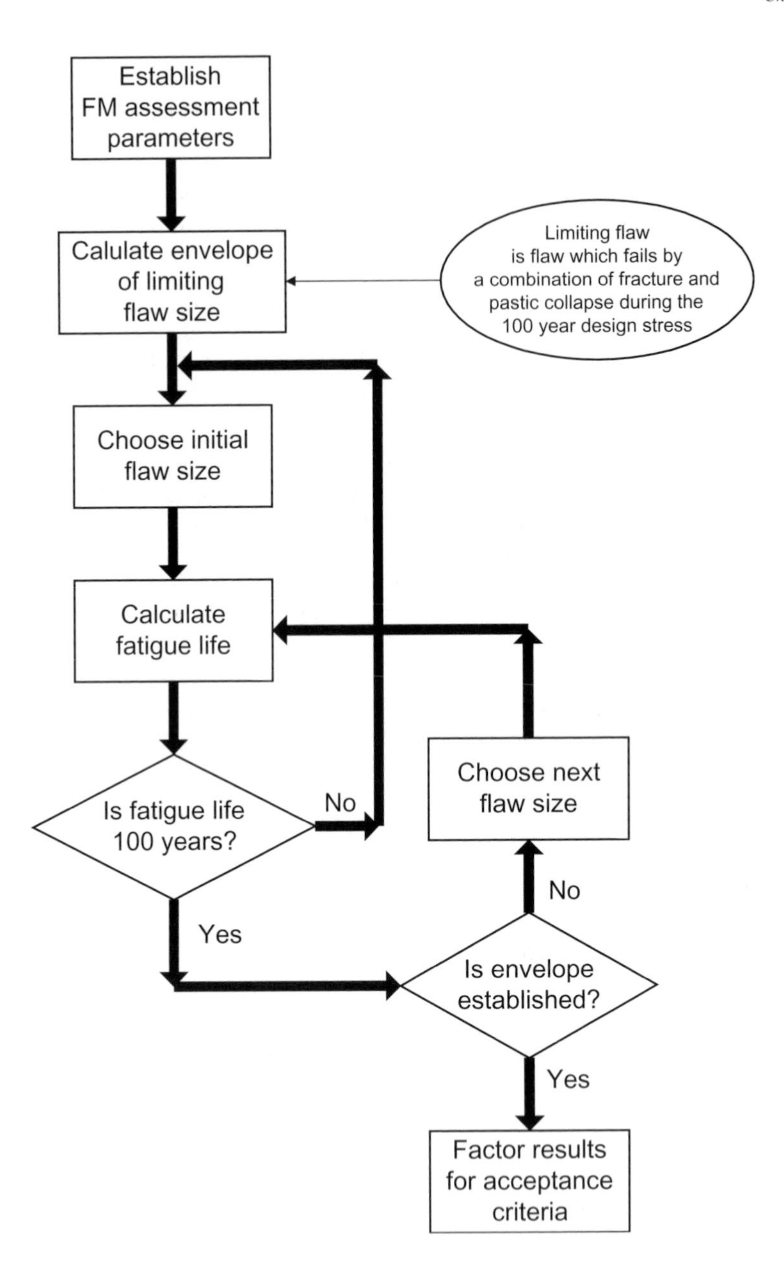

Figure 9.72 Schematic of example acceptance criteria development procedure

the deepwater SCR design for floating vessels is relatively a new technology. Hence it may be some time before the sufficient statistical data is available on SCR. However, procedures to determine component and system reliabilities have been investigated as part of the Integrated Mooring and Riser Design JIP, and are described in a Technical Bulletin (1990). A major step forward is also provided by DNV's "Dynamic Risers" which provides an LRFD format with reliability-based calibration of partial safety factors.

Development of long-term response distribution and comprehensive reliability assessment is possible but far from being standard analysis for risers. Nevertheless, limited methods and examples have been demonstrated for flexible risers [Farnes and Moan, 1993; Larsen and Olufsen, 1992; Trim, 1992] and more recently for an SCR hung off a ship-shaped vessel in the GOM [Gopalkrishnan, et al 1998].

The key advantage of the reliability/long-term methods is their consistency; i.e. the fact that exceedance probabilities are used which account for all environmental conditions arising in the long-term. This is exemplified in Corr, et al (2000), which reports 100-yr responses 20% lower than those obtained using conventional combination of collinear 100-yr wind, wave and current conditions. In this method the joint statistics of environmental inputs were developed and combined with results of representative dynamic simulations to produce a response-surface (a response which is a function of several environmental variables).

It should be cautioned that use of such methods cannot be assumed to always result in reduced response predictions, as that depends on the "conventional" methods to which they are compared. Nevertheless, their consistency and resulting high levels of confidence point the way to safer and more economic design.

9.10 Design Verification

The purpose of design verification is to provide the designer with an independent review and confirmation that the design adequately addresses the key issues outlined below:

- Functional and operational requirements in the client's specifications and documentation.
- Structural integrity.
- Resistance to fatigue and other forms of long-term degradation.
- Stable overall configuration; no detrimental interference with adjacent risers, umbilicals, moorings.
- Compatibility with fluids being transported.

The design verification process should also include riser appurtenances. In cases where the installation process results in significant effects on the riser; e.g. reeling, it will also be appropriate to include the installation operations, limits and contingency procedures in the scope of the review. The process is one of the confirmation for the client/designer and is not intended to replace the more formal independent design review required by certifying authorities.

It is appropriate when addressing the state of the art technology applied to critical equipment to consider two levels of design verification: (i) a systematic review of key documentation – specifications, design bases, design reports – to confirm the adequacy of the design process and documentation; (ii) an independent analysis of selected key load cases, preferably by a consultant with access to different analytical software to that used by the designer.

Sources of uncertainty, as far as the current SCR design is concerned, include compression in the TDZ, riser/soil interaction and riser fatigue due to VIV. Model testing to confirm key design issues and assumptions may be considered as part of Design Verification, particularly where assumptions relate to safety-critical features of the design. It should be realised that the modelling process has fundamental shortcomings when used to address the behaviour of an integrated riser/host/mooring system in that the model scaling requirements for different parts of the system cannot be satisfied in a single model.

9.11 Design Codes

The main design codes and standards used for riser design are:

- "Recommended Practice for Design of Risers for Floating Production Systems and TLPs", First Edition, API RP 2RD, June 1998.
- "Dynamic Risers", DNV-OS-F201, 2001.
- "Submarine Pipeline Systems", DNV-OS-F101, 2000.
- "Recommended Practice for Design, Construction, Operation, and Maintenance of Offshore Hydrocarbon Pipelines (Limit State Design)", API RP 1111, 3rd Edition, July 1999.
- "Guidance on Methods for Assessing the Acceptability of Flaws in Fusion Welded Structures", BS PD 6493, August 1991.
- "Fatigue Strength Analysis of Mobile Offshore Units", DNV Classification Note No.30.2, August 1984.
- "Offshore Installations: Guidance on Design, Construction and Certification", HSE Books, 1995 (supersedes same title from UK Dept. of Energy, HMSO, 1984 and takes precedence over "Code of Practice for Fatigue Design and Assessment of Steel Structures", BS7608:1993, with or without amendment of February 1995).
- "Recommended Practice RP B401: Cathodic Protection Design: 1993", DNV.

Regarding the Fatigue Design Codes, the reader is referred to an industry design codes, which provide guidance on the appropriate selection of S–N curves to apply to girth welds and other components under cyclic fatigue loading. Factors, which the designer may need to consider are:

- Project-specific conditions (materials, production chemistry, welding procedures) etc. which may cause the riser fatigue performance to depart from published curves

- Environmental conditions – in air, in seawater, in seawater with cathodic protection etc.
- Presence of mean stress for non-welded components
- Ovality and mis-match causing hi-lo conditions at the weld
- Compressive stress cycles

Acknowledgement

We acknowledge that Dr. A. D. Trim edited part of the material included in the section of Steel Catenary Riser of this chapter.

References

API RP 1111 (1999). "Design, construction, operation, and maintenance of offshore hydrocarbon pipelines (Limit State Design)", (3rd ed.). July.

API RP 16Q (1993). "Recommended practice for design, selection, operation, operation and maintenance of marine drilling riser systems", American Petroleum Inst., Wash, D.C.

API RP 2A-LRFD (1993). "Recommended practice for planning, designing and constructing fixed offshore platforms – load and resistance factor design", (1st ed.). July.

API RP 2RD (1998). "Recommended practice for design of risers for floating production systems (FPSs) and Tension Leg Platforms (TLPs)", (1st ed.). June.

API (1994). "Bulletin on formulas and calculations for casing, rubing, drill pipe, and line pipe properties", API Bulletin 5C3, (6th ed.). American Petroleum Inst., Wash, D.C., October.

ASME 31.4. American Society Of Mechanical Engineers: Gas Transmission and Distribution Piping Systems.

ASME 31.8. American Society Of Mechanical Engineers: Liquid Transportation Systems for Hydrocarbons, Liquid Petroleum Gas, Anhydrous Ammonia, and Alcohols.

Baxter, C., Pillai, S., and Hutt, G. (1997). "Advances in titanium risers for FPSOs", OTC 8409.

Bell, M. and Daly, R. (2001). "Reeled PIP SCR", OTC 13184. May.

Blevins, R. D. (1977). *Flow-Induced Vibrations*, Van Nostrand Reinhold Company, New York, N.Y.

Brekke, J. N., Chou, B., Virk, G. S., and Thompson, H. M. (1999). "Behavior of a drilling riser hung off in deep water", *Deep Offshore Technology Conference.*

Brooks, I. H. (1987). "A pragmatic approach to vortex-induced vibrations of a drilling riser", *Proceedings of the Offshore Technology Conference*, OTC 5522, Houston, TX.

BS7910:1999 (later BS version of above PD 6493).

BS8010 (1973). "Code of practice for pipelines: Pt.3: Pipelines subsea: design, construction and installation".

BS7608 (1995). "Code of practice for fatigue design and assessment of steel structures", 1993, with amendment, February.

Buitrago, J. and Zettlemoyer, N. (1998). "Fatigue design of critical girth welds for deepwater applications", OMAE 98-2004, Lisbon, Portugal, July.

Campbell, M. (1999). "Complexities of fatigue analysis of deepwater risers", *Deepwater Pipeline Technology Conference*, March.

Chaudhury, G. and Kennefick, J. (1999). "Design, testing and installation of steel catenary risers", *Proceedings of the Offshore Technology Conference*, OTC 10980, Houston, Texas, May 3–6.

Clausen, T. and D'Souza, R. (2001). "Dynamic risers key component for deepwater drilling, floating production", *Offshore Magazine*, pp. 89–93, May.

Connelly, L. M. and Zettlemoyer, N. (1993). "Stress concentration at girth welds of tubulars with axial wall misalignment", *Proc. 5th Intl. Symp. Tubular Structures*, Nottingham, UK, August.

DNV Classification Note 30.2, (1984). "Fatigue strength snalysis of mobile offshore units", August.

DNV Classification Notes 30.5, Environmental Conditions and Environmental Loads.

DNV (1981). "Rules for submarine pipeline systems", Hφvik. Norway.

DNV (1996). "Rules for submarine pipeline systems", December.

DNV-OS-F101 (2000). "Submarine pipeline systems".

DNV-OS-F201 (2001). "Dynamic Risers".

Erb, P. R., Ma, Tien-Chi, and Stockinger, M. P. (1983). "Riser collapse – A unique problem in deep-water drilling", IADC/SPE 11394, Society of Petroleum Engineers.

Farnes, K. A. and Moan, T. (1993). "Extreme response of a flexible riser system using a complete long-term approach", *Proc. ISOPE.*

Finn, L. (1999). "Reliable riser systems for spars" *Journal of Offshore Mechanics and Arctic Engineering*, Vol. 121, pp. 201–206, November.

Fumes, G. K., Hassanein, T., Halse, K. H., and Eriksen, M. (1998). "A field study of flow induced vibrations on a deepwater drilling riser", *Proceedings of the Offshore Technology Conference*, OTC 8702, Houston, TX, pp. 199–208.

Gardner, T. N. and Cole, M. W. (1982). "Deepwater drilling in high current environment", *Proceedings of the Offshore Technology Conference*, OTC 4316, Houston, TX.

Garrett, D. L., Gu, G. Z., and Watters, A. J. (1995). "Frequency content selection for dynamic analysis of marine systems", *Proceedings OMAE*, Copenhagen, Vol. 1-B, pp. 393–399.

Gopalkrishnan, R., Kopp, F., Rao, V. S., Swanson, R. C., Yu, X., Zhang, J. Q., Jones, W. T., and Zhang, H. (1998). "Development of the dynamic riser system for a ship-shaped production host in the deepwater Gulf of Mexico", *10th Deep Offshore Technology Conference*, New Orleans, November.

HSE Books (1995). "Offshore installations: guidance on design, construction and certification", (supersedes same title from UK Dept. of Energy, HMSO, 1984).

Integrated Mooring and Riser Design JIP (1999). "Technical bulletin: analysis methodology", MCS International and Noble Denton Europe, September.

IP6: Institute of Petroleum Pipeline Safety Code (UK).

Jiao, G. and Moan, T. (1990). "Probabilistic analysis of fatigue due to Gaussian load processes", *Probabilistic Engineering Mechanics*, Vol. 5, No. 2.

Johns, T. G., Meshlokh, R. E., and Sorenson, J. E. (1976). "Propagating buckle arrestors for offshore pipelines", OTC 2680.

Kavanagh, K. Dib, M., and Balch, E. (2004). "New revision of drilling riser recommended practice" (API RP 16Q), OTC 14263, May.

Krolikowski, L. P. and Gay, T. A. (1980). "An improved linearization technique for frequency domain riser analysis", *Proceedings of the Offshore Technology Conference*, OTC 3777, Houston, TX.

Langner, C. G. (1975). "Arrest of propagating collapse failures in offshore pipelines", Shell Deepwater Pipeline Feasibility Study.

Larsen, C. E. and Lutes, L. D. (1991). "Predicting the fatigue life of offshore structures by the single-moment spectral method", *Probabilistic Engineering Mechanics*, Vol. 6, No. 2, pp. 96–108.

Larsen, C. M. and Olufsen, A. (1992). "Extreme response estimation of flexible risers by use of long term statistics", *Proc. ISOPE.*

Lee, L., Allen D. W., Henning, D. L., and McMullen, D. (2004). "Damping characteristics of fairings for suppressing vortex-induced vibrations", OMAE 2004-51209.

Lutes, L. D. and Larsen, C. E. (1990). "Improved spectral method for variable amplitude fatigue prediction", *J. Struct. Engng.*, Vol. 116, No. 4, pp. 1149–1164, April.

McIver, D. B. and Olson, R. J. (1981). "Riser effective tension – now you see it, now you don't!", *37th Petroleum Mechanical Engineering Workshop and Conference*, ASME, Dallas, TX, September.

Mekha, B. B. (2001). "New frontiers in the design of steel catenary risers for floating production systems", *Journal of Offshore Mechanics and Arctic Engineering*, Vol. 123, pp. 153–158, November.

Miller, J. E. and Young, R. D. (1985). "Influence of mud column dynamics on top tension of suspended deepwater drilling risers", *Proceedings of the Offshore Technology Conference*, OTC 5015, Houston, TX.

Moe, G., Teigen, T., Simantiras, P., and Willis, N. (2004). "Predictions and model tests of a SCR undergoing VIM in flow at oblique angles", *Proceedings of Offshore Mechanics and Arctic Engineering*, OMAE2004-51563, Vancouver, Canada.

Mork, K. J., Chen, M. Z., Spolton, S., and Baxter, C. (2001). "Collapse and buckling design aspects of titanium alloy pipes", *20th Int. Conf. OMAE*, Rio de Janeiro, June.

Murphy, C. E. and Langner, C. G. (1985). "Ultimate pipe strength under bending, collapse and fatigue".

NTNF Research Programme, "Handbook of hydrodynamic coefficients of flexible risers", FPS 2000/Flexible Risers and Pipes, Report 2.1-16.

Offshore magazine, "Drilling & production riser systems and components for floating units", Poster distributed with May 2001.

Ortiz, K. and Chen, N. K. (1987). "Fatigue damage prediction for stationary wideband stresses", *5th Int. Conf. on Application of Statistics and Probability in Soil and Structural Engng.*

Rothbart, Harold A. (Ed), (1964). *Mechanical Design and Systems Handbook*, McGraw-Hill Book Company.

Siewert, T. A., Manahan, M. P., McCowan, C. N., Holt, J. M., Marsh, F. J., and Ruth, E. A. (1999). "The history and importance of impact testing", Pendulum Impact Testing: A Century of Progress, ASTEM STP 1380, T. A. Siewert and M. P. Manahan, Sr., (Editors), American Society for Testing and Materials, West Conshohocken, PA.

SHEAR7 User Manual for use with Version 4.1.

Sparks, C.P. (1983). "The influence of tension, pressure and weight on pipe and riser deformations and stresses", *Proc. 2nd Int. OMAE Symposium*, Houston.

Spolton, S. and Trim, A. D. (2000). "Aspects of steel-titanium riser design", *5th Int. Conf. On Advances in Riser Technologies*, Aberdeen, June.

Stahl, M. J. (2000). "Controlling recoil in drilling risers following emergency disconnect", *Proceedings of ETCE/OMAE2000 Joint Conference*, ETCE2000/DRILL-10105, New Orleans, LA.

Timoshenko and Gere (1961). "Theory of elastic stability", (2nd ed.). McGraw-Hill.

Triantafyllou, M., Triantafyllou, G., Tien, D., and Ambrose, B. D. (1999). "Pragmatic riser VIV analysis", *Proceedings of the Offshore Technology Conference*, OTC 10931, Houston, TX.

Triantafyllou, M. S. (2001). "User guide for VIVA", February.

Trim, A. D. (1992). "Extreme responses of flexible risers", *Marine Structures: Special Issue on Flexible Risers (Part II)*, Vol. 5, No. 5.

Vandiver, J. K. (1998). "Research challenges in the vortex-induced vibration prediction of marine risers", *Proceedings of the Offshore Technology Conference*, OTC 8698, Houston, TX, pp. 155–159.

Venkataraman, G. (2001). "Reeled risers: deepwater and dynamic considerations", OTC 13016, May.

Wirsching, P. and Light, M. C. (1980). "Fatigue under wide band random stresses", *Journal of the Structural Division*, ASCE, Vol. 106, No. ST7, pp. 1593–1607, July.

See http://www.matweb.com/, web site for material properties.

Handbook of Offshore Engineering
S. Chakrabarti (Ed.)
© 2005 Elsevier Ltd. All rights reserved.

Chapter 10

Topside Facilities Layout Development

Kenneth E. Arnold and Demir I. Karsan
Paragon Engineering Services Inc., Houston, TX, USA

Subrata Chakrabarti
Offshore Structure Analysis, Inc., Plainfield, IL, USA

10.1 Introduction

The most important factors governing an offshore platform topside facilities layout and design are its purpose and whether it will be manned or unmanned. A manned facility will require accommodation quarters for the personnel and will be subject to additional safety requirements. A manned facility will also require special transportation, landing and evacuation facilities for the personnel. These requirements will necessitate additional deck space.

Based on the equipment and personnel requirements for the topside facilities, first a deck layout plan should be developed. The layout plan is based on the operational workability and maintainability of the equipment and the health and safety requirements for the personnel who will operate it. The layout plan may be accommodated in a single deck level or may require multiple deck levels depending on the type of the offshore structure. For example, a Floating Production Storage and Offtake System (FPSO), which may be supported by a new built or converted ship shaped vessel, would normally have ample space available on its deck to accommodate most equipment and personnel on a single deck level. On the other hand, a fixed jacket, SPAR, or TLP topsides would have a smaller footprint and the production equipment may be laid in multiple levels.

This chapter describes the general considerations for the layout and design of the topside facilities for offshore platforms. The effect of the environment on the deck design; the types of topside deck structures and the split of the construction, hookup and commissioning (HUC) activities between the onshore and offshore sites depending on the deck type; and the control and safety requirements, including fuel and ignition sources, firewall and fire equipment are presented. The practical limitations of the topside design are described. As examples, two different layout systems are compared and the topside design of the North

Sea Britannia platform is presented. Much of the material presented in this chapter has been derived from course notes prepared by Mr. Ken Arnold, CEO of the Paragon Engineering Services, Inc.

10.2 General Layout Considerations

The following items require special attention in the topsides facilities design:

- Prevailing Wind Direction
- Firewalls and Barrier Walls
- Process Flow
- Safe Work Areas
- Storage
- Ventilation
- Escape Routes
- Fire Fighting Equipment
- Thermal Radiation
- Vapour Dispersion
- Future Expansion(s)
- Simultaneous Operations Provisions (such as producing while drilling or working over wells)

A number of offshore platform topside deck layouts have evolved in response to operational requirements and the fabrication infrastructure and installation equipment availability. Operational requirements dictate the general deck size and configuration (number of deck levels and their layout, etc.). For example, the need for a fully integrated drilling and production system would dictate vertical and horizontal layering of the deck structure in such a manner as to provide an efficient operation while also providing an acceptable level of human and environmental safety.

If fabrication facilities and skilled labour are not available in the area; the economics may dictate building the deck in smaller pieces and modules and assembling them offshore using the low capacity offshore lifting equipment available. This approach may result in increased steel weight, and offshore construction time and cost, while avoiding the expense of investing in a major fabrication yard.

Alternatively, the owner may design the deck as an integrated single piece structure or as a Module Support Frame (MSF) supporting a few large modules, which can be built at a location where fabrication infrastructure and equipment are readily available. The "integrated deck" may then be installed on site using high capacity lifting cranes, or if not available, a float-over deck installation approach. In a float-over deck installation approach, the fully integrated and pre-commissioned deck (or a large module) is loaded out onto a large transportation vessel(s) and transported to the installation site as a single piece. At the installation site, the deck is floated over and then lowered onto the support structure by either ballasting the vessel or using quick drop mechanisms. Alternatively, for the case of a floating support structure, the support structure may be de-ballasted to pick the deck up.

An integrated deck may be divided into a number of levels and areas depending on the functions they support. Typical levels are:

- *Main (upper) deck*, which supports the drilling/production systems and several modules (drilling, process, utilities, living quarters, compression, etc.)
- *Cellar deck,* which supports systems that need to be placed at a lower elevation and installed with the deck structures, such as pumps, some utilities, pig launchers/ receivers, Christmas trees, wellhead manifolds, piping, etc.
- *Additional deck levels*, if needed. For example, if simultaneous drilling and production operations are planned, some process equipment may be located in a mezzanine deck.

An example of such a topside layout is the Diana SPAR design [Milburn and Williams, 2001]. In the Diana Spar topsides design, the upper deck is called the "Drilling Deck", and has the production and temporary quarters buildings, drill rig, chemical tote tank storage and communications and radar satellite dishes. The mid-level (mezzanine) deck is called the "Production Deck" and contains the majority of oil and gas separation, processing, treating, compression equipment, power generation equipment, the MCC/Control Room and many of the utilities. The lower "Cellar Deck" contains other utility systems (cooling water, fresh water, firewater, flare scrubber, etc.) as well as oil and gas sales meters, pipeline pig launchers and receivers, manifolds and shutdown valves.

A subcellar Deck, which is a partial deck suspended below the cellar deck could also be installed to contain the gravity drain sumps and pumps. Because this deck is usually small it could be designed to withstand impact from the wave crest and transport the lateral loads to the rest of the structure.

A modular deck may be divided into a number of pieces and modules depending on the functions they support and the installation equipment available. Typical modular deck components are:

- Module Support Frame (MSF), which provides a space frame for supporting the modules and transferring their load to the jacket/tower structure. The MSF may also be designed to include a number of platform facilities, such as the storage tanks, pig launching and receiving systems, metering/proving devices and the associated piping systems,
- Modules. These provide a number of production and life support systems, such as the –

  Living Quarters Module (generally supporting a heliport, communication systems, hotel, messing, office and recreational facilities).

  Utilities Module (generally supporting power generation and electrical and production control systems, including a control room).

  Wellhead Module (generally supporting the wellheads, well test and control equipment).

  Drill Rig Module (containing the drill tower, draw-works, drillers and control rooms, drill pipe and casing storage racks and pipe handling systems). Drill rig module is located over and supported by the wellhead module.

Production Module (containing the oil/gas/water separation and treatment systems, other piping, control systems and valves for safe production, metering and transfer of the produced liquids and gas to the offloading system).
Compression Module, if gas compression for injection to the formation and/or high-pressure gas pumping to the shore is needed. Since compression may be needed at later production stages, this module may be installed on the deck at a later date or on a nearby separate platform (generally bridge connected to the deck). Similarly, water injection and pumping modules may be added if these functions are needed at later field development stages.

In general, integrated decks result in more efficient and lighter structural systems than modular decks, since additional module steel, which is only needed for installation reasons, is avoided. For demonstration purposes, the following paragraphs will elaborate only on the components and design of a mid-sized single deck structure. The design of MSF and modules follow similar design principles and methods.

10.2.1 General Requirements

It is advantageous to design the topsides facilities using a Three (*3*) *D*imensional *C*omputer *A*ided *D*esign (3D CAD) model. If the 3D model has a high degree of accuracy on all structural, piping and equipment layouts; the possibility of encountering "clashes" in fabrication between piping, fittings, structural members, instrumentation, electrical cable trays and conduits can be minimised. In addition, the use of virtual reality with a 3D model allows an operator to "move" around inside the deck structure and identify potential clashes, ensure the correct orientation of valves, study that ample access to equipment exists, and the equipment could be easily removed for maintenance or replaced.

In general, there are two broad categories of equipment. One of these may be termed the "fuel sources" and the other the "ignition sources". The primary goal in a deck layout should be to prevent hydrocarbon ignition and fire escalation by separating the fuel sources from the ignition sources. Any layout is a compromise that balances the probability of occurrence of these undesirable events against their consequences.

Modern platform designs incorporate the learnings and recommendations from past disasters. Many of the decisions on selection and layout of the process and its control and safety systems are derived through quantified safety/risk analysis processes to ensure a low occurrence of accidental events, and in the event of an incident, to ensure safe evacuation of the personnel on board within acceptable risk levels.

The cost of having to scaffold offshore for access to equipment can be very high. Therefore the designers should site equipment at deck level, wherever possible, or adjacent to access platforms.

During design and development of the Process and Instrument Drawings (P&IDs) and plant layout [Croft-Bednarski and Johnston, 1999] the designer should ensure that the control valves are situated in easily accessible places so that start up, shutdown, isolation and maintenance can be carried out efficiently and safely. Another important aspect of the layout design is to identify areas of the plant, which would require frequent maintenance and ensure easy access to these areas.

The design of the control rooms in offshore platform topsides is very important. The control room operator must be able to control and manage safety and production critical emergencies efficiently and effectively to ensure that the platform can be shutdown and vented safely, ensure that fire control and mitigation efforts are initiated and that evacuation can be accomplished if necessary. On large platforms it may be necessary to have an emergency control room separate from the main control room to serve as a backup for these functions, if the main control room is not available. This room also duplicates as the emergency command centre.

The layout and design of the personnel accommodation facilities is also very important. The operators must provide input to the layout of these facilities. Offshore personnel who normally work in areas such as the galley and sickbay should be brought in to work with the architects to achieve an optimal accommodation facilities layout. Designs should be based around natural colours and wherever possible, areas should give a feeling of comfort and security to the personnel who will be lodged in these facilities.

10.2.2 Deepwater Facility Considerations

Deepwater floating facilities [Milburn and Williams, 2001] require a number of considerations during design that are not normally found in conventional fixed offshore structure. The motions of a floating facility must be taken into consideration during the design of its topside. Structural details that will be subjected to inertial loadings due to platform motions have to be checked for one or more load cases such as the operating, survival, transportation and installation conditions (for more details see Chapter 6, Fixed Offshore Platform Design).

The sea state at which the facility should continue its normal operation must be determined. The process vessels must be designed to meet these conditions. For a horizontal vessel, the motion effect might require special internal designs. Normally, the longer the vessel, the greater is the need for special care. Slugging and tight emulsions from subsea wells caused by the long vertical risers and cold sea temperatures must also be considered in the design of the production equipment. The expected motions in the design sea state should be supplied to all process vendors to insure their understanding of the operating conditions. Other systems that require careful consideration of motions are the drains and mechanical rotating equipment (turbines, compressors, generators, etc.).

Deepwater subsea production presents a number of "flow assurance" problems for an offshore host facility. These include hydrates, wax, multi-phase flow, slugging and low temperatures. The host platform topside facilities may be required to provide for methanol storage or methanol recovery and regeneration for hydrate inhibition, equipment for heating flowlines or for recirculating hot fluids, pig launchers and receivers for management of wax, slug catching capacity and valving for slugs and inlet heaters to increase temperature of the fluids for further processing.

During design and fabrication, careful consideration must be given to regulatory authority requirements and the classing of the vessel. For SPARs in the US, the vessel is classed with the American Bureau of Shipping (ABS). The United States Coast Guard (USCG) and Mineral Management Service (MMS) are the principal regulatory authorities. Usually, considerable USCG and ABS oversight occurs on topside systems. The systems which require such oversight include: primary topside structure, quarters and buildings, firewater system, life saving systems, compressed air supporting marine systems, diesel system, fuel gas systems supporting power generation, helicopter refuelling system, any systems for bulk storage and handling of liquids in the hull, deck drainage system, potable water system, sewage system and freshwater wash down system.

In the USA, the USCG also reviews platform safety (including access/egress); lifesaving, fire protection, personal protection, ventilation and marine transfer facilities. Electrical systems and equipment include the alarm system, aids to navigation, communication system, area classification, power generation, emergency generator, electrical switchgear/MCC, lighting systems and fire detection.

10.2.3 Prevailing Wind Direction

In locating the equipment on the deck, it is important to consider the effect of the prevailing wind direction. An example "wind rose" summarising the wind data is shown in fig. 10.1. Certain equipment and elements should be placed upwind as much as possible. These components are quarters and control buildings, helidecks, air intakes of fired vessels, engines, turbines, air compressors and HVAC equipment. Similarly, certain components should be placed downwind, such as, vents, storage tanks, compressors, wellheads, etc. These will minimise the probability of escaping vapours being carried toward ignition sources and personnel. The main boat landing should be located on the leeward side, which will shelter the boat landing and keep vessels from hitting the platform. The main crane should be located on the boat landing side. Where high current is not aligned with the wind, the relative effects of each must be considered in the design of the component placement.

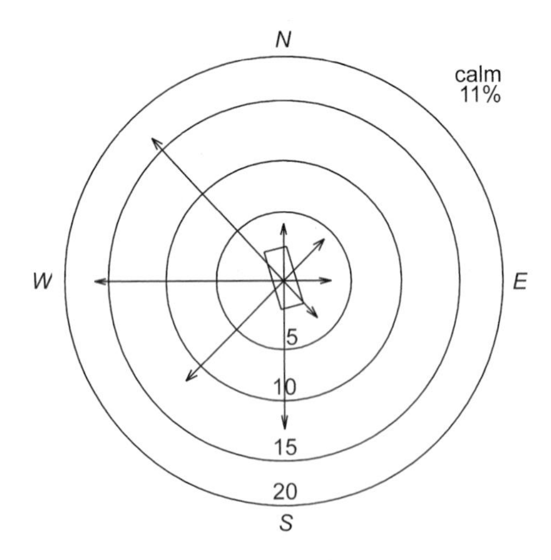

Figure 10.1 Typical average yearly wind rose shown as percent occurrence per year

The main escape areas (such as the safe gathering or mustering areas, helidecks, primary escape routes, stairs to boat landing, etc.) should be located upwind, wherever possible. However, rarely can it be guaranteed that "prevailing" wind conditions will occur at the time of the accident. Thus, secondary means of escape should be located downwind.

10.2.4 Fuel and Ignition Sources

Typical equipment found on the topside may be categorised as either "fuel" or "ignition" sources. Table 10.1 below shows these two categories of equipment and other major topside fuel or ignition sources as listed by API 14J (API, September 1993). The desired topside locations for the fuel and ignition sources listed in table 10.1 are given in table 10.2.

Table 10.1 Equipment – fuel sources vs. ignition sources (From API 14J)

Fuel Sources	Ignition Sources
Wellheads	Fired Vessels
Manifolds	Combustion Engines (including gas turbines)
Separators and Scrubbers	Electrical Equipment (including offices and buildings)
Coalescers	Flares
Oil Treaters	Welding Machines
Gas Compressors	Grinding Machinery
Liquid Hydrocarbon Pumps	Cutting Machinery or Torches
Heat Exchangers	Waste Heat Recovery Equipment
Hydrocarbon Storage Tanks	Static Electricity
Process Piping	Lightning
Gas-Metering Equipment	Spark Producing Hand Tools
Risers and Pipelines	Portable Computers
Vents	Cameras
Pig Launchers and Receivers	Cellular Phones
Drains	Non-Intrinsically Safe Flashlights
Portable Engine-Driven Equipment	
Portable Fuel Tanks	
Chemical Storage	
Laboratory Gas Cylinders	
Sample Pots	

Table 10.2 Recommended topside location objectives for fuel and ignition sources listed in table 10.1 (From API 14J (API, September 1993))

Area	Location Objective	Example Equipment Types	Source Type
Wellhead	Minimise sources of ignition and fuel supply Protect from mechanical damage and exposure to fire	Wellheads, Chokes, Manifolds, Headers	Fuel
Unfired Process	Minimise sources of ignition	Manifolds and Headers, Separators, Gas Sales Station, Pig Traps, Heat Exchangers, Water Treating Equipment, Pumps, Compressors, LACT Units	Fuel
Hydrocarbon Storage	Minimise sources of ignition	Storage Tanks, Gunbarrel Tanks, Sump Tanks, Produced Water Treating Tanks	Fuel
Direct Fired Process	Minimise fuel supply	Fired Treaters, Line Heaters, Glycol Reboilers	Ignition and fuel
Machinery	Minimise fuel supply	Generators, Electric Hoisting Equipment, Air Compressors, Engines, Turbines	Ignition
Quarters/ Utilities Building	Personnel safety Minimise sources of fuel	Office, Control Room, Switchgear/MCC, Warehouse, Maintenance Areas/Building	Ignition
Pipeline	Minimise sources of ignition Protect from mechanical damage and exposure to fire	Pig Launchers, Pig Traps, Valve Stations, Meter Stations	Fuel
Flares	Minimise fuel sources	Discharge Point	
Vents	Minimise ignition sources	Discharge Point	

10.2.5 Control and Safety Systems

The control and safety systems on platform facilities [Milburn and Williams, 2001] generally include:

- Either local or central operational control systems
- Data acquisition systems
- Manual operator interface
- Local equipment control and shutdown systems
- Well control and shut down systems
- Emergency Shut Down (ESD) System
- Fire detection systems
- Combustible gas monitoring systems

10.2.6 Firewalls, Barrier Walls and Blast Walls

For safety reasons, adequate barrier and firewalls should be considered for areas where it is desirable to attempt to isolate certain areas where explosion, spillage or fire is possible. Barrier walls impede escaping gas or liquid leaks from entering an area with ignition sources. Firewalls provide a heat shield to allow personnel escape and protect potential fuel sources. Blast walls contain an overpressure from an explosion in a confined space from causing secondary damage on the other side of the wall.

The disadvantages of firewalls, barrier walls and blast walls are that they restrict ventilation, hamper escape and can, in themselves, help create overpressure in explosions. Thus, any decision to include one of these walls in the layout must balance the potential detriments against the potential benefits. Careful consideration is required for location of Shut Down Valves on lines that penetrate walls.

10.2.7 Fire Fighting Equipment

Fire fighting equipment should be easily accessible in any location on the deck. This is particularly true for manned platforms. Hose stations should be located so that two hoses can reach any point of the deck. Firewater pumps, fire fighting chemicals and hose stations should be accessible and removed from locations where fire might occur. Ramps should be provided for wheeled chemical units. Spray systems should cover the entire area and point upwards at the wellheads, rather than downward. Automatic fire suppression systems can be considered for enclosures containing an ignition source, which cannot be isolated from a fuel source. In designing fire-fighting systems, consideration should be given to providing two separate pumps on opposite sides of the platforms so that damage to one would not likely cause the other to be inoperable. Firewater mains should be isolated, so that if a main is severed in an explosion, pressured water can still be delivered to the intact system.

10.2.8 Process Flow

A well-developed process flow diagram is necessary to define the parameters for design of individual pieces of equipment. In laying out the equipment, a logical and orderly flow path is desirable from wellhead to sales meter. This minimises the required piping while reducing the inventory of potential fuel, which could feed a fire. Locating the equipment solely to minimise piping for main process streams is often not the best answer because of other safety issues involved. The needs to separate fuel and ignition sources, to consider prevailing wind, and to allow for ease of maintenance may be overriding. In addition, equipment items have many connections in addition to the main process flow and the most efficient piping arrangement may not necessarily follow the main flow pattern.

The amount of high temperature piping should be minimised as well to reduce heat loss and insulation requirements. High-pressure piping should be kept away from high traffic areas and moving equipment. Long piping runs should be avoided where pressure drop is critical. The need for gravity flow may dictate relative vertical positioning.

10.2.9 Maintenance of Equipment

Provide adequate room for operations and maintenance. This includes the following:

- Pulling fire tubes from fired heaters
- Pulling tube bundles or plates from heat exchangers
- Removing compressor cylinders
- Replacing turbines, engines, generators, compressors and pumps
- Pulling vertical turbine or can-type pumps
- Removing plate packs from plate coalescers
- Pig insertion and removal
- Changing filter elements and filter media
- Removing and installing bulk storage containers
- Opening and removing inspection plates and manways

Supplementary overhead cranes or lifting devices should be provided where necessary. Most injuries are due to falls from high places and handling of heavy loads. Layout should consider access to ladders and landings for maintenance purposes.

10.2.10 Safe Work Areas and Operations

Provide safe welding and cutting areas for minor construction or routine maintenance. Floors should be solid and adequate ventilation and separation from fuel sources should be provided. Isolate the work areas from drains containing live hydrocarbons with liquid seals. Attention should be given to equipment handling requirements and weather protection.

Operations should be planned for production, drilling, completion, wireline, pumpdown, snubbing unit work, construction activities, surface preparation and painting, removal or installation of wellhead equipment and installation of conductor pipe. Planning is required for equipment used in all phases of work anticipated. Adequate space and handling equipment are needed for consumables and support operations.

10.2.11 Storage

Storage areas should be provided for diesel fuel, treating chemicals (e.g. corrosion inhibitors, demulsifiers, hydrate inhibition, glycols, biocides, etc.) and waste fluids. Storage for spare

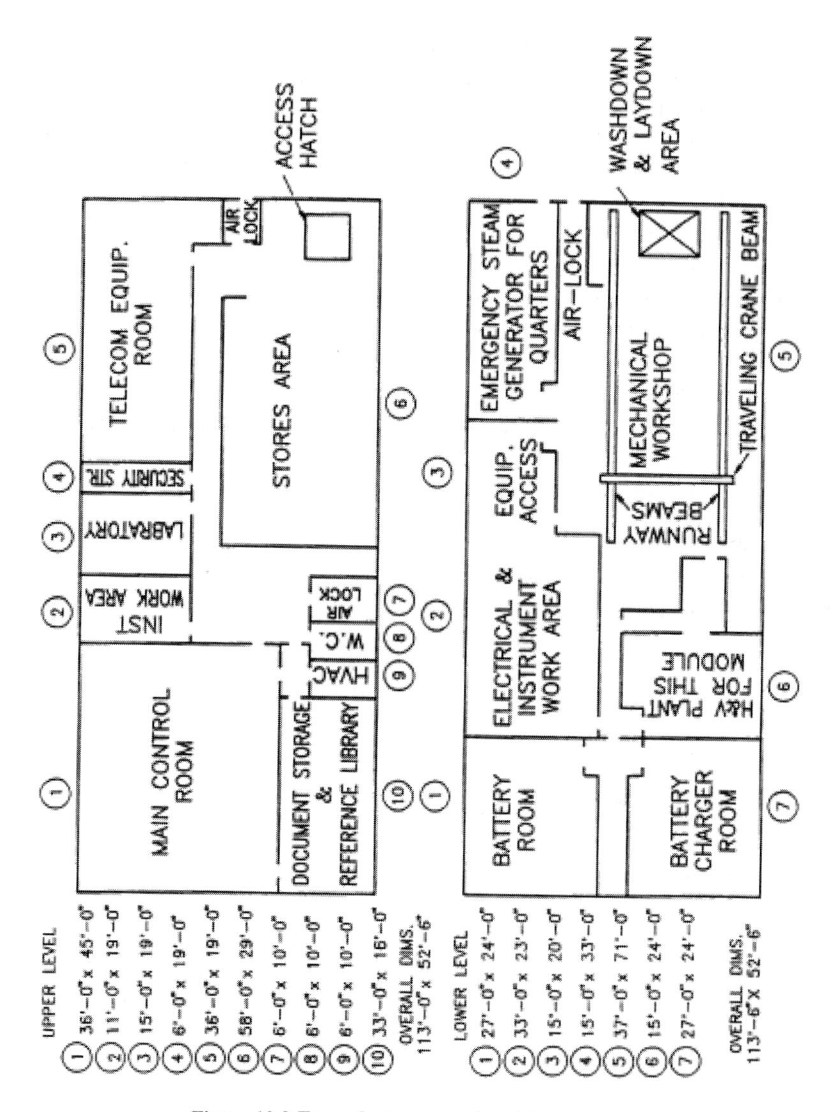

Figure 10.2 Example storage and maintenance area

parts and solid consumables is normally provided in buildings, shops or warehouses. An example of a storage and maintenance area is shown in fig. 10.2.

10.2.12 Ventilation

Due to the possibility of accidental flammable gas and flashing liquid discharges (leak, incorrect valve opening, sampling, etc.), adequate ventilation is a critical safety consideration. Provide ventilation where necessary to disperse hazardous vapours and prevent their accumulation in gas traps. Enclosed buildings, which contain ignition sources, should be pressurised to prevent entry of flammable gas to the external atmosphere. Locate any air intakes in a safe area. Completely enclosed structures which house hydrocarbon fuel sources should have an air circulation and exhaust system to replace accumulated vapours with fresh air. Provide space for ducting. Fire and blast walls reduce natural ventilation. Try to keep at least two sides of the wellhead area open to natural ventilation.

10.2.13 Escape Routes

Provide two independent escape routes from each location. Maintain escape ways with a minimum clearance width of three feet, free of obstructions and with adequate headroom. Two stairs between all levels at opposite ends of a platform are preferred. Enclosed areas with fuel sources should have two exit doors, which open to independent escape routes.

Evacuation can be provided through use of boat landing, survival craft and helideck (helideck may be difficult to use in case of a hydrocarbon related emergency).

10.3 Areas and Equipment

The following are items to be considered in the design and layout of the deck for the different areas described in table 10.1 (Section 10.2.4).

10.3.1 Wellhead Areas

Potential for uncontrolled flow and high pressures exists. The following considerations should be given to the wellhead areas:

- Provide adequate ventilation.
- Protect from sources of ignition, other large inventories of fuels, machinery and dropped objects and traffic.
- Protect equipment and instrumentation from drilling and completion fluid spillage.
- Provide unobstructed access to and egress from wellheads and separate them from the living quarters.

For the wellhead area, size is a function of the drilling or work over rig and the number and spacing of the wells. Tight spacing makes access and escape paths difficult. On large platforms, the wells are usually isolated on one end of the structure. A firewall is sometimes placed to isolate the wellhead area from the production equipment.

10.3.2 Unfired Process Areas

Unfired process areas are a potential source of fuel. Considerations should be given to the vertical placement of equipment. Liquid leaks from this area could ignite on hot surfaces or ignition sources below. Gas leaks could ignite on hot surfaces or ignition sources above. The unfired process area is usually located near the wellhead area. The area should be protected from dropped objects. An estimation of space required for these process vessels can be based on the following assumptions for piping twelve inches in diameter and smaller:

- Piping for horizontal vessels will take up an area of about four feet wider and two feet longer than the vessel itself,
- Piping for vertical vessels will take up an area of about two feet wider and four feet longer than the vessel diameter and
- Additional space is needed for walkways.

10.3.3 Hydrocarbon Storage Tanks

Hydrocarbon storage tanks can provide a large inventory on the platform to feed a fire. There is a potential for a tank roof to fail, if subjected to overpressure. Also, tanks can be easily punctured. Therefore, careful considerations should be given to separate tanks from ignition sources and from other equipment, which can add fuel to a fire, such as, wellheads, pipelines and risers. Protect other equipment from liquids spilled from tanks and provide containment, where necessary. Protect from movement of equipment, which can puncture the tank. On offshore platforms, space can be saved using rectangular tanks.

Locate oil tanks on the upper level, if possible, since it is very likely that the roof will fail if the liquid in the tank catches fire or if the tank is over pressured. Separate storage tanks for diesel or lube oil can be avoided by using the interiors of deck legs or crane pedestals. In sectionalised tanks, it is sometimes desirable to store non-flammable liquids between the stored fuel and the potential ignition sources to act as a safety buffer.

Atmospheric tanks containing crude oil must be vented. If the vented gas is not to be recovered, it should be routed to a vent stack on the downwind side. Level gauges, controls and access will normally require about three feet on one side of a tank. On the sides of tanks without piping, only a walkway will be necessary.

10.3.4 Fired Process Equipment

Direct fired process equipment is a source of ignition. If it contains flammable liquid (crude, gas, glycol), then it is also a potential fuel source. Air intakes should be from the perimeter of the platform on the upwind side to avoid sucking in hydrocarbon fuel with the air. The hot exhaust stack should be isolated from potential oil spills, since the pipe may be hot enough to ignite the spilled oil. Consider firewalls to protect surrounding equipment. Suggested clearance around fired process equipment is at least 15 ft. Maintenance space must be provided for pulling the fire tube. Fire tube maintenance will also require lifting equipment.

10.3.5 Machinery Areas

Be aware of oil leaks from above or gas leaks from below the sources of ignition (especially hot surfaces) in the machinery areas. Failure of mechanical seals or packing in compressors and pumps can provide fuel. The probability of failures of piping and connections in machinery areas are higher than normal due to vibration.

Machinery areas, which do not contain flammable fluid, can be located near quarters or office/warehouse/auxiliaries as both are ignition sources. If flammables are present, then the machinery area is a potential source of fire and should be separated from wellheads, pipelines, risers and tanks, which could escalate the fire, and separated from quarters due to hazard to personnel. Consider enclosing turbine and engine driven equipment and providing the enclosure with fire and gas detection and suppression equipment. A positive pressure could be maintained to exclude migration of gases into the area and disperse leaking gases. Isolate turbine inlets from ingesting gas with the air.

AC motor driven pumps and compressors with proper electrical classification can be installed in process areas. DC motors are an ignition source, which when used on pumps or compressors should be in an enclosure with gas detectors and positive pressure from a safe intake source.

Adequate space and hoisting capability should be provided. Generally, three feet of space on each side of a skid plus special clearances are needed. Heaviest and largest parts can be moved to an area accessible by crane. Noise should be taken into account as well.

10.3.6 Quarters and Utility Buildings

Protection from external fires, noise and vibration is needed for these areas where there is a concentration of personnel. Consider fire resistant construction materials for the quarters. Potential sources of ignition from cooking, smoking and electrical equipment should be studied. Isolate quarters from potential gas leaks. Try to locate the quarters away from sources of noise and vibration. A firewall may be advantageous if the building cannot be safely located away from hazardous equipment. Minimise windows, which face the process area. Pay attention to escape routes and minimise exposure of personnel to radiation from potential flame sources. Try to locate utilities near the quarters building to minimise piping and conduit runs and minimise the external exposure to the quarters. In locating the quarters, consider the proximity to electrical generation, sewage treatment, heating, ventilation and air conditioning and potable water supply.

10.3.7 Pipelines

Potential uncontrolled flows from pipeline risers and pig traps and launchers should be separated from quarters, control buildings and wellheads. Consider automatic Shut Down Valves (SDVs) and protect them from blast, fire or dropped objects by location of firewalls. Do not install instruments, vent valves or drain valves outside of SDVs. Risers should be protected from boat impact and dropped objects. Provide space for access to risers and space and work platforms for access to pig traps and launchers for pig removal. Consider the need for lift equipment for large diameter pigs.

10.3.8 Flares and Vents

A vent is a potential gas fuel source and a potential liquid fuel source due to carryover from the vent. This is also true for a flare, if the pilot fails. A flare, or a vent, which has been ignited by lightning, is a potential ignition source. Liquid carryover from a flare is a potential ignition source as well. A flare may become a potential source of dangerously high SO_2 levels and a vent may become a potential source of dangerously high H_2S levels, if H_2S is present in the gas. If the pilot fails, a flare may become a potential source of dangerously high H_2S levels.

The potential danger from radiant heat from a flare as well as from accidental ignition of a vent should be taken into account for all potential wind directions. The normal flow from the flare provides continuous exposure to radiation, while a short-term radiation exposure is given from the emergency relief. Vents can give a short-term exposure from the normal flow if accidentally ignited. Dispersion of gases from vents must not create a problem with helicopter and boat approaches, air intakes for turbines and drilling derricks. An adequate scrubber should be provided for all flares and vents to minimise the possibility of liquid carryover.

10.4 Deck Impact Loads

The deck of an offshore structure is generally positioned at an elevation above the maximum water level that may be reached by a statistically probable wave crest, which may be experienced throughout the structure's operation. This elevation is determined as part of the design process (see Chapters 3, and 6 of this handbook) from probability-based models aimed at predicting the largest wave in a particular return period. The maximum lateral pressure exerted on a structure by a wave occurs at or near a wave crest. The preference is to position the deck at an elevation sufficiently high to avoid an impact between the wave crest and a large area of the structure. For various reasons, situations may arise where the probability of a wave impact on at least a small portion of the deck is high enough that an estimate of the wave impact load on the deck is required (see Chapter 4).

Offshore structure designers traditionally use the design wave method to establish the ultimate design load. The design wave approach considers the largest wave that appears in a random wave time series. Estimation of wave impact load is generally based on a numerical model, which is based on empirical factors. These factors are derived from scaled model tests in which deck structures are modeled. One such test set up is shown in fig. 10.3 in which the deck of a jacket structure subjected to a high wave in a random sea time series was floated on a load cell to measure this impact load. The jacket platform in the picture is the Vermilion 46A platform, located approximately 30 miles offshore South of

Figure 10.3 Wave impact load on a jacket platform deck model

New Orleans. The model platform, which has 40 legs, consists of two identical 20-leg jackets connected together at the base. The deck of each platform was separately instrumented to measure the two-component horizontal loads produced due to the wave impact.

The associated errors in the empirical factors are difficult to quantify. That is why an attempt is made to position the horizontal members and floor beams which make up the lowest level of the deck above the maximum expected wave crest including wave run up on the vertical members of the structure supporting the deck. The distance between the design crest elevation and the lowest elevation of the significant area of horizontal steel is called the "air gap". API recommends five-foot air gap for Gulf of Mexico Platforms to protect equipment from splash damage as well as provide a safety factor against the calculation of wave crest elevation. For further details see Section 6.2.3.2.

10.5 Deck Placement and Configuration

Almost every offshore structure includes some type of a deck. The size and design of topside depends on the type of structure and its function. Fixed structures can be a jacket type structure, which is piled to the foundation with a deck on top. A typical fixed jacket platform and deck structure is shown in fig. 6.11. Platforms could also be steel or concrete gravity structures, which have a deck on top; or wood, concrete or steel slabs with piles and cap beams. Platforms can also be designed to be floated into place and "jacked up" as are mobile jack-up rigs. Other structure types, such as articulated towers, guyed towers, Semi Submersible Vessels (SSVs), SPARs or Tension Leg Platforms (TLPs) will have a deck which may either be an integrated part of the structure itself or installed on top, as with a fixed jacket. A ship shaped FPSO will have a module support frame, which supports the equipment 10 or more feet above the ship's deck. The module support frame performs the same function as a deck.

10.5.1 Horizontal Placement of Equipment on Deck

From a structural efficiency standpoint, it is beneficial to place heavy equipment near truss supports and to try to balance the vertical load on each leg. Adequate room for future equipment additions should be provided on the top deck and along the perimeter of the deck. Provide clearances for pad eyes and lifting slings. The need to keep deck equipment weight within capacities of lift barges may necessitate that some of the equipment be installed in separate lifts, require two or more decks side by side or necessitate a float over system (Sections 10.6.4 and 6.2.1.1). Rotating equipment should be oriented with its long axis transverse to the platform floor beams for increased stiffness. Allocate space on the top deck or around the perimeter for future equipment.

10.5.2 Vertical Placement of Equipment

Allow adequate height for piping (e.g. relief valves, gas outlets) and maintenance. It may be necessary to have a tall piece of equipment penetrate the deck above due to its height. Place the equipment to take advantage of gravity flow. Pumps with a high negative suction head requirements may be located at a low elevation. Provide hatches in the upper decks or porches in lower decks for crane access. Locate heavy equipment as low as possible to

Table 10.3 Typical load intensities on deck

Load Type	Item	Load Intensity
Dead Load	Floor beams and plate	50 psf
Live Loads for Floor Beam Design	Production Equipment	350 to 500 psf
	Drilling Equipment	1000 psf
	Mud Tanks	1250 psf
	Derrick Load	2500 kip
	Living Quarters	150 psf
Deck Truss, Jacket and Piling	Carry Down to Lower Levels	70% of live loads

lower the vertical centre of gravity, which will optimise stability, minimise dynamic response and aid in deck transportation.

Open gravity drains must flow to a low point sump, which could be located in a subcellar deck.

If specific loads are unknown, the deck should be designed based on the typical load intensities for the different types of loads shown in table 10.3.

10.5.3 Installation Considerations

Installation procedures should consider the availability of lift equipment vs. minimising offshore hook-up time. Evaluation of the alternatives should take into account lift weight, available lift equipment and time required for the installation. Types of lifts that should be examined are single hook lift, lift with a spreader bar or a two-point lift (fig. 10.4). The deck on barges, concrete towers, etc. is normally installed as a unit.

10.5.4 Deck Installation Schemes

There are several types of decks that may be installed on a structure based on their construction and transportation method. An Integrated Deck is one in which the equipment is pre-installed on the deck at an onshore yard. In this case, the deck structure supports the skids directly as well as provides any lateral support needed. A multi-piece Integrated Deck is used where the complete deck is too costly to lift in one piece and the platform has eight or more legs. A Modular Deck will have its equipment installed in modules. The module support frame may be an integral part of the tower or installed on the platform as a separate unit. Modular decks allow the fabrication workload to be spread out among several yards so that different yards may work on different modules in parallel, thus potentially decreasing the construction time. Lastly, Skidded Equipment containing some piping, valves and controls, but generally smaller than individual modules, may be used to minimise construction effort at the yard assembling an integrated deck or large module. Skids are generally transportable by truck and thus can be bid to a number of smaller fabricators for increased competition, lower project cost and shorter construction

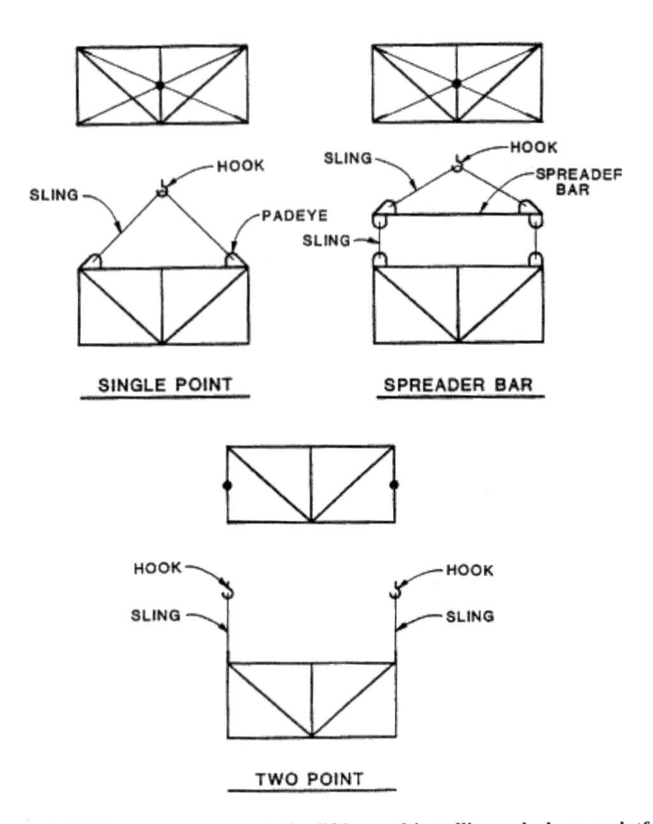

Figure 10.4 Different arrangements for lifting and installing a deck on a platform

time. Skids may be installed inside a module or an integrated deck. They do tend to increase the structural weight, however.

The different installation schemes described above are depicted in fig. 10.5. The top figure shows integrated modules or deck installed on a steel tower with a few legs. On a shallow water concrete barge, the deck may be composed of skidded equipment lifted onto a steel or concrete deck or an integrated structure placed on steel or concrete columns. The bottom figure shows an integrated deck on a multi-piled platform. The cap beams are shown here.

Various installation schemes on piled jacket structures are shown in fig. 10.6. In the top figures, modules are placed on the deck by using offshore lifting cranes. For this purpose a module support frame should be designed which accommodates the modules. In the bottom left figure the deck is brought on a barge, which is positioned in between the jacket legs. The deck structure is then lowered on the jacket legs and secured in place by use of lowering mechanisms or de-ballasting the barge. In the bottom right figure, two four pile integrated decks are installed on an eight-pile jacket with spanning integrated deck insert added next.

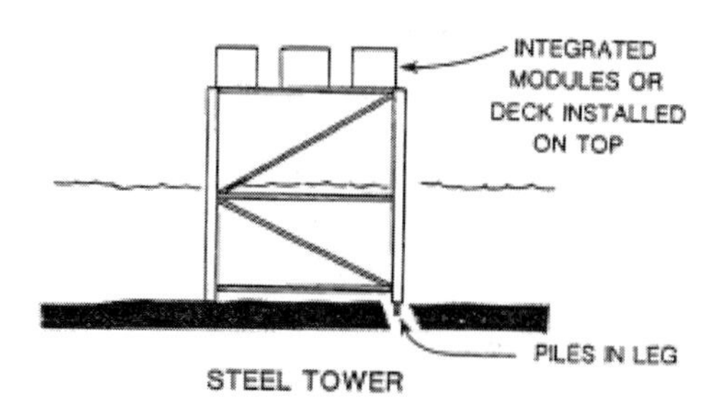

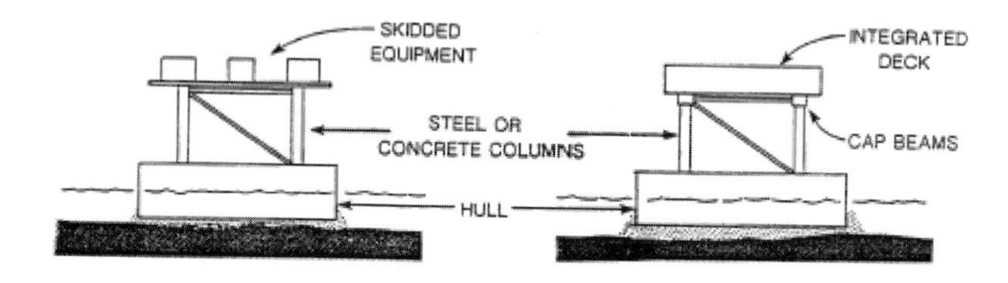

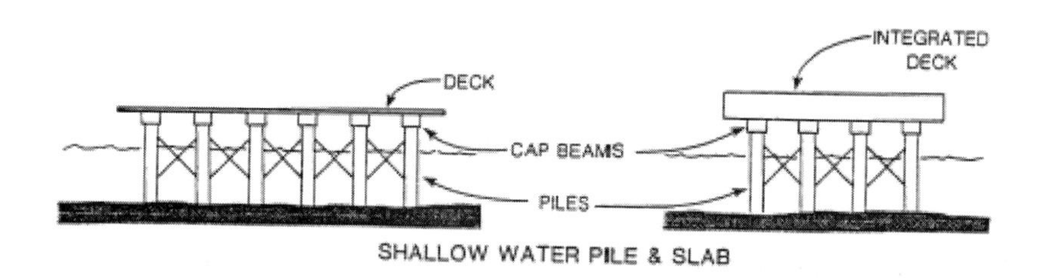

Figure 10.5 Deck installation scheme on piled jacket structures or floating barge

Figure 10.7 shows the layout of a deck installed on an FPSO. In this case the helideck and the personnel quarters are separated from the process equipment by a sufficient distance. Such a separation is possible because the ship-shaped structure provides ample space on its deck. The flare is placed far away from the personnel and fuel sources.

10.6 Floatover Deck Installation

Deck installation using the floatover deck concept [Salama, et al 1999 and Section 6.1.1 of this handbook] in lieu of the traditional crane vessel lifting is a well-accepted method. This method provides an attractively cost effective way to install decks, especially when the deck weight exceeds available crane lifting capacity. The floatover method eliminates the use of

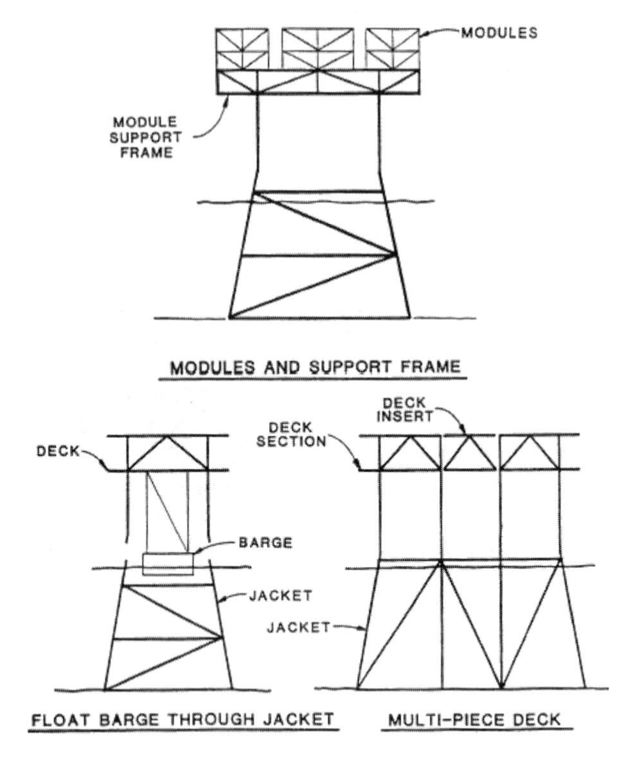

Figure 10.6 Several methods for installing decks on piled jacket structures

heavy lift crane vessels since it uses the cargo vessel itself as an installation vessel. This method has been successfully used offshore in numerous deck installations. If environmental conditions are favourable and a protected deepwater site is available, a catamaran type floatover installation is also possible. In this method, the deck is transported with two vessels and lowered over the platform, which may be a fixed or floating structure.

The floatover installation can be accomplished in two different ways. In the first method the installation barge enters inside the jacket (as shown earlier), moored to it and then lowered down by de-ballasting, gently transferring the deck load onto the jacket legs. The barge is then retrieved from under the installed deck. This method is called floatover installation by barge ballasting. The second method is similar to the first except that hydraulic jacks are used to lift the deck prior to entrance in the jacket and then used again to rapidly transfer the deck load to the piles. This method is called floatover installation by jacking system. This second case is mainly used when the deck is transported with a low centre of gravity above installation barge deck and is then raised prior to entrance in the jacket to maintain sufficient clearance for the entry. Variations of this method are required in areas such as West Africa where ocean swells could cause damaging barge impact loads on the jacket if the barge ballasting method is used.

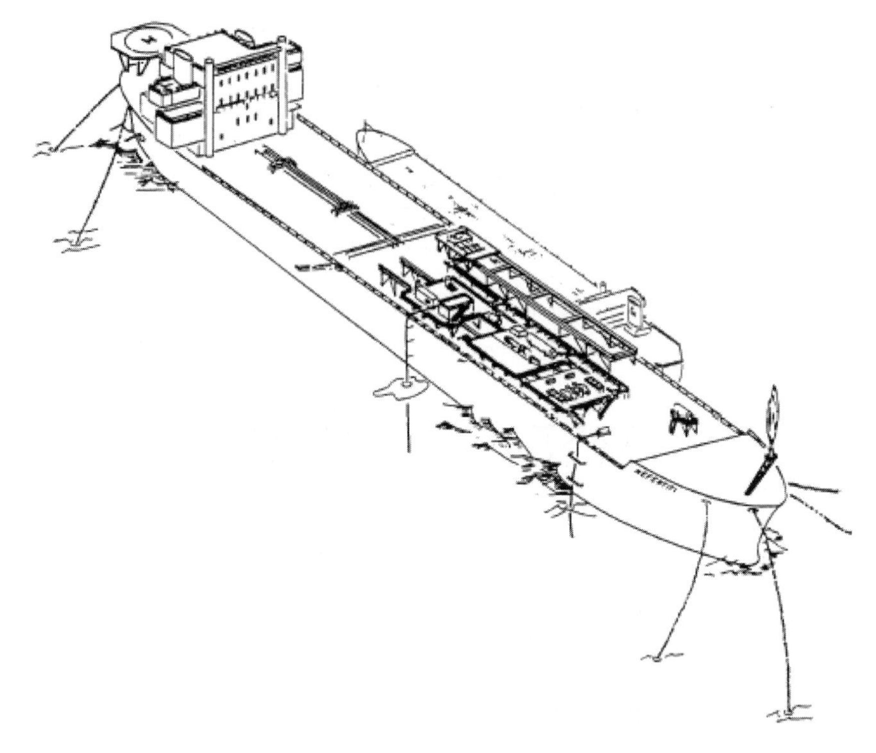

Figure 10.7 Module support frame and equipment installed on an FPSO deck

The selection of the size of the installation barge so that its width would fit the space between the jacket legs while providing adequate stability during transportation, is a vital issue for the whole operation, since the whole philosophy of the installation is based on this selection. Additionally, a fendering/shock absorbing system between the barge and the jacket legs is required in order to prevent steel-to-steel contact at any stage of the installation.

The selection of the floatover installation method will have a major impact on topside layout and design and the selection of the installation barge. The floatover method may look simple at a first glance, but would require considerable preparation for a successful operation.

The Crane Vessel Lifting Method is compared against the Floatover method in table 10.4.

10.7 Helideck

The helideck can be the roof of the quarters building. However, installing the helideck as a separate level over the roof of the quarters building at an additional expense has the advantage of isolating vibration. The size of the helideck is based on its intended use and

Table 10.4 Comparison of floatover and crane vessel lifting methods

Parameter	Floatover Method	Derrick Lifting Method
Cargo barge	Required (generally larger)	Required
Heavy lift crane vessel	Not required	Required
Sensitivity to weather	More weather sensitive	Less weather sensitive
Hook up and commissioning	Most done onshore (using cheaper and more productive man-hours)	Most done offshore (using expensive and less productive man-hours)
Jacket design requirements	Requires bigger and usually heavier jacket. Jacket design governed by the installation method.	Does not need a special jacket design. Jacket design governed by topside layout.
Engineering analysis involved	Requires several types of marine and structural analyses	Only lift analysis required
Weight limitations	Cargo barge capacity is the limit	Derrick lift barge capacity is the limit
Tug boats requirements for the operation	Requires 3 or more tugs	Generally one tug needed
Installation aids/equipment requirements	Cargo barge rigged with installation equipment. Fendering and mooring system required	Lifting gear and spreaders required
Hookup and commissioning	Single piece completed onshore. Efficient layout and piping runs.	May have to be installed in two or more pieces; significantly increases HUC time and cost.
Deck structural strength requirements	Deck strength generally not governed by installation loads	Deck strength may be governed by installation loads
System requirements	Requires an accurate cargo barge ballasting system	May require a specific crane vessel. Installation dependent on vessel availability
Installation time	Weather dependent. Usually longer	Less weather dependent
Installation of secondary items inside the jacket perimeter	After deck installation	Before deck installation
Fendering system requirements	Three different types of fendering required for the deck and the jacket	Bumpers and guides required
Risks during installation	Low	Relatively high

the type of helicopters that will be landing. The surface area of the helideck must exceed that of the helicopter's rotor diameter for proper ground cushion effect. The perimeter safety shelf may be solid for increased ground cushion area or open netting. The landing/departure paths for the helicopter should be provided. All tall objects should be marked with a contrasting paint scheme. Gas should not be vented near a helideck. Gas injected with ambient air can cause the helicopter turbines to overspeed. For further details and recommended practice, see API Recommended Practice 2L, Planning, Designing and Constructing Heliports for Fixed Offshore Platforms (API, January 1983).

10.8 Platform Crane

The main function of the platform crane is to load and off-load material and supplies from boats. The crane is usually located on the top deck over the boat landing area. It is recommended that an open laydown/storage area be located near the crane on each deck level. Loading porches should be provided on the lower deck for easier access. Hatches may be required through the main deck to access equipment on lower levels. The crane is also used for routine equipment maintenance, including handling such items as compressor cylinders, pumps, generators and fire tubes in fired vessels. Localised hoists or monorails may be needed in an area not accessible with the platform crane. Two cranes may be required on large platforms or in areas with rough seas. For further guidance and details please refer to API Recommended Practices 2D "Operation and Maintenance of Offshore Cranes (API, March 1983 and June 2003)".

10.9 Practical Limitations

The layout of equipment, facility and operation is always a compromise as it is not possible to fully separate all equipment from each other. Trade-offs are required. The first step in laying out the equipment for a specific layout and installation concept is to draw a wind rose. The wellheads are then located. These may be pre-determined because they have to be in platform legs or must be accessible from a rig. If a platform rig is required, it is normally laid out next. As a guideline, the production equipment is laid out with preference given roughly to the following hierarchy:

- Isolate quarters and helideck on windward side.
- Place vent or flare on leeward side and locate cranes.
- Separate ignition sources from fuel sources where possible.
- Locate rotating machinery for access to cranes.
- Put utilities and water handling equipment near quarters.
- Optimise placement of equipment to minimise piping.

10.10 Analysis of Two Example Layouts

Two different deck layouts are compared in this section. The examples are taken from an API Recommended Practice, which is no longer in print (API RP 2G 1974). The layout

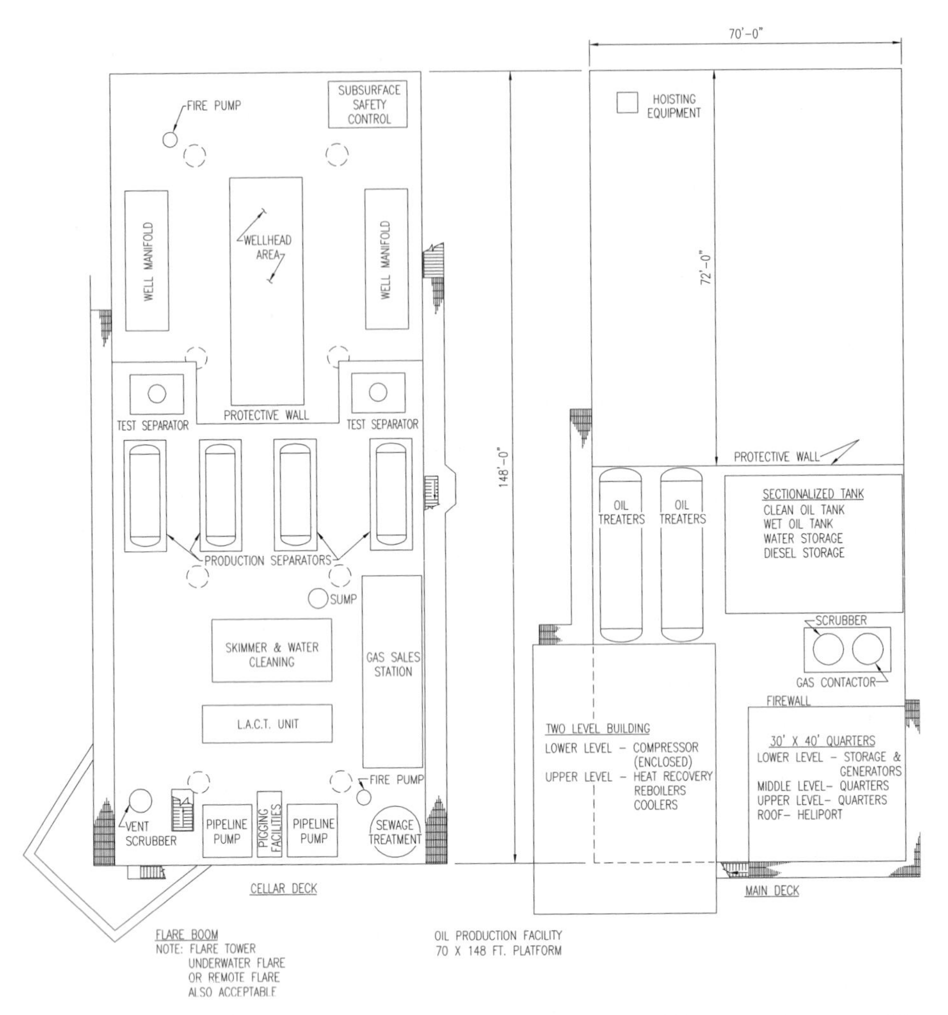

Figure 10.8 First example layout, oil production facility (API, January 1974. Reproduced courtesy of the American Petroleum Institute).

in the first example given in fig. 10.8 shows a preliminary oil production facility layout on a two level $70' \times 148'$ platform deck. The prevailing wind direction is from the northwest.

The following is a brief discussion of the positive points of this layout:

1. Protective walls on both the cellar and main deck effectively isolate the hazardous wellhead area from the remainder of the platform.
2. The quarters building is located as far away from the wellhead area as possible and is protected with firewalls on the sides facing the process equipment.

3. The two escape routes of the main deck are located near the quarters building and are partially shielded by the building itself.
4. The cellar deck is well laid out with the separators conveniently located near the wells. The skimmer and water cleaning skid, the oil automatic custody transfer (LACT) unit and the gas sales station are located between and isolate the fuel source of the separators from the potential ignition source of the engine driven pipeline pumps.
5. The cellar deck firewater pump has a duplicate backup pump, which is located away from the main pump in the event of a local area problem.
6. The enclosed compressor on the main deck is vertically isolated from the reboilers and heat recovery units by using a two level building.
7. Good vertical isolation has been obtained for the major fuel sources of the main and cellar decks by placing the oil treaters and storage tank directly above the separators.

Some of the negative points concerning this layout are:

1. There is only one escape route around the protective wall on the main deck.
2. The main deck is quite congested and access around the oil treaters is restricted.
3. The large fuel sources represented by the oil treaters and the storage tank on the main deck are a major hazard to the quarters building if a fire should occur.
4. The compressors are located adjacent to the quarters and the generators beneath the quarters presenting noise and vibration problems.
5. The flare boom is located near the helideck.
6. The following sources of potential high-pressure gas leaks are located near the quarters: contact tower, compressor and gas sales.
7. The platform crane cannot be used to maintain either the compressor or generator. No provision is made for crane access to lower deck.

The layout in the second example shown in fig. 10.9 shows another oil production facility layout on a two level $72' \times 150'$ platform deck.

A discussion of the positive points of this layout follows:

1. Quarters building is located as far from the wellhead area as possible.
2. Quarters building is additionally protected by using firewalls on the inboard sides of the building.
3. Locating the potable water compartment of the sectionalised tank adjacent to the quarters serves as a safety buffer between the personnel and a large concentration of clean oil.
4. Oil treater and glycol reconcentrator utilise waste heat from compressors and are located directly above for compact, efficient arrangement.
5. The cellar deck is well laid out with adequate space between skids.
6. The platform crane can be used to maintain the compressor and aid in the maintenance of the generator.

Some of the negative points concerning this layout are:

1. No escape routes are shown off the cellar deck to the boat landing.

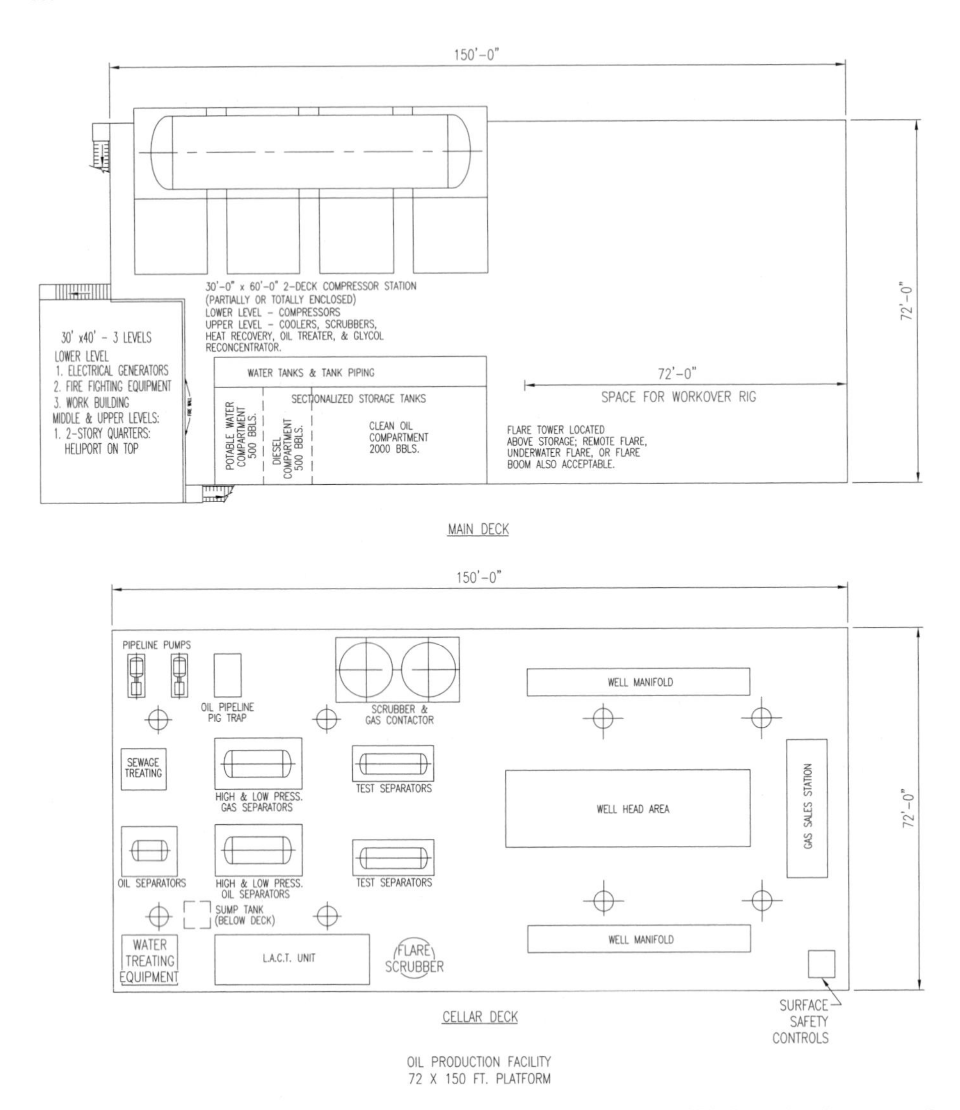

Figure 10.9 Second example layout oil production facility (API, January 1974. Reproduced courtesy of the American Petroleum Institute).

2. The stair at the corner of the main deck should be rotated 180 degrees to put the top of the stair closer to the quarters building in case of an emergency.
3. The large clean oil tank on the main deck has an exposed side near the workover rig area. There is a possibility that this tank may be punctured; a protective wall should be added to protect the tank.

4. The gas sales station on the cellar deck is in an awkward location from a piping standpoint. In addition, the LACT unit and pipeline pumps are on opposite sides of the platform.
5. No provision is made for crane access to lower deck to help in maintaining pumps, etc.
6. The helideck is located close to the vertical flare tower.
7. The compressors and generators are located close to the quarters presenting potential noise and vibration problems.
8. The compressor is close to the quarters providing a potential source of high-pressure gas leaks.
9. There may be insufficient vertical clearance for the gas contactor.
10. Potable water in sectionalised tank is subject to possible contamination from diesel fuel in the adjacent compartment.

10.11 Example North Sea Britannia Topside Facility

The development of the North Sea SPAR Britannia topside facility was described by Garga (1999). Key factors that dictated the choice of facilities on the production platform were:

- Approximately 30 wells were drilled at the platform location in order to reach the vast aerial extent of the reservoir. The use of extended reach, near horizontal wells with measured depths of 31,000 ft (true vertical depths of 13,000 ft). Although 10 wells were to be pre-drilled, the remaining wells required a single, full drilling rig and services facility with a hook load of 510 ton, rotary torque of 60,000 lb ft 5000 psi surface rated equipment and mud circulation rates of 1500 usgpm at 7500 psi at 1.3SG,
- Basic separation of wet gas from the gas condensate reservoir into rich gas, condensate liquids and water. Produced gas to be dried to water dryness of 1 lb of water/MMSCF and to remain in dense phase at a pressure higher than 110 bars at the on-shore terminal end of the gas export pipeline. The condensate to be stabilised to 125 TVP (true vapor pressure) and boosted to an inlet pressure of 180 barg into the condensate export pipeline,
- Production is intended from a large number of subsea wells. The control of production required a heating medium system, chemicals and their own dedicated test separator,
- Blast walls were used between hazardous and non-hazardous areas of the plant to control the effects of blast over-pressures and assist in the safe evacuation of personnel.

The basic topside layout of the Britannia platform segregates the hazardous areas (containing hydrocarbons – process, well bay, gas compressor and condensate export) from the non-hazardous areas (containing no hydrocarbons – utilities, control room, accommodation and lifeboats) by means of solid blast resisting walls and floors. The hazardous areas are themselves compartmentalised by blast walls in order to contain the consequences of likely events. This aspect of Britannia facilities is shown in fig. 10.10, in which the blast walls are shown as thick, black lines. Solid blast walls (shown as dark lines) separate hazardous process facilities from wellbay and from non-hazardous utilities.

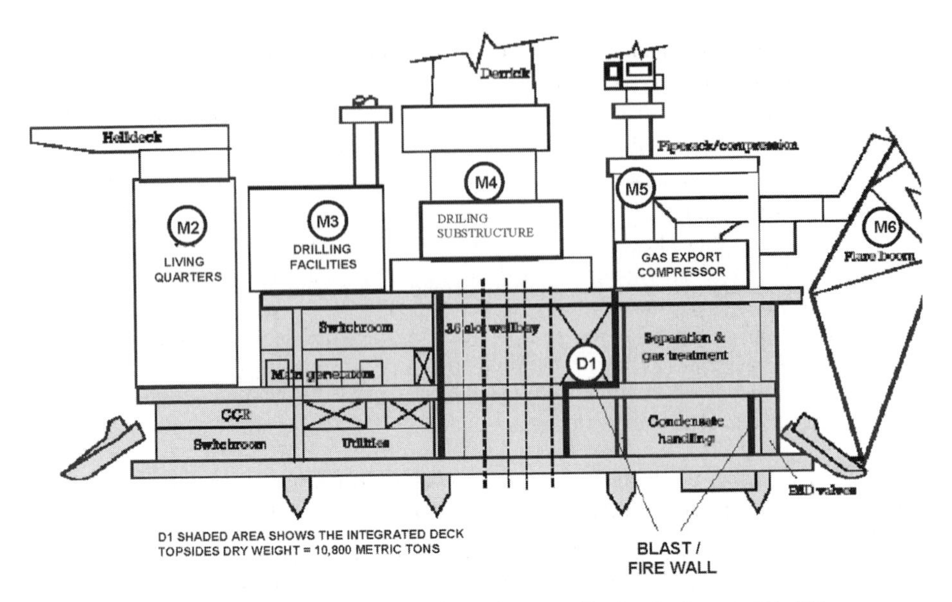

Figure 10.10 Side elevation of Britannia platform topside showing layout of facilities

The safety design of the topside, apart from incorporating the usual active and passive fire protection systems, fire and gas detection systems and multiple safe evacuation devices, took a structured approach towards increasing safety with respect to fire and explosion events. This was achieved by incorporating – inherent safety via reduction of likelihood of leaks, measures to reduce ignition probabilities, improved ventilation and explosion venting, good detection of leaks and rapid isolation and blast protection for personnel, emergency equipment and critical plant. The notable design features that resulted from the above structured approach were:

- The knock-on benefit of minimum plant/minimum sparing in reducing hydrocarbon inventories in vessels, equipment and pipe work,
- Critical review and elimination of breaks or entries into pipework normally associated with maintenance or control needs. This review, undertaken with operations and maintenance personnel, was significantly aided by having all the key influencing parties in an alliance, working to common objectives, – open, high vertical height module spaces to provide good ventilation (over 100 air changes per hour) and explosion venting. Note that lowering module packing densities by enlarging roof heights goes against the philosophy of tighter packing to reduce weight and cost, but significantly reduces blast overpressures with its resultant benefits in lowering the weight and cost of blast resisting structures,
- Blast walls between process and wellbays (see fig. 10.10) designed to resist between 2 and 4 bars of overpressure. The blast wall between wellbay and utilities areas of double skin construction, also used as a potable water tank,

- A large flare relief system of 830 MMSCFD, allowing blowdown to 7 barg in 13 min and atmospheric pressure in 30 min,
- High specification on cable entries and terminations into equipment in hazardous areas and isolation of power at 50% gas LEL (lower explosion limit),
- Extensive use of the latest technology in fire and gas detection and control systems,
- Use of electrically driven combined seawater and firewater duty pumps providing a more reliable, faster and high deluge water volume response,
- Critical systems (fire water, control cables) and their supports designed to survive high blast overpressures, – the adoption of self-verification methods, with their requirement to define "safety critical systems" and the attention to detail of the critical elements within those systems (in design, manufacture, testing, operations and maintenance).

References

API (January 1974). "Production facilities on offshore structures", Recommended practice API-2G, (1st ed.), American Petroleum Institute, Washington DC.

API (January 1983). "Planning, designing and constructing heliports for fixed offshore platforms", Recommended practice API-2L, (2nd ed.), American Petroleum Institute, Washington DC.

API (March 1983). "Offshore cranes", Specification API-2C, (3rd ed.), American Petroleum Institute, Washington DC.

API (September 1993). "Recommended Practice for design and hazard analysis for offshore production facilities", Recommended practice API-14J, (RP 14J), American Petroleum Institute, Washington DC.

API (June 2003). "Operation and maintenance of offshore cranes", Recommended practice API-2D, (5th ed.), American Petroleum Institute, Washington DC.

Croft-Bednarski, S. and Johnston, K. (1999). "Britannia topsides: a low cost, safe and productive north sea facility", *Offshore Technology Conference*, OTC 11018, Houston, Texas.

Edwards, C. D., Cox, B. E., Geesling, J. F., Harris, C. T., Earles, G. A., and Webre, Jr., D. (1997). "Design and construction of the mars TLP deck", *Offshore Technology Conference*, OTC 8372, Houston, Texas, 5–8 May.

Garga, P. K. (1999). "Britannia topsides: a safe and productive north sea facility at lowest cost", *Offshore Technology Conference*, OTC 11015, Houston, Texas.

Milburn, F. H. and Williams, R. H. (2001). "Hoover/Diana: topsides", *Offshore Technology Conference*, OTC 13083, Houston, Texas.

Salama, K. S., Suresh, P. K., and Gutierrez, E. C. (1999). "Deck installation by floatover method in the arabian Gulf", *Offshore Technology Conference*, OTC 11026, Houston, Texas.

Handbook of Offshore Engineering
S. Chakrabarti (Ed.)
© 2005 Elsevier Ltd. All rights reserved.

Chapter 11

Design and Construction of Offshore Pipelines

André C. Nogueira and David S. Mckeehan
INTEC Engineering, Houston, Texas

11.1 Introduction

During the sixties, offshore pipeline design saw the vigour and strength for a young structured engineering field, as solutions to the practical problems demanded innovation and vision. Such initial vigour and strength is documented in numerous scientific papers and research reports of this era. For example, in the 1960s Shell Research and Development carefully studied and advanced the water depth of pipeline. Dixon and Rutledge (1968) published the stiffened catenary solution for offshore pipelines. In the mid 1960s, a straight stinger was used to lay pipe in the North Sea [Berry, 1968]. A patent for the articulated stinger was issued in 1969 [Broussard, et al 1969]. The articulated stinger provided major technology advancement in the feasibility of laying pipe in ever-deeper waters. For the historically inclined reader, Timmermans (2000) presents an interesting overview of the development and achievements of the offshore pipeline design and construction discipline worldwide.

The objective of this chapter is to serve as a reference and guide to the offshore pipeline engineer during the design process. The following aspects of offshore pipeline design are discussed: the establishment of a design basis, aspects of route selection, guidance in sizing the pipe diameter, wall thickness requirements, on-bottom pipeline stability, bottom roughness analysis, external corrosion protection, crossing design and construction feasibility. These topics encompass the majority of issues regarding offshore pipelines. Some issues not covered herein are expansion analysis, curve stability, risers and steel catenary risers (SCRs), analysis of the installation of in-line appurtenances, fracture analysis of weldments and subsea connections.

11.2 Design Basis

The first step in offshore pipeline design is establishing a concise design basis document (DBD). For consistency, every project requires, in its early phase, the establishment of the DBD. This is to be used as a reference by the design team for the different aspects of offshore pipeline design. The DBD provides basic project-dependent information, and enables consistency and correctness of project calculations, reports, bid specifications, contract documents, installation procedures, etc, with respect to the fundamental parameters of the project.

For a major project, a DBD should include the following sections:

- *Development overview*: describes the project location and basic layout.
- *Reservoir and well information*: provides reservoir characteristics, fluid rheology and production rates.
- *Environmental*: defines geotechnical properties along the proposed route (shear strength, weight, etc.), meta-ocean data (waves and currents), and seawater temperatures/chemistry.
- *Flow assurance*: provides information on flowline parameters, e.g. operating pressures, temperatures and velocities; identify and address flow hazards such as hydrates, wax, scale, corrosion, slugging.
- *Wellbore, drilling and completion information*: provides safety valve philosophy, downhole chemical injection, completion design, downhole monitoring, general rig description, well servicing and intervention process, etc.
- *Equipment design philosophy*: describes the design and selection approach, standardisation of components and interfaces, equipment design life, quality programme.
- *Subsea trees and flowline/pipeline sleds*: describes subsea trees, flowline/pipeline sled characteristics, tie-in jumpers, completion/workover system, equipment marking, corrosion protection.
- *Production control system and umbilicals*: establishes codes and standards, system overview, subsea instrumentation, redundancy, emergency shut-down valve (ESV) requirements, surface equipment, subsea equipment, power and hydraulic umbilical, methanol distribution umbilical (if required), intelligent well completions, metering.
- *Pipelines*: provides general characteristics (grade, size, water depth), route selection, applicable codes and regulation, system design requirements (design life, cathodic protection system, etc.), risers and tie-ins, maximum shut-in pressure, corrosion allowance.
- *Host facilities*: gives general description of the host, process design, major equipment list, interface definitions.
- *Operation and maintenance*: outlines normal production parameters, start-up and shutdown procedures, routine testing requirements, pigging, system maintenance, abandonment philosophy.

In general, the DBD is a living document that goes through several revisions during the course of a project. However, after the front end engineering design (FEED) phase of a project, the DBD must define the majority of the project requirements. After FEED, a successful project will incorporate change only by an established "management of change" process, which provides evaluation of the proposed change, and its implication with regard to safety, cost and schedule.

11.3 Route Selection and Marine Survey

In the FEED phase of a project, typically the seafloor bathymetry data is available to the so-called "regional survey" level. This means that coarse surface tow and swath bathymetry survey data are available for preliminary route selection, but not to a level of detail required for a finalisation. At this point, the pipeline lead engineer should select the base case route based on the regional survey data.

If a challenging bathymetry is present, alternative routes should also be defined; environmental sensitivity zones should be avoided, as well as excessive span areas. During the detailed marine survey, a pipeline engineer should be on-board to perform a real-time bottom roughness analysis. Frequent communications should take place between the on-board pipeline engineer and the design office to assure a successful marine survey, which will suffice for purposes of supporting a final route selection as well as the required geohazard survey report.

11.4 Diameter Selection

The selection of diameter is a process where the initial capital expenditure (capex) and operational expenditure (opex) are evaluated leading to an optimised design by minimising total cost through the life of the project. The main criterion for selection of pipeline diameter is the ability to carry fluids at the design flow rates, within the allowable pressure. Figure 11.1 depicts the processes and logic involved in the selection of an initial diameter, and the flow assurance work needed to guarantee operability of the system.

To provide some reference point regarding diameter, flowrate and operating pressures, table 11.1 summarises the pipe diameters used in selected offshore developments. Key parameters are given, as well as references, for the interested reader to obtain more details regarding diameter selection and flow assurance. Diameters are provided as nominal outside diameter (OD).

11.4.1 Sizing Gas Lines

The practice for selecting a pipe diameter is a detailed hydraulic analysis; especially for multi-phase flow with untreated gas. However, a quick way to estimate the size of dry, single phase, gas lines is to use the simplified equation (11.1) [McAllister, 1993]. For small gathering lines, the answer will have an accuracy within 10% of that obtained by more complex formulas.

$$Q = \frac{500\, ID^3 \sqrt{P_1^2 - P_2^2}}{\sqrt{L}} \tag{11.1}$$

where Q = cubic ft of gas per 24 h, ID = internal pipe diameter in inches, P_1 = psia at starting point, P_2 = psia at ending point, L = length of line in miles.

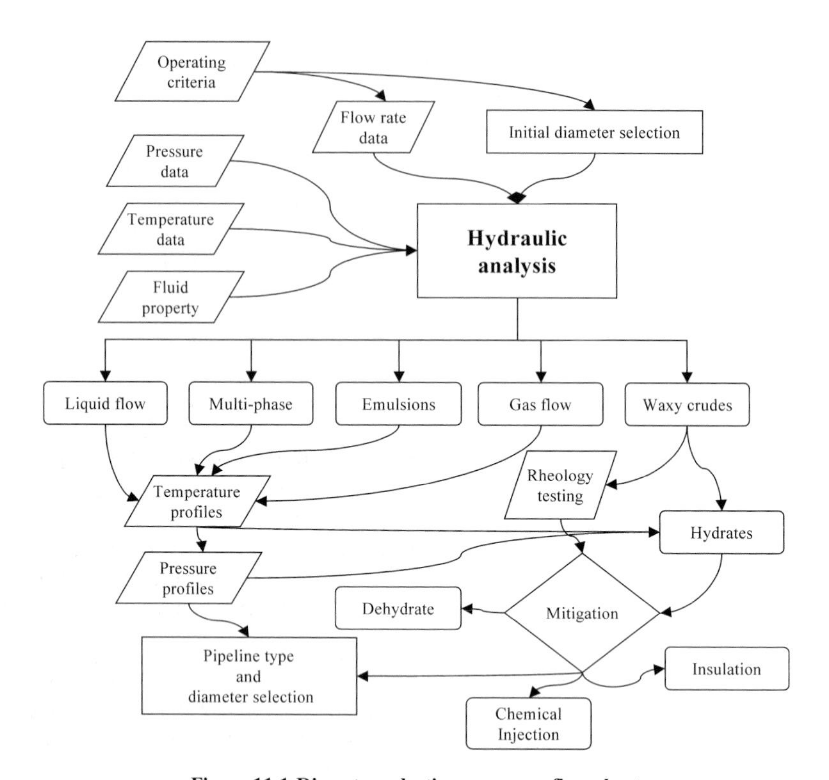

Figure 11.1 Diameter selection processes flow chart

Table 11.1 Diameter for selected offshore projects

Project	Contents and peak flow rate	Maximum operating pressure	Pipeline length	Nominal OD
Canyon Express[1]	Gas-condensate at 500 MMSCFD	4200 psig at 40°F	55 miles	2 × 12 in.
Northstar Export Line[2]	42° API Oil at 65,000 barrels per day	1480 psig at 100°F max. temperature	6.0 miles	10 in.
Northstar Gas Line[2]	Processed gas 100 MMSCFD	1480 psig	6.0 miles	10 in.
Scarab/Saffron[3]	Gas-condensate at 600-1800 MMSCFD	3750 psig	56 miles	2 × 20 in. 1 × 24 in. 1 × 36 in.

[1]Wallace, et al (2003)
[2]Lanan, et al (2001)
[3]Choate, et al (2002)

For example, given an 8 in. ID line, 9 miles long, if the pressure at the staring point is 485 psi and the pressure at the downstream termination is 283 psi, the total gas flow is estimated at:

$$Q = \frac{500(8^3)\sqrt{(485+15)^2-(285+15)^2}}{\sqrt{9}} = 34.1 \text{ million cubic ft per day (MMCFD)}$$

11.4.2 Sizing Oil Lines

The sizing of oil lines is more complex than gas lines due to the different viscosities and specific gravities of crude oil. However, the table 11.2 from McAllister (1993) provides guidance in selecting line size, and pressure drop for oil of approximately 40° API gravity and 60 SSU viscosity.

11.5 Wall Thickness and Grade

API 5L grade X-65 has become the steel grade of choice for deepwater offshore pipelines. The main reasons for this choice are cost-effectiveness and adequate welding technology. A lower grade, X-60, is typically used for SCRs, to ensure easier welding overmatch for these structures, and an improved fatigue life. For buried offshore pipelines in the Arctic, a more ductile, X-52 grade has proven the best choice for limit state design and the need for high toughness material that could sustain the high strain base design [Lanan, et al 2000; Nogueira, et al 2000; Lanan, et al 2001].

To calculate the required wall thickness for an offshore pipeline, three different failure modes must be assessed:

- Internal pressure containment (burst) during operation and hydrotest.
- Collapse due to external pressure.
- Local buckling due to bending and external pressure.

Table 11.2 Crude oil sizing guidance

For	Use	
Throughputs (bbl per day)	Pipe outside diameter (in.)	Pressure drop (psi per mile)
0 to 2000	3½	–
2000 to 3000	4½	32
3000 to 7500	6⅝	16
7500 to 16,500	8⅝	10.5
16,500 to 23,500	10¾	8.5
23,500 to 40,000	12¾	7

A fourth failure mode may be used to calculate the required wall thickness in deep water:

- Buckle propagation and its arrest.

Designing for each of these failure modes is discussed in each of the sub-sections below. A numerical design example, covering each failure mode, is given in Section 11.7.

11.5.1 Internal Pressure Containment (Burst)

Pipelines to be installed in the Gulf of Mexico, or in any place within the jurisdiction of the Minerals Management Service of the United States, must comply with the appropriate Code of Federal Regulation (CFR). Three parts of these regulations are applicable for offshore pipelines:

- Title 30, part 250 of the CFR [30 CFR 250, 2002] entitled "Oil and gas and sulphur operations in the outer continental shelf (OCS)", and in particular subpart J entitled "Pipelines and Pipeline Rights-of-way". This defines the so-called Department of Interior's (DOI) jurisdiction, or a DOI pipeline. Per 30 CFR 250.1001: "DOI pipeline refers to a pipeline extending upstream from a point on the OCS where operating responsibility transfers from a producing operator to a transporting operator". This is applicable to pipelines from wells to platforms.
- Pipelines in the OCS, which are not DOI pipelines and are used in the transportation of hazardous liquids or carbon dioxide, must follow the Department of Transportation standards presented in 49 CFR 195 (2002), subpart A. These provisions are applicable for oil pipelines from platforms to shore, or other tie-in points into existing pipeline transportation systems.
- Pipelines in the OCS, which are not DOI pipelines, and are used in the transportation of gas, must follow the Department of Transportation standards presented in 49 CFR 192 (2002), subpart A. These provisions are typically applicable for gas pipelines from platforms to shore, or other tie-in points into existing pipeline transportation systems.

For simplicity, the following wall thickness design requirements are based on the provisions of 30 CFR 250.1002, entitled "Design requirements for DOI pipelines". The other two CFRs contain very similar design requirements.

30 CFR 250.1002 adopts an allowable stress design format. That is, the basic (burst) design equation sets the internal design pressure, P_{id}, to a value such that the resulting hoop stress is a fraction of the pipeline yield stress. The relationship between P_{id} and the (nominal) wall thickness is given by:

$$P_{id} = \frac{2S_Y t}{D} FET \tag{11.2}$$

Equation (11.2) above is given by 30 CFR 250.1002(a), which also defines the following terms (definition below are transcribed verbatim):

P_{id} = internal design pressure,
t = nominal wall thickness,
D = nominal outside diameter of pipe,
S_Y = specified minimum yield stress,

Table 11.3 Temperature de-rating factor, *T*, for steel pipe

Temperature (°F)	Temperature De-rating Factor, T
250 or less	1.000
300	0.967
350	0.933
400	0.900
450	0.867

F = construction design factor of 0.72 for the submerged component and 0.60 for the riser component,

T = temperature de-rating factor obtained from Table 841.1C of ANSI B31.8, see table 11.3.

E = longitudinal joint factor. Obtained from Table 841.1B of ANSI B31.8 (see also Section 811.253(d)) – see table 11.4.

According to 30 CFR 250, all pipelines should be hydrostatically tested with water at a stabilised pressure of at least 1.25 times the maximum allowable operating pressure (MAOP) for at least 8 h. The test pressure should not produce a stress in the pipeline in excess of 95% of the specified minimum-yield strength of the pipeline.

The relationship between the maximum hydrotest pressure and the (nominal) wall thickness is similar to equation (11.2), and is given by:

$$P_{\text{max-hyd}} = \frac{2S_Y t}{D} FET \qquad (11.3)$$

where $P_{\text{max-hyd}}$ = maximum hydrostatic test pressure, and F = construction design factor, 0.95 for hydrotest.

11.5.2 Collapse Due to External Pressure

During installation, offshore pipelines are typically subjected to conditions where the external pressure exceeds the internal pressure. The differential pressure acting on the pipe wall due to hydrostatic head may cause collapse of the pipe. Several design codes present formulation addressing the design against this failure mode. Amongst these codes, the most prominent are API RP 1111 (1999) and DNV OS-F101 (2000).

The elastic collapse pressure [equation (11.5b)] and the plastic collapse pressure [equation (11.5c)] bound the problem. Timoshenko and Gere (1961) proposed a bi-linear transition between the two equations [see Timoshenko and Gere, 1961, figs. 7–9], which adequately bridges the two equations. API RP 1111 (1999) adopts a transition between the two equations [equation (11.5a)], which is simpler than the cubic interaction equation proposed by DNV. A comparison between the API and DNV collapse pressures, P_c, normalised by the plastic collapse pressure, P_y, is given in fig. 11.2. P_c as calculated by equation (5.18) in the DNV OS-F101 (2000), uses ovalisation parameter $f_o = 1\%$, as defined in DNV OS-F101

Table 11.4 Longitudinal joint factor, E

Spec No.	Pipe class	E factor
ASTM A 53	Seamless	1.0
	Electric resistance welded	1.0
	Furnace butt welded – continuous weld	0.6
ASTM A 106	Seamless	1.0
ASTM A 134	Electric fusion arc welded	0.8
ASTM A 135	Electric resistance welded	1.0
ASTM A 139	Electric fusion welded	0.8
ASTM A 211	Spiral welded steel pipe	0.8
ASTM A 333	Seamless	1.0
	Electric resistance welded	1.0
ASTM A 381	Double submerged-arc-welded	1.0
ASTM A 671	Electric fusion welded	
	Classes 13, 23, 33, 43, 53	0.8
	Classes 12, 22, 32, 42, 52	1.0
ASTM A 672	Electric fusion welded	
	Classes 13, 23, 33, 43, 53	0.8
	Classes 12, 22, 32, 42, 52	1.0
API 5L	Seamless	1.0
	Electric resistance welded	1.0
	Electric flash welded	1.0
	Submerged arc welded	1.0
	Furnace butt welded	0.6

(2000) equation (5.21). No factor of safety has been applied to either formulation. It can be seen that both codes yield very similar results for the collapse pressure. Due to their simplicity, the API equations are recommended for wall thickness design against collapse due to external pressure.

Following API RP 1111 (1999), Section 4.3.2.1, the pipe collapse pressure P_c (i.e. pipe collapse capacity) must be greater than the net external pressure (i.e. effective applied external pressure) everywhere along the pipeline, as specified by equation (11.4) below:

$$(P_o - P_i) \leq f_o \cdot P_c \tag{11.4}$$

where f_o = safety factor: 0.7 for seamless or ERW pipe and 0.6 for cold expanded pipe, P_o = external pressure and P_i = internal pressure.

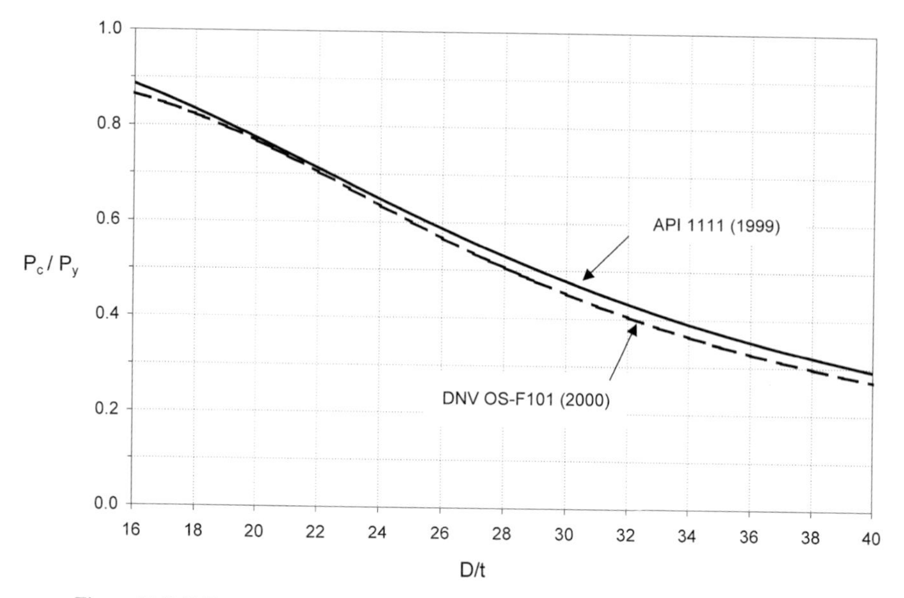

Figure 11.2 Collapse pressure vs. D/t per API 1111 (1999) and DNV OS-F101 (2000)

The collapse pressure is determined by equations (11.5a)–(11.5c):

$$P_c = \frac{P_e P_y}{\sqrt{P_e^2 + P_y^2}} \quad \text{collapse pressure} \tag{11.5a}$$

$$P_e = \frac{2E}{1-\nu^2}\left(\frac{t}{D}\right)^3 \quad \text{elastic collapse pressure} \tag{11.5b}$$

$$P_y = 2S_y\frac{t}{D} \quad \text{plastic collapse pressure} \tag{11.5c}$$

where E = modulus of elasticity of steel, and ν = Poisson's ratio, 0.3 for steel.

11.5.3 Local Buckling Due to Bending and External Pressure

This failure mode is typically most severe during installation when bending and external pressure effects are critical. However, local buckling also applies for the installed pipeline, in case of depressurisation. API RP 1111 (1999) and DNV OS-F101 (2000) have adequate formulations that address this failure mode, which are based exclusively on empirical data fitting.

Once again, due to its simpler treatment of the subject, the API RP 1111 (1999) is the one presented herein. For D/t upto 50, the following interaction equation needs to be satisfied following API RP 1111 (1999), Section 4.3.2.2.

$$\frac{\varepsilon}{\varepsilon_b} + \frac{(P_o - P_i)}{P_c} \le g(\delta) \tag{11.6}$$

where

ε = critical strain (maximum compressive strain at onset of buckling)

$\varepsilon_b = \frac{t}{2D}$ = critical strain under pure bending

$g(\delta) = (1+20\,\delta)^{-1}$ = collapse reduction factor

$\delta = \dfrac{D_{\max} - D_{\min}}{D_{\max} + D_{\min}}$ = ovality

$D_{\max}$ = maximum diameter at any given cross section

$D_{\min}$ = minimum diameter at any given cross section

Equation 11.6 can be rewritten as:

$$\varepsilon \leq \left[g(\delta) - \frac{(P_o - P_i)}{P_c} \right] \frac{t}{2D} \tag{11.7}$$

The bending strains shall be limited as follows:

$$f_1 \varepsilon_1 \leq \varepsilon \tag{11.8a}$$

$$f_2 \varepsilon_2 \leq \varepsilon \tag{11.8b}$$

where ε_1 = maximum installation bending strain, ε_2 = maximum in-place bending strain, f_1 = safety factor for installation bending plus external pressure, and f_2 = safety factor for in-place bending plus external pressure.

A value of 2.0 for safety factors f_1 and f_2 is suggested by API RP 1111 (1999). Safety factor f_1 may be larger than 2.0 for cases where installation bending strain could increase significantly due to off-normal conditions, or smaller than 2.0 for cases where bending strains are well defined (e.g. reeling).

11.5.4 Rational Model for Collapse of Deepwater Pipelines

The above API formulation is based on empirical data fitting [Murphey and Langner, 1985]. Palmer (1994) pointed out that "it is surprising to discover that theoretical prediction (of tubular members collapse under combined loading) has lagged behind empirical prediction, and that many of the formula have no real theoretical backup beyond dimensional analysis". Recently, this situation has changed with the rational model formulation presented by Nogueira and Lanan (2001). The rational model has been derived from first principles, e.g. equilibrium of forces and moments; and its predictions have been shown to correlate very well with test results.

The cornerstone of the rational model is the recognition that when a pipe is subjected to bending moment, the longitudinal stresses generate transverse force components due to the pipe curvature. As a pipe bends, components of the longitudinal bending stresses act into the cross-section. This, in turn, generates a transverse moment, which ovalises the pipe cross section, or ring, until it collapses. A pipe under bending will collapse when its cross section (or ring) loses stiffness due to plastic hinges mechanism formation at the onset of local buckling. Therefore, when rings of the pipe lose their stiffness, the ovalisation (initially uniform along the pipe length) will concentrate at the weakest point along the pipe (e.g. a thinner ring) and a local buckle will form. If in addition to bending, pressure is applied, its effects are taken into account by noticing that it contributes to reduce the ring capacity to resist bending. This is due to the effects of the compressive hoop stress.

Since this model has sound theoretical basis, it provides explanation to some intriguing issues in pipe collapse. For example, the rational model includes, by derivation, in its formulation the anisotropy ratio $N = \sigma_{oH}/\sigma_{oL}$, where σ_{oH} is the pipe yield stress in the hoop direction, and σ_{oL} is the pipe yield stress in the longitudinal direction. Tam, et al (1996) reported that when the anisotropy ratio is included in their model, its predictions fit more precisely the experimental results. However, Tam, et al (1996) could not attribute a physical meaning to the ratio N. Following the rational model derivation, the explanation is that greater values of the longitudinal yield stress σ_{oL} (which generates the applied ring load) result in a greater applied transversal load, and lower values of the hoop yield stress σ_{oH} (which characterises the ring load capacity) result in a reduction in the ring capacity. Of course, this effect is numerically captured in the anisotropy ratio N, as defined above. The ratio N can be less than one especially for pipe manufactured by the UOE method.

For explanations of other issues and for complete derivation of the equations of the rational model, see Nogueira and Lanan (2001). The model equations are given below.

Interaction equation $$\frac{8}{\pi} N P_R \Delta_{B\&P} + \varepsilon_T(1 + 2\Delta_{B\&P}) = \varepsilon_{co}(1 - P_R)(1.31 P_R + 1) \tag{11.9}$$

Anisotropy ratio $$N = \frac{\sigma_{oH}}{\sigma_{oL}} \tag{11.10}$$

Normalised pressure $$P_R = \frac{p}{P_y} \tag{11.11}$$

Yield pressure $$P_y = \frac{2\sigma_{oH}}{(D/t)} \tag{11.12}$$

Ovality (due to bending and pressure) $$\Delta_{B\&P} = \frac{f(\varepsilon_T)}{1 - p/P_C} \tag{11.13}$$

Bending ovality $$\Delta = f(\varepsilon_T) = \begin{cases} \dfrac{\pi}{8}\dfrac{\sigma_{oL}}{E}(D/t)^2 \varepsilon_T + \Delta_I; & \text{for } \varepsilon_T \le \varepsilon_{TY} \\ \Delta_Y \left[\dfrac{S-1}{S - (\varepsilon_T/\varepsilon_{TY})}\right]^{(S-1)}; & \varepsilon_T \ge \varepsilon_{TY} \end{cases} \tag{11.14}$$

Reference strain $$\varepsilon_{TY} = \frac{8N}{\pi 3\sqrt{3}(D/t)} \frac{1}{(1 + 2\Delta_I)} \tag{11.15}$$

Reference ovality $$\Delta_Y = \frac{N}{3\sqrt{3}} \frac{\sigma_{oL}}{E} \frac{(D/t)}{(1 + 2\Delta_I)} + \Delta_I \tag{11.16}$$

Initial ovality $$\Delta_I = \frac{D_{max} - D_{min}}{D_{max} + D_{min}} \tag{11.17}$$

Hyperbolic ratio $$S = \varepsilon_{co}/\varepsilon_{TY} \tag{11.18}$$

Critical strain approximation $$\varepsilon_{co} = \frac{2\sqrt{3}N}{\pi(D/t)} = \frac{1.1N}{(D/t)} \tag{11.19}$$

In the above equations, p is the maximum applied external pressure, and ε_T is the rational model's critical strain. For example, given p, the solution of the above equations will yield ε_T. The data required to solve the above equations are:

- Pipe diameter (D)
- Pipe wall thickness (t)
- Pipe yield stress in the hoop direction (σ_{oH})
- Pipe yield stress in the longitudinal direction (σ_{oL})

In the case of external pressure only, the term with ε_T on the left hand-side of equation (11.9) vanishes. In order to obtain the correct results, equation (11.9) is re-arranged as shown by equation (11.20). This will lead to correct critical pressures for the perfect circular pipe.

$$\frac{8}{\pi}NP_R\Delta = \varepsilon_{co}(1 - P_R)(1.31P_R + 1)(1 - p/p_c) \tag{11.20}$$

The collapse pressure predictions of equation (11.20) are shown in figs. 11.3 and 11.4, for pipe with different initial ovalities, compared to experimental results reported by Murphey and Langner (1985). The rational model shows slightly conservative results. The main

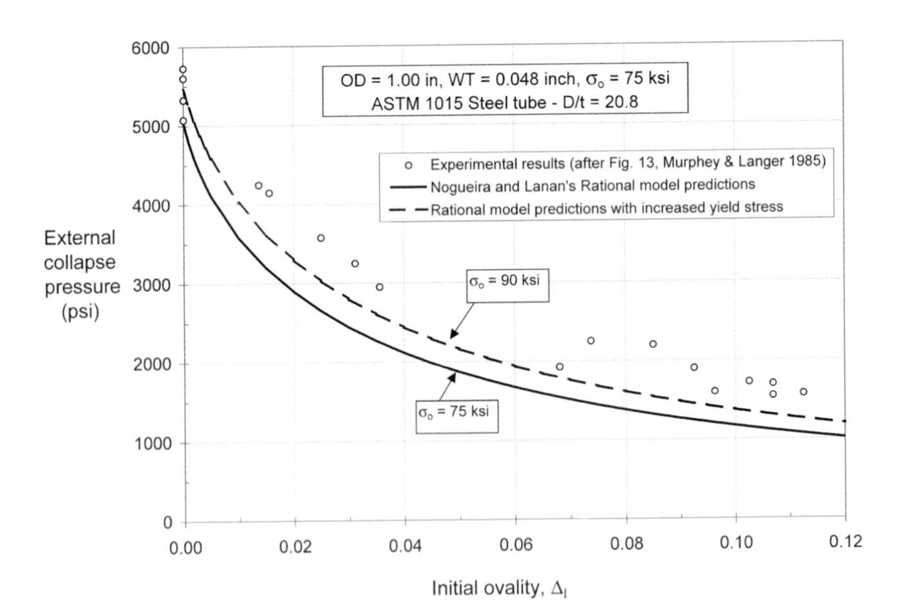

Figure 11.3 Rational model prediction of collapse pressure vs. initial ovality, compared to experimental results for pipe with D/t = 20.8

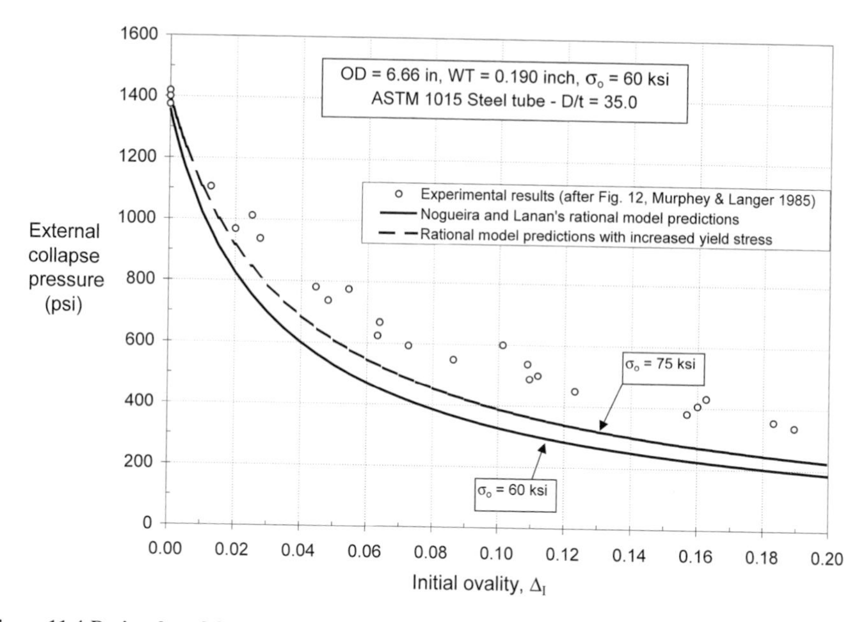

Figure 11.4 Rational model prediction of collapse pressure vs. initial ovality, compared to experimental results for pipe with D/t = 35

reason for this conservatism is the model's elasto-perfectly-plastic material assumption, and that the yield stress given by the authors probably underestimates the actual yield plateau. It is common that the actual yield plateau to be 10–20 ksi higher than the yield stress. Therefore, the collapse pressures predictions are also given for higher yield stresses, as indicated in the figures. Predictions become closer to the experimental values, but are still conservative.

The interaction equation (11.9) can be solved by means of spreadsheets. Results of critical pressure vs. critical strains predicted by the rational model are shown in fig.11.5, which also shows the interaction equation results of API RP 1111 (1999) and DNV OS-F101 (2000), for a pipe with diameter-to-thickness, D/t, ratio equal to 20. The API formulation is conservative for the most part of the interaction plot, except that the rational model is more conservative at very low levels of bending strain. The DNV formulation as well as the rational model produces somewhat similar predictions for higher bending strains, with the rational model formulation predicting the highest capacity for small levels of external pressure.

Figure 11.6 shows rational model interaction equation predictions and compares them with experimental results presented in Table 5 of Fowler (1990). The rational model predictions are based on the average pipe properties. The rational model interaction equation predictions are very close to the experimental values.

As a final comparison, results from the Oman-to-India pipe collapse programme [Stark and McKeehan, 1995] are shown in fig. 11.7, together with the rational model predictions.

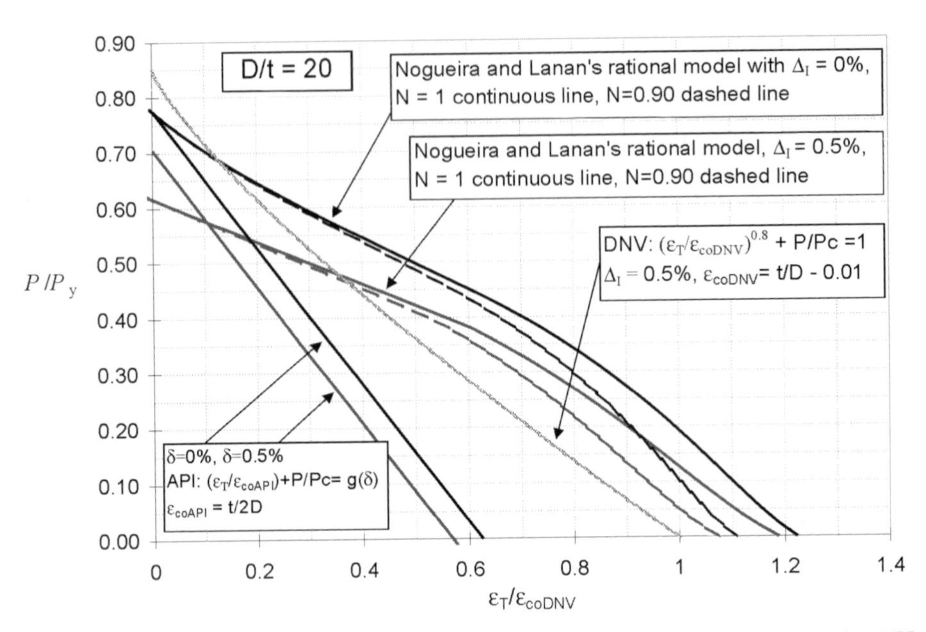

Figure 11.5 Nogueira and Lanan's rational model pressure vs. bending strain prediction compared to API 1111 and DNV OS-F101 for pipe with D/t = 20

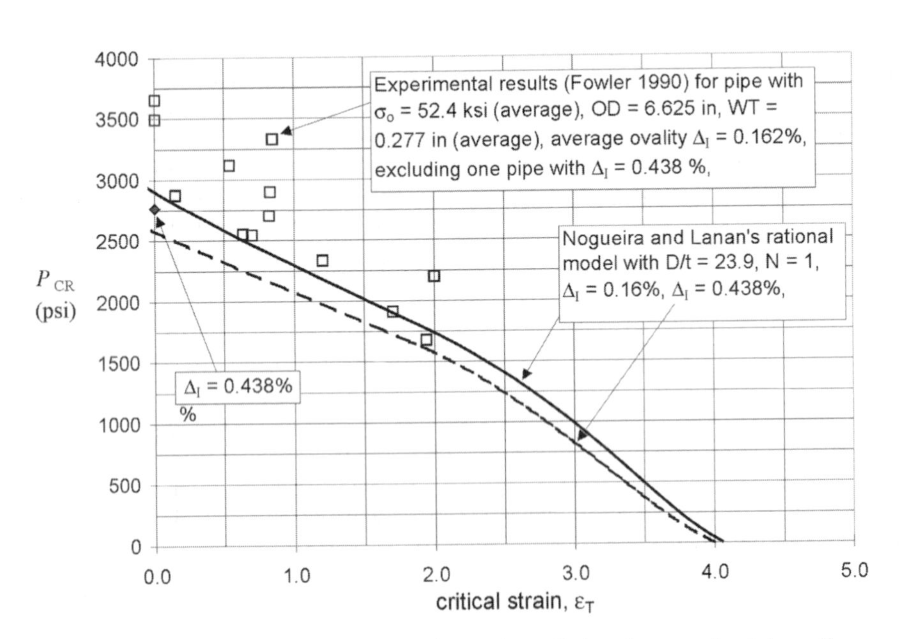

Figure 11.6 Rational model pressure vs. bending strain predictions (average pipe data used) vs. experimental results for pipes with D/t = 23.9 (average)

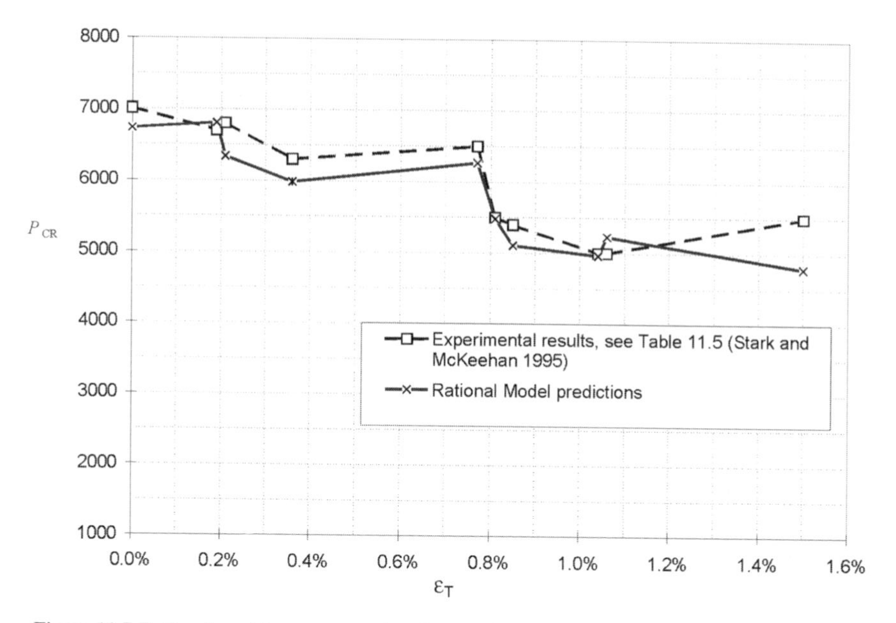

Figure 11.7 Rational model pressure vs. bending strain prediction vs. experimental results after Stark and McKeehan (1995)

In this case, the individual pipe characteristics were taken into account, including hoop and longitudinal yield data (not shown). It can be seen that excellent agreement is also obtained for all test results. For completeness, the original data published by Stark and McKeehan (1995) is provided herein in table 11.5. The hoop yield stress σ_{oH} was obtained by compressive uniaxial test, which is more difficult to obtain, as it requires stiff testing conditions, an accurately machined specimen, and alignment of the line of loading with the axis of the specimen. For this reason, an abnormal hoop stress result for specimen ZFV18 was discarded and substituted by the hoop stress average for all specimens.

11.6 Buckle Propagation

In the unlikely event of a local buckling, the external pressure may cause a buckle to propagate along the pipeline. As long as the external pressure is less than the propagation pressure threshold, a buckle cannot propagate. A number of empirical relationships have been published for determining the minimum pressure, P_p, at which buckle propagation can occur for a given pipe diameter, wall thickness and steel grade. The mechanics of buckle propagation is explained by Nogueira (1998a,b,c).

The API RP 1111 (1999) equation for calculating the propagation pressure is as follows:

$$P_p = 24S\left(\frac{t_{\text{nom}}}{D}\right)^{2.4} \qquad (11.21)$$

Table 11.5 Test data and results for 26 in. OD, 1.625 in. WT, Grade X60, after Stark and McKeehan (1995)

Pipe designation	ZFV8	ZFV16	ZFV21	ZFV18	ZFV19	ZFV22	ZFV7	ZFV23	ZFV20	ZFV13
Applied (live) critical strain (%)	0.00	0.19	0.21	0.36	0.77	0.81	0.85	1.04	1.06	1.50
Collapse pressure (psi)	7009	6700	6800	6300	6500	5500	5400	5000	5000	5500
σ_{oH} (ksi)	60.5	71.5	66.9	63.9[1]	73.4	63.3	61.5	60.8	65.0	61.1
σ_{oL} (ksi)	73.0	80.4	79.4	75.2	73.1	76.3	74.2	75.2	75.0	74.0
OD (in.)	26	26	26	26	26	26	26	26	26	26
WT (in.)	1.619	1.620	1.623	1.618	1.620	1.622	1.616	1.628	1.621	1.640
D/t	16.06	16.05	16.02	16.07	16.05	16.03	16.09	15.97	16.04	15.85
Δ_I(%)	0.0038	0.13	0.16	0.14	0.19	0.16	0.16	0.19	0.18	0.17

[1]The measured hoop stress of specimen ZFV18 was reported to be 56.1 ksi, which was deemed low, due to testing error. The average hoop stress of all specimens was used herein.

If the following equation is satisfied, with the buckle propagation safety factor, f_p, of 0.80, then buckle arrestors are not required. Since, in this case a buckle cannot propagate along the pipeline.

$$P_o - P_i \leq f_p \cdot P_p \tag{11.22}$$

In deepwaters, it is not economically feasible to have the wall thickness satisfy the buckle propagation criteria of equation (11.22). The wall thickness can be chosen to be less than the minimum calculated in equation (11.21), provided that buckle arrestors are recommended to mitigate the risk of buckle propagation. Buckle arrestors can be designed by the formulations presented by Park and Kyriakides (1997) or Langner (1999).

11.7 Design Example

This section provides an example of offshore pipeline design.

Figure 11.8 shows schematically an example gas pipeline; corresponding design data is listed in table 11.6. The example gas pipeline runs from a subsea well at 8000 ft water depth to a shallow water host platform at a water depth of 500 ft. At the platform, a riser segment brings the gas to the topside piping. The pipeline is assumed to be 10.75 in. diameter and it has wall thickness (WT) break at 3000 ft water depth. 30 CFR 250 (2002) applies, and each wall thickness will be calculated, or verified, to prevent the three failure modes identified as internal pressure containment (burst) during operation and hydrotest, collapse due to external pressure, and local buckling due to bending and external pressure.

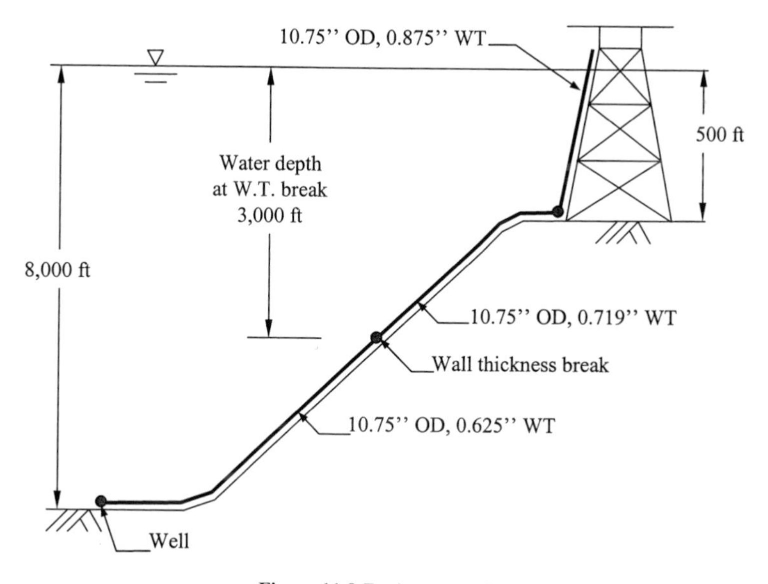

Figure 11.8 Design example

Table 11.6 Summarises the design data for this example

Parameter	Value
Pipe material	API 5L X-65 seamless
Pipe OD[1]	10.75 in.
Corrosion allowance, CA	1/16 in.
Specified minimum yield strength	65,000 psi
Minimum ultimate tensile strength	77,000 psi
Modulus of elasticity	29×10^6 psi
Steel density	490 lb/ft^3
Poisson's Ratio	0.3
Pipeline content	Natural gas
Gas density	14 lb/ft^3
Gas maximum temperature	200°F
Gas minimum temperature	40°F
Minimum internal pressure	0 psig
Maximum Source Pressure (MSP) at well-pipeline interface	6400 psig
Seawater density	64 lb/ft^3
Maximum water depth	8000 ft
Minimum water depth	500 ft
Maximum installation bending strain	0.2%
Maximum in-place bending strain	0.2%
100-year ARP[2] bottom current velocity	1.3 ft/s

[1]OD = outside diameter
[2]ARP = average return period

11.7.1 Preliminary Wall Thickness for Internal Pressure Containment (Burst)

The first step to determine the wall thickness is to satisfy the required design pressure, P_{req}, at the shut-in condition. The general equation for the required design pressure is:

$$P_{\text{req}} = MSP - P_{pgas} - P_o \qquad (11.23)$$

where

MSP = maximum source pressure, which equals the wellhead pressure at the well–pipeline interface.

P_{pgas} = internal gas weight (at shut-in condition) from source elevation to elevation of interest.

Table 11.7 Required design pressure for maximum source pressure

Water depth (ft)	P_o (psig)	P_{pgas} (psig)	P_{req} (psig)
0	0	778	5622
500	222	729	5449
3000	1333	486	4581
8000	3556	0	2844

Table 11.7 gives the required design pressure, per equation (11.23), for the significant water depths assuming a seawater weight of 64 pcf. These values will be used to check that the required design pressure does not exceed the maximum allowable operating pressure (MAOP) of the pipeline system.

Based on the numbers given on table 11.7, the minimum hydrostatic test pressure equals 1.25 times the maximum required design pressure, or $1.25 \times 5622 = 7028$ psig. This assumes that the entire pipeline system will be subject to a single hydrostatic test. The nominal hydrostatic test pressure is set at a slightly higher value, as follows:

- Nominal hydrostatic test pressure, $P_{\text{nom-hyd}} = 7100$ psig @ mean water level (MWL)
- Leading to 80% of hydrostatic test pressure = 5680 psig @ MWL

The wall thickness must be selected such that equation (11.2) [given as a design inequality by equation (11.24)] and equation (11.3) [given as a design inequality by equation (11.25)] are satisfied at every point along the pipeline and riser system, as follows:

$$P_{\text{req}} \leq P_{id} = \frac{2S_Y(t - CA)}{D} FET \tag{11.24}$$

$$P_{\text{nom-hyd}} \leq P_{\text{max-hyd}} = \frac{2S_Y t}{D} FET \tag{11.25}$$

Recall that F = construction design factor = 0.6 for riser component, = 0.72 for submerged component, and = 0.95 for hydrotest, E = longitudinal joint factor, E = 1 for API 5L seamless line pipe, and T = temperature de-rating factor, T = 1 for maximum temperature of 200°F.

Also recall that 30 CFR 250 defines that the maximum allowable operating pressure (MAOP) to be the least of the following:

- internal design pressure of the pipeline, valves, flanges and fittings,
- 80% of hydrostatic test pressure,
- MAOP of the receiving pipeline.

The preliminary nominal wall thickness for each pipeline section is calculated to satisfy equations (11.24) and (11.25). The API 5L (2000) wall thickness selection is shown in table 11.8 (see also fig. 11.8). To verify that the design equations are satisfied, table 11.8 presents the numerical values at each pertinent location along the pipeline elevation. Note that:

- For each pipeline segment the controlling (higher) required design pressure, P_{req}, is at the shallower water depth.

Table 11.8 Wall thickness design checks for burst

Water depth (ft)	Wall thickness (in.)	P_{req} (psig)	P_{id}[1] (psig)	$P_{max\text{-}hyd}$[2] (psig)	MAOP[3] (psig)
0	0.875	5622	5895	10052	5680
500	0.875	5449	5895	10052	5680
500	0.719	5449	5716	8260	5680
3000	0.719	4581	5716	8260	5680
3000	0.625	4581	4897	7180	4897
8000	0.625	2844	4897	7180	4897

[1] P_{id} must be greater than P_{req} [equation (11.24)]
[2] $P_{max\text{-}hyd}$ must be greater than $P_{nom\text{-}hyd} = 7100$ psig in this example [equation (11.25)]
[3] MAOP is the least of (a) $0.8 \times P_{nom\text{-}hyd}$, (b) P_{id}, (c) internal design pressure of valves, flanges and fittings; which in this example will be assumed greater than the values shown and (d) MAOP of receiving pipelines.

- An API wall thickness thinner than 0.625 in., which is 0.562 in., would not be adequate since it would not satisfy equation (11.25), even though it would satisfy equation (11.24) for water depths greater than 3670 ft.
- There is room to revise the wall thickness break to shallower water depth, thus making the 0.625 in. segment longer, which would lead to project savings. To calculate the water depth, X, at the wall thickness break, use equation (11.23) with $P_{req} = P_{id}$. In this example such an equation is: $6400 - X(64/144) - (8000 - X)(14/144) = 4897$, which leads to $X = 2088$ ft. Of course the prudent offshore pipeline engineer will always give some allowance for installation tolerances, and the revised water depth for the WT break would be set at 2200 ft. However, for purposes of the ensuing numerical examples, the water depth of 3000 ft will be maintained.

11.7.2 Collapse Due to External Pressure

Equations (11.5a–11.5c) yield the collapse pressure. The factored collapse pressure $f_o \cdot P_c$ must exceed the net external pressure everywhere along the pipeline, as shown in equation (11.4). The collapse reduction factor in this example for seamless pipe is $f_o = 0.7$. The net external pressure can be determined as the hydrostatic water pressure at maximum water depth of each pipeline section assuming zero internal pressure. The net external pressures are all within the allowable collapse pressure as summarised in table 11.9.

In this example the nominal wall thickness of each pipeline section is used. The authors judge that for collapse as well as local buckling, it is over-conservative to deduct the corrosion allowance in such calculations. While burst will occur at the maximum stress which occurs at the thinnest wall thickness (justifying the corrosion allowance deduction on burst limit state check), collapse and local buckling involves the formation of a four-hinge collapse mechanism, with maximum bending moments at four hinges 90° apart around the pipeline cross section [see Nogueira and Lanan 2001; Timoshenko and Gere, 1961, Section 7.5]. Given that the pipe mill's average wall thickness is

Table 11.9 Collapse pressure vs. external pressure

Water depth (ft)	Wall thickness (in.)	Collapse pressure, P_c (psig)	$f_o \cdot P_c$ (psig)	External pressure, P_o (psig)
0–500	0.875	10113	7079	222
500–3000	0.719	7911	5538	1333
3000–8000	0.625	6471	4530	3556

Table 11.10 Factored strain vs. limiting bending strain

Water depth (ft)	Wall thickness (in.)	Factored installation bending strain $f_1\varepsilon_1$ (%)	Factored in-place bending strain $f_2\varepsilon_2$ (%)	$\frac{P_o - P_i}{P_c}$	Critical bending strain [equation (11.7)] ε (%)
0–500	0.875	0.4	0.4	0.022	3.30
500–3000	0.719	0.4	0.4	0.168	2.22
3000–8000	0.625	0.4	0.4	0.550	0.82

typically 10% greater than the nominal wall thickness, around the line pipe circumference, the nominal wall thickness is recommended to be used in collapse and local buckling. Of course, this is a project decision that should be clearly stated within the project DBD.

11.7.3 Local Buckling Due to Bending and External Pressure

Equation 11.6 gives the limiting bending strain, to avoid the local buckling limit state. By rearranging it, equation (11.7) is obtained, which yields the maximum bending strain. Therefore, the installation and in-place bending strains shall be limited per equations (11.8a) and (11.8b). In this design example, factors of safety adopted are $f_1 = f_2 = 2.0$. The ovality is conservatively set at $\delta = 1\%$, which leads to $g(\delta) = 0.833$. The critical bending strains per equation (11.7) are shown in the right hand column of table 11.10 (with $P_i = 0$) and are all greater than the factored bending strains.

11.7.4 Buckle Propagation

The pipeline propagation pressure value, per API 1111 (1999), is given by equations (11.21) and (11.22). Assuming zero internal pressure the results are shown in table 11.11.

From the results presented in table 11.11, buckle arrestors are required along the 0.625 in. WT segment, when the external water pressure is greater than 1352 psi. This corresponds to 3042 ft and greater water depth. Buckle arrestor design guidelines can be found in Park and Kyriakides (1997) and Langner (1999).

Table 11.11 Buckle arrestor requirement

Water depth (ft)	Wall thickness (in.)	Propagation pressure P_{pr} (psig)	$f_o \cdot P_{pr}$ (psig)	Net external pressure (psig)	Buckle arrestor
500	0.875	3789	3031	222	not required
3000	0.719	2365	1892	1333	not required
8000	0.625	1690	1352	3556	required

11.8 On-Bottom Stability

This section addresses stability analysis of offshore pipelines on the seabed under hydrodynamic loads (wave and current). On-bottom stability is checked for the installation case with the pipe empty using the 1-yr return period condition and for lifetime using the 100-yr storm. Additionally, a minimum pipeline specific gravity of 1.20 during installation is desired.

Hydrodynamic stability analysis involves the following steps:

1. Define environmental criteria for the 1-yr and 100-yr condition:
   - Water depth
   - Significant wave height (H), wave period (T) and the angle of attack (β)
   - Steady current velocity (U_c) and angle of attack (β)
   - Wave only particle velocity (U_w), maximum water particle velocity due to wave and current (U_m) and steady current ratio ($U_R = U_c/U_m$)
   - Soil submerged weight (γ), soil friction factor () or undrained shear strength (S_u)
   - Seabed slope (δ) measured positive in downward loading
2. Determine hydrodynamic coefficients: drag (C_D), lift (C_L) and inertia (C_I). These may be adjusted for Reynolds number, Keulegan–Carpenter number, ratio of wave to steady current and embedment.
3. Calculate hydrodynamic forces drag (F_D), lift (F_L) and inertia (F_I).
4. Perform static force balance at time step increments and assess stability and calculate concrete coating thickness for worst combination of lift, drag and inertial force.

Hydrodynamic stability is determined using Morison's equation, which relates hydraulic lift, drag and inertial forces to local water particle velocity and acceleration. The coefficients used, however, vary from one situation to another. For example, the lift and drag coefficients of 0.6 and 1.2, which is representative of a steady current, is not appropriate for oscillating flow in a wave field. Additionally, these coefficients are reduced if the pipe is not fully exposed because of trenching or embedment.

To determine wave particle velocity, the theory used depends on wave height, water depth and wave period. For most situations, linear theory is adequate as bottom velocities and accelerations do not vary significantly between theories. However, as the wave height to water depth ratio increases, Stoke's fifth order theory becomes appropriate. For shallow water or very high wave heights, a solitary theory should be used to predict particle velocity and accelerations [Sarpkaya and Isaacson, 1981]. For breaking waves, or large diameter pipe that may affect the flow regime, other analysis methods may be appropriate. In general, pipelines should be trenched within the breaking wave (surf) zone.

Experimental and theoretical researches [Ayers, et al 1989; DNV RP E305, 1988] have shown that the traditional static analysis methods have been conservative in most cases. In the 1980s, two research groups developed theoretical and experimental models to assess pipe stability. Findings of these groups (American Gas Association in USA and PIPESTAB in Europe) resulted in the development of program LSTAB, which accounts for the effects of embedment. The commercially available computer program LSTAB, with the American Gas Association, is the state-of-art tool for assessing on-bottom stability of pipelines. It is comprehensive and easy to use.

What follows is a summary of the most important factors for an on-bottom stability analysis and relevant references.

11.8.1 Soil Friction Factor

The friction factor is defined as the ratio between the force required to move a section of pipe and the vertical contact force applied by the pipe on the seabed. This simplified model (Coulomb) is used to assess stability and requires an estimate of the friction factor, . Strictly speaking, the friction factor, , depends on the type of soil, the pipe roughness, seabed slope and depth of burial; however, the pipe roughness is typically ignored.

For stability analysis, a lower bound estimate for soil friction is conservatively assumed, whereas for pulling or towing analysis, an upper bound estimate would be appropriate. The following lateral friction factors [Lyons, 1973; Lambrakos, 1985] are given as a guideline for stability analysis in the absence of site-specific data:

- Loose sand: tan ϕ (generally $\phi = 30°$)
- Compact sand: tan ϕ (generally $\phi = 35°$)
- Soft clay: 0.7
- Stiff clay: 0.4
- Rock and gravel: 0.7

These coefficients are adequate for generalised soil types and do not include safety factors. Small-scale tests [Lyons, 1973] and offshore tests [Lambrakos, 1985] have shown that the starting friction factor in sand is about 30% less than the maximum value, which occurs after a very small displacement of the pipe builds a wedge of soil; past this point, the friction factor levels off. The values given above account for the build-up of this wedge of soil, which has been shown to take place.

11.8.2 Hydrodynamic Coefficient Selection

Hydrodynamic coefficients have been the subject of numerous theoretical and experimental investigations and are often subject to controversy. Selection of C_D, C_L and C_I are dependent on one of the following situations:

- Steady current only
- Steady current and waves

For steady current conditions acting on a pipeline resting on the sea floor, $C_D \approx 0.7$ and $C_L \approx 0.9$. However, these coefficients are dependent on the Reynold's number ($R_e = U_c D/\nu$, with $\nu = 1.7 \times 10^{-5}$ ft^2/s), and if more precision is warranted Jones (1976) may be consulted. For steady-current conditions, a conservative stability check may be performed by subtracting lift from the submerged weight, calculating the available friction force and verifying that the drag force is smaller than the available friction force.

For waves and currents, these parameters are dependent on the Keulegan–Carpenter number ($K_e = U_w T/D$, where D = pipe outside diameter), pipe roughness and the steady current ratio. Bryndum, et al (1983, 1988) include guidelines in selecting these parameters.

11.8.3 Hydrodynamic Force Calculation

The drag force, lift force and inertia force are given by the Morrison's equations:

- Drag force: $F_D = \frac{1}{2} C_D \rho D U_m |U_m|$
- Lift force: $F_L = \frac{1}{2} C_L \rho D U_m^2$
- Inertia force: $F_I = C_I \rho \left(\frac{\pi D^2}{4} \right) \dot{U}_m$

11.8.4 Stability Criteria

The last step of the simplified on-bottom stability analysis consists in assessing stability using a simple lateral force equilibrium equation. In the following equation the symbols are as defined in Section 11.8 and W_s is the pipeline submerged weight:

$$\mu(W_s \cos\delta - F_L) \geq \xi(F_D + F_I + W_s \sin\delta) \tag{11.26}$$

This formulation assumes a Coulomb friction model as described above and is over-conservative if the pipe is embedded. A preliminary conservative approach, however, is to consider no embedment. The drag force in equation (11.26) may include the effects of the angle of attack, in case that the design wave and current are not expected to be perpendicular to the pipeline alignment. The safety factor (ξ) in equation (11.26) is designed to account for uncertainties in:

- Actual soil friction factor
- Actual environmental data (wave, current)
- Actual particle velocity and acceleration
- Actual hydrodynamic coefficients

Recommended safety factors are:

- $\xi = 1.05$ for installation
- $\xi = 1.1$ for operation

11.9 Bottom Roughness Analysis

The objective of bottom roughness analysis is to identify possible free spans that exceed the maximum allowable span length that may occur during pipeline installation, hydrotest and

Table 11.12 Allowable pipeline stresses

Condition	Longitudinal stress (% SMYS)	Total Von-Mises stress (% SMYS)
Hydrotest[1]	95	95
Operation[2]	80	90
Installation[3]	80	90

[1]Based on 30 CFR 250 stress limit during hydrotest
[2]ASME B31.4 and ASME B31.8 requirements
[3]Assumed identical to the ASME limits for the operating case

operation. The bottom roughness analysis can be performed using computer software such as OFFPIPE, which is an industry recognised finite element tool used for the analysis of offshore pipelines (see www.offpipe.com for software information). The OFFPIPE model assumes a linear elastic foundation under the pipeline with supports at regular intervals. Due to this regular support interval, actual span lengths may be shorter than the calculated span lengths.

One of the criteria to establish the maximum allowable span, is to limit the maximum pipeline stresses under static conditions. This is done by limiting both the total Von-Mises stress and the longitudinal stress as shown in table 11.12.

In addition, the pipeline span lengths cannot exceed the maximum span lengths at which in-line vortex-induced-vibration (VIV) will occur. Therefore, the determination of the pipeline allowable span lengths must consider the following five criteria:

- onset of in-line VIV,
- onset of cross-flow VIV,
- maximum allowable equivalent stress,
- maximum allowable longitudinal stress,
- fatigue life due to in-line VIV (optional criteria).

The spans from the first four criteria for each segment are considered when calculating the maximum allowable span, evaluating the bottom roughness analysis, or evaluating the pipeline crossing analysis. The last criterion involves performing a fatigue analysis to increase the span length due to in-line VIV as explained here.

When free spans occur due to seabed irregularities or pipeline crossings along the pipeline route, the presence of bottom current may cause dynamic effects. The fluid interaction with the pipeline can cause the free span to oscillate due to vortex shedding. Two distinct forms of oscillation can be observed due to vortex shedding: in-line and cross-flow. In-line VIV occurs when the pipeline vibrates parallel to the direction of flow in a constant current. The amplitude of the in-line vibrations is typically less than 20% of the outside diameter of the pipe and is significantly smaller than (only about 10% of) the amplitudes for cross-flow vibrations. In-line VIV occurs at lower flow velocities and shorter spans than cross-flow VIV. It is the industry practice to allow span lengths to exceed the in-line VIV criteria, provided a fatigue analysis is done, that demonstrates adequate design life.

DNV Guideline 14 (1998) presents a complete treatment of the subject of oscillations both in-line and cross-flow, including current and wave effects. What follows is a simplified approach for current-dominated oscillations.

11.9.1 Allowable Span Length on Current-Dominated Oscillations

Several parameters are used to assess the allowable span length, for a given current velocity, that will lead to the onset of in-line VIV. For this analysis, the stability parameter (K_s) and the reduced velocity (V_r) are used. The dimensionless stability parameter is calculated using equation (11.27).

$$K_s = \frac{2M_e\delta}{\rho D^2} \tag{11.27}$$

where K_s = stability parameter, M_e = the effective mass, $M_e = M_p + M_c + M_a$, M_p = pipe mass, M_c = mass of pipe contents, M_a = added mass, $M_a = \rho\pi D^2/4$, δ = logarithmic decrement of structural damping, for steel pipe, $\delta = 0.125$ and ρ = mass density of the fluid around the pipe, for seawater $\rho = 2$ slugs/ft^3.

The reduced velocity, V_r, can be determined as a function of the stability parameter by:

$$V_r = \begin{cases} K_s < 0.25, 1 \\ 0.25 < K_s \le 1.2,\ 0.188 + 3.6K_s - 1.6K_s^2 \\ K_s > 1.2,\ 2.2 \end{cases} \tag{11.28}$$

The reduced velocity is then used to determine the critical frequency at which the onset of in-line VIV can occur. Calculation of the critical frequency is shown in equation (11.29).

$$f_{\mathrm{cr}} = \frac{V}{V_r D} \tag{11.29}$$

where f_{cr} = the critical frequency, and V = the current design velocity.

To determine the span length at which the onset of in-line VIV can occur for the design current velocity, the natural frequency of the span is set equal to the critical frequency and solved for the corresponding span length. Equation (11.30) is used to calculate the natural frequency of the span. Equation (11.31) shows how the critical span length is calculated from the critical frequency.

$$f_n = \frac{C}{2\pi}\sqrt{\frac{EI}{M_e L^4}} \tag{11.30}$$

$$L_{\mathrm{cr}} = \sqrt{\frac{1}{f_{cr}}\frac{C}{2\pi}\sqrt{\frac{EI}{M_e}}} \tag{11.31}$$

where f_n = natural frequency of the span, C = the end condition constant ($1.25^2\pi^2$ for pinned-fixed), E = the modulus of elasticity of the pipeline, I = the moment of inertia of the pipeline, L = a given span length and L_{cr} = the critical span length.

Cross-flow VIV occurs when the pipeline vibrates perpendicular to the direction of flow due to vortex shedding in a constant current. The response amplitudes for cross-flow VIV are much greater than for in-line VIV. Span lengths for the onset of cross-flow VIV are to be avoided.

The parameters used to assess the potential of cross-flow VIV are the Reynolds Number, R_e, and the reduced velocity, V_r. For cross-flow VIV, the reduced velocity can be estimated as a function of the Reynolds Number by:

$$R_e = \frac{VD}{\nu} \tag{11.32}$$

where R_e = Reynolds Number, V = flow velocity, D = pipe outside diameter and ν = kinematic viscosity of the fluid, for seawater $\nu = 1.26 \times 10^{-5}\ \text{ft}^2/\text{s}$.

$$V_r = \begin{cases} R_e < 5 \times 10^4, \ 5 \\ 5 \times 10^4 < R_e \leq 3 \times 10^6, \ c_1 - c_2 R_e + c_3 R_e^2 + c_4 R_e^3 + c_5 R_e^4 \\ R_e > 3 \times 10^6, \ 3.87 \end{cases} \tag{11.33}$$

where V_r = reduced velocity for onset of cross-flow VIV, $c_1 = 5.07148$, $c_2 = 1.61569 \times 10^{-6}$, $c_3 = 8.73792 \times 10^{-13}$, $c_4 = 2.11781 \times 10^{-19}$ and $c_5 = 1.89218 \times 10^{-26}$.

Using the reduced velocity for cross-flow VIV, the critical frequency and critical span length are determined in the same manner as for in-line VIV.

Note that a conventional riser along a platform (as shown in Fig.11.8) must be designed such that in-line and cross-flow VIV does not occur. This check must consider wave and current. Clamps to the platform must be designed to avoid this critical design case in risers. Failure in risers in the Gulf of Mexico are rare, but have been reported during hurricanes, thus the extreme case combination for the environmental loads must be taken into account.

11.9.2 Design Example

The design example (table 11.13) calculates the allowable span based on VIV criteria already described. From table 11.6, the bottom current used is 1.3 ft/s. The example assumes the pipeline is in the operating condition, i.e. the pipeline is filled with product. The allowable span lengths for both the in-line motion and cross-flow motion are obtained.

11.10 External Corrosion Protection

The external corrosion of the offshore pipelines is usually controlled by ways of an external corrosion coating and a sacrificial anode-cathodic protection system. The corrosion coating for the offshore pipelines is normally fusion-bonded epoxy (FBE) coating of about 16 mil. The design of the sacrificial anode-cathodic protection system is typically performed using the design guidelines given by DNV RP B401 (1993).

The surface areas to receive cathodic protection should be calculated separately for areas where the environmental conditions or the application of coatings imply different current requirements. All components to be connected to the system should be included in the surface area calculations. This may include various types of appurtenances or outfitting to be installed along the pipeline.

Table 11.13 Allowable span design example. Segment 3 pipe: OD = 10.75 in., WT = 0.625 in.

Step 1	Value	Unit
Pipe mass, M_p	2.103	slug/ft
Mass of contents, M_c	0.214	slug/ft
Added mass, M_a	1.254	slug/ft
Equivalent mass, M_e	3.571	slug/ft
Step 2		
Stability parameter, Ks	0.559	–
Step 3		
Reynolds number, Re	92,427	–
Step 4		
Reduced velocity, Vr, in-line	1.701	–
Reduced velocity, Vr, cross-flow	4.93	–
Step 5		
Critical frequency for in-line VIV	0.853	1/s
Critical span length for in-line VIV	104.4	ft
Step 6		
Critical frequency for cross-flow VIV	0.294	1/s
Critical span length for cross-flow VIV	177.8	ft

Surface area demand involves assumptions of coating breakdown factors. Offshore engineers designing pipelines in the Gulf of Mexico, typically use coating breakdown factors smaller and more realistic than those recommended by DNV RP B401 (1993). For example, see Britton (1999) who suggests initial coating breakdown factor of 3%, and final coating breakdown factor of 5% for a 20-yr design life. Thus, coating breakdown factor established for a project shall always be documented very clearly in the DBD, so that the project team consistently uses the project-specific values.

11.10.1 Current Demand Calculations

The current demand I_c to achieve polarisation during the initial and final lives of the cathodic protection system, and the average current demand to maintain cathodic protection throughout the design life should be calculated separately.

The surface area A_c to be cathodically protected should be multiplied with the relevant design current density i_c and the coating breakdown factor f_c:

$$I_c = A_c \cdot f_c \cdot i_c \tag{11.34}$$

where I_c = current demand for a specific surface area, i_c = design current density, selected from tables 11.14 and 11.15, which follow guidance provided by DNV RP B401 (1993)

Table 11.14 Initial and final design current densities for various climatic regions and depths – adapted from table 6.3.1 of DNV RP B401 (1993)

Water depth (ft)	Design current densities (initial/final) in A/ft^2							
	Tropical (>20°C)		Subtropical (12–20°C)		Temperate (7–12°C)		Arctic (<7°C)	
	Initial	Final	Initial	Final	Initial	Final	Initial	Final
0–100	0.0139	0.0084	0.0158	0.0102	0.1860	0.0121	0.0232	0.0158
>100	0.0121	0.0074	0.0139	0.0084	0.0167	0.0102	0.0204	0.0121

Table 11.15 Average (Maintenance) design current densities for various climatic regions and depths – adapted from table 6.3.2 of DNV RP B401 (1993)

Water depth (ft)	Design current densities (initial/final) in A/ft^2			
	Tropical (>20°C)	Subtropical (12–20°C)	Temperate (7–12°C)	Arctic (<7°C)
0–100	0.0065	0.0074	0.0093	0.011
>100	0.0056	0.0065	0.0074	0.0093

Section 11.3, f_c = coating breakdown factors. See DNV RP B401 (1993) table 11.4.1 and Sections 6.5.3 and 6.5.4 for guidance on offshore pipelines.

For items with major surface areas of bare metal, the current demands required for initial polarisation, I_c (initial), and for re-polarisation at the end of the design life, I_c (final), should be calculated, together with the average current demand I_c (average) required to maintain cathodic protection throughout the design period. For pipelines and other items with high-quality coatings, the initial current demand can be deleted in the design calculations.

11.10.2 Selection of Anode Type and Dimensions

The type of anode to be used is largely dependent on fabrication, installation and operational parameters. The anode type is determining for which anode resistance formulas and anode utilisation factors are used in further calculations. For pipeline bracelet anodes that are mounted flush with the coating, the thickness of the coating layer will be decisive to the anode dimensions.

11.10.3 Anode Mass Calculations

The total net anode mass M (kg) required to maintain cathodic protection throughout the design life t_r (yr) should be calculated from the average current demand I_c:

$$M = \frac{I_c\,(\text{average}) \cdot t_r \cdot 8760}{u \cdot \varepsilon_{LT}} \tag{11.35}$$

where 8760 is the number of hours per year, u is the utilisation factor, and ε_{LT} (A-h/lb) is the electrochemical efficiency of the anode material, which is 950 A h/lb for aluminum-based anode material type – see Section 6.6, DNV RP B401 (1993).

11.10.4 Calculation of Number of Anodes

For the anode type selected, the number of anodes, anode dimensions and anode net mass should be selected to meet the requirements for initial/final current output (A) and the current capacity ($A \cdot h$), which relate to the protection current demand of the protection object. The anode current output I_a is calculated from Ohm's law:

$$I_a = \frac{E_c^0 - E_a^0}{R_a} \tag{11.36}$$

whereE_a^0 (V) is the design closed circuit potential of the anode, typically -1.05 V (relative to Ag/AgCl/seawater), see Section 6.65 of DNV RP B401 (1993). E_c^0 (V) is the design protective potential, which is chosen to be -0.80 V (relative to Ag/AgCl/seawater). R_a (ohm) is the anode resistance, is given by DNV RP B401 (1993), table 6.7.1; which for bracelet anode is:

$$R = 0.315\rho/\sqrt{A} \tag{11.37}$$

where A is the anode surface area, and ρ is the environmental resistivity; for which guidance can be found in Section 6.8 of DNV RP B401 (1993). For the Gulf of Mexico, typically, $\rho = 30$ ohm-cm.

Anode dimensions and net weight are to be selected to match all requirements for current output (initial/final) and current capacity for a specific number of anodes. This is an iterative process and a simple computer spreadsheet may be helpful. Calculations should be carried out to demonstrate that the following requirements are met:

$$C_a = n \cdot c_a \geq I_c \cdot t_r \cdot 8760 \tag{11.38}$$

$$n \cdot I_a(\text{initial/final}) \geq I_c(\text{initial/final}) \tag{11.39}$$

To summarise, the cathodic protection design should optimise anode spacing and weight. The selected anode characteristics must meet two requirements:

- The anode mass must be sufficient to meet the current demand over its design life.
- The anode surface area at the end of its design life must be sufficient to provide the required current. At the end of its design life, the anode's surface area is assumed to be the product of the pipe circumference and the anode's length.

11.10.5 Design Example

The following shows an example of the pipeline cathodic protection design. The data is from the deep segment (assumed 100,000 ft long) of the design example shown in Section 11.7. The rows on table 11.16 are numbered so that the calculation may be easily followed, thus C1 in column B, row 3, refers to the numerical value shown in column C, row 1. Cross reference to the above said equations is also provided.

11.11 Pipeline Crossing Design

Pipeline crossing design basically involved protecting the crossed pipeline using articulated concrete mattresses. Typically, in the US Gulf of Mexico (GOM), two 9 in. thick articulated concrete mats are used, for a 18 in. separation between the pipelines. The single lift equations (11.41)–(11.43) below [see Troitsky, 1982; Section 11.6.5] can be used to calculate the crossing loads, which were relatively small, so that the crossed pipeline could transfer such forces to the underlying seafloor. Typically, no intermediate supports for the crossing pipeline were needed.

However, recent GOM Federal regulations require that crossing pipelines be covered by mats from touchdown to touchdown, for water depths less than 500 ft. This leads to a crossing arrangement depicted in figs. 11.9 and 11.10, which shows with relative scale a 12 in. pipeline crossed by a 24 in. pipeline and 9 in. thick concrete mats. The capping mats impose a load on top of the crossing pipeline, which is transferred to the crossed pipeline as a concentrated load on a short ring of pipe (shown with a length L in fig. 11.10). Finally, the crossed pipeline transfers this load to the seafloor. Thus, crossing design needs to be evaluated as follows.

The first step in the crossing analysis is to estimate the crossing load. Concrete mattresses' submerged weight is approximately 6000 lb for a 9 in. thick mat with 8 ft by 20 ft dimensions, which leads to a submerged load of about $w = 38$ psf. A 4.5 in. thick mat of same dimensions has a 3600 lb total submerged weight, for a load of about $w = 23$ psf. The load imposed by the capping mattress is estimated by assuming an average drape angle of 30° (fig. 11.10), and the corresponding maximum linear load, q, at the crown of the crossing is given by:

$$q = w[2 \times 1.16(d + OD_{\mathrm{TOP}}) + OD_{\mathrm{TOP}}] \qquad (11.40)$$

where $1.16 \cong 1/\cos 30$, d is the crossing pipe prop height, which equals distance from adjacent seafloor (mudline) to bottom of crossing pipe, OD_{TOP} is the diameter of the crossing pipe.

Given an existing pipeline with $OD_{\mathrm{BOT}} = 20.00$in., $WT_{\mathrm{BOT}} = 0.812$, embedded 3 in, two 9 in. separation mattresses, a crossing pipeline with $OD_{\mathrm{TOP}} = 12.75$ in., $WT_{\mathrm{TOP}} = 0.750$, and a 9 in. capping mattress; then: $d = 35$ in. and $q = 391.2$ plf. The crossing pipeline has water filled submerged weight of 83.7 plf (= 96.2 plf steel weight in air, 44.2 plf water contents at 64 pcf, minus 56.7 plf buoyancy. When the mattress load is added, the total maximum pipeline load is $q_T = 475$ plf, or 5.7 times that of the crossing pipeline during hydrotest.

A pipeline on a prop with height d from the adjacent seafloor (see fig. 11.11), with Young's modulus E, moment of inertia I and total submerged load q, will have a distance from the prop point to touchdown l given by:

$$l^4 = \frac{72EId}{q} \qquad (11.41)$$

The total prop force F_P and maximum bending moment at the centre of the span, M_P, are given by:

$$F_P = 4ql/3 \qquad (11.42)$$

$$M_P = ql^2/6 \qquad (11.43)$$

Table 11.16 Cathodic protection design example

Row number	Item	Source/Equation	Value	Unit
	A	B	C	D
1	Pipeline diameter	Input data	10.75	in.
2	Pipeline length	Input data	100,000	ft
3	Total surface area	π*C1*C2/12	281,434	sq ft
4	Mean coating breakdown factor	Project specific	5	%
5	Final coating breakdown factor	Project specific	9	%
6	Mean bare area	C4*C3	14,072	sq ft
7	Final bare area	C5*C3	25,329	sq ft
8	Required current density	Project specific	0.00837	A/ft^2
9	Required mean current	C6*C8	118	A
10	Required final current	C7*C8	212	A
11	Electrochemical efficiency	Input data	950	A-h/lb
12	Design Life	Input data	20	yr
13	Anode efficiency	Input data	0.85	

14	Weight required	Eq. 11.35: C9*C13* 8760/(C10*C14)	25,602	Lb
15	Anode spacing	Input data	480	Ft
16	Number of anodes required	INT(C2/C16) + 1	209	
17	Minimum anode weight	C15/C17	122.5	lbs
18	Nominal net anode weight	Project specified	130	lbs
19	Anode specific weight	Vendor specified	180	lb/ft^3
20	Anode thickness	Project defined	1.5	in.
21	Average anode diameter	C1 + C22	12.25	in.
22	Estimated anode length	1728*C19/(C21*π*C23*C2)	22	in.
23	Environmental resistivity	Project specified	30	ohm-cm
24	Anode Area	π*C21*C22	847	$in.^2$
25	Anode resistance	Equation (11.37): $0.315*(C24/2.54)/C25^{0.5}$	0.1278	ohm
26	Anode potential	Project specified	−1.05	V
27	Cathode potential	Project specified	−0.80	V
28	Anode current	Equation (11.36): (C28–C27)/C26	1.956	A
29	Total current available	Verify Equation (11.39): C29*C17	408	A

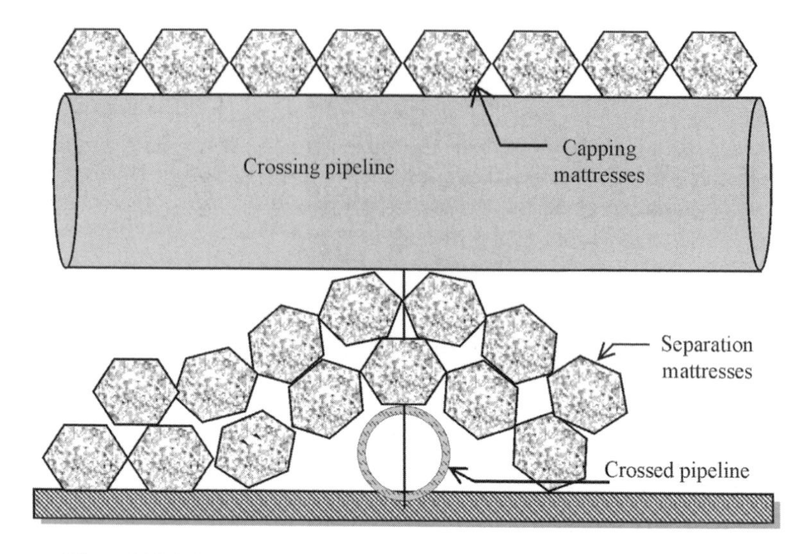

Figure 11.9 Schematic of crossing arrangement – Side view of crossing pipeline

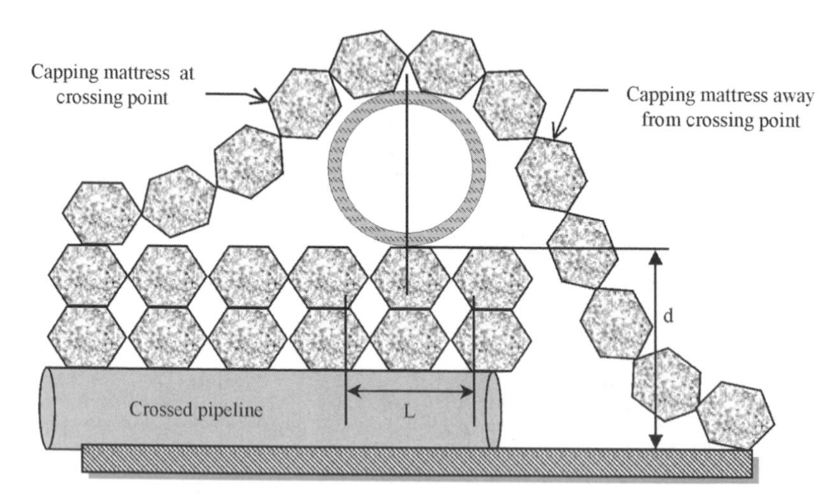

Figure 11.10 Schematic of crossing arrangement – Side view of crossed pipeline

In order to estimate the crossing load and make an initial assessment of the crossing integrity, a conservative analysis may be done as follows: Assume that the maximum load q_T is applied on the crossing pipeline along the entire crossing span. Calculate the prop force F_P as a function of the prop height d, for several values of the crossed pipeline additional embedment, A_e. A simplified yet conservative reactive force can be calculated by assuming a soil reaction acting on the entire crossed pipeline outside diameter, thus leading to a soil reaction of $q_s = 3.4\ S_u\ OD_{\mathrm{BOT}}$, where S_u is the undrained shear strength, and the

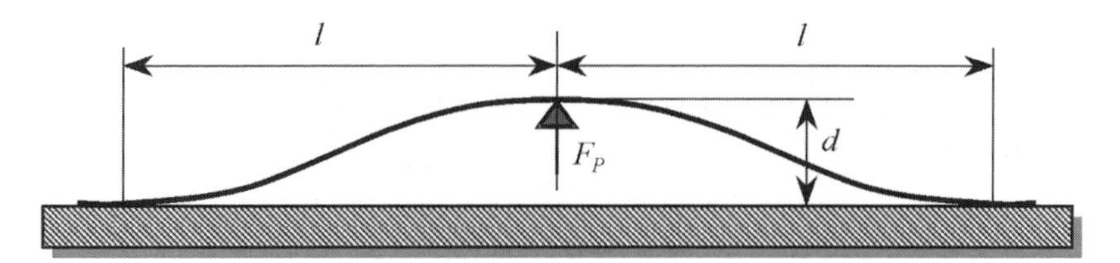

Figure 11.11 Schematics of a propped pipeline

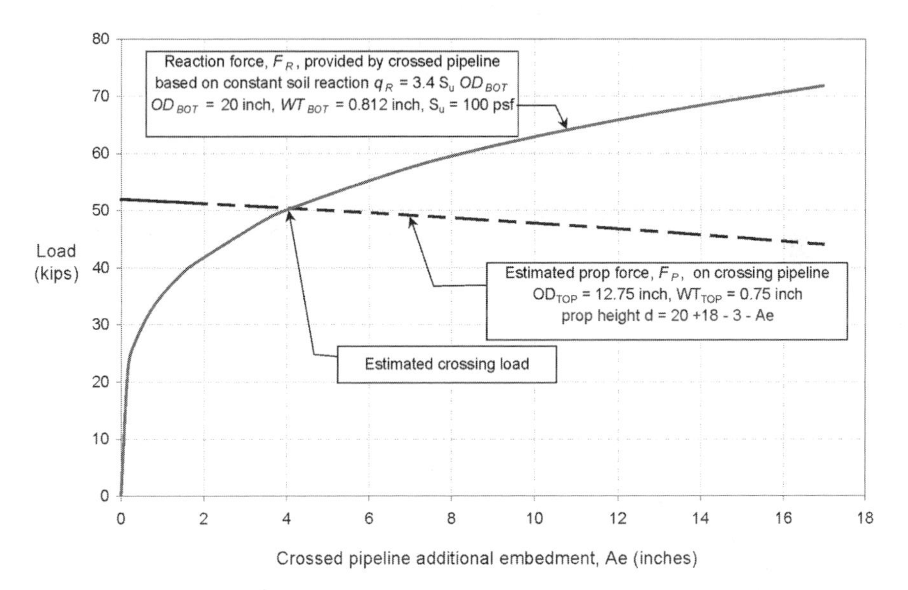

Figure 11.12 Propping and reactive load for a pipeline crossing

factor 3.4 accounts for the round pipeline shape as a foundation. With this linear soil reaction, the same equations of pipeline on a prop can be used for the crossed pipeline to calculate the total reaction provided by the soil, F_R, as a function of additional embedment A_e. Both forces F_P and F_R can be plotted as a function of the additional embedment and their intersection will provide the crossing load and corresponding lengths to touchdown for each pipe. Such plots are shown in fig. 11.12 for the pipeline with the characteristics given in the example, and assuming that the crossed pipeline has an initial embedment (before crossing installation) of 3 in.

In this case, the resultant crossing load is $F_P = F_R = 50.4$ kip, at an additional embedment of 4.1 in. (for a total embedment of 7.1 in.). The constant soil reaction on the crossed pipeline is $q_R = 567$ plf, as a result of the undrained shear strength value of 100 psf adopted. The prudent offshore engineer needs to adopt an upper bound for the shear strength, since this will lead to higher crossing loads. The distances to touchdown for the crossing (top) and crossed (bottom) pipeline are $l_{TOP} = 79.5$ ft and $l_{BOT} = 66.6$ ft. Therefore, the bending moments at the crossing point applied at each pipelines are $M_{TOP} = 500.4$ kip-ft and $M_{BOT} = 418.9$ kip-ft. Note that, if no capping mattresses were present, the crossing load would be about 14 kip, leading to $l_{TOP} = 127$ ft, and a smaller bending moment $M_{TOP} = 225$ kip-ft would be present.

Of course, the assumption of constant maximum load on the crossing pipeline results in an estimated crossing force higher than actual. Similarly, the maximum soil reaction acting along the entire crossed pipeline span results in the estimated reactive force higher than actual. A more precise finite element analysis considering the soil as hyperbolic non-linear springs and varying the applied load on the crossing pipeline may be performed using a finite element program, if warranted, thus leading to somewhat smaller crossing loads.

With the applied crossing load and resultant bending moments, a checking against pipe capacity must be performed. Two checks are required: a longitudinal bending moment check as well as local collapse check (e.g. a ring of pipe being crushed, or excessive ovalised, at the crossing point). A limit state design is proposed where each failure mechanism is checked against the corresponding limit state. A factor of safety of 1.5 is suggested.

The plastic bending moment capacity, M_P, is given by $M_P = S_Y\,(D - t)^2 t$. Adopting $S_Y =$ 65 ksi for both pipelines in the example above, $M_{P\text{-TOP}} = 585$ kip-ft and $M_{P\text{-BOT}} =$ 1619 kip-ft. Therefore, the crossing pipeline has a factor of safety, FS = 585/500.4 = 1.17, and does not meet the safety criteria proposed above. The crossed pipeline has a factor of safety, FS = 1619/418.9 = 3.9 and it is adequate.

For local collapse to occur, a three-hinge collapse mechanism must take place. This is shown schematically in fig. 11.13, where a free-body of a pipe ring at the crossing location is depicted. The effective ring of pipe has length L (see also fig. 11.10). The total soil reaction acting along the effective ring is 3.4 S_u OD L, which is less than the crossing load. The shear at each end of the effective ring (shown in fig. 11.12 as V at each side of the ring) provides the force necessary for equilibrium. Collapse will occur when the total applied moment equals the total plastic capacity of the upper half of the ring [Baker and Heyman, 1969]. The total applied moment on the upper half of a ring due to a load F at 12 o'clock is

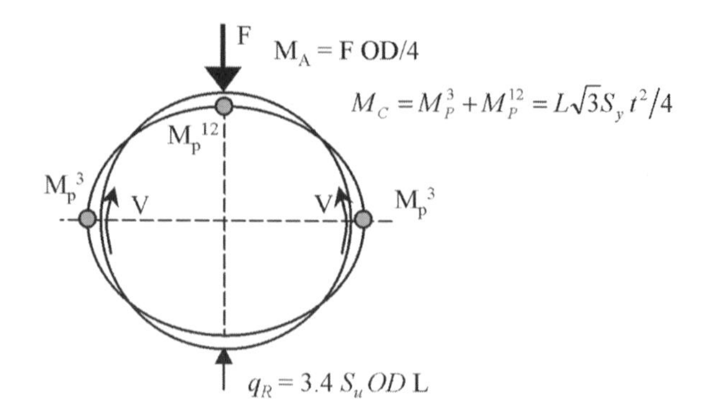

Figure 11.13 Free body of crossed pipeline at crossing point

$M_A = FOD/4$ (similar to a centred point load in a simply-supported beam). The total plastic moment capacity, M_C, is the summation of the plastic capacity of the hinges at 3 o'clock and 12 o'clock, which is $M_C = L\sqrt{3}S_y t^2/4$ [see Nogueira and Lanan, 2001 for derivation details]. The effective ring length is assumed to be equal to the pipeline diameter, $L = OD$; while this assumption is adequate for pipes with $D/t < 25$, it needs to be validated for thinner pipes. The inequality $M_A \leq M_C$ leads to the local ring collapse, or denting load, F_D, on a pipeline, as given in equation (11.44):

$$F_D = 1.73 S_y t^2 \tag{11.44}$$

The data of the example above leads to, for both pipes, $F_{D\text{-}TOP} = 1.73(65)(0.75^2) = 63.3$ kip and $F_{D\text{-}BOT} = 1.73(65)(0.812^2) = 74.1$ kip. Therefore the crossing pipeline has a factor of safety, FS = 63.3/50.4 = 1.26 and does not meet the safety criteria against local collapse. The crossed pipeline has a factor of safety, FS = 74.1/50.4 = 1.47, which also does not meet the safety criteria. Regarding the ring collapse limit state, changing the capping mats to 4.5 in. thick (which leads to a crossing load of $F_L = 38.2$ kip), would lead to a crossing design within the safety guidelines suggested herein.

In this example, extra supports adjacent to the crossed pipeline are needed for the safety criteria suggested herein to be achieved. Note that *both* pipelines must be checked, since the free body shown in fig. 11.13 also applies to the crossing pipeline; except that the point load would be inverted: the higher force would be applied at 6 o'clock (the crossed pipeline reaction) and the smaller force at the 12 o'clock is due to the pipeline self-weight and capping mats.

The above equations neglect the effects of external pressure. While this is an adequate assumption for water depths less than 500 ft, such effects can be readily addressed by using the rational model for pipeline collapse. This way, crossing capacity in deep water can be more precisely estimated. Nogueira and Lanan (2001) showed that external pressure has the effect of adding to the total applied moments and also decreasing the ring collapse hinge capacity due to additional compressive hoop stress around the pipe. Thus, advantage can be taken of the terms given in equation (24) of Nogueira and Lanan (2001) to add the effects of external water pressure.

The example above illustrates the fact that a new US GOM regulation, while with the goal of decreasing the risk of pipeline being dragged or damaged by fishing gear, has the effect of increasing substantially the stresses on pipeline crossings. Therefore, the offshore pipeline engineer needs to be aware that what used to be a traditionally trivial design matter, now requires renewed attention. Actually, any change on status quo in any area of engineering always needs to be carefully considered by knowledgeable and careful engineers, to assess all implications.

11.12 Construction Feasibility

Pipelines are installed on the seafloor by one of the four typical installation methods: J-lay, S-lay, Reel-lay and Tow. The J-lay and the S-lay method are shown schematically in figs. 11.14 and 11.15 (the shape each pipe assumes justifies the corresponding name). The reel-lay method includes one or more pipe spools on board the vessel, and the pipeline is

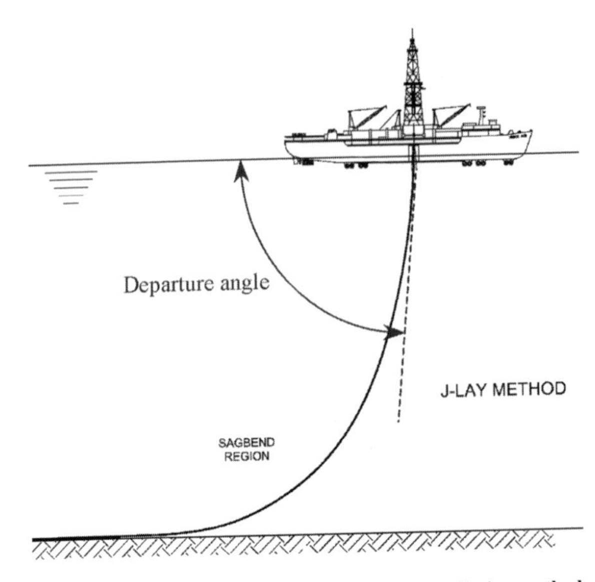

Figure 11.14 Schematic depiction of the J-lay installation method

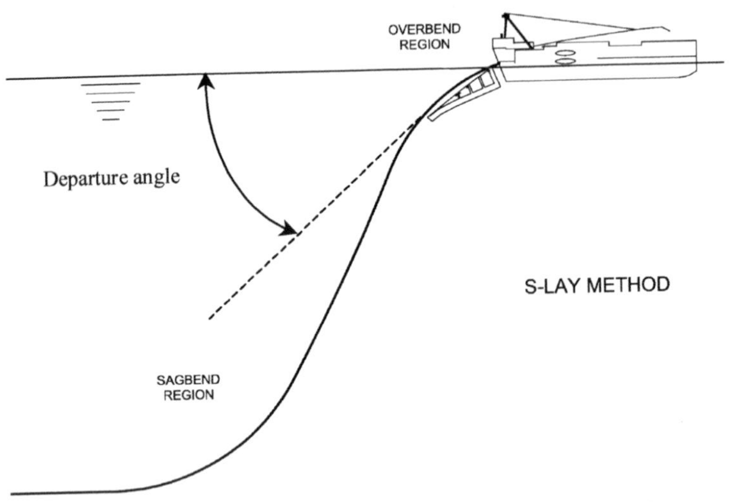

Figure 11.15 Schematic depiction of the S-lay installation method

un-spooled during offshore works. It departs the vessel in a J-lay or S-lay configuration, depending on the vessel method employed. By J-lay mode it is meant a large departure angle, thus the J-lay tower can assume a large departure angle to the horizontal, leading the pipe to a single curvature, or J-shape. Conversely, the S-lay mode has a smaller departure angle and the pipe has a double curvature, or S-shape.

With the exception of the Tow method, all others use a self-contained laybarge to store pipe (with additional supply barges, as required). Some laybarges use anchored mooring system to mantain position, such as the Castoro 10 (at this time, owned by Saipem); others use thrusters in the dynamically positioned station-keeping mode. Station-keeping is very important during pipelay, since unexpected movement away from the planned laying route may severely bend the pipeline either in a sagbend or in an overbend, and the pipe may buckle or kink. The Allseas S-lay barge Lorelay (fig. 11.17) was first to apply dynamic positioning system to pipelay. The McDermott DB50 (J-lay) is also dynamically positioned. Both vessels use an integrated control system, which tracks the relative position of the touchdown point and the vessel.

At the time of this writing, the Canyon Express flowlines in the Gulf of Mexico achieved the world's deepest pipeline installation, which took place in the Summer 2002 at a maximum water depth of 7300 ft or 2225 m [de Reals, et al 2003; Nogueira and Stearns, 2003]. This successful project consisted of 110 miles of 12 in. pipe, with a number of in-line structures, which had to land at precise locations on the seabed, with tight tolerances. Previously, the Blue Stream pipeline at 7050 ft (2150 m) water depth was installed across the Black Sea [McKeehan and Kashunin, 1999] consisting of about 390 km of 24 in. pipe. The installation contractor, Saipem, used the J-lay vessels during the installation of these projects: the Canyon Express project was installed with the vessel FDS and the Blue Stream project used the semi-submersible S-7000.

The Horn Mountain 10-in. pipeline has been installed by Allseas, using the S-lay method with the vessel Solitaire, during the Winter 2002 in the Gulf of Mexico at a water depth of about 5500 ft. Allseas was able to achieve an impressive maximum lay rate of 5.6 miles in one day. The interested reader may consult Langner (2000) for a more comprehensive description of recent projects installed in the Gulf of Mexico.

For the offshore pipeline engineer, it is interesting to know the availability of the vessel fleet, as well as its pipe storage capacity and lay rates. This information is important to help establish the potential cost of a project and, therefore, its feasibility. Table 11.17 lists all the major pipelay contractors and their addresses in Houston, Texas. This table will allow information to be obtained directly from the contractors, who frequently are upgrading their vessel fleet. For example, recent additions are the lay barges Deep Blue (Coflexip Stena reel ship) and the Q-4000 (Cal Dive).

11.12.1 J -lay Installation Method

The J-lay installation method is a relatively new type of installation method specifically aimed at deepwater and ultra-deepwater projects. This method is characterised by a steep ramp, typically 65° or higher departure angle, so that the pipe has a suspended J-shape. While fig. 11.14 depicts this schematically, fig. 11.16 shows the Balder J-laying pipe with the aid of a side tower. The stresses and strain close to the top are minimised, as well as the horizontal tension component at the top and the horizontal tension at the mudline [Langner and Ayers, 1985]. The main advantages and disadvantages of the J-lay method are described in table 11.18.

Typically, to assess the technical feasibility, analysis is performed using commercially available software packages, such as OFFPIPE. Alternatively, a simplified analysis may

Table 11.17 Major offshore pipeline installation contractors in Houston, Texas

Installation Contractor and website	Address in Houston, Texas, USA
DSND Horizon www.dsnd.com www.subsea7.com	2500 City West Blvd Suite 300 Houston, Texas 77042 713-267-2246
Saipem Inc. www.saipem.it	15950 Park Row Houston TX 77084 281-552-5706
Stolt Comex Seaway www.stoltoffshore.com	900 Town & Country Lane Suite 400 Houston, Texas 77024
Oceaneering International www.oceaneering.com	11911 FM 529 Houston Texas 77041 713-329-4500
CalDive International www.caldive.com	400 N. Sam Houston Parkway E. Suite 400 Houston, Texas 77060 281-618-0400
Heerema Marine Contractors www.heerema.com	17154 Butte Creek, Suite 200 Houston, Texas 77090 281-880-1600
Allseas www.allseas.com	333 N. Sam Houston Pkwy. E. Suite 750 Houston, Texas 77060 281-999-3330
Coflexip Stena Offshore www.technip-coflexip.com	7660 Woodway, Suite 390 Houston, Texas 77063 713-789-8540
J. Ray McDermott, Inc. www.jraymcdermott.com	200 WestLake Park Blvd. Houston, TX 77079-2663 281-870-5235
Global Industries www.globalind.com	5151 San Felipe, Suite 900 Houston, Texas 77056 713-479-7911
Torch Offshore Inc. www.torchinc.com	11757 Katy Freeway, Suite 1300 Houston, Texas 77079 713-781-7990

be performed using the stiffened catenary equations, which can yield very accurate results for the J-lay configuration [Langner, 1984]. Such analysis will provide top tension, bottom tension and pipeline stresses and strains along the suspended catenary. With these parameters, the pipeline wall thickness can be checked, as well as the required vessel forces.

Figure 11.16 Heerema's balder in J-lay mode – Courtesy Heerema marine contractors

Table 11.18 Advantages and disadvantages of J-lay

Adv.	Best suited for ultra deep-water pipeline installation.
Adv.	Suited for all diameters.
Adv.	Smallest bottom tension of all methods, which leads to the smallest route radius, and allows more flexibility for route layout. This may be important in congested areas.
Adv.	Touchdown point is relatively closer to the vessel, thus easier to monitor and position.
Adv.	Can typically handle in-line appurtenances with relative ease, with respect to landing on the seafloor, but within the constraints of the J-lay tower.
Disadv.	Some vessels require the use of J-lay collars to hold the pipe.
Disadv.	If shallower water pipeline installation is required in the same route, the J-lay tower must be lowered to a less steep angle. Even then, depending on the water depth, it may be not feasible to J-lay the shallow end with a particular vessel and a dual (J-lay/S-lay) installation may be required. Such was the case of the Canyon Express project [de Reals, et al 2003].

Figure 11.17 Allseas Lorelay S-lay vessel – Courtesy C. Langner

11.12.2 S-lay

S-lay is utilised to install the vast majority of all offshore pipelines. Allseas have configured its flagship, the Solitaire, with a stinger that can reach very steep departure angles. As a result, it was able to install, a 10-in. pipeline at 5400 ft water depth.

S-lay is a very efficient lay method, since all welding is done with pipe in an horizontal position. The main advantages and disadvantages of the S-lay method are presented in table 11.19.

Table 11.19 Advantages and disadvantages of S-lay

Adv.	All welds are done on horizontal position, making for efficient productivity of multiple stations (typically 5–6).
Adv.	Suited for all diameter lines.
Adv/Disadv.	Can typically handle smaller, more compact in-line appurtenances with ease, but larger in-line structures may be too large to go through the stinger.
Disadv.	Buckle arrestors will induce concentrated higher strains in their vicinity within the stinger.
Disadv.	Typically, pipeline will rotate axially during installation [Endal and Verley, 2000; Endal, et al 1995].
Disadv.	Requires a relatively high component of horizontal tension.

Table 11.20 Advantages and disadvantages of reel-lay

Adv.	Almost all welds are done on-shore, minimising offshore welding.
Adv.	Well suited for smaller diameter lines and smaller diameter-to-thickness ratios.
Adv.	If all pipeline can stored on-board, a very fast installation campaign is achieved, making this method very cost effective.
Disadv.	If the route is too long or the diameter is relatively large, all the pipes may not be able to be stored on-board and a number of recharging trips to the spooling base may be necessary to re-load, thus offsetting the high lay rate.
Disadv.	Very high pipeline strains (of the order of 3–5%) are applied into the pipeline.
Disadv.	Due to high strains, welding methods and acceptance criteria are more stringent.
Disadv.	Pipeline will rotate during installation and may coil on the seafloor.
Disadv.	In-line structures are typically more difficult to handle and install.

11.12.3 Reel-lay

The reel method was patented in the USA by Gurtler (1968), who makes reference to a British Patent of 1948. The patent [Gurtler, 1968] has very detailed drawings of a horizontal reel, as well as a pre-bending apparatus and straightener. The main advantages and disadvantages of the reel-lay method are described in table 11.20.

11.12.4 Towed Pipelines

In this installation method, the pipeline is constructed onshore and towed into place. There are different ways to tow the pipeline string to site: surface tow, mid-depth tow or bottom tow. In the surface tow the pipe is positively buoyant, towed to location on the surface, and sunk in position by flooding. Wave action is a factor; therefore this method is used typically where rough seas are not likely. In the mid-depth tow typically the pipe or pipe bundle is negatively buoyant, suspended above the seabed and towed by a lead tug with a tail tug at the end of the pipe string. If the pipe is positively buoyant, mid-depth tow may be achieved by incorporating the use of drag chains at specified intervals along the pipe string, so that the pipe string assumes an equilibrium position above the seabed. For the bottom tow method, the pipeline rests on the seabed, and a tug pulls it. The length of the towed string is limited to about ten miles in the most favourable conditions.

The tow methods are challenging due to the effects of the environment such as waves action, oscillations during pull or abrasive effects of the seabed during bottom tow. However, the onshore construction may significantly reduce cost when compared to the installation methods described in the previous sections. Several failures of pipe bundles during tow attest to the precautions that the offshore pipeline engineer must take when using the tow method of installation.

References

American Petroleum Institute – Recommended Practice 1111 (1999). "Design, construction, operation, and maintenance of offshore hydrocarbon pipelines (Limit State Design)", (3rd ed.).

American Petroleum Institute – Specification 5L (2000). "Specification for line pipe", (42nd ed.).

American Society of Mechanical Engineers – B31.4 (2002). "Pipeline transportation systems for liquid hydrocarbons and other liquids".

American Society of Mechanical Engineers B31.8 (1999). "Gas transmission and distribution piping systems".

Ayers, R. R., et al (1989). "Submarine on-bottom stability: recent AGA research", *Eighth Int. Conf. on Off. Mech. and Arctic Eng.*, March 19–23, The Hague.

Baker, J. and Heyman, J. (1969). "Plastic design of frames", Cambridge University Press, Cambridge, UK.

Berry, W. H. (1968). "Pipelines from the North Sea block 49/26 to the Norfolk coast". *Journal of Petroleum Inst.*, Vol. 54, No. 532, pp. 104–106.

Britton, J. (1999). "External corrosion control and inspection of deepwater pipelines", *Deepwater Pipeline Tech. Conf.*, organized by Pipes and Pipelines Int., New Orleans, Louisiana.

Broussard, D. E., Barry, D. W., Kinzbach, R. B., and Kerschner, S. G. (1969). "Pipe laying barge with adjustable pipe discharge ramp". U.S. Patent no. 3,438,213.

Bryndum, M. B. et al (1983). "Hydrodynamic forces from wave and current loads on marine pipelines", *Off. Tech. Conf.*, Paper 4454, Houston, Texas.

Bryndum, M. B., Jacobsen, V., and Tsahalis, D. T. (1988). "Hydrodynamic forces on pipelines: model tests", *Seventh Int. Conf. on Off. Mech. and Arctic Eng.*

Choate, T. G. A., Davis, H., and Gaber, M. (2002). *Mediterranean Off. Conf.*, Alexandria, Egypt.

Code of Federal Regulations, Title 30, Part 250, Subpart J (2002). "Part 250 – oil and gas and sulphur operations in the outer continental shelf, subpart J – pipelines and pipeline rights-of-way". 7-01–02 Ed., U.S. Gov. Printing Office, Washington, D.C.

Code of Federal Regulations, Title 49, Part 192, Subpart A (2002). "Part 192 – transportation of natural and other gas by pipeline: minimum federal safety standards, subpart A – General". 10-01-02 Ed., U.S. Gov. Printing Office, Washington, D.C.

Code of Federal Regulations, Title 49, Part 195, Subpart A (2002). "Part 195 – transportation of hazardous liquids by pipeline, subpart A – General". 10-01-02 Ed., U.S. Gov. Printing Office, Washington, D.C.

DNV Guideline 14 (1998). "Free spanning pipelines", Det Norske Veritas, Norway.

DNV RP B401 (1993). "Cathodic protection design", Det Norske Veritas, Norway.

Det Norske Veritas RP E305 (1988). "On-bottom stability design of submarine pipelines", Det Norske Veritas, Norway.

Det Norske Veritas OS F101 (2000). "Submarine pipeline systems", Det Norske Veritas, Norway.

Dixon, D. A. and Rutledge, D. R. (1968). "Stiffened catenary calculations in pipeline laying problem". *Journal of Engineering for Ind.*, Vol. 90, pp. 153–170.

Endal, G., Ness, O. B., Verley, R., Holthe, K., and Remseth, S. (1995). "Behavior of offshore pipelines subjected to residual curvature during laying", *14th Int. Conf. of Off. Mech. and Arctic Eng.*

Endal, G. and Verley, R. (2000). "Cyclic roll of large diameter pipeline during laying", *19th Int. Conf. of Off. Mech. and Arctic Eng.*, New Orleans, LA, USA.

Fowler, J. R. (1990). "Large scale collapse testing", *Proceedings of Collapse of Offshore Pipelines*, Pipeline Research Committee – American Gas Association, Houston, Texas.

Gurtler, H. (1968). Method of Laying Pipeline, U.S. Patent Office, Patent number 3,372,461, patented March 12, 1968, New Orleans, LA, USA.

Jones, W. T. (1976). "On-bottom pipeline stability in steady water currents", *Off. Tech. Conf.*, Paper 2598, Houston, Texas.

Lambrakos, K. F. (1985). "Marine pipeline soil friction coefficients from in-situ testing", *Ocean Engineering*, Vol. 12, No. 2, pp. 131–150.

Lanan, G. A., Ennis, J. O. S., Egger, P. S., and Yockey, K. E. (2001). "Northstar offshore Arctic pipeline design and construction", *Off. Tech. Conf.*, Paper 13133, Houston, Texas.

Lanan, G. A., Nogueira, A. C, McShane, B. M., and Ennis, J. O. (2000). "Northstar development project pipeline description and environmental loadings", ASME, *Int. Pipeline Conf.*, Calgary, Canada.

Langner, C. G. (1984). "Relationships for deepwater suspended pipe spans", *Third Int. Conf. on Off. Mech. and Arctic Eng.*, New Orleans, LA.

Langner, C. G. (1999). "Buckle arrestors for deepwater pipelines", *Off. Tech. Conf.*, Paper 10711, Houston, Texas.

Langner C. G. (2000). "Technical challenges for deepwater pipelines in the Gulf of Mexico – update 2000", Marine Pipeline Engineering Course, Houston, Texas.

Langner, C. G. and Ayers, R. R. (1985). "The feasibility of laying pipelines in deep water", *Fourth Int. Conf. on Off. Mech. and Arctic Eng.*

Lyons, C. G. (1973). "Soil resistance to lateral sliding of marine pipelines", *Off. Tech. Conf.*, Paper 1876, Houston, Texas.

McAllister, E. W. (1993). "Pipe line rules of thumb handbook", (3rd ed.), Gulf Publishing Co., Houston, Texas.

McKeehan, D. S. and Kashunin, K. A. (1999). "The blue stream project-A large diameter deepwater pipeline in the Black Sea", *Proc. of the Non-governmental Ecological Vernadsky Foundation*, the Black Sea Regional Energy Centre, April 1999.

Murphey, C. and Langner, C. (1985). "Ultimate pipe strength under bending, collapse and fatigue", *4th Int. Conf. Off. Mech. and Arctic Eng.*

Nogueira, A. C. (1998a). "A new model for understanding buckle propagation: Link-beam", *Deepwater Pipeline Tech. Conf.*, organized by Pipes and Pipelines Int., New Orleans, LA.

Nogueira, A. C. (1998b). "Link-beam model for pipeline buckle propagation", *Off. Tech. Conf.*, Paper 8673, Houston, Texas.

Nogueira, A. C. (1998c). "Link-beam model for dynamic buckle propagation in pipelines", *Eighth Int. Off. And Polar Eng. Conf. (ISOPE)*, Montreal, Canada.

Nogueira, A. C., Lanan, G. A., Even, T. M., Fowler, J. R., and Hormberg, B. A. (2000). "Northstar development: Limit state design and experimental program", *Int. Pipeline Conf.*, Calgary, Canada.

Nogueira, A. C. and Lanan, G. A. (2000). "Rational modeling of ultimate pipe strength under bending and external pressure", *Int. Pipeline Conf.*, Calgary, Canada.

Nogueira, A. C. and Lanan, G. A. (2001). "Application of a rational model for collapse of deepwater pipelines", *4th Deepwater Pipeline and Riser Tech. Conf.*, Huston, Texas.

Nogueira, A. C. and Stearns, J. P. (2003). "World record breaking: design and installation of the Canyon Express flowlines 7,300 foot deep", *Off. Pipeline Tech. Conf.*, Houston, Texas.

Palmer, A. C. (1994). "Deepwater pipelines: improving the state-of-the art", *Off. Tech. Conf.*, Paper 7541, Houston, Texas.

Park, T. D. and Kyriakides, S. (1997). "On the performance of integral buckle arrestors for offshore pipelines", *International Journal of Mech. Sciences*, Vol. 29, No. 6.

de Reals, T. B., Lomenech, H., Nogueira, A. C., and Stearns, J. P. (2003). "Canyon express flowline system: design and installation", *Off. Tech. Conf.*, Paper 15096, Houston, Texas.

Sarpkaya, T. and Isaacson, M. (1981). Mechanics of Wave Forces on Offshore Structures, Van Nostrand Reinhold Company, New York.

Stark, P. R. and McKeehan, D. S. (1995). "Hydrostatic collapse research in support of the Oman India gas pipeline", *Off. Tech. Conf.*, Paper 7705, Houston, Texas.

Tam, C., Raven, P., Robinson, R., Stensgaard, T., Al-Sharif, A. M., and Preston, R. (1996). "Oman India pipeline: development of design methods for hydrostatic collapse in deep water", *Off. Pipeline Tech. Conf.*, Amsterdam.

Timmermans, W. J. (2000). "The past and future of offshore pipelines". *Off. Pipeline Tech. Conf.*, Oslo, Norway.

Timoshenko and Gere (1961). Theory of Elastic Stability, McGraw-Hill, New York.

Troitsky, M. S. (1982). "Tubular steel structures – theory and design", (2nd ed.). The James F. Lincolin Are Welding Foundation, Cleveland, Ohio.,

Wallace, B. K., Gudimetla, R., Nelson, S., and Hassold, T. A. (2003). "Canyon express subsea multiphase flow metering system: principles and experience", *Off. Tech. Conf.*, Paper 15098, Houston, Texas.

Handbook of Offshore Engineering
S. Chakrabarti (Ed.)
© 2005 Elsevier Ltd. All rights reserved.

Chapter 12

Design for Reliability: Human and Organisational Factors

Robert G. Bea
University of California, Berkeley, CA

12.1 Introduction

Very advanced technology has been developed to assist offshore engineers in the design of platforms, floating structures, pipelines and ships. Those who have used and are using this technology have much to be proud of. Today there is a vast infrastructure of these structures located on the world's continental shelves and slopes. In the main, this infrastructure has had a remarkable record of success. This chapter is about a part of this technology that is focused on people and their organisations and how to design offshore structures to achieve desirable reliability. The objective of this chapter is to provide the engineer – designer-oriented guidelines to help reap success in the design of offshore structures. The application of these guidelines is illustrated with two examples: (1) design of a "minimum" offshore structures and (2) design of an innovative deepwater structure.

This chapter will address the following topics:

- Recent experiences of designs gone bad
- Design objectives: life cycle quality, reliability and minimum costs
- Approaches to achieve successful designs
- Instruments to help achieve design success
- Example applications

12.2 Recent Experiences of Designs Gone Bad

As a result of studying more than 600 "well documented" (these are difficult to find) major recent failures of offshore structures, some interesting insights have been developed [Bea, 2000a]:

(1) Approximately 80% of the major failures (cost more than U.S. $ 1 million) are directly due to human and organisational factors (HOF) and the malfunctions that develop as a result of these factors (e.g. platform fails due to explosion and fire). These causes will be identified as *exherent causes*. Only about 20% of these failures can be regarded as being natural or representing residual risk (e.g. platform fails due to hurricane forces). These causes will be identified as *inherent causes*.

This finding is a tribute to the engineers and technology that has been used to design these structures. The primary causes of failures are not associated directly with the technology concerned with design for the conditions traditionally addressed by offshore engineers.

- Of the 80% of the major failures that are due to exherent causes, about 80% of these occur during operations and maintenance activities; frequently, the maintenance activities interact with the operations activities in an undesirable way. Frequently, the structure cannot be operated as intended and short-cuts and adaptations must be made in the field.
- It is important to define failure. In this chapter, failure is defined as realising undesirable and unanticipated compromises in the quality of the offshore structure. Quality is the result of four attributes: (1) serviceability (fitness for purpose), (2) safety (freedom from undue exposure to harm or injury), (3) durability (freedom from unanticipated degradation in the quality attributes), and (4) compatibility (meets business and social objectives – on time, on budget and happy customers, including the environment). The probability of failure is defined as the likelihood that the quality objectives are not realised during the life cycle of the offshore structure. Reliability is the likelihood that these quality objectives are realised.
- Of the failures due to the exherent causes that occur during operations and maintenance, more than half (50%) of these can be traced back to seriously flawed engineering design; offshore structures may be designed according to the accepted industry standards and yet are seriously flawed due to limitations and imperfections that are embedded in the industry standards and/or how they are used. Offshore structures are designed that cannot be built, operated and maintained as originally intended; the structures cannot be built as intended and changes must be made during the construction process to allow the construction to proceed; flaws can be introduced by these changes or flaws can be introduced by the construction process itself. When the structure gets to the field, modifications are made in an attempt to make the structure workable or to facilitate the operations, and in the process additional flaws can be introduced. Thus, during the operations and the maintenance phases, operations personnel are faced with a seriously deficient or a defective structure that cannot be operated and maintained as intended.
- Of the 20% of failures that do not occur during the operations, the percentages of failures developing during the design and the construction phase are about equal. There are a large number of "quiet" failures that develop during these phases that are

increasingly frequently ending up in legal proceedings. Recently, there have been several of these failures that have had costs exceeding U.S. $ 1 billion. Initially the causes of the failures were identified to be due to flaws in the engineering design processes. However, the causes of these failures were ultimately traced to flaws in EPCO (Engineering, Procurement, Construction, Operating) contracting, organisational and management processes.

- The failure development process can be organised into three categories of events or stages: (1) initiating, (2) contributing and (3) propagating. The dominant initiating events are developed by "operators" performing erroneous acts of commission or interfacing with the system components that have "embedded pathogens" that are activated by such acts of commission (about 80%); the other initiating events are acts or developments involving acts of omissions (something important left out).

  The dominant contributing events are organisational; these contributors act directly to encourage or cause the initiating events. In the same way, the dominant propagating events are also organisational; these propagators are generally responsible for allowing the initiating events to unfold into a failure. A taxonomy (classification system) will be developed for these malfunctions later in this section. It is also important to note that these same organisational aspects very frequently are responsible for the development of "near-misses" that do not unfold into failures.

- It is important to define what constitutes an offshore structure "system". In this work, a system has been defined as composed of seven primary components: (1) the structure (provides support for facilities and operations), (2) the hardware (facilities, control systems, life support), (3) the procedures (formal, informal, written, computer software), (4) the environments (external, internal, social), (5) the operators (those that interface directly with the system), (6) the organisations (institutional frameworks in which operations are conducted), and (7) the interfaces among the foregoing. Systems have a life cycle that consists of concept development, design, construction, operation, maintenance and decommissioning. Failures must be examined in the framework of the components that comprise an offshore structure system and contexts of the life cycle activities.
- Most failures involve never being exactly repeated sequences of events and multiple breakdowns or malfunctions in the components that comprise an offshore structure system. These events are frequently dubbed "incredible" or "impossible". After many of these failures, it is observed that if only one of the protective "barriers" had not been breached, then the accident would not have occurred. Experience has adequately shown that it is extremely difficult, if not impossible, to accurately recreate the time sequence of the events that actually took place during the period leading to the failure. Unknowable complexities generally pervade this process because a detailed information on the failure development is not available or is withheld. Hindsight and confirmational bias are common, as are distorted recollections. Stories told from a variety of viewpoints involved in the development of a failure seem to be the best way currently available to

capture the richness of the factors, elements and processes that unfold in the development of a failure.

- The discriminating difference between the "major" and the "not-so-major" failure involves the "energy" released by and/or expended on the failure. A not-so-major failure generally involves only a few people, only a few malfunctions or breakdowns, and only small amounts of energy that frequently is reflected in the not-so-major direct and indirect, short-term and long-term "costs" associated with the failure. The major failures are characterised with the involvement of many people and their organisations, a multitude of malfunctions or breakdowns, and the release and/or expenditure of major amounts of energy; this seems to be because it is only through the organisation that so many individuals become involved and the access provided to the major sources of this energy (money is a form of energy). Frequently, the organisation will construct "barriers" to prevent the failure causation to be traced in this direction. In addition, until recently, the legal process has focused on the "proximate causes" in failures; there have been some major exceptions to this focus recently, and the major roles of organisational malfunctions in an accident causation have been recognised in court. It is important to realise that the not-so-major accidents, if repeated very frequently, can lead to major losses.
- To many engineers who design offshore structures, the human and organisational factor part of the challenge of designing high quality and reliability systems is "not an engineering problem"; frequently, this is believed to be a "management problem". The case histories of these recent major failures clearly indicate that engineers have a critical role to play if the splendid histories of successes are to be maintained or improved. Engineers can learn how to use existing technology to reach such a goal. The challenge is to wisely apply what is known. To continue to ignore the human and organisational issues in design engineering of offshore structures is to continue to experience things that we do not want to happen and whose occurrence can be reduced.

An experience-based (heuristic) classification system (taxonomy) was developed to describe the causes of the recent failures (compromises in quality) that were studied [Bea, 2000a]. The taxonomies go beyond human and organisational malfunctions (errors) [Reason, 1990, 1997] and include the structure-hardware malfunctions, the procedure malfunctions, and the environmental influences. This encourages examination of the "parts" in the context of the whole – the offshore structure "system". The taxonomies define the hows of malfunctions; the generic categories of actions or activities that result in flaws and malfunctions.

12.2.1 Operator Malfunctions

There are many different ways to define, classify and describe operator (individual) malfunctions that develop during design, construction, operation and maintenance of offshore structures [Wenk, 1986; Reason, 1990; Kirwan, 1994; Gertman and Blackman, 1994; Center for Chemical Processing Safety, 1994]. Operator malfunctions can be defined as actions taken by individuals that can lead an activity to realise a lower quality and reliability than intended. These are malfunctions of commission. Operator malfunctions

Table 12.1 Taxonomy of operator malfunctions

Communications – ineffective transmission of information
Slips – accidental lapses
Violations – intentional infringements or transgressions
Ignorance – unaware, unlearned
Planning and preparation – lack of sufficient program, procedures, readiness, and robustness
Selection and training – not suited, educated or practiced for the activities
Limitations and impairment – excessively fatigued, stressed and having diminished senses
Mistakes – cognitive malfunctions of perception, interpretation, decision, discrimination, diagnosis and action

also include actions not taken that can lead an activity to realise a lower quality than intended. These are malfunctions of omission. Operator malfunctions might best be described as action and inaction that result in lower than acceptable quality. Operator malfunctions also have been described as mis-administrations and unsafe actions.

Frequently, the causes of failures are identified as the result of "human errors". This identification is seriously flawed because errors are results, not causes [Woods, 1990; Reason, 1997]. This is an important distinction if one is really interested in understanding how and why malfunctions develop. Operator malfunctions can be described by types of error mechanisms. These include slips or lapses, mistakes and circumventions. Slips and lapses lead to low-quality actions where the outcome of the action was not what was intended. Frequently, the significance of this type of malfunction is small because these actions are not easily recognised by the person involved and in most cases easily corrected. A taxonomy of operator malfunctions based on the study of the failures of offshore structures is given in table 12.1.

Mistakes can develop where the action was intended, but the intention was wrong. Circumventions (violations, intentional shortcuts) are developed where a person decides to break some rule for what seems to be a good (or benign) reason to simplify or avoid a task. Mistakes are perhaps the most significant because the perpetrator has limited clues that there is a problem. Often, it takes an outsider to the situation to identify mistakes. A taxonomy of operator mistakes is given in table 12.2.

It is important to note that the study of failures involving offshore structures clearly indicates that the single leading factor in operator malfunctions is communications. Communications can be very easily flawed by "transmission" problems and "reception" problems. Feedback, that is so important to validate the communications, frequently is not present or encouraged. Language, culture, societal, physical problems and environmental influences can make this a very malfunction-prone process. Also note the importance of violations, ignorance (failure to use the existing technology or develop the necessary

Table 12.2 Taxonomy of mistakes

Perception – unaware, not knowing
Interpretation – improper evaluation and assessment of meaning
Decision – incorrect choice between alternatives
Discrimination – not perceiving the distinguishing features
Diagnosis – incorrect attribution of causes and or effects
Action – improper or incorrect carrying out activities

technology), planning and preparation, and selection and training. Engineers are frequently asked or required to do things that they are not sufficiently trained to do, and in some cases, are not capable of doing. But, they try.

12.2.2 Organisational Malfunctions

The analysis of the history of failures of offshore structures provides many examples in which organisational malfunctions have been primarily responsible for the failures. Organisational malfunction is defined as a departure from the acceptable or the desirable practice on the part of a group of individuals that results in unacceptable or undesirable results. Based on the study of case histories regarding the failures of offshore structures, studies of High Reliability Organisations (HRO) [Roberts, 1989, 1993; Weick, 1999], and managing organisational risks [Reason, 1997; Haber, et al 1991; Wu, et al 1989], a classification of organisational malfunctions is given in table 12.3.

Table 12.3 Taxonomy of organisational malfunctions

Communications – ineffective transmission of information
Culture – inappropriate goals, incentives, values and trust
Violations – intentional infringements or transgressions
Ignorance – unaware, unlearned
Planning and preparation – lack of sufficient program, procedures, readiness
Structure and organisation – ineffective connectedness, interdependence, lateral and vertical integration, lack of sufficient robustness
Monitoring and controlling – inappropriate awareness of critical developments and utilisation of ineffective corrective measures
Mistakes – cognitive malfunctions of perception, interpretation, decision, discrimination, diagnosis and action

Frequently, the organisation develops high rewards for maintaining and increasing production; meanwhile the organisation hopes for quality and reliability (*rewarding "A" while hoping for "B"*) [Roberts, 1993]. The *formal and informal* rewards and incentives provided by an organisation have a major influence on the performance of operators and on the quality and reliability of offshore structures. In a very major way, the performance of people is influenced by the incentives, rewards, and disincentives provided by the organisation. Many of these aspects are embodied in the "culture" (shared beliefs, artefacts) of an organisation. This culture largely results from the history (development and evolution) of the organisation. Cultures are extremely resistant to change; particularly if they have been "successful".

Several examples of organisational malfunctions recently have developed as a result of efforts to down-size and out-source as a part of *re-engineering* organisations [Bea, et al 1996]. The loss of corporate memories (leading to repetition of errors), creation of more difficult and intricate communications and organisational interfaces, degradation in morale, unwarranted reliance on the expertise of outside contractors, cut-backs in quality assurance and control, and provision of conflicting incentives (e.g. cut costs, yet maintain quality) are examples of activities that have lead to substantial compromises in the intended quality of systems. Much of the down-sizing ("right-sizing") outsourcing ("hopeful thinking") and repeated cost-cutting ("remove the fat until there is no muscle or bone") seems to have its source in modern business consulting. While some of this thinking can help promote "increased efficiency" and maybe even lower CapEx (Capital Expenditures), the robustness (damage and defect tolerance) of the organisation and the systems it creates are greatly reduced. Higher OpEx (Operating Expenditures) and more "accidents" can be expected; particularly in the long-run – if there is one, before the system is scraped or sold.

Experience indicates that one of the major factors in organisational malfunctions is the culture of the organisation. Organisational culture is reflected in how action, change, and innovation are viewed; the degree of external focus as contrasted with internal focus; incentives provided for risk-taking; the degree of lateral and vertical integration of the organisation; the effectiveness and honesty of communications; autonomy, responsibility, authority and decision making; trust; rewards and incentives; and the orientation towards the quality of performance contrasted with the quantity of production. In some organisations, the primary objective becomes "looking good", not doing good. The culture of an organisation is embedded in its history.

One of the major cultural elements is how managers in the organisation react to suggestions for a change in the management. Given the extreme importance of the organisation and its managers on quality and reliability, it is essential that these managers see suggestions for change (criticism?) in a positive manner. This is extremely difficult for some managers because they do not want to relinquish or change the strategies and processes that had made them managers.

Another major cultural element is how organisations react to failures. Often the focus is on blame and shame; the author calls this focus "kill the victims". Often the view is one that localises the failure; the fostered belief is that the failure was caused by a few misguided, poorly motivated or trained people. These reactions tend to stop the learning that can be developed by truly understanding the factors and situations that result in failures. These reactions tend to suppress the early warning signs that developing failures can provide.

12.2.3 Structure, Hardware, Equipment Malfunctions

Human malfunctions can be initiated by or exacerbated by poorly designed offshore structures and procedures that invite errors. Such structures are difficult to construct, operate and maintain. Table 12.4 summarises a classification system for hardware- (equipment, structure) related malfunctions. New technologies compound the problems of latent system flaws (structural pathogens) [Reason, 1997]. An excessively complex design, a close coupling (the failure of one component leads to the failure of other components) and severe performance demands on systems increase the difficulty in controlling the impact of human malfunctions, even in well operated systems. The field of ergonomics (people–hardware interfacing) has much to offer in helping create "people-friendly" offshore structures [ABS, 1998]. Such structures are designed for what people will do, not what they should be able to do. Such structures facilitate construction (constructability), operations (operability), and maintenance (maintainability, repairability).

The issues of offshore structure system robustness (defect or damage tolerance), design for constructability [Bea, 1992] and design for IMR (inspection, maintenance, repair) are critical aspects of engineering systems that will be able to deliver acceptable quality. The design of the system to assure robustness is intended to combine the beneficial aspects of configuration, ductility and excess capacity (it takes all three!) in those parts of the structure system that have high likelihoods and consequences associated with developing defects and damage. The result is a defect and damage tolerant system that is able to maintain its quality characteristics in the face of HOF. This has important ramifications with regard to engineering system design criteria and guidelines. A design for constructability is a design to facilitate construction, taking account of worker's qualifications, capabilities, and safety, environmental conditions, and the interfaces between the equipment and workers. A design for IMR has similar objectives. A reliability-centered maintenance (RCM) has been developed to address some of these problems, and particularly the unknowable and the HOF aspects [Jones, 1995].

Table 12.4 Taxonomy of structure and equipment malfunctions

Serviceability – inability to satisfy purposes for intended conditions
Safety – excessive threat of harm to life and the environment, demands exceed capacities
Durability – occurrence of unexpected maintenance and less than expected useful life; unexpected degradation in other quality characteristics
Compatibility – unacceptable and undesirable economic, schedule, environmental and aesthetic characteristics

It is becoming painfully clear that our engineering design guidelines for the creation of sufficient robustness – damage – defect tolerance in offshore structure systems is not sufficient. Our thinking about sufficient damage stability and damage tolerance needs rethinking. Our thinking about designing for the "maximum incredible" events needs more development. While two offshore structures can both be designed to "resist the 100-yr conditions" with exactly the same probabilities of failure, the two structures can have very different robustness or damage tolerance during the life cycle of the structures. The "minimum" CapEx offshore structure may not have a configuration, excess capacity or ductility to allow it to weather the inevitable defects and the damage that should be expected to develop during its life. Sufficient damage tolerance almost invariably results in increases in CapEx; the expectation and the frequent reality are that OpEx will be lowered. But, one must have a "long-term" view for this to be apparent.

Recent work has clearly shown that the foregoing statements about structure and hardware robustness apply equally well to organisations and operating teams. Proper configuration, excess capacity and ductility play out in organisations and teams in the same way that they do in the structure and hardware [Bea, 2000a, b]. It is when the organisation or an operating team encounters defects and damage – and is under serious stress, that the benefits of robustness become evident. A robust organisation or an operating team is not a repeatedly downsized (lean and mean?), out-sourced and financially strangled organisation. A robust organisation is a Higher Reliability Organisation (HRO).

12.2.4 Procedure and Software Malfunctions

Based on the study of procedure and software-related problems that have resulted in failures of offshore structures, table 12.5 summarises a classification system for procedure or software malfunctions. These malfunctions can be embedded in engineering design guidelines and computer programs, construction specifications and operations manuals. They can be embedded in contracts (formal and informal) and subcontracts. They can be embedded in how people are taught to do things. With the advent of computers and their integration into many aspects of the design, construction, and operation of oil and gas structures, software errors are of particular concern because the computer is the ultimate fool [Knoll, 1986; Rochllin, 1997].

Software errors in which incorrect and inaccurate algorithms were coded into computer programs have been the root cause of several recent failures of offshore structures

Table 12.5 Taxonomy of procedure and software malfunctions

Incorrect – faulty
Inaccurate – untrue
Incomplete – lacking the necessary parts
Excessive complexity – unnecessary intricacy
Poor organisation – dysfunctional structure
Poor documentation – ineffective information transmission

[Bea, 2000a, b]. Guidelines have been developed to address the quality of computer software for the performance of finite element analyses. Extensive software testing is required to assure that the software performs as it should and that the documentation is sufficient. Of particular importance is the provision of independent checking procedures that can be used to validate the results from analyses. High-quality procedures need to be verifiable based on first principles, results from testing and field experience. This has particular importance in the quality assurance and quality control (QA/QC) in design.

Given the rapid pace at which significant industrial and technical developments have been taking place, there has been a tendency to make design guidelines, construction specifications and operating manuals more and more complex. Such a tendency can be seen in many current guidelines used for the design of offshore structures. In many cases, poor organisation and documentation of software and procedures has exacerbated the tendencies for humans to make errors [Rochlin, 1997]. Simplicity, clarity, completeness, accuracy and good organisation are desirable attributes in procedures developed for the design, construction, maintenance and operation of offshore structures.

12.2.5 Environmental Influences

Environmental influences can have important effects on the performance characteristics of individuals, organisations, hardware and software. These include:

- External (e.g. wind, temperature, rain, fog, time of day),
- Internal (lighting, ventilation, noise, motions) and
- Sociological factors (e.g. values, beliefs and morays).

All three of these environmental influences can have extremely important effects on human, operating team and organisational malfunctions.

12.3 Design Objectives: Life Cycle Quality, Reliability and Minimum Costs

12.3.1 Quality

In this development, the term "quality" is defined as freedom from unanticipated defects in offshore structures. Quality is fitness for purpose. Quality is meeting the requirements of those that own, operate, design, construct and regulate offshore structures. These requirements include those of serviceability, safety, compatibility and durability [Matousek, 1990]. Quality is freedom from unanticipated defects in the serviceability, safety, durability and compatibility of the offshore structure system.

Serviceability is suitability for the proposed purposes, i.e. functionality. Serviceability is intended to guarantee the use of the system for the agreed purpose and under the agreed conditions of use. Safety is the freedom from excessive danger to human life, the environment and property damage. Safety is the state of being free of undesirable and hazardous situations. The capacity of a structure to perform acceptably during extreme demands and other hazards is directly related to and most often associated with safety. Compatibility assures that the structure does not have unnecessary or excessive negative impacts on the environment and society during its life cycle. Compatibility also is the ability of the structure to meet economic, time, political, business and environmental requirements.

Durability assures that serviceability, safety and environmental compatibility are maintained during the intended life of the structure. Durability is freedom from unanticipated maintenance problems and costs. Experience with offshore structures has shown that durability is one of the most important characteristics that must be achieved; if insufficient durability is developed, then there are unanticipated and often undetected degradations in the other quality characteristics, and many times these degradations have disastrous results.

This is a holistic definition of the key objective of engineering design processes; to achieve adequate and acceptable quality [Hessami, 1999]. In recent years, a wide variety of processes, procedures and philosophies intended to improve and achieve adequate quality in goods and services have been developed and implemented including Total Quality Management [Demming, 1982], QA/QC and the International Standards Organization Quality Standards [ISO, 1994a, 1994b, 1994c]. These components are the building blocks of a quality management system. Engineers have learned that it is important to recognise that these processes, procedures and philosophies are all related to the same objective; they represent complementary parts of activities that are intended to help achieve adequate and acceptable quality in offshore structures. The challenge has been to learn how to use these processes, procedures and philosophies wisely, effectively and efficiently. Also, it is important to note that the traditional "business" objectives (e.g. serviceability, compatibility) have been merged with the traditional "safety" objectives; quality can be good for business and vice versa.

12.3.2 Reliability

Reliability is defined as the probability (likelihood) that a given level of quality will be achieved during the design, construction and operating life cycle phases of an offshore structure. Reliability is the likelihood that the structure system will perform in an acceptable manner. Acceptable performance means that the structure system has desirable serviceability, safety, compatibility and durability. The reliability, Ps (likelihood of success), can be expressed analytically as:

$$Ps = P[C \geq D] \tag{12.1}$$

where $P[\,]$ is read as the likelihood that [ ], D is the demand/s imposed on the system, and C is the capacity/ies of the system to successfully withstand the imposed demand/s.

The complement of reliability is the likelihood or probability of unacceptable quality; the probability of failure, Pf.

$$Ps = P[D \geq C] = 1 - Ps \tag{12.2}$$

This definition has linked the concepts of probability, uncertainty and reliability with the holistic definition of quality to reflect upon the likelihoods of achieving acceptable quality in offshore structures.

Compromises in quality of a structure system can occur in the structure itself and/or in the facilities it supports. These failures can be rooted in malfunctions developed by individuals (operators) in design, construction, operation, and/or maintenance. Individuals, the people who design, construct, operate and maintain the systems have direct influence on malfunctions developed in these phases. However, the malfunctions developed by the individuals can be and often are caused (contributing factors) or compounded (propagating factors) by malfunction-inducing influences from organisations, hardware, software (procedures) and environment (external, internal). It is the combination of the individuals, organisations, procedures, environments, hardware, structure and interfaces between the foregoing that constitutes an offshore structure system. An offshore structure can only be understood realistically in the context of all of the components or elements that comprise the structure system and influence its life cycle performance characteristics.

The calculation of reliability or its complement, the likelihood of failure can be done in a variety of ways. The most straightforward method is to numerically integrate two distributions:

$$Pf = \sum F_C[d] f_D[d] \Delta d \tag{12.3}$$

where F_c is the cumulative distribution for the capacities (probability that capacity is equal to or less than a given demand, d), f_D is the density distribution for the demands (probability that the demand is in the interval Δd, and Δd is the integration interval. This expression can be used for any form of the distributions of demands and capacities. This expression can incorporate dependency (correlation) between the demands and capacities (e.g. as demand increases, capacity decreases) through the means used to define the cumulative distribution for the capacities.

Given that the distributions of demands and capacities can be reasonably characterised as normal (Gaussian) and independent, then Pf can be computed directly from:

$$\beta = \frac{\overline{C} - \overline{D}}{\sqrt{\sigma_C^2 + \sigma_D^2}} \tag{12.4}$$

where β is defined as the safety index, $\overline{C}$ is the mean (average) capacity, $\overline{D}$ is the mean demand, σ_C is the standard deviation of the distribution of capacities, and σ_D is the standard deviation of the demands. If the demands and capacities are not independent, then the safety index can be determined from:

$$\beta = \frac{\overline{C} - \overline{D}}{\sqrt{\sigma_C^2 + \sigma_D^2 - 2\rho_{DC}\sigma_C\sigma_D}} \tag{12.5}$$

where ρ_{DC} is the correlation coefficient ($-1 \leq \rho \leq 1$)that recognises the dependency of the magnitudes of the demands and capacities.

Given that the distributions of demands and capacities can be reasonably characterised as Lognormal (normal distribution of logarithms) and independent, then Pf can be computed directly from:

$$\beta = \frac{\ln(C_{50}/D_{50})}{\sqrt{\sigma_{\ln C^2} + \sigma_{\ln D^2}}} \tag{12.6}$$

where β is defined as the safety index, C_{50} is the median (50th percentile) capacity, D_{50} is the median demand, $\sigma_{\ln C}$ is the standard deviation of the Lognormal distribution of capacities and $\sigma_{\ln D}$ is the standard deviation of the Lognormal distribution demands. If the demands and capacities are not independent, then the safety index can be determined from:

$$\beta = \frac{\ln(C_{50}/D_{50})}{\sqrt{\sigma^2_{\ln C} + \sigma^2_{\ln D} - 2\rho_{DC}\sigma_{\ln C}\sigma_{\ln D}}} \tag{12.7}$$

The likelihood of failure is determined from the safety index as:

$$Pf = 1 - \Phi\ (\beta) \tag{12.8}$$

where $\Phi\ (\beta)$ is the standard cumulative normal distribution for the value of the safety index. For values of β between 1 and 3:

$$Pf \approx 0.475 \text{ exp } -(\beta)^{1.6} \tag{12.9}$$

even more approximately:

$$Pf \approx 10^{-\beta} \tag{12.10}$$

The safety index is like a factor-of-safety; as β gets larger, Pf gets smaller.

Note in equations (12.6) and (12.7) the ratio C_{50}/D_{50} is like the traditional factor-of-safety; it is the ratio of the median capacity to the median demand (load). This ratio is referred to as the median factor-of-safety. As the factor-of-safety gets larger, the safety index gets larger, and the likelihood of failure gets smaller. Also, as the uncertainty in the demand and capacity increases (reflected in the standard deviations), the safety index gets smaller, and the likelihood of failure gets larger. Greater uncertainties lead to greater likelihoods of failure.

A very useful "normalised" characterisation of the uncertainty characteristics is the coefficient of variation (COV, ratio of standard deviation to mean value of variable $X = V_X$). The COV of a variable, X, is related to the standard deviation of the logarithm of the distribution of X as:

$$\sigma_{\ln X} = \sqrt{\ln(1 + V_X^2)} \tag{12.11}$$

for small values of V_X (less than about 40%), $\sigma_{\ln X} \approx V_X$.

It is important to recognise that the variables used in designing offshore structures are often "conservative". Thus, there can be a source or sources of "bias" that must be eliminated or recognised quantitatively in analyses of *Ps* or *Pf*. The actual mean or median values of demands and capacities are required to develop realistic evaluations of *Ps* or *Pf*.

Also, it is important to recognise that there are different types of uncertainties that determine the resultant uncertainties associated with demands and capacities. One type of uncertainty (Type 1) is natural or inherent; this type of uncertainty is "information insensitive and random". A second type of uncertainty (Type 2) is associated with modelling, parametric and state uncertainties; this type of uncertainty is "information sensitive" and systematic. A third type (Type 3) of uncertainty is related to HOF. The focus of this chapter is on the third type of uncertainty. However, many of the thinking and analytical processes that have been used to address the Type 1 and the Type 2 uncertainties associated with designing offshore structures are adaptable to the Type 3 uncertainties. This adaptation will be illustrated later in this chapter.

It is very important to properly identify and characterise the Type 2 uncertainties. One approach is to express the Type 2 uncertainties as a "Bias" where this term is defined as the ratio of the actual or true value of the variable to the predicted or nominal (design) value of the variable. A variety of methods can be used to characterise the bias including field test data, laboratory test data, numerical data, and "expert" judgement. Often it is not possible to develop unambiguous separations of the Type 1 and Type 2 uncertainties and it is important not to include them twice.

There are "advanced" approaches to calculating *Ps* and *Pf* that are "distribution free" in the sense that a particular type of likelihood distribution (e.g. Normal, Lognormal) does not have to be assumed. These approaches have been termed first order reliability methods (FORM) and second-order reliability methods (SORM). While they are more advanced, they require much more complicated numerical methods to perform the calculations, and they too involve approximations. Because of these properties, the author suggests the use of equations (12.3), (12.6) and (12.7) to perform the majority of reliability analyses. The Lognormal distributions can be "fitted" to the important parts of the parameter distributions of concern (this takes some knowledge) and develop results that are very close to those from the more advanced approaches. The advantage of this "simplified" approach is that it is relatively transparent compared with the advanced approaches (calculations can be readily performed) and it can be used by design engineers that have a knowledge of the fundamentals of statistics and probability.

12.3.3 Minimum Costs

Providing quality in the design of an offshore structure can result in lower life cycle costs, be safer, and minimise unrealised expectations during the life cycle of the facility. Quality can result in significant benefits to minimise costs and increase income through maximised serviceability and availability (durability). In this development, the focus will be solely on costs; however, even greater benefits can be developed if the maximised serviceability and availability effects on income are recognised.

Achieving adequate levels of quality and reliability is not quick, easy, or free. It can be costly in terms of the initial investments of manpower, time and other resources required to achieve it (fig. 12.1). But, if it is developed and maintained, it can result in significant savings in future costs. In addition, initial costs can be reduced by discarding ineffective

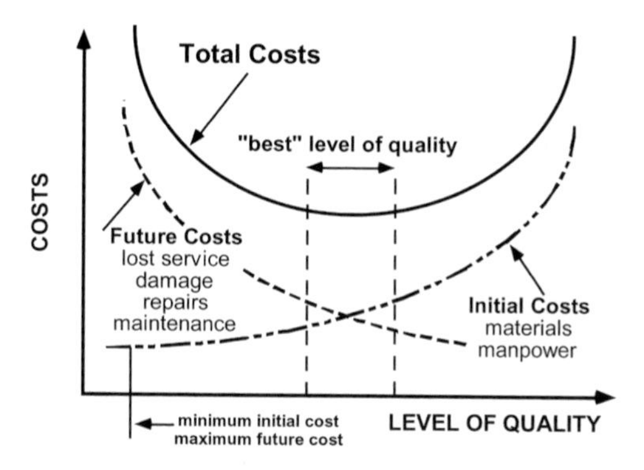

Figure 12.1 Consideration of initial and future costs associated with various levels of quality

and inefficient programs that are currently in use. A basic objective is to find ways to reduce both initial and future costs and thereby provide both a short-term and long-term financial incentive to implement improved quality and reliability programs. The objective is to find the level or degree of quality that will minimise the total of initial and future costs.

Different levels of quality are needed for different levels of criticality of elements in a system. If a system element or component is particularly critical to the quality and reliability of a system, then even though it may have identical initial costs, it may have very different future costs (fig. 12.2). Higher levels of quality and more intense QA/QC (Quality Assurance and Quality Control) measures should be relegated to those elements and components that have higher levels of criticality.

The costs to correct insufficient quality are a function of when the deficiencies are detected and corrected (fig. 12.3). The earlier the deficiencies are caught and fixed, the lesser the costs. The most expensive time to fix quality deficiencies is after the system is placed in service. This places a large premium on early detection and correction of errors. Not only

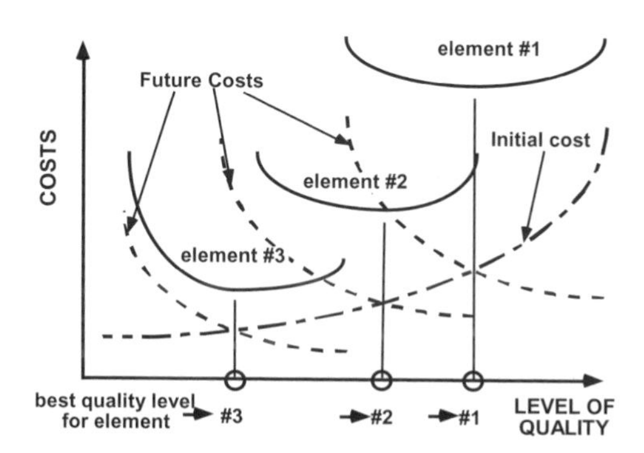

Figure 12.2 Criticality should determine the level of quality

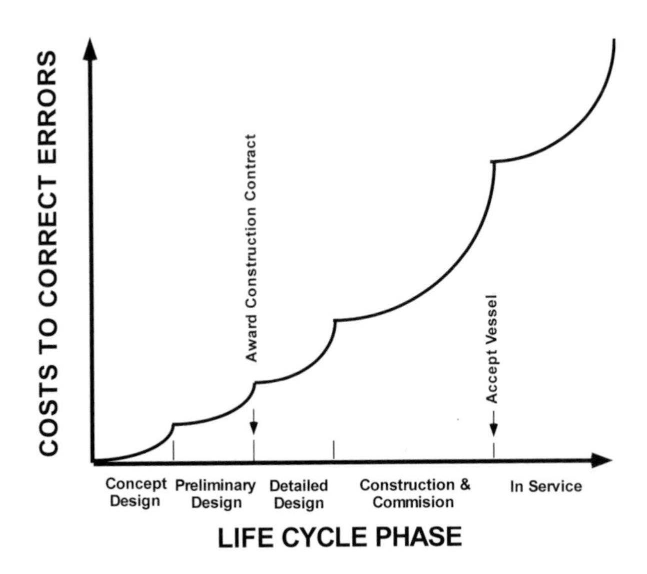

Figure 12.3 Life cycle costs to correct errors

are there large direct future costs associated with fixing errors, but also large indirect costs associated with loss of business and loss of image.

The present value of the total life cycle cost, C, associated with the performance of a system can be expressed as:

$$C = C_O + (C_S + C_I + C_M + C_R) = C_O + C_F \tag{12.12}$$

where the subscript O refers to the initial cost, S refers to loss of serviceability cost, I refers to inspection cost, M refers to structural maintenance costs, R refers to structural repair costs and F to the total future costs associated with the maintenance of the system.

Assuming a continuous discounting, each of the individual costs can be expressed as:

$$C_X = \sum C_x \exp(-r\, T_x) = \sum C_x(\mathrm{PVF}_x) \tag{12.13}$$

where the uppercase subscript X refers to a type of cost, the lowercase subscript x refers to the specific cost, the summation is taken over the occasions or time for the category of cost, r is the net discount rate, T is the time that the expense is incurred, and PVF is the resultant net present value function.

All of these categories of costs are variable and uncertain. Likelihoods (or probabilities) can be entered into the process in several ways. A traditional approach has been to focus on expected costs in which the estimated cost is multiplied by the probability, P, of experiencing that cost:

$$E[C_X] = C_X P_X \tag{12.14}$$

The total expected cost can be written as:

$$E[C] = \sum C_X P_X \tag{12.15}$$

The expected initial cost includes the costs associated with the system capacity, durability (degree of corrosion and fatigue protection provided including materials, redundancy and robustness integrated into the system), and construction (including degree of QA/QC provided). The expected future cost includes the costs associated with loss of serviceability of the system (C_S) and the costs associated with a given inspection, maintenance, repair (IMR) program (C_I, C_M, C_R).

Often, it is useful to provide the decision making process with an expression of the uncertainties associated with the expected costs. The uncertainties associated with each of the cost and the probability variables can be estimated on the basis of analyses, data and experience. Based on a first-order, second moment approximation that utilises the mean values and coefficients of variation (COV = V = ratio of standard deviation to mean) for each of the cost and components ($\overline{C_X}$, $\overline{P_X}$, V_{Cx}, V_{Px}) the mean cost and coefficient of variation in that cost can be estimated as:

$$\overline{C} = \sum (\overline{C_X}\,\overline{P_X}) \tag{12.16}$$

$$V_C = \sqrt{\sum V_{C_X}^2 + \sum V_{P_X}^2} \tag{12.17}$$

The process of defining what constitutes desirable quality and reliability can be expressed as a utility maximisation process. The objective of the utility maximisation process can be expressed as an expected total cost minimisation:

$$E[C]_{\min} = \left[\sum C_X P_X\right]_{\min} \tag{12.18}$$

The expected value costs associated with an alternative is the average monetary result per decision that would be realised if the decision makers accepted the alternative over a series of identical repeated trials. The expected value concept is a philosophy for consistent decision making, which if practiced consistently, can bring the sum total of the utilities of the decision to the highest possible level.

The expected value is not an absolute measure of a monetary outcome. It is incorrect to believe that the expected value is the most probable result of selecting an alternative. If one wanted to determine the probabilities of different magnitudes of utilities, then likelihoods could be assigned to each of the cost elements and these likelihoods propagated through the cost and likelihood evaluations to develop probability distributions of the potential utilities.

Given that the initial costs associated with a given quality alternative can be related linearly to the logarithm of P_F:

$$E[C_O] = P_O(C_O + \Delta C_O \log_{10} P_F) \tag{12.19}$$

C_O is the initial cost versus P_F intercept, ΔC_O is the slope of the initial cost curve, and P_O is the probability that the estimated initial cost will be realised. Given that the inspection, maintenance, and repair costs do not vary significantly with P_F, differentiating the sum of initial and future cost with respect to P_F to find the point of zero slope gives the P_F that produces the lowest total cost (P_{fo}):

$$P_{fo} = \frac{0.435}{R_c \, \mathrm{PVF}} \tag{12.20}$$

R_c (cost ratio) is the ratio of the present valued future cost, C_F, to the expected cost needed to decrease P_F by a factor of 10, ΔC_O:

$$R_c = \frac{C_F}{\Delta C_O} \tag{12.21}$$

PVF is a present value discount function. Based on continuous discounting and replacement:

$$\mathrm{PVF} = \left[1 - (1 + r)^{-L}\right]/r \tag{12.22}$$

For a continuous discount function and long-life system (life $\geq$ 10 yr), PVF $\approx r^{-1}$ where r is the monetary net discount rate (investment rate minus inflation rate). For short-life systems (life $\leq$ 5 yr), PVF $\approx$ L, where L is the life in years.

As shown in fig. 12.4, as the costs associated with the development of insufficient quality increases, the reliability must increase. As the initial costs to achieve quality increases, the optimum reliability decreases. The optimum reliability is based on the quality that will develop the lowest total initial and future costs. The marginal probability of insufficient quality is double the optimum quality probability. It is the quality in which the incremental investment to achieve quality equals the incremental future benefit (cost/benefit = 1.0). Reliability of a system element, component and system is a function of its criticality expressed by the product of the cost ratio and present value function.

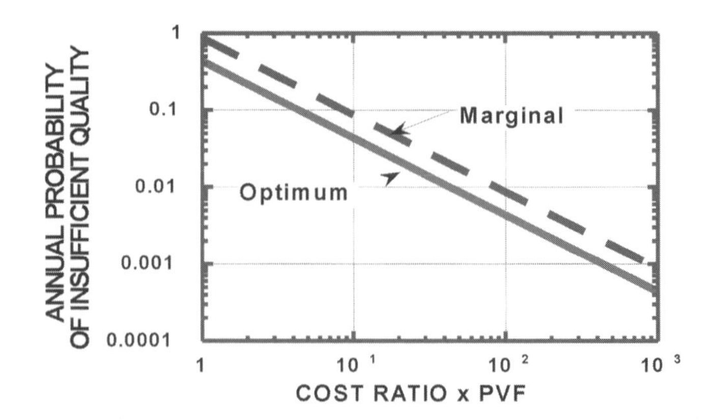

Figure 12.4 The economics and likelihood of insufficient quality

Quality can be a substantial competitive aspect in industrial activities. If a purchaser or user recognises the benefits of adequate quality and is able and willing to pay for it, then quality can be a competitive advantage. If a purchaser or user does not recognise the benefits of adequate quality or is unable or unwilling to pay for it, then quality can be a competitive disadvantage. Purchaser/owner quality goals must be carefully defined so that uniformity can be developed in the degrees of quality offered in a product or service sector. Once these goals have been defined, then the purchaser/owner must be willing to pay for the required quality.

12.4 Approaches to Achieve Successful Designs

An important starting point in addressing HOF in the quality and reliability of offshore structures is to recognise that while human and organisational malfunctions and errors are inevitable, their occurrence can be reduced and their effects mitigated by improving how structures are designed, constructed, operated, maintained and decommissioned. Engineering can improve the processes and products of design, construction, operations, maintenance and decommissioning to reduce the malfunction promoting characteristics, and to increase malfunction detection and recovery characteristics. Engineering can help develop systems for what people will do, not for what they should do. Engineering also can have important influences on the organisation and management aspects of these systems.

Engineering organisations have important and pervasive influences on the reliability of offshore structure systems. High reliability organisations (HROs) have been shown to be able to develop high reliability systems that operate relatively error free over long periods of time and in many cases, in very hazardous environments. The HROs go beyond Total Quality Management and the International Standards Organization certifications in their quest for quality and reliability. They have extensive process auditing procedures to help spot safety problems and have reward systems that encourage risk-mitigating behaviours, They have high quality standards and maintain their risk perception and awareness. Most importantly, such organisations maintain a strong command and control system that provides for organisational robustness or defect tolerance.

There are three fundamental, complimentary and interactive approaches to achieving adequate and acceptable quality and reliability in offshore structures:

- Proactive (activities implemented before malfunctions occur),
- Reactive (activities implemented after malfunctions occur) and
- Interactive or real-time (activities implemented during occurrence of malfunctions).

In the context of these three approaches, there are three primary strategies to be employed:

- Reduce incidence of malfunctions,
- Increase detection and correction of malfunctions and
- Reduce effects of malfunctions.

One approach frequently considered by engineers is to combat the potential effects of human and organisational malfunctions by increasing the structure's factors-of-safety. This has not proven to be an effective approach because making the structure stronger for the design loadings does not necessarily make the structure more reliable for extrinsic hazards. It has proven to be much more effective to implement measures directed at the source of the unreliability – people and their organisations.

12.4.1 Proactive Approaches

The proactive approach attempts to understand a system even before it fails (unacceptable quality) in an attempt to identify how it could fail in the future. Measures can then be put in place to prevent the failure or failures that have been anticipated. Proactive approaches include well-developed qualitative methods such as HazOp (Hazard Operability) and FMEA (Failure Mode and Effects Analyses) and quantitative methods such as SRA (Structural Reliability Analyses), PRA (Probabilistic Risk Analyses) and QRA (Quantified Risk Analyses) [Center for Chemical Process Safety, 1989; Spouge, 1999; Moan, 1997; Soares, 1998; Vinnem, 1998]. Each of these methods have benefits and limitations [Groeneweg, 1994; Molak, 1997; Apostolakis, et al 1990; Aven and Porn, 1998; Bier, 1999].

Proactive approaches also include organisational – management improvements and strategies that are intended to develop higher reliability organisations (HROs). Such organisations are able to operate over long periods of time conducting relatively error free operations and to consistently make good decisions regarding quality and reliability. The creation of HROs is perhaps the most important proactive approach.

Another proactive approach that has not received the attention that it deserves is the creation of "robust" offshore structures and similarly robust design organisations. Robustness is defined here as damage or defect tolerance. Robustness in a structure or an organisation means that it can continue to operate satisfactorily without surrendering fundamental quality and reliability objectives until repairs and/or modifications can be made. These are "human friendly" structures in the sense that they can tolerate high probability defects and damage that have sources in human and organisational malfunctions. Studies of robustness in offshore structures [Avigutero and Bea, 1998; Bea, 2000a] have shown that it takes the combination of four attributes to create a robust structure system:

- configuration,
- ductility,
- excess capacity and
- appropriate correlation

Configuration relates to the topology of the structure system; how elements and materials are arranged. Frequently, this has been called "redundancy"; referring to the degree of static indeterminancy. But, configuration goes beyond redundancy so that as elements or members are damaged or defective, that the structure system is still able to perform acceptably until repairs and modifications can be made. Ductility relates to the ability of the structure to shift the paths of demands or loads imposed on the elements and system. It relates to the ability of the structure materials and elements to deform nonlinearly without undue loss in capacity. Excess capacity relates to the ability of the structure system to carry normal loadings and over-loadings even though some of its elements may be damaged or defective. This means that some elements must be intentionally "over-designed" relative to the normal loading patterns and distributions so that these elements can carry the loadings that are transferred to them when other members or elements are damaged, defective or fail. Appropriate correlation refers to the dependence between the strengths of paired elements. In systems comprised of parallel elements, independence is desirable. In systems comprised of elements in series, dependence or high correlation is desirable. This is fail-safe or intrinsically safe design. Robust systems are not created by over zealous value improvement programs (VIP), excessive downsizing and outsourcing and excessive initial cost cutting (reduced CAPEX at the expense of future OPEX).

A Recent work with an HRO has clearly shown that development of robustness in engineering organisations is a very desirable proactive measure. Such organisations can tolerate defects and damage and still perform acceptably. This work also has shown that it takes the same three fundamental attributes: configuration, ductility and excess capacity. Such organisations are not downsized, out-sourced or cost-cut to the point that the organisation cannot tolerate daily and abnormal demands. Some organisation "fat" is a good thing when it allows the organisation to perform well when distressed.

The author has been an active protagonist and practitioner of the proactive reliability analysis-based approach to help improve the quality of offshore structures for more than three decades [Bea, 1974, 1975, 2000a; Marshall and Bea, 1976]. He believed that this approach provided an ability to forecast how systems could go bad. Very sophisticated analytical models could be developed to help foster this belief. Results from these analyses seemed to have value and to enhance his abilities to address some types of variability and uncertainty. This approach was workable as long as he dealt with systems in which the interactions of people with the systems were minimal or minimised. However, the problem changed radically when people began to exert major influences on the quality of the systems and in many cases on the physical aspects of the systems [Bea, 1996a, b]. In this case, his lack of knowledge of the physics and mechanics of the complex behaviours of people that in the future would design, construct, operate and maintain the system defined an "unpredictable", or certainly one with very limited predictability. The author's analytical models addressed systems that were essentially static and mechanical. Yet the real systems were dynamic, constantly changing, and more organic than mechanical. The analytical models generally failed to capture the complex interactions between people and the systems that they designed, constructed, operated and maintained.

The author found most data on the reliability of humans in performing tasks to be very limited [Kirwan, 1994; Gertman and Blackman, 1994; Dougherty and Fragola, 1986; Center for Chemical Process Safety, 1994]. Existing databases failed to capture or adequately characterise influences that had major effects on human reliability [Wu, et al 1989; Haber, et al 1991]. Yet, when the numbers were supplied to the very complex analytical models and the numbers were produced, the results were often mistaken for "reality". There was no way to verify the numbers. If the results indicated that the system was "acceptable", then nothing was done. If the results indicated that the system was "not acceptable", then generally the equipment and the hardware fixes were studied in an attempt to define a fix or fixes that would make the system acceptable or ALARP (As Low As Reasonably Practicable) [Melchers, 1993]. When the author went to the field to compare his analytical models with what was really there, he found little resemblance between his models and what was in the field [Bea, 1996b].

The author does not advocate discarding the analytical–quantitative proactive approach. He advocates using different types of proactive approaches to gain insights into how systems might fail and what might be done to keep them from failing [Weick, 2000; Bea, 2000a, b]. The marked limitations of the analytical models and the quantitative methods must be recognised or major damage can be done to the cause of the quality and reliability of offshore structures. The potential for engineers to be "hyper rational" and attempt to extend the applicability of SRA/PRA/QRA methods beyond their limitations must be recognised and countered. On the other hand, qualitative methods (e.g. HazOp, FMEA), in the hands of qualified and properly motivated assessors (both internal and external) can do much to help the causes of quality and reliability [Center for Chemical Process Safety, 1989, 1994]. Experience, judgement and intuition of the assessors needs to be properly recognised, respected and fully integrated into the proactive qualitative and quantitative approaches. Much headway has been made recently in combining the powers of qualitative methods with quantitative Risk Assessment and Management (RAM) methods [Bea, 2000a, b]. The qualitative methods are able to capture more fully the dynamic, changing, organic, complex interactions that cannot be analysed [Weick, 2000; Haber, et al 1991; Groeneweg, 1994]. Given an input from the qualitative methods, the quantitative methods are able to provide numbers that can be used to assist the development of judgements about when, where and how to better achieve quality and reliability in offshore structures. But, even at this level of development, the proactive RAM methods are very limited in their abilities to truly provide quality and reliability in offshore structures. Other methods (e.g. interactive RAM) must be used to address the unknowable and the unimaginable hazards.

It is the author's experience in working with and on offshore structure systems for more than four decades, that many if not most of the important proactive developments in the quality and reliability of these systems were originated in a cooperative, trust-based venture of knowledgeable "facilitators" working with seasoned veterans that have daily responsibilities for the quality of these systems. This cooperative venture includes design, construction/decommissioning, operations and maintenance/inspection personnel. Yet, it is also the author's experience, that many engineering and many well-meaning reliability – risk analysis "experts" are not developing a cooperative environment. This is very disturbing. The conduct of each operation during the life cycle of an engineered system should be regarded as the operations of "families". Knowledgeable, trained, experienced and sensitive outsiders can help, encourage and assist "families" to become "better". But, they cannot make the families better. Families can only be changed from within by the family members. Proactive measures based on casual or superficial knowledge of a system or of an operation of that system should be regarded as tinkering. And, tinkering can have some very undesirable effects and results [Wenk, 1986; Woods, 1990; Weick, 1995; Bea, 1996, 2001a].

The crux of the problem with the proactive analytical approaches is with the severe limitations of such approaches in their abilities to reasonably characterise human and organisational factors and their effects on the performance of a system [Center for Chemical Process Safety, 1994; Reason, 1997; Groeneweg, 1994; Haber, et al 1991; Wu, et al

1981; Rasmussen, et al 1987]. Quantitative analytical approaches rely on an underlying fundamental understanding of the physics and mechanics of the processes, elements and systems that are to be evaluated. Such understanding then allows the analyst to make projections into the future about the potential performance characteristics of the systems. And, it is here that the primary difficulties arise. There is no fundamental understanding of the physics and mechanics of the future performance – behaviour characteristics of the people that will come into contact with a system and even less understanding of the future organisational influences on this behaviour. One can provide very general projections of the performance of systems including the human and organisational aspects based on extensive assumptions about how things will be done, but little more. The problem is that engineers and managers start believing that the numbers represent reality.

To the author, the true value of proactive approaches does not lie in their predictive abilities. The true value lies in the disciplined process such approaches can provide to examine the strengths and weaknesses in systems; the objective is detection and not prediction. The magnitudes of the quantitative results, if these results have been generated using reasonable models and input information, can provide insights into where and how one might implement effective processes to encourage development of acceptable quality and reliability. The primary problems that the author has with the quantitative reliability analysis proactive approach are with how this method is used and what it is used to do. Frequently the results from the approach are used to justify meeting or not meeting regulatory/management targets and, in some cases not implementing clearly justified – needed improvements in the quality – reliability of an engineered system.

Perhaps the most severe limitation to the proactive approaches regards "knowability". One can only analyse what one can or does know. Predictability and knowability are the foundation blocks of the quantitative analytical models [Apostolakis, et al 1990; Rasmussen, 1996; Center for Chemical Process Safety, 1989; Spouge, 1999]. But, what about the unknowable and the unpredictable? Can we really convince ourselves that we can project into the future of offshore structure systems and perform analyses that can provide sufficient insights to enable us to implement the measures required to fully assure their quality and reliability? Or are some other processes and measures needed? This fundamental property of the unknowability has some extremely important ramifications with regard to application of the ALARP principle [Melchers, 1993; Hessami, 1999]. We can ALARP only what we recognise and this has proven to be extremely limited when it comes to very low probability – high consequence events that have their sources in human and organisational factors.

The author has concern for some proactive reliability based analyses that have been and are being used to define IMR (Inspection, Maintenance, Repair) programs for offshore structures [Bea, 1992]. Such analyses can only address the knowable and predictable aspects that influence IMR programs (e.g. fatigue damage at brace joints). Such analyses are frequently used to justify reductions in IMR program frequencies, intensities and costs [Faber, 1997; Soares, 1998; Marine Technology Directorate, 1989, 1992]. But what about the unknowable and the unpredictable elements that influence the IMR programs? We look for cracks where we do not find them and we find them where we do not look for them [Bucknell, 2000]. What about the host of major "biases" (differences between reality and the calculated results) that exert major influences on the results that come from such analyses [Xu, et al 1999]? These elements are frequently referred to as being founded in "gross errors" [Marine Technology Directorate, 1989; Bea, 1992]. Experience has adequately demonstrated that a very large amount, if not the majority of the defects and damages we encounter in offshore structures are not in any reasonable or practical sense "predictable" [Marine Technology Directorate, 1994; Winkworth and Fisher, 1992; Bucknell, 2000; De Leon and Heredia-Zavoni, 2001]. Other approaches (e.g. inductive information based) must be used to address the unknowable – unpredictable aspects that still must be managed in the operations of offshore structures.

Studies of the HROs (Higher Reliability Organisations) have shed some light on the factors that contribute to errors made by organisations and risk mitigation in HRO. The HROs are those organisations that have operated nearly error-free over long periods of time. A wide variety of HROs have been studied over long periods of time. The HRO research has been directed to define what these organisations do to reduce the probabilities of serious errors. The work has shown that the reduction in error occurrence is accomplished by the following [Roberts, 1989, 1993; Weick, 1995; Weick, et al 1999]: (1) command by exception or negation, (2) redundancy (robustness – defect and damage tolerance), (3) procedures and rules, (4) selection and training, (5) appropriate rewards and punishment and (6) ability of management to "see the big picture".

Command by exception (management by exception) refers to the management activity in which the authority is pushed to the lower levels of the organisation by managers who constantly monitor the behaviour of their subordinates. Decision-making responsibility is allowed to migrate to the persons with the most expertise to make the decision when unfamiliar situations arise (employee empowerment).

Redundancy involves people, procedures and hardware. It involves numerous individuals who serve as redundant decision-makers. There are multiple hardware components that will permit the system to function when one of the components fails. The term redundancy is directed towards the identification of the need for organisational "robustness" – damage and defect tolerance that can be developed, given proper configuration (deployment), ductility – ability and willingness to shift demands, excess capacity (ability to carry temporary overloads) and appropriate correlation (low for parallel elements, high for series elements).

Procedures that are correct, accurate, complete, well organised, well documented and are not excessively complex are an important part of an HRO. Adherence to the rules is emphasised as a way to prevent errors, unless the rules themselves contribute to error.

The HROs develop constant and high quality programs of personnel selection and training. Personnel selection is intended to select people that have natural talents for performing the tasks that have to be performed. Training in the conduct of normal and abnormal activities is mandatory to avoid errors. Training in how to handle unpredictable and unimaginable unraveling of systems is also needed. Establishment of appropriate rewards and punishment that are consistent with the organisational goals is critical; incentives are a key to performance.

An HRO's organisational structure is defined as one that allows the key decision-makers to understand the big picture. These decision-makers with the big picture perceive the important developing situations, properly integrate them, and then develop high reliability responses.

In a recent organisational research performed by Libuser (1994), five prominent failures were addressed including the Chernobyl nuclear power plant, the grounding of the Exxon Valdez, the Bhopal chemical plant gas leak, the mis-grinding of the Hubble Telescope mirror and the explosion of the space shuttle, Challenger. These failures were evaluated in the context of five hypotheses that defined risk mitigating and non-risk mitigating organisations. The failures provided support for the following five hypotheses:

- Risk mitigating organisations will have extensive process auditing procedures. Process auditing is an established system for ongoing checks designed to spot expected as well as unexpected safety problems. Safety drills would be included in this category as would be equipment testing. Follow-ups on problems revealed in prior audits are a critical part of this function.
- Risk mitigating organisations will have reward systems that encourage risk mitigating behaviour on the part of the organisation, its members and constituents. The reward system is the payoff that an individual or organisation gets for behaving in one way or another. It is concerned with reducing risky behaviour.
- Risk mitigating organisations will have quality standards that exceed the referent standard of quality in the industry. Risk mitigating organisations will correctly assess the risk associated with the given problem or situation. Two elements of risk perception are involved. One is whether or not there was any knowledge that risks existed at all. The second is if there was knowledge that risk existed, the extent to which it was understood sufficiently.
- Risk mitigating organisations will have a strong command and control system consisting of five elements: (a) migrating decision making, (b) redundancy, (c) rules and procedures, (d) training and (e) senior management has the big picture.

These concepts have been extended to characterise how organisations can organise to achieve high quality and reliability. Effective HROs are characterised by [Weick, et al 1999; Weick and Quinn, 1999; Weick and Sutcliffe, 2001]:

- Preoccupation with failure – any and all failures are regarded as insights on the health of a system, thorough analyses of near-failures, generalise (not localise) failures, encourage self-reporting of errors and understand the liabilities of successes.
- Reluctance to simplify interpretations – regard simplifications as potentially dangerous because they limit both the precautions people take and the number of undesired consequences they envision, respect what they do not know, match external complexities with internal complexities (requisite variety), diverse checks and balances, encourage a divergence in analytical perspectives among members of an organisation (it is the divergence, not the commonalties, that hold the key to detecting anomalies).
- Sensitivity to operations – construct and maintain a cognitive map that allows them to integrate diverse inputs into a single picture of the overall situation and status (situational awareness, "having the bubble"), people act thinkingly and with heed,

redundancy involving cross-checks, doubts that precautions are sufficient, and wariness about claimed levels of competence, exhibit extraordinary sensitivity to the incipient overloading of any one of it members, sensemaking.

- Commitment to resilience – capacity to cope with unanticipated dangers after they have become manifest, continuous management of fluctuations, prepare for inevitable surprises by expanding the general knowledge, technical facility, and command over resources, formal support for improvisation (capability to recombine actions in repertoire into novel successful combinations), and simultaneously believe and doubt their past experience.
- Under-specification of structures – avoid the adoption of orderly procedures to reduce error that often spreads them around, avoid higher level errors that tend to pick up and combine with lower level errors that make them harder to comprehend and more interactively complex, gain flexibility by enacting moments of organised anarchy, loosen specification of who is the important decision maker in order to allow decision making to migrate along with problems (migrating decision making), move in the direction of a garbage can structure in which problems, solutions, decision makers and choice opportunities are independent streams flowing through a system that become linked by their arrival and departure times and by any structural constraints that affect which problems, solutions and decision makers have access to which opportunities.

The other side of this coin is LROs (Lower Reliability Organisations). The studies show that these non-HROs are characterised by a focus on success rather than failure, and efficiency rather than reliability [Weick, et al 1999; Weick and Sutcliffe, 2001]. In a non-HRO, the cognitive infrastructure is underdeveloped, failures are localised rather than generalised, and highly specified structures and processes are put in place that develop inertial blind spots that allow failures to cumulate and produce catastrophic outcomes. The LROs have little or no robustness, have little or no diversity; they have focused conformity.

Efficient organisations practice stable activity patterns and unpredictable cognitive processes that often result in errors; they do the same things in the face of changing events, these changes go undetected because people are rushed, distracted, careless or ignorant [Weick and Quinn, 1999]. In the non-HRO expensive and inefficient learning and diversity in problem solving are not welcomed. Information, particularly "bad" or "useless" information is not actively sought, failures are not taken as learning lessons and new ideas are rejected. Communications are regarded as wasteful and hence the sharing of information and interpretations between individuals is stymied. Divergent views are discouraged, so that there is a narrow set of assumptions that sensitise it to a narrow variety of inputs.

In the non-HRO, success breeds confidence and fantasy, managers attribute success to themselves, rather than to luck, and they trust procedures to keep them appraised of developing problems. Under the assumption that success demonstrates competence, the non-HRO drifts into complacency, inattention, and habituated routines, which they often justify with the argument that they are eliminating unnecessary effort and redundancy. Often downsizing and out-sourcing are used to further the drives of efficiency and insensitivity is developed to overloading and its effects on judgement and performance. Redundancy (robustness or defect tolerance) is eliminated or reduced in the same drive resulting in the elimination of cross-checks, assumption that precautions and existing levels of training and experience are sufficient, and dependence on claimed levels of competence. With outsourcing, it is now the supplier, not the buyer that must become preoccupied with failure. But, the supplier is preoccupied with success, not failure, and because of low-bid contracting, often is concerned with the lowest possible cost success. The buyer now becomes more mindless and if novel forms of failure are possible, then the loss of a preoccupation with failure makes the buyer more vulnerable to failure. The non-HROs tend to lean towards anticipation of "expected surprises", risk aversion and planned defences against foreseeable accidents and risks; unforeseeable accidents and risks are not recognised or believed.

Reason (1997) in expanding his work from the individual [Reason, 1990] to the organisation, develops another series of important insights and findings. Reason observes that all technological organisations are governed by two primary processes: production and protection. Production produces the resources that make protection possible. Thus, the needs of production will generally have priority throughout most of an organisation's life, and consequently, most of those that manage the organisation will have skills in production, not protection. It is only after an accident or a near-miss that protection becomes for a short time period paramount in the minds of those that manage an organisation. Reason observes that production and protection are dependent on the same underlying organisational processes. If priority is given to production by management and the skills of the organisation are directed to maximising production, then unless other measures are implemented, one can expect an inevitable loss in protection until significant accidents cause an awakening of the need to implement protective measures. The organisation chooses to focus on problems that it always has (production) and not on problems it almost never has (major accidents). The organisation becomes "habituated" to the risks it faces and people forget to be afraid: "chronic worry is the price of quality and reliability" [Reason, 1997].

12.4.2 Reactive Approaches

The reactive approach is based on the analysis of the failure or near failures (incidents, near-misses) of a system. An attempt is to made to understand the reasons for the failure or near-failures, and then to put measures in place to prevent future failures of the system. The field of worker safety has largely developed from the application of this approach.

This attention to accidents, near-misses and incidents is clearly warranted. Studies have indicated that generally there are about 100+ incidents, 10–100 near-misses, to every accident [Hale, et al 1997; Rassmussen, et al 1987]. The incidents and near-misses can give early warnings of potential degradation in the safety of the system. The incidents and near-misses, if well understood and communicated provide important clues as to how the system operators are able to rescue their systems, returning them to a safe state and to the potential degradation in the inherent safety characteristics of the system. We have come to understand that responses to accidents and incidents can reveal much more about maintaining adequate quality and reliability than responses associated with successes.

Well-developed guidelines have been developed for investigating incidents and performing audits or assessments associated with near-misses and accidents [Center for Chemical Process Safety, 1992; Hale, et al 1997]. These guidelines indicate that the attitudes and beliefs of the involved organisations are critical in developing successful reactive processes and systems, particularly doing away with "blame and shame" cultures and practices. It is further observed that many if not most systems focus on "technical causes" including equipment and hardware. Human system failures are treated in a cursory manner and often from a safety engineering perspective that has a focus on outcomes of errors (e.g. inattention, lack of motivation) and statistical data (e.g. lost-time accidents) [Reason, 1997; Fischoff, 1975].

Most important, most reactive processes completely ignore the organisational malfunctions that are critically important in contributing to and compounding the initiating events that lead to accidents [Reason, 1997]. Finding "well-documented" failures is more the exception than the rule. Most accident investigation procedures and processes have been seriously flawed. The qualifications, experience and motivations of the accident assessors are critical; as are the processes that are used to investigate, assess and document the factors and events that developed during the accident. A wide variety of biases "infect" the investigation processes and investigators (e.g. confirmational bias, organisational bias, reductive bias) [Reason, 1997; Fischoff, 1975].

In the author's direct involvement with several major failures of offshore structures (casualties whose total cost exceeds U.S. $1 billion each), the most complete information develops during the legal, regulatory induced and insurance investigation proceedings. Many of these failures are "quiet". Fires and explosions (e.g. Piper Alpha), sinkings (e.g. Petrobras P-36) and collisions/groundings (e.g. Exxon Valdez) are "noisy" and often attract media, regulatory and public attention. Quiet failures on the other hand are not noisy; in fact, many times overt attempts are made to "keep them quiet". These quiet failures frequently are developed during the design and/or construction phases. These represent offshore structure "project failures".

The author recently has worked on two major quiet failures that involved the international EPC (Engineering, Procurement, Construction) offshore structure project failures that developed during construction. A third major failure involved an EPCO (add Operation) project that failed when the system was not able to develop the quality and reliability that had been contracted for. In both these cases, the initial "knee jerk" reaction was to direct the blame at "engineering errors" and a contended "lack of meeting the engineering standard of practice". Upon further extensive background development (taking 2 and 3 yr of legal proceedings), the issues shifted from the engineering "operating teams" to the "organisational and management" issues. Even though "partnering" was a primary theme of the formation of the contractors and contracting, in fact partnering was a myth. Even though ISO certifications were required and provided, the ISO QA/QC guidelines were not followed. The international organisations involved in the work developed severe "cultural conflicts" and communication breakdowns. Promises were made and not honoured; integrity was compromised. Experienced personnel were promised and not provided ("bait and switch"). There was a continually recurring theme of trying to get something/everything for nothing or next to nothing. As ultimately judged in the courts, these failures were firmly rooted in organisational malfunctions, not engineering malfunctions. The problem with most legal proceedings is that it is very rare that the results are made public. Thus, the insights important to the engineering profession is largely lost, and in some cases, seriously distorted.

A primary objective of incident reporting systems is to identify recurring trends from the large numbers of incidents with relatively minor outcomes. The primary objective of near-miss systems is to learn lessons (good and bad) from operational experiences. Near-misses have the potential for providing more information about the causes of serious accidents than accident information systems. The near-misses potentially include information on how the human operators have successfully returned their systems to the safe states. These lessons and insights should be reinforced to better equip operators to maintain the quality of their systems in the face of unpredictable and unimaginable unraveling of their systems.

A root cause analysis is generally interpreted to apply to systems that are concerned with the detailed investigations of accidents with major consequences. The author has a fundamental objection to root cause analysis because of the implication that there is a single cause at the root of the accident (reductive bias) [Center for Chemical Process Safety, 1994]. This is rarely the case. This is an attempt to simplify what is generally a very complex set of interactions and factors, and in this attempt, the lessons that could be learned from the accident are frequently lost. Important elements in a root cause analysis include an investigation procedure based on a model of accident causation. A systematic framework is needed so that the right issues are addressed during the investigation [Hale, et al 1997; Bea, et al 1996]. There are high priority requirements for comprehensiveness and consistency. The comprehensiveness needs to be based on a systems approach that includes error tendencies, error inducing environments, multiple causations, latent factors and causes and organisational influences. The focus should be on a model of the system factors so that error reduction measures and strategies can be identified. The requirement for consistency is particularly important if the results from multiple accident analyses are to be useful for evaluating trends in underlying causes over time.

There is no shortage of methods to provide a basis for a detailed analysis and the reporting of incidents, near-misses and accidents. The primary challenge is to determine how such methods can be introduced into the life cycle risk assessment and management (RAM) of offshore structures and how their long-term support can be developed (business incentives).

Inspections during construction, operation, and maintenance are a key element in reactive RAM approaches. Thus, development of IMR (Inspection, Maintenance, Repair) programs is a key element in the development of reactive management of the quality and reliability of offshore structures [Bea, 1992]. Deductive methods involving mechanics-based SRA/PRA/QRA techniques have been highly developed [Faber, 1997; Spouge, 1999; Soares, 1998]. These techniques focus on "predictable" damage that is focused primarily on durability; fatigue and corrosion degradations. The inductive methods involving discovery of defects and damage are focused primarily on "unpredictable" elements that are due primarily to unanticipated HOE such as weld flaws, fit-up or alignment defects, dropped objects, ineffective corrosion protection and collisions. Reliability centre maintenance (RCM) approaches have been developed and are continuing to be developed to help address both predictable and unpredictable damage and defects [Jones, 1995]. Some very significant forward strides have been made in the development and implementation of life cycle IMR database analysis and communications systems. But, due to expense and cost concerns, and unwillingness or inability of the organisation to integrate such systems into their business systems, much of this progress has been short lived.

The reactive approach has some important limitations. It is not often that one can truly understand the causes of accidents. If one does not understand the true causes, how can one expect to put the right measures in place to prevent future accidents? Further, if the causes of accidents represent an almost never-to-be repeated collusion of complex actions and events, then how can one expect to use this approach to prevent future accidents? Further, the usual reaction to accidents has been to attempt to put in place hardware and equipment that will help prevent the next accident. Attempts to use equipment and hardware to fix what are the basic HOF problems generally have not proven to be effective [Reason, 1997]. It has been observed that progressive application of the reactive approach can lead to decreasing the accepted "safe" operating space for operating personnel through increased formal procedures to the point where the operators have to violate the formal procedures to operate the system.

12.4.3 Interactive Approaches

Experience with developing acceptable and desirable quality and reliability of offshore structures indicates that there is a third important approach that needs to be recognised and further developed. Until recently, it was contended that there were only proactive and reactive approaches [Rasmussen, 1996; Rasmussen, et al 1987]. The third approach is interactive (real-time) engineering and management in which danger or hazards builds up in a system and it is necessary to actively intervene with the system to return it to an acceptable quality and reliability state. This approach is based on the contention that many aspects that influence or determine the failure of offshore structures in the future are fundamentally unpredictable and unknowable. These are the incredible, unbelievable, complex sequences of events and developments that unravel a system until it fails. We want to be able to assess and manage these evolving disintegrations. This approach is based on providing systems (including the human operators) that have enhanced abilities to rescue themselves. This approach is based on the observation that people more frequently return systems to safe states than they do to the unsafe states that result in accidents.

Engineers can have important influences on the abilities of people to rescue systems and on the abilities of the systems to be rescued by providing adequate measures to support and protect the operating personnel and the system components that are essential to their operations. Quality assurance and quality control (QA/QC) is an example of the real-time approach [Matousek, 1990]. QA is done before the activity, but QC is conducted during the activity. The objective of the QC is to assure that what was intended is actually being carried out.

Two fundamental approaches to improving interactive performance are: (1) providing people support and (2) providing system support. People-support strategies include such things as selecting personnel well suited to address challenges to acceptable performance, and then training them so they possess the required skills and knowledge. Re-training is important to maintain skills and achieve vigilance. The cognitive skills developed for interactive RAM degrade rapidly if they are not maintained and used [Weick, 1995; Klein, 1999; Knoll, 1986; Weick and Sutcilffe, 2001].

Interactive teams should be developed that have the requisite variety to recognise and manage the challenges to quality and reliability and have developed teamwork processes so the necessary awareness, skills and knowledge are mobilised when they are needed. Auditing, training and re-training are needed to help maintain and hone skills, improve knowledge and maintain readiness [Center for Chemical Process Safety, 1993]. The interactive RAM teams need to be trained in problem "divide and conquer" strategies that preserve situational awareness through organisation of strategic and tactical commands and utilisation of "expert task performance" (specialists) teams [Klein, 1999]. Interactive teams need to be provided with practical and adaptable strategies and plans that can serve as useful "templates" in helping manage each unique crisis. These templates help reduce the amount and intensity of cognitive processing that is required to manage the challenges to quality and reliability.

An improved system support includes factors such as improved maintenance of the necessary critical equipment and procedures so they are workable and available as the system developments unfold. Data systems and communications systems are needed to provide and maintain accurate, relevant and timely information in "chunks" that can be recognised, evaluated and managed. Adequate "safe haven" measures need to be provided to allow interactive RAM teams to recognise and manage the challenges without major concerns for their well being. Hardware and structure systems need to be provided to slow the escalation of the hazards, and re-stabilise the system.

One would think that the improved interactive system support would be highly developed by engineers. This does not seem to be the case [Kletz, 1991]. A few practitioners recognise its importance, but generally it has not been incorporated into general engineering practice or guidelines. Systems that are intentionally designed to be stabilising (when pushed to their limits, they tend to become more stable) and robust (sufficient damage and defect tolerance) are not usual. Some provisions have been made to develop systems that slow the progression of some system degradations.

Effective early warning systems and "status" information and communication systems have not received the attention they deserve in providing system support for interactive RAM. Systems need to be designed to clearly and calmly indicate when they are nearing the edges of safe performance. Once these edges are passed, multiple barriers need to be in place to slow further degradation and there should be warnings of the breaching of these barriers. More work in this area is definitely needed.

Reason (1997) suggested that latent problems with insufficient quality (failures, accidents) in technical systems are similar to diseases in the human body:

> *"Latent failures in technical systems are analogous to resident pathogens in the human body which combine with local triggering factors (i.e. life stresses, toxic chemicals and the like) to overcome the immune system and produce disease. Like cancers and cardiovascular disorders, accidents in defended systems do not arise from single causes. They occur because of the adverse conjunction of several factors, each one necessary but*

not sufficient to breach the defenses. As in the case of the human body, all technical systems will have some pathogens lying dormant within them".

Reason developed eight assertions regarding the error tolerance in complex systems in the context of offshore structures:

- The likelihood of an accident is a function of the number of pathogens within the system.
- The more complex and opaque the system, the more pathogens it will contain.
- Simpler, less well-defended systems need fewer pathogens to bring about an accident.
- The higher a person's position within the decision-making structure of the organisation, the greater is his or her potential for spawning pathogens.
- Local pathogens or accident triggers are hard to anticipate.
- Resident pathogens can be identified proactively, given adequate access and system knowledge.
- Efforts directed at identifying and neutralising pathogens are likely to have more safety benefits than those directed at minimising active failures.
- Establish diagnostic tests and signs, analogous to white cell counts and blood pressure, that give indications of the health or morbidity of a high hazard technical system.

The single dominant cause of structure design-related failures has been errors committed, contributed, and/or compounded by the organisations that were involved in and with the designs. At the core of many of these organisation-based errors was a culture that did not promote quality and reliability in the design process. The culture and the organisations did not provide the incentives, values, standards, goals, resources and controls that were required to achieve adequate quality.

Loss of corporate memory also has been involved in many cases of structure failures. The painful lessons of the past were lost and the lessons were repeated with generally even more painful results. Such loss of corporate memory are particularly probable in times of down-sizing, out-sourcing and mergers.

The second leading cause of structure failures is associated with the individuals that comprise the design team. Errors of omission and commission, violations (circumventions), mistakes, rejection of information and incorrect transmission of information (communications) have been the dominant causes of failures. Lack of adequate training, time and teamwork or back-up (insufficient redundancy) has been responsible for not catching and correcting many of these errors [Bea, 2000b].

The third leading cause of structure failures has been errors embedded in procedures. The traditional and established ways of doing things when applied to structures and systems that "push the envelope" have resulted in a multitude of structure failures. There are many cases where such errors have been embedded in design guidelines and codes and in computer software used in design. Newly developed, advanced and frequently very complex design technology applied in the development of design procedures and the design of offshore structures have not been sufficiently debugged and failures (compromises in quality) have resulted.

This insight indicates the priorities of where one should devote attention and resources if one is interested in improving and assuring sufficient quality in the design of offshore structures [Bea, 2000b]: (1) organisations (administrative and functional structures), (2) operating teams (the design teams), (3) procedures (the design processes and guidelines), (4) robust structures and (5) life cycle engineering of "human friendly" structures that facilitate construction, operation, maintenance and decommissioning.

Formalised methods of QA/QC take into account the need to develop the full range of quality attributes in the offshore structure including serviceability, safety, durability and compatibility. QA is the proactive element in which planning is developed to help preserve desirable quality. QC is the interactive element in which planning is implemented and carried out. QA/QC measures are focused both on error prevention and error detection and correction [Harris and Chaney, 1969]. There can be a real danger in excessively formalised QA/QC processes. If not properly managed, they can lead to a self-defeating generation of paperwork, waste of scarce resources that can be devoted to QA/QC and a minimum compliance mentality.

In design, adequate QC (detection, correction) can play a vital role in assuring the desired quality is achieved in an offshore structure. Independent, third-party verification, if properly directed and motivated, can be extremely valuable in disclosing embedded errors committed during the design process. In many problems involving insufficient quality in offshore structures, these embedded errors have been centred in fundamental assumptions regarding the design conditions and constraints and in the determination of loadings or demands that will be placed on the structure. These embedded errors can be institutionalised in the form of design codes, guidelines and specifications. It takes an experienced outside viewpoint to detect and then urge the correction of such embedded errors [Klein, 1999]. The design organisation must be such that identification of potential major problems is encouraged; the incentives and rewards for such detection need to be provided.

It is important to understand that adequate correction does not always follow detection of an important or significant error in the design of a structure. Again, QA/QC processes need to adequately provide for correction after detection. Potential significant problems that can degrade the quality of a structure need to be recognised at the outset of the design process and measures provided to solve these problems if they occur. A study of the offshore structure design errors and the effectiveness of QA/QC activities in detecting and correcting such errors leads to the checking strategies summarised in table 12.6.

The structure design checking studies performed by Knoll (1986), the series of studies performed by Stewart and Melchers (1988), and the studies performed during this research indicate that there is one part of the design process that is particularly prone to errors committed by the design team. That part of the process is the one that deals with the definition of design loadings that are imposed on and induced in the structure. This recognition has several implications with regard to managing HOF in design. The first implication regards the loading analysis procedures themselves. The second implication regard the education and training of structure design engineers in the development and performance of loading analyses. Given the complexities associated with performing

Table 12.6 Structure design QA/QC

• **What to check?**	• **How to check?**
– high likelihood of error parts (e.g. assumptions, loadings, documentation) – high consequence of error parts • **When to check?** – before design starts (verify process, qualify team) – during concept development – periodically during remainder of process – after design documentation completed	– direct towards the important parts of the structure (error intolerant) – be independent from circumstances which lead to generation of the design – use qualified and experienced engineers – provide sufficient QA/QC resources – assure constructability and IMR • **Who to check?** – the organisations most prone to errors – the design teams most prone to errors – the individuals most prone to errors

loading analyses, the complexities associated with the loading processes and conditions and the close coupling between the structure response and the loading environment, it is little wonder that loading analyses are probably the single largest source of structure design errors. What is somewhat disturbing is that many designers of offshore structures do not understand these complexities nor have been taught how to properly address them in structure design.

The third implication regards the need for independent (of the situations that potentially create errors), third-party QA and QC checking measures that are an integral part of the offshore structure design process. This checking should start with the basic tools (guidelines, codes, programs) of the structure design process to assure that "standardised errors" have not been embedded in the design tools. The checking should extend through the major phases of the design process, with a particular attention given to the loading analysis portions of that process. Computer programs used to perform analyses for design of critical parts of the structure should be subjected to verifications and these analyses repeated using independently developed programs.

The intensity and the extent of the design-checking process needs to be matched to the particular design situation. Repetitive designs that have been adequately tested in operations to demonstrate that they have the requisite quality do not need to be verified and checked as closely as those that are "first-offs" and "new designs" that may push the boundaries of current technology.

The elements of organisational sensemaking are critical parts of an effective QA/QC process, and in particular, the needs for requisite variety and experience. There is a need for background and experience in those performing the QA/QC process that matches the complexity of the design being checked. Provision of adequate resources and motivations are also necessary, particularly the willingness of management and engineering to provide integrity to the process and to be prepared to deal adequately with "bad news".

12.5 Instruments to Help Achieve Design Success

Two instruments will be discussed in the remainder of this chapter that have been developed recently to help promote more effective application of proactive, reactive and interactive processes during the life cycle of offshore structures. A development of these two instruments have been concentrated on taking full advantage of the progress cited in this chapter while addressing some of the major limitations that have been recognised.

The first instrument (computer program, application protocol) is identified as a Quality Management Assessment System (*QMAS©*); this is fundamentally a qualitative process to help guide assessment teams to examine the important parts of offshore structure systems at different times during their life cycle. These assessment teams include members of the offshore structure design engineering team system being assessed. The instrument has been designed to elicit the insights and information that only these people can have.

The second instrument (computer program, application protocol) is a System Risk Analysis System (*SYRAS©*); this is a PRA/QRA/SRA instrument to help develop quantitative results that are often required by engineers and managers. Traditional event tree and fault tree analysis methods have been used in *SYRAS*. The analytical templates in *SYRAS* enable the analyses of each of the life cycles of an offshore structure and address each of the quality attributes.

A "link" has been developed between the results from the *QMAS* and the input required for the *SYRAS* instrument. This link is based on translating the "grades" developed from application of *QMAS* to performance shaping factors (PSF) that are used to modify normal rates of human/operator team malfunctions. The link has been developed, verified and calibrated from the *QMAS–SYRAS* analyses of failures and successes of offshore structure systems during their different life cycle phases [Bea, 2000a, b].

12.5.1 Quality Management Assessment System

The *QMAS* is a method that is intended to provide a level of detail between the qualitative/less-detailed methods (e.g. HazOps, FMEA) and the highly quantitative/very detailed methods (PRA, QRA). The *QMAS* encompasses two levels of safety assessment: coarse and detailed qualitative. The objective of the *QMAS* is with the least effort possible, to identify those factors that are not of concern relative to quality and reliability, to identify those mitigation measures that need to be implemented to improve quality and reliability and to identify those factors that are of concern that should be relegated to more detailed quantitative evaluations and analyses.

Components

The *QMAS* system is comprised of three primary components: (1) a laptop computer program and documentation that is used to help guide platform assessments and record their results, (2) an assessor qualification protocol and training program, and (3) a three-stage assessment process that is started with information gathering and identification on factors of concern (FOC), then proceeds to observe operations, and is concluded with a final assessment and set of recommendations.

The surveying instrument is in the form of a laptop computer program that contains interactive algorithms to facilitate development of consistent and meaningful evaluations of existing facilities. The instrument includes evaluations of the categories of facility factors defined earlier: the operating personnel, organisations, hardware (equipment, structure), procedures (normal, emergency), environments, and interfaces between the categories of factors. A standardised and customised written, tabular and graphical output reporting and routines are provided. This instrument is intended to help identify alternatives for how a given facility might best be upgraded so that it can be fit for the intended purposes.

The *QMAS* process has been developed so that it can be used effectively and efficiently by those that have daily involvement and responsibilities for the quality and reliability of offshore structures. The *QMAS* system is intended to help empower those that have such responsibilities to identify important potential quality and reliability degradation hazards, prioritise those hazards, and then define warranted or needed mitigation measures.

Evaluation Steps

There are five major steps in the *QMAS*. Step #1 is to select a system for assessment. This selection would be based on an evaluation of the history of quality and reliability degradation events and other types of high-consequence accidents involving comparable systems, and the general likelihood and consequences of potential quality and reliability degradations.

Step #2 is to identify an assessment team. This team is comprised of qualified and trained *QMAS* assessors indicated as designated assessment representatives (DARs). These DARs normally come from the organisation/s and operation/s being assessed, regulatory or classification agencies, and/or consulting engineering service firms. DAR appointment is based on technical and operations experience. Integrity, credibility and deep knowledge are the key DAR qualification attributes. The DARs are qualified based on *QMAS* specific training and experience that includes development of in-depth knowledge of human and organisational factors and their potential influences on the quality and reliability of offshore structure systems. To avoid conflicts of interest, the DARs are allowed to request replacement by when such conflicts arise. It is desirable that the assessment teams include members of management and operations/engineering. The DAR teams include experienced "outsiders" (counsellors) who have extensive HOF background and *QMAS* applications experience.

Step #3 consists of a coarse qualitative assessment of the seven categories of elements that comprise an offshore structure system. This assessment is based on the general history of similar types of facilities and operations and details on the specific system. These details would consist of current information on the structure, equipment, procedures (normal operations and maintenance, and emergency/crisis management), operating personnel (including contractors), and organisations/management. Discussions would be held with representatives of the operator/system organisation and the operating/engineering teams.

The product of Step #3 is identification of the FOC that could lead to degradations in quality and reliability of an offshore structure. As a part of the assessment process that will be described later, the assessment team records the rationale for identification of the FOC. The assessment may at this stage also identify suggested mitigation. The results are reported in user-selected standard textural and graphical formats and in user-defined textural and graphical formats (that can be stored in the computer or produced each time). For some systems, the information at this stage may be sufficient to allow the system to exit the *QMAS* with the implementation of the mitigation, recording the results and scheduling the next assessment.

If it is deemed necessary, the *QMAS* proceeds to Step #4; development of scenario/s to express and evaluate the FOC. These scenarios or sequences of events are intended to capture the initiating, contributing and compounding events that could lead to degradations in quality and reliability. These scenarios help focus the attention of the assessors on specific elements that could pose high risks to the system. Based on the FOC and the associated scenarios, Step #5 proceeds with a detailed qualitative assessment. Additional information is developed to perform this assessment and includes more detailed information on the general history of the structure system, its details, results from previous studies, and management and operating personnel interviews. In recording results from the interviews, provisions are made for anonymous discussions and reporting.

The product of Step #5 is a detailing of the mitigation measures suggested for mitigation of the FOC confirmed in Step #5. The rationale for the suggested mitigation are detailed together with projected beneficial effects on the FOC. As for the results of Step #3, the results of Step #4 are reported in standard and user-defined formats. At this point, the assessment team could elect to continue the *QMAS* in one of the two ways. The first option would be to return to the FOC stage and repeat Step #5-based "new" FOC and the associated scenarios. The second option would be to proceed with some of the FOC and the associated scenarios into coarse quantitative analyses and evaluations. If the assessment team is elected, the *QMAS* could be terminated at the end of Step #5. The results would be recorded, and the next assessment scheduled.

Evaluation Processes

The *QMAS* evaluation is organised into three sections or "Levels" (fig. 12.5). The first level identifies each of the seven structure system components: 1.0 – operators, 2.0 – organisations, 3.0 – procedures, 4.0 – equipment, 5.0 – structure, 6.0 – environments and

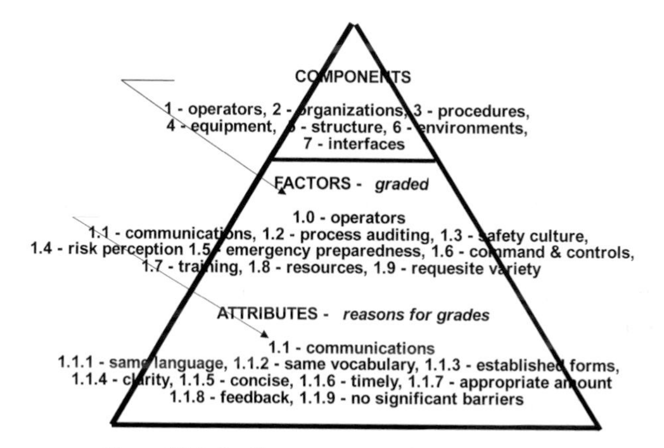

Figure 12.5 Quality components, factors and attributes

7.0 – interfaces. These seven components comprise "modules" in the *QMAS* computer program. The structure and equipment factors are modified to recognise the unique characteristics of different offshore structures.

The second level identifies the factors that should be considered in developing assessments of the components. For example, for the operators (1.0), seven factors are identified: communications (1.1), selection (1.2), knowledge (1.3), training (1.4), skills (1.5), limitations/impairments (1.6) and organisation/coordination (1.7). If in the judgement of the assessment team, additional factors should be considered, then they can be added. Using a process that will be described later, the assessment team develops grades for each of these factors.

The third level identifies attributes associated with each of the factors. These attributes are observable (behaviours) or measurable. These attributes provide the basis or rationale for grading the factors. For example, for the communications factor (1.1) six attributes are included: clarity (1.1.1), accuracy (1.1.2), frequency (1.1.3), openness/honesty (1.1.4), verifying or checking feedback (1.1.5) and encouraging (1.1.6). Again, if in the judgement of the assessment team, additional attributes are needed, they can be added to the *QMAS*.

The factors and the attributes for each of the system components have been based on results from current research on these components with a particular focus on the HOF-related aspects. This approach avoids many of the problems associated with the traditional "question-based" instruments that frequently involve hundreds of questions that may be only tangentially applicable to the unique elements of a given structure system.

Factors Grading

The *QMAS* assessment team assigns grades for each component factor and attribute. Three grades are assigned: the most likely, the best, and the worst. These three grades help the assessors express the uncertainties associated with the gradings. Each of the attributes for a given factor are assessed based on a seven-point grading scale (fig. 12.6). An attribute or factor that is average in meeting referent standards and requirements is given a grade of 4. An attribute or factor that is outstanding and exceeds all referent standards and requirements is given a grade of 1. An attribute or factor that is very poor and does not meet any referent standards or requirements is given a grade of 7. Other grades are used to express characteristics that are intermediate to these. The reasons for the attribute and factors grades are recorded by the assessment team members. This process develops a consensus among the system or domain experts, allowing for expressions of dissenting opinions.

The grades for the attributes are summed and divided by the number of attributes used to develop a resultant grade for the factor. Weightings of the factors and attributes can be introduced by the assessors. The assessors review the resultant grades and if they are acceptable, the grades are recorded. If it is not, they are revised and the reasons for the revisions noted. The uncertainties associated with the grades for the attributes are propagated using a first order statistical method.

In the same manner, the grades for the factors are summed and divided by the number of factors to develop a resultant grade for the component. Again, the assessors review this resultant grade and if it is acceptable, the grade is recorded. If it is not, it is revised

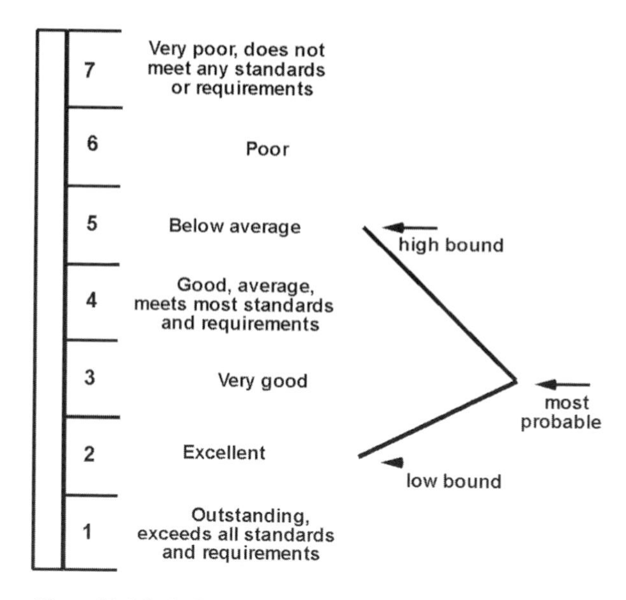

Figure 12.6 Scale for grading attributes, factors and components

and reasons for the revision noted. The uncertainties associated with the grades for the factors are propagated using a first-order statistical method.

A "Braille" chart is then developed that summarises the mean grades (and, if desired, their standard deviations) developed by the assessment team for each of the factors (fig. 12.7). The "high" grades (those above 4) indicate components and the associated factors that are candidates for attention and mitigation.

Assessors

The most important element in the *QMAS* system is the team of assessors. It does not matter how good the *QMAS* assessment instruments and procedures are, if the personnel

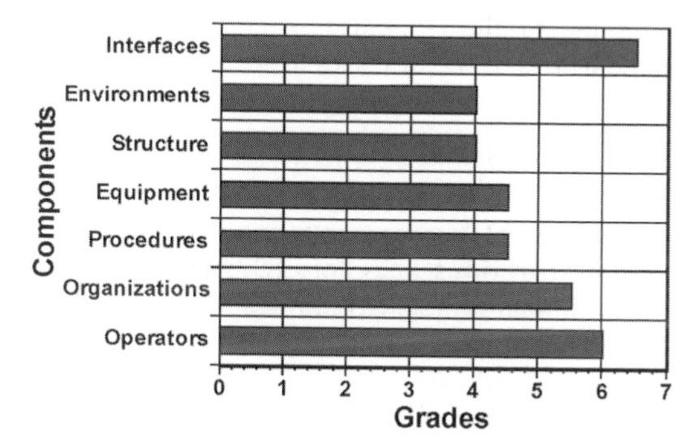

Figure 12.7 Example component mean grading results

using the instrument do not have the proper experience, training and motivations. The *QMAS* assessors must have experience with the system being assessed, quality auditing experience and training in human and organisational factors. The assessor team is comprised of members from the system (operators, engineers, managers, regulators) and *QMAS* "counselors" who have extensive experience with the *QMAS* system and operations – facilities similar to those being assessed.

An important aspect of the qualifications of assessors regards their aptitude, attitude, trust and motivation. It is very desirable that the assessors be highly motivated to learn about the human and organisational factors and safety assessment techniques, have high sensitivity to quality hazards ("perverse imaginations"), be observant and thoughtful, have good communication abilities and have a willingness to report "bad news" when it is warranted. It is vital that both the assessors and the *QMAS* counselor have the trust and respect of the system operators and managers.

An assessor "just-in-time" training program has been developed as part of the *QMAS* instrument. This program includes training in human and organisational factors and the *QMAS* assessment process. Example applications are used to illustrate applications and to help reinforce the training. A final examination is used to help assure that the assessor has learned the course material and can apply the important concepts.

The assessor training program has two parts: (1) informational and (2) practical exercises. The informational part contains background on the *QMAS* assessment process and computer instrument, failures involving offshore structures and other types of engineered structures, human and organisational performance factors and evaluations.

The second part of training is the hands-on use of the computer software. Training exercises are performed to demonstrate the use of the *QMAS* instrument. Software demonstrations using offshore structures as case studies are walked through. Then the assessors assess another system on their own. Following this, the assessments are compared and evaluated. The assessors are asked for feedback on the *QMAS*.

This approach is identified as a "participatory ergonomics" approach. The people who participate in the daily activities associated with their portion of the life cycle of a system are directly involved in the evaluations and assessments of that system. These people know their system better than any outsider ever can. Yet, they need help to recognise the potential threats to the quality and reliability of their system. These people provide the memory of what should be done and how it should be done. These are the people who must change and must help their colleagues change so that desirable and acceptable system quality and reliability are developed. This is a job that outsiders can never do or should be expected to do.

The *QMAS* has been applied to a wide variety of offshore structure systems including marine terminals, offshore platforms and ships. *QMAS* has been applied in proactive assessments (before operations conducted), in reactive assessments (after operations conducted), and in interactive assessments (during conduct of operations). Multiple assessment teams have been used to assess the same system; the results have shown a very high degree of consistency in identification of the primary factors of concern and potential mitigation measures. *QMAS* has proven to provide a much more complete and realistic understanding of the human and organisational elements that comprise offshore structure systems than the traditional PRA/QRA/SRA approaches [Weick, 2000]. Frequently, RAM of an offshore structure system can be conducted solely on the basis of results developed from *QMAS*, factors important to quality and reliability can be defined and characterised sufficiently to enable effective actions to achieve these objectives.

12.5.2 System Risk Assessment System

The System Risk Assessment System (*SYRAS*) has been developed to assist engineers in the assessment of system failure probabilities based on the identification of the primary or major tasks that characterise a particular part of the life cycle (design, construction, maintenance, operation) of an offshore structure (fig. 12.8). This RAM instrument has been applied in the study of tradeoffs regarding "minimum" platforms, in quality assurance and quality control (QA/QC) of the design of innovative deepwater structures, and the effects of Value Improvement Programs for several major offshore structures [Shetty, 2001; Bea, 2000a]. The *SYRAS* instrument consists of a computer program and an applications protocol [Bea, 2000b].

Failures to achieve the desirable quality in an offshore structure can develop from intrinsic (I) or extrinsic (E) causes. Intrinsic causes include factors such as extreme environmental conditions and other similar inherent, natural and professional uncertainties. Extrinsic causes are due to human and organisational factors – identified here as "human errors". The probability of failure of the structure to develop quality attribute (i), $P(F_{Si})$, is

$$P(F_{Si}) = P(F_{SiI} \cup F_{SiE}) \tag{12.23}$$

where ($\cup$) is the union of the failure events. The probability of failure of any one of the quality attributes (i) due to inherent randomness is $P(F_{SiI})$. The probability of failure of any

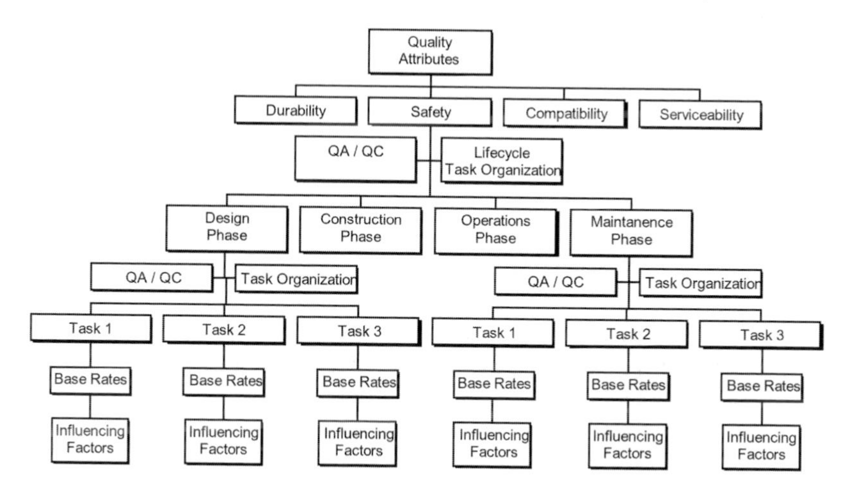

Figure 12.8 *SYRAS* components

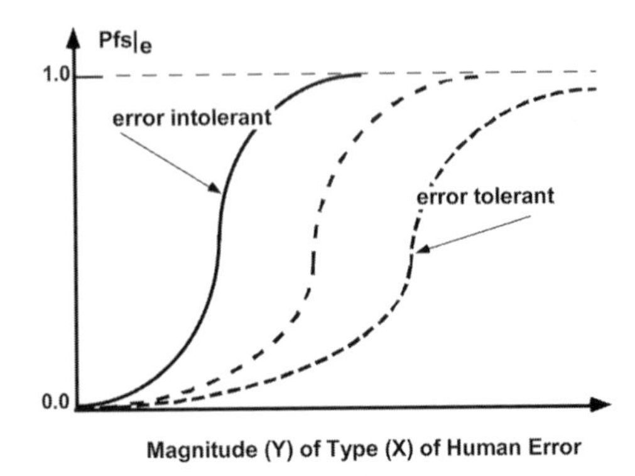

Figure 12.9 Likelihood of unsatisfactory quality

one of the quality attributes (i) due to the occurrence of human error is $P(F_{SiE})$. The probability of human error in developing a quality attribute (i) in the structure is $P(E_{Si})$. Then:

$$P(F_{Si}) = P(FS_{iI}|E_{Si})P(E_{Si}) + P\left(F_{SiI}|\not{E}_{Si}\right)P(\not{E}_{Si}) + P(F_{SiE}|E_{Si})P(E_{Si}) \quad (12.24)$$

The first term addresses the likelihood of structure failure due to inherent causes given a human error (e.g. structure fails in a storm due to damage from a boat collision). The second term addresses the same likelihood given no human error. This is the term normally included in structural reliability analyses. The third term addresses the likelihood of structure failure directly due to human error (e.g. structure fails due to explosions and fire).

The probability of failure given HOE, $P(F_S|E)$, characterises the "robustness" or defect and damage tolerance of the structure to human errors. The shape of the fragility curve (fig. 12.9) can be controlled by engineering. This is explicit design for robustness or defect (error) tolerance and fail-safe or intrinsically safe design. For the intensities (magnitude) and types of malfunctions that normally can be expected, the structure should be configured and designed so that it does not fail catastrophically (or have unacceptable quality) when these types and magnitude of malfunctions occur. The fragility curve for a particular system is determined using off-line analyses or experimental results and the results input to *SYRAS*.

The probability of no human error is:

$$P(\not{E}_{Si}) = 1 - P(E_{Si}) \quad (12.25)$$

The probability of insufficient quality in the structure due to HOE, $P(F_{SiE})$, can be evaluated in the (j) life cycle activities of design ($j=1$), construction ($j=2$), operations ($j=3$), and maintenance ($j=4$) as

$$P(F_{SiE}) = P\left(\bigcup_{j=1}^{4} F_{SiEj}\right) \quad (12.26)$$

or

$$P(F_{SiE}) = \sum_{j=1}^{4} P\big(F_{Sij}|E_{Sij}\big)P(E_{Sij}) \tag{12.27}$$

Each of the life cycle activities ($j = 1 - 4$) can be organised into (n) parts ($k = 1 - n$):

$$P(F_{SiEj}) = P\left(\bigcup_{k=1}^{n} F_{SiEjk}\right) \tag{12.28}$$

This task-based formulation addresses the major functions that are involved in the principal activities that occur during the life cycle of an offshore platform.

For example, the system design activity ($j = 1$) can be organised into four parts ($n = 4$): configuration ($k = 1$), system demand analyses ($k = 2$), system capacity analyses ($k = 3$) and documentation ($k = 4$). The likelihood of insufficient quality in the system due to human error during the design activity is

$$P(F_{SiE1}) = P\left(\bigcup_{k=1}^{4} F_{SiE1k}\right) \tag{12.29}$$

If desirable, the primary functions or tasks can be decomposed into subtasks to provide additional essential details.

The base rates of human errors of type "*m*", $P(E'_{Sijkm})$, are based on the published information on human task performance reliability (fig. 12.10) [Center for Chemical Process Safety, 1994; Swain and Guttman, 1983; Kirwan, 1994; Gertman and Blackman, 1994; Kontogiannis and Lucas, 1990; Haber, et al 1991). Performance Shaping Factors (PSF) are used to modify the base or "normal" rates of human errors, $P(E'_{Sijkm})$, to recognise the effects of organisations, structure, equipment, procedures, environments and interfaces:

$$P(E_{jkm}) = P(E'_{jkm}) \cdot \prod \mathrm{PSF}\varepsilon_{jkm} \leq 1 \tag{12.30}$$

As discussed earlier, gradings from the *QMAS* component evaluations ($G_{\varepsilon jkm}$) are developed on a seven-point scale (fig. 12.6). The mean value and the coefficient of variation of each of the categories of PSF are developed based on an average of the mean values and coefficients of variation of each of the *QMAS* categories. Evaluation of each of the seven categories of PSF results in a final overall grading ($\overline{G_{\varepsilon jkm}}$) and coefficient of variation ($V_{G\varepsilon jkm}$) on this grading that can be used to quantify a specified PSF.

Each of the seven PSF ($\mathrm{PSF}_{\varepsilon jkm}$) can act to increase or decrease the base rates of human errors. *SYRAS* allows the user to specify the base rates and then scale the base rates by multiplying the base rates by the PSF identified by the user. The scales allow the base rates to be increased or decreased by three orders of magnitude. When quantification of the PSF is based on the use of the *QMAS* instrument and protocol, the PSF is computed from (fig. 12.11):

$$\log \mathrm{PSF}_{\varepsilon jkm} = \left(\overline{G_{\varepsilon jkm}} - 4\right) \tag{12.31}$$

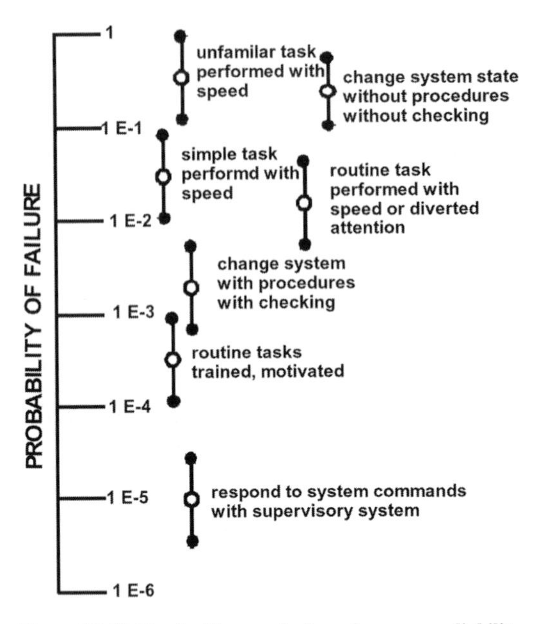

Figure 12.10 Nominal human task performance reliability

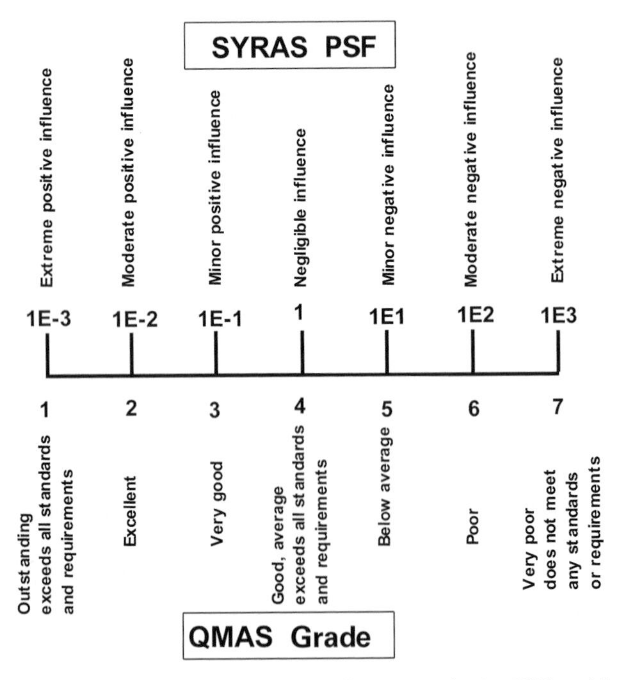

Figure 12.11 *QMAS* qualitative grading translation to quantitative PSF used in *SYRAS*

The resultant PSF that modifies the base rate of error is computed from the product of the seven mean PSF:

$$\mathrm{PSF}_{\varepsilon} = \prod_{i=1}^{7} \mathrm{PSF}_{\varepsilon jkm} \tag{12.32}$$

The resultant coefficient of variation of the PSF is computed from the square root of the sum of the squares of the PSF coefficients of variation:

$$V_{\mathrm{PSF}}^{2} = \sum_{i=1}^{7} V_{\mathrm{PSF}i}^{2} \tag{12.33}$$

The PSF provides the important link between the qualitative *QMAS* assessment process and the quantitative PRA-based *SYRAS* analysis process [Bea, 2000a]. Results from *QMAS* are then "translated" to input that can be used in the traditional PRA/QRA approach embodied in *SYRAS*. The *QMAS–SYRAS* link has been based on a repetitive calibration process involving applications of *QMAS* and *SYRAS* to offshore structures that have failed (very high probabilities of failure) and succeeded (very low probabilities of failure) [Bea, 2000b]. As would be expected, due to the natural variability in human and organisational performance and the uncertainties associated with the evaluations of such a performance, the PSF have very large coefficients of variation (in range of 100–200%) [Bea, 2000a, 2000b].

The *QMAS* grades, FOC and system quality improvement recommendations are intended to help capture the processes that cannot be incorporated into highly structured quantitative analyses; these are the dynamic organic processes that characterise most real offshore structure systems. Frequently, the intensive application of the *QMAS* instrument and the underlying organisational philosophies provide the insights essential to help achieve desirable and acceptable quality and reliability. The coupling of the results from *QMAS* with the *SYRAS* probabilities are intended to provide engineers and managers with quantitative assessments of systems so that the effects of potential mitigation measures can be examined and the effects of VIP assessed. Of course, this means that potentially much of the richness of insights provided by *QMAS* can be lost or obscured by intense attention to the numerical results provided by *SYRAS*. The best experiences have been those in which both instruments are diligently applied; thus, capturing both qualitative and quantitative insights.

Once the tasks are organised into the task structure for the life cycle phase, correlation among elements is assessed. In order to facilitate the calculation of the likelihood of failure, the elements can be designated as either perfectly correlated or perfectly independent.

After determining the overall system task structures, the user has the option of analysing the effects of Quality Assurance and Quality Control (QA/QC) on the overall system probability. This is done in an "overlay edit-mode". This means that the user is able to go back into the task structures and add in the QA/QC procedures as independent tasks with corresponding influences. The user is presented with both the original system *Pf* and the QA/QC-modified *Pf*.

Consequently, the next step in the *SYRAS* development addresses the HOF malfunction detection (D) and correction (C). This is an attempt to place parallel elements in the quality system so that failure of a component (assembly of elements) requires the failure of more than one weak link. Given the high positive correlation that could be expected in such a system, this would indicate that QA/QC efforts should be placed in those parts of the system that are most prone to error or likely to compromise the intended quality of the system.

Conditional on the occurrence of type (m) of HOE, E_m, the probability that the error gets through the QA/QC system can be developed as follows: The probability of detection is $P(D)$ and the probability of correction is $P(C)$. The compliments of these probabilities (not detected and not corrected) are:

$$P(\not{D}) = 1 - P(D), \quad \text{and} \quad P(\not{C}) = 1 - P(C) \tag{12.34}$$

The undetected and uncorrected error event, UE_m, associated with a human error of type m is:

$$\mathrm{UE}_m = \bigcup_{m=1}^{8} (E_m \cap \not{D}_m \cap \not{C}_m) \tag{12.35}$$

The probability of the undetected and corrected HOE of type m is:

$$P(\mathrm{UE}) = \sum_{m=1}^{8} P(E_m | \not{D}_m \cap C_m)(P(\not{D}_m | \not{C}_m)P(\not{C}_m) \tag{12.36}$$

Assuming independent detection and correction activities or tasks, the probability of the undetected and corrected HOE of type m is

$$P(\mathrm{UE}_m) = P(E_m)[P(D_m)P(\not{C}_m) + P(\not{D}_m)] = P(E_m)[1 - P(D_m)P(C_m)] \tag{12.37}$$

The probability of error detection and the probability of error correction play important roles in reducing the likelihood of human malfunctions compromising the system quality. Introduction of QA/QC considerations into the developments into the earlier developments is accomplished by replacing $P(E_{Sijkm})$ with $P(\mathrm{UE}_{Sijkm})$ into the desirable parts of the *SYRAS* analysis.

12.6 Example Applications

12.6.1 Minimum Structures

Results from a joint industry – government sponsored project that addressed the system reliability levels of three minimum structures compared with a standard four-pile jacket recently have become publicly available [Shetty, 2000]. The study considered extreme storm, fatigue and ship collision conditions and considered the potential effects on reliability from errors due to human and organisational factors that develop during design, construction and operation of such structures [Bea and Lawson, 1997]. The structural concepts considered were a three-pile Monotower, Vierendeel Tower, Braced Caisson and a conventional four-pile Jacket (fig. 12.12).

The structures were designed using a common design criteria, analysis and design procedure, and for operation at the same field and to support the same topside operations (RAMBØLL, 1999). Key members were designed to have utilisation ratios close to 0.8 under the 100-yr return period environmental loading.

Welded joints were designed to have minimum fatigue lives of five times the service life (20 yr) for the three minimum structures and three times the service life for the four-pile jacket. The in-place operational conditions, vortex shedding, and on-bottom stability requirements were considered; reinforcements were made to joint cans and braces to ensure that the structures were able to fully mobilise their capacity during ship impacts. These "minimum" structures were designed to be much more robust (damage–defect tolerant) than their counterparts for the Gulf of Mexico [Bea, et al 1998].

The performance of the four structures under extreme conditions was studied by performing deterministic pushover and system reliability analyses [Gierlinski and Rozmarynowski, 1999; MSL Engineering, 1999] (fig. 12.13). Based on the joint probability distributions of wave heights, periods and current parameters, and the ultimate capacity of the structures based on results from the pushover analyses, and accounting for the uncertainties in the calculated hydrodynamic loads and capacities of the four structures, system probabilities of failure were evaluated for each structure. Reliability characteristics for extreme storm conditions also were evaluated for other locations.

The reliability under fatigue conditions were evaluated based on the failure of individual joints and the sequences of two, three and four joints assuming that the initial joint failures were not detected and repaired. The impact of fatigue failure of joints was evaluated by calculating the conditional probability of collapse due to environmental overload given the initial failure of one or more joints by fatigue, and multiplying this with the probability of the fatigue failure sequence occurrence.

Time domain, non-linear, ship-structure collision analyses were performed to study the performance of the structures against collisions from supply vessels [MSL Engineering, 1999]. Analyses were carried out for a number of vessel mass and velocity combinations, which were considered as credible for operations in the Southern and Central North Sea fields. Following the impacts, a post-impact pushover analysis was performed to determine the reduction in capacity as a result of ship impact damage.

A methodology to evaluate the potentials for and effects of human and organisational malfunctions were developed and implemented in the form of two computer programs – instruments previously identified earlier in this chapter as the *QMAS* and the *SYRAS*. Based on the results from a questionnaire circulated to the operators of structures similar to those studied, a review of the world-wide accident database for marine structures, and reported incidents of damage to offshore structures in the North Sea, five error scenarios were identified [Bea and Lawson, 1999]. These scenarios addressed errors that could develop during design (fatigue due to pile driving not considered), fabrication (fit-up, welding

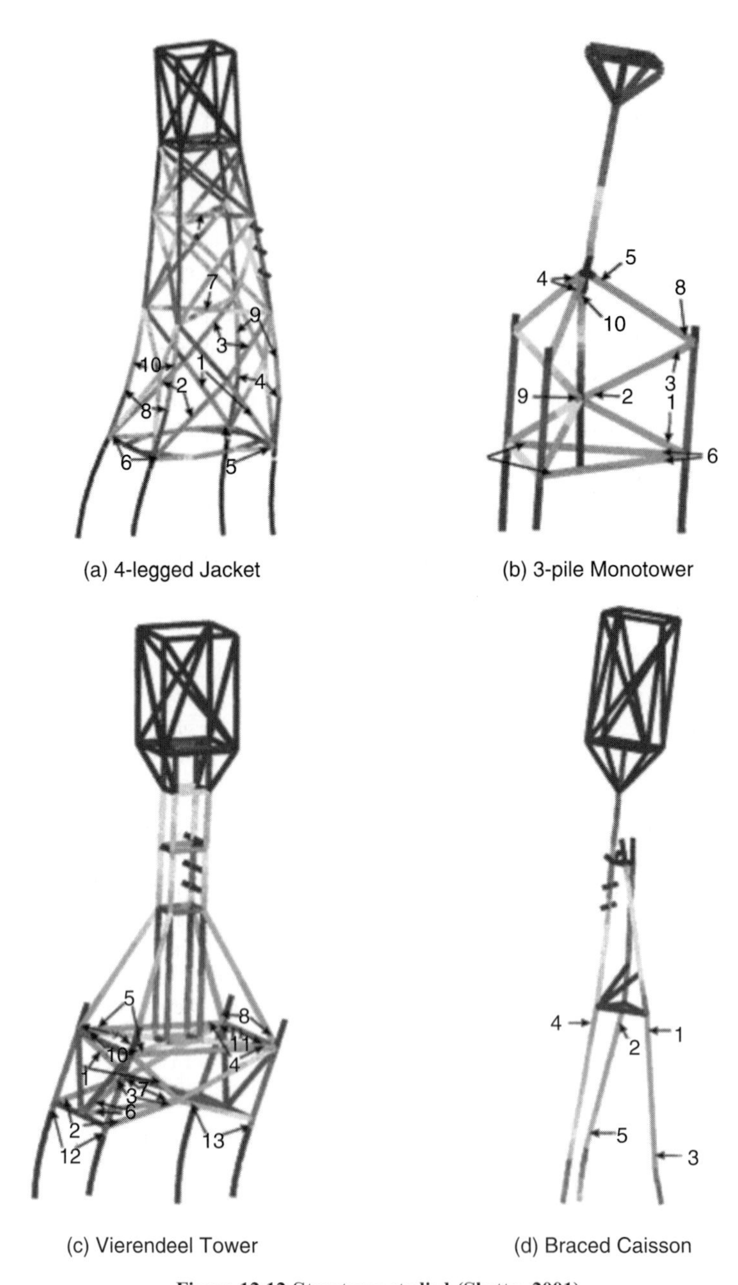

Figure 12.12 Structures studied (Shetty, 2001)

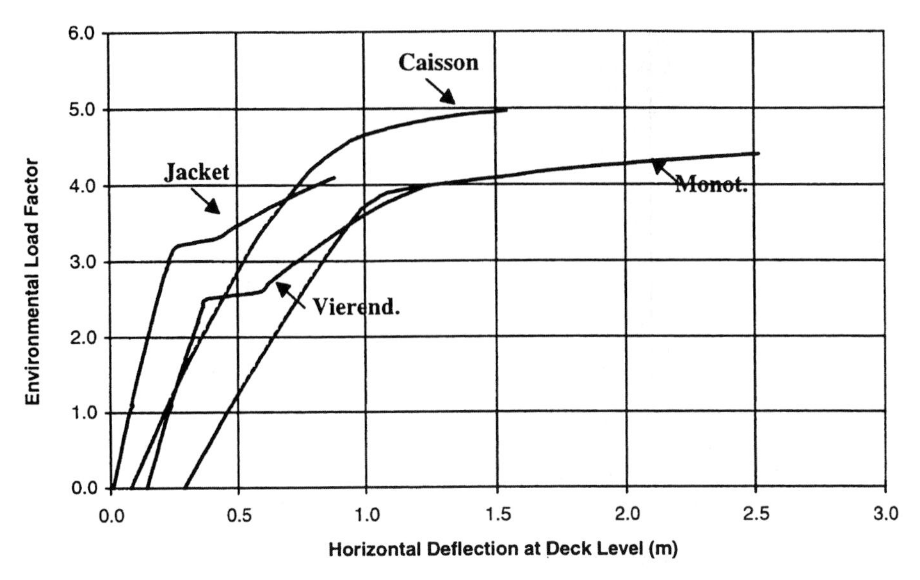

Figure 12.13 Results from static non-linear pushover analyses (Shetty, 2000)

defects), installation (dented braces due to pile stabbing) and operation (dropped production package, supply boat collision) phase of the structures. For each scenario, the damage to the structures were determined and their reliabilities under the damaged condition were evaluated considering fatigue, extreme storm and ship collision conditions.

An example of the evaluation process that was used in this study is that associated with the design phase and the omission of consideration of pile installation–driving-induced stresses. The source of the less-than-desired fatigue durability is due to pile driving stress-induced fatigue in the joints that connect the pile sleeves or guides to the structure. The stress is due to the difficulties associated with maintaining proper alignment of the piles in the underwater pile sleeves or in the caisson pile sleeve connections during installation of the piles. The structures were not designed to sustain the pile driving stresses nor were provisions developed to allow more precise alignment of the piles during pile driving.

The structure is fabricated as specified. During installation of the platform, the pile driving-induced stresses cause fatigue cracking to be initiated in the joints of the vertical diagonal braces that connect the pile sleeves/guides to the primary structure elements. This damage leads to through-thickness cracking of several joints. In the case of the three-pile and four-pile monopods, through-thickness cracks are developed in the pile sleeves to vertical diagonal braces that connect to the central column. In the case of the braced caisson, fatigue cracks are developed at the connection between the caisson and the diagonal braces – piles that are driven through the connection. In the case of the four-leg jacket, the piles can be aligned in the legs and driven without imparting significant fatigue damage.

The probabilities associated with each of the eight potential causes of this malfunction by the design team during this phase are: communications 5E-4, selection and training 2E-3, planning and preparations 5E-4, limitations and impairments 6E-4, violations 1E-4, slips 1E-4, lack of knowledge 5E-3 and mistakes 1E-3. These probabilities reflect influences from the organisations (direction not provided by owner/operator, design contractor, regulatory), procedures (effects not included in design guidelines), hardware (no significant influences) and environments (no significant influences).

The *SYRAS* analyses indicated that the probability of this HOE scenario is $P_{E1.1} = 8.9E\text{-}3$. The dominant causes of the potential malfunctions are ignorance (56%) and selection and training of the members of the design team (22%). The ignorance source error was influenced primarily by lack of organisational communications and defined design procedures to address this problem.

There can be sources of correlation between the sources or causes of malfunctions. Such correlation can be developed through organisational influences that embed a specific "culture" in an organisation and result in "group think" biases. In the analysis of HOE, the *SYRAS* user is able to introduce correlation between the sources of HOE embedded in a task structure. In this case, for the case of perfect positive correlation between the sources or causes of HOE in the design process and team, the probability of the design error would be $P_{E1.1} = 5.0E\text{-}3$. The values of $P_{E1.1} = 5.0E\text{-}3$ and $P_{E1.1} = 8.9E\text{-}3$ could be viewed as "bounds" on the possible likelihoods of this specific HOE scenario.

For this scenario, two QA/QC alternatives were considered. The first was QA/QC conducted during the design process. Two design process QA/QC alternatives were evaluated. One was a conventional checking of the design analysis calculations. The other was the verification of the design analysis processes by experienced "third-party" design and construction engineers. Based on the results of design process checking cited earlier, the probabilities of detection ($P_D = 0.10$) and correction ($P_C = 0.80$) for the first alternative were determined to be $P_{DC} = 0.08$. The probability of not detecting and correcting the design HOE was therefore $P_{NDC} = 0.92$. In the second instance, the probabilities of detection ($P_D = 0.80$) and correction ($P_C = 0.90$) were determined to be $P_{DC} = 0.72$. Thus, the probability of not detecting and correction were determined to be $P_{NDC} = 0.28$.

The resulting probabilities of design HOE with additional QA/QC measures were thus determined to be $P_{E1.1A} = 4.6E\text{-}3$ to 8.2E-3 for the first QA/QC alternative, and $P_{E1.1B} = 1.4E\text{-}3$ to 2.5E-3 for the second QA/QC alternative. Results for the QA/QC alternatives are summarised in table 2.

Table 12.7 summarises the likelihoods associated with each of the five life cycle scenarios. The base rate likelihoods refer to the condition where the currently specified QA/QC measures were employed. Likelihoods were also developed for additional QA/QC measures representing significant (Alternative A) and major (Alternative B) improvements in these processes.

The base rate likelihoods range from 1E-3 to 9E-3. The ranges in the likelihoods represents the potential effects of "correlation" between the causes and tasks involved in the HOE scenarios (fig. 12.14). These likelihoods are in good agreement with the database results developed by the Marine Technology Directorate (1994) on platforms in the North Sea. In some cases, the additional QA/QC measures are able to substantially reduce the base error rates, reducing them by a factor of 10 when the QA/QC measure is highly effective.

Table 12.7 Summary of HOE scenarios likelihoods with and without additional QA/QC measures

Phase, HOE, scenario ID	Base rate likelihood	QA/QC alt. A likelihood	QA/QC alt. B likelihood
Design, omit install fatigue	5.0–8.9E-3	4.6–8.2E-3	1.4–2.5E-3
Fabrication, fit-up/welding defects	0.7–2.1E-3	1.0–2.8E-4	1.3–3.8E-4
Installation, dented braces	2.0–3.7E-3	2.0–3.7E-4	4.0–7.4E-4
Production, dropped package	1.0–4.0E-3	1.4–5.7E-4	1.9–7.8E-4
Production supply boat collision	3.0–8.7E-3	0.6–1.7E-3	0.9–2.6E-3

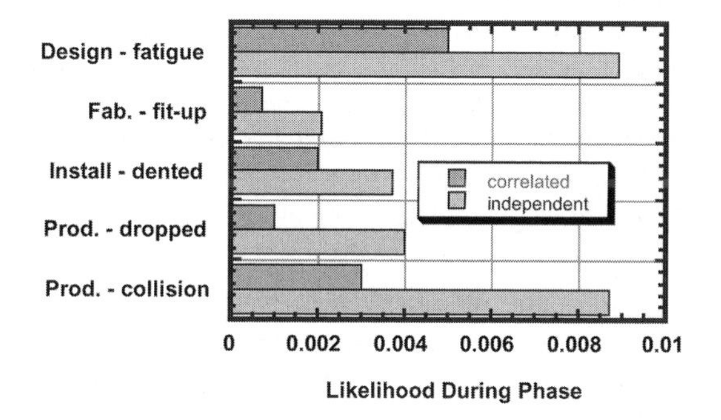

Figure 12.14 Life cycle malfunctions scenario likelihoods

In some cases, the QA/QC measures are not very effective in reducing the likelihood of the HOE effects.

The primary results from the system reliability analyses of the four structures are summarised in fig. 12.15. The intrinsic (error free) probabilities of system failure under extreme conditions and combined fatigue and extreme condition loadings are given in the first row of table 12.7. The probabilities of system failure as a result of extrinsic or HOE causes are added to the "error free" intrinsic probabilities of system failure to obtain the total probability of system failure.

Based on the results from the analyses, the first two malfunction scenarios: (1) omission of pile driving stresses during design and not making adequate provisions for alignment of piles during driving and (2) fit-up and welding flaws introduced during fabrication, both of which affect fatigue strength, have the most significant influence in degrading the system reliability of the three-pile Monotower and Vierendeel Tower structures. These two structures are less robust under these HOE scenarios. The four-pile Jacket shows only a marginal influence due to HOE scenario (2) while the Braced Caisson shows practically no influence from these HOE scenarios. Under HOE scenario (4), only the three-pile

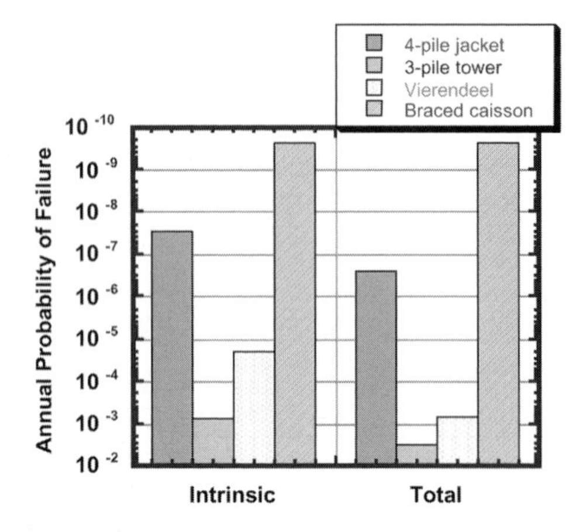

Figure 12.15 Intrinsic and total annual probabilities of system failure

Monotower shows a significant reduction in reliability as a result of ship impact damage, while the other three structures show high levels of robustness. The HOE scenarios (3) and (5) involving damage to one of the braces do not show a significant impact on the system reliability of any of the four structures.

The three-pile Monotower and the Vierendeel Tower structures were shown to be particularly susceptible to potential HOE, which affect the fatigue strength of critical welds. The implication is that effective QA/QC measures should be employed to safeguard these structures against such defects. In addition, designing the critical welds to longer fatigue lives and thorough inspection of welds at the fabrication yard and after installation are implicated to help minimise the risk of these scenarios actually developing.

12.6.2 Deepwater Structure Design Project

A review has been performed of a deep-water structure design project that involved the use of very innovative design methods and technology. The assessment team was given full access to the design management organisation, the engineering organisation and the classification–verification organisation. This included reviews of design documentation, specifications and background information, and interviews – discussions with the members of each of the organisations.

In an attempt to reduce initial costs, the design approach involved very advanced and innovative design procedures and technology. Specific "target" reliabilities were defined by the owner/operator for the structure. A Value Improvement Program (VIP) was instituted. The goal of the VIP was to reduce the initial cost of the project by 25%.

At the time of this review, the design had been underway for two years. The work had included extensive analyses of alternatives, development of computer programs and performance of experimental work on several of the critical components. A leading classification society was involved in an on-going QA/QC program that included a failure modes and effects analysis of the structure system.

The assessment team included representatives of the management, engineering, and classification organisations (participatory ergonomics approach). They participated in training workshops that focused on the HOF aspects of the engineering design process and on the HOF considerations in developing successful platform life cycles.

The consensus results from the first round of analyses indicated significant concerns for the design procedures, design personnel and management, technology and quality incentives. The concerns for:

- Design procedures were focused on the very sophisticated and complicated methods that were involved in the analysis of a very complex interaction of the structure, the foundation and the oceanographic environment.
- Design personnel and management were focused on the low level of experience of the lead design engineers and on their on-going debates with the project management's requirements for verifications and validations of the results from the design analyses.
- Technology was focused in the first-time nature of the engineering methods and analytical tools being used in the design (based on limiting strains and deformations).
- Quality incentives regarded the VIP and the lack of specific guidelines on the effects of VIP on the quality and reliability of the structure.

After the first round evaluations were completed, a second team of assessors was organised that included representatives of the design organisation's management, engineering and verification teams. The results are summarised in fig. 12.16. The resultant uncertainties are indicated for each of the components (best estimate, $\pm 1\sigma$ standard deviation).

The high grades (indicating below average quality attributes) for interfaces, procedures and operators reflected the same primary issues indicated by the qualitative assessment. Examination of the factors and attributes associated with the gradings indicated that the primary reasons for the high grading of the interfaces referred to the lack of appropriate interfacing between the design and management teams. There was a contention between engineering and management. Engineering felt that once an analysis was completed and verified, then the results should be implemented in the design. Management felt that

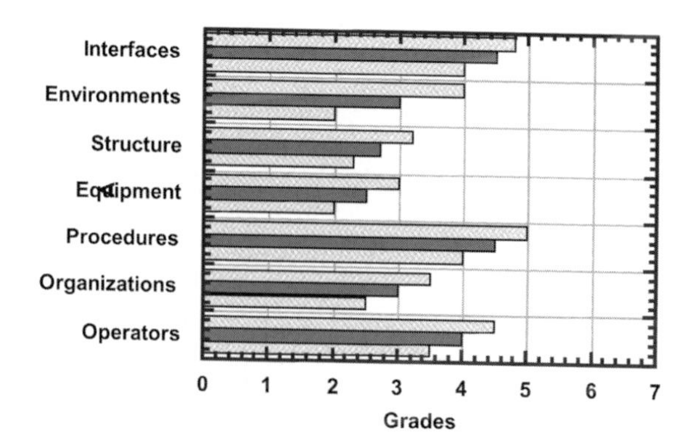

Figure 12.16 Grades from interactive application of *QMAS* to design team, process, and organisation

interpretation and judgement needed to be used as screens to assure that the results "made sense" before they were used.

The reasons for the high grading of the procedures referred to the lack of first principles and experimental verifications of the computer programs that were being used in the design and the lack of any specific guidelines to determine the effects on the structure reliability of the VIP.

The reason for the average grading of the operators (the design team) was the relatively low level of structure design experience in the design team and the lack of in-depth construction and operations experience in the team.

The review included five recommendations to improve the *QMAS* gradings:

(1) Develop and implement definitive guidelines to evaluate the quantitative effects of the VIP alternatives and measures on life cycle costs and reliability of the structure (these guidelines would be consistent with the background that had been used to develop the reliability targets),
(2) Develop and implement a "challenge" process in the design procedures that would assure that all results from engineering analyses were validated by alternative analyses, experimental field data and experienced judgement (the ongoing QA/QC process would be replaced),
(3) Assign additional experienced structural design engineers to the design team (less experienced personnel would be assigned to other projects),
(4) Temporarily assign construction and operations engineering personnel to the design team to review the construction, operations and maintenance characteristics being used in the design (these personnel were representatives of the organisations that would build and operate the structure) and
(5) Develop a structural robustness program and design guidelines that would assure fail-safe design (intrinsic safety) for all of the critical structural and equipment components through the life cycle of the structure (explicit design for damage and defect tolerance).

The *SYRAS* instrument was used to evaluate the reliability and life cycle cost implications of the VIP alternatives [Bea, 2000la, 2001]. The structural quality profiling instrument and the *QMAS* instrument proved to be effective and efficient. The recommendations developed during the assessment process were implemented by the management, engineering and classification–verification organisations. The recommendations proved to be practical and cost-effective.

12.7 Summary and Conclusions

Those responsible for the development and creation of offshore structures, the associated regulatory agencies, their engineers, managers and the operating staffs have much to be proud of. There is a vast international infrastructure of offshore structures that supply much needed goods and services to the societies they serve. This chapter addresses the issues associated with helping achieve desirable quality and reliability of offshore structures during their life cycles. The primary challenge that is addressed is not associated with the traditional engineering technologies that have been employed in the creation of these structures. History has shown that this is not the challenge. Rather, the primary challenge that is addressed is associated with the human and organisational aspects of these systems.

A colleague recently stated: "most engineers want to believe that the planet is not inhabited". It is clear that human and organisational factors are the primary challenges in developing offshore structure systems that have desirable and acceptable quality and reliability. Also, it is clear that there is a significant body of knowledge about how to address this challenge. The problem is wise implementation of this knowledge on a continuing basis.

Two instruments have been advanced to enable improved recognition of HOF in the design of offshore structures. Qualitative insights into potential performance characteristics of offshore structures are provided by the *QMAS* instrument; the primary focus of this instrument is on the HOF that influences the quality and reliability of these structures. Quantitative insights are provided by the *SYRAS* instrument. A "calibrated link" has been developed to enable the insights developed with application of *QMAS* to be translated into "reasonable" quantitative results that include explicit analyses of HOF. The combination of *QMAS* and *SYRAS* have been applied in several industry projects that have studied the considerations associated with "minimum" offshore structures, and in a variety of operating settings including design QA/QC, construction and operations.

It should be apparent to all engineers that HOF is of fundamental importance in the development of offshore structures that will have acceptable and desirable quality and reliability during their life cycles. Design engineers have a fundamental and primary responsibility in addressing HOF as an integral part of the design engineering process.

It should also be apparent to all concerned with the quality and reliability of offshore structures that organisations (industrial and regulatory) have pervasive influences on the assessment and management of threats to the quality and reliability of offshore structures. Management's drives for greater productivity and efficiency need to be tempered with the need to provide sufficient protections to assure adequate quality and reliability.

The threats to adequate quality and reliability in offshore structures emerge slowly in the design office. It is this slow emergence that generally masks the development of the threats to words quality and reliability. Often, the participants do not recognise the emerging problems and hazards. They become risk habituated and loose their wariness. Often, emerging threats are not clearly recognised because the goals of quality and reliability are subjugated to the goals of production and profitability. This is a problem, because there must be profitability to have the necessary resources to achieve quality and reliability. Perhaps, with the present high costs of lack of quality and reliability, these two goals are not in conflict. Quality and reliability can help lead to production and profitability. One must adopt a long-term view to achieve the goals of quality and reliability, and one must wait for production and profitability to follow. However, often we are tempted for today, not tomorrow.

The second important thing that we have learned about approaches to help achieve management desirable quality and reliability is organising the "right stuff" for the "right job". This is much more than job design. It is selecting those able to perform the daily tasks of the job within the daily organisation required to perform that job. Yet, these people must be able to re-organise and re-deploy themselves and their resources as the pace of the job changes from daily to unusual (it improves time). Given most systems, they must be team players. This is no place for "super stars" or "aces". The demands for highly developed cognitive talents and skills is great for successful crisis management teams. In its elegant simplicity, Crew Resource Management has much to offer in helping identify, train and maintain the right stuff. If properly selected, trained and motivated, even "pick-up ball teams" can be successful design engineering teams.

The final part of the 15-yr stream of research and development on which this chapter is based addresses the issues associated with implementation [Bea, 2000a]. A case-based reasoning study of a dozen organisations that had tried the implementation for a significant period of time identified five key attributes associated with successful implementation:

- *Cognisance* – of the threats to quality and reliability,
- *Capabilities* – to address the HOF and HRO aspects to improve quality and reliability,
- *Commitment* – to a continuing process of improvement of the HOF and HRO aspects,
- *Culture* – to bring into balance the pressures of productivity and protection and to realise trust and integrity, and
- *Counting* – financial and social, positive and negative, ongoing incentives to achieve adequate and desirable quality and reliability.

It is interesting to note that of the seven organisations that tried implementation, only two succeeded. It is obvious that this is not an easy challenge, and that at the present time, failure is more the rule than success. It is also interesting to note that the two organisations that succeeded recently have shown signs of "backsliding". Organisational–management evolution has resulted in a degradation in the awareness of what had been accomplished and why it had been accomplished. The pressures of doing something "new", downsizing, outsourcing, merging, and other measures to achieve higher short-term profitability have resulted in cutbacks in the means and measures that had been successfully implemented to reduce the costs associated with lack of adequate and acceptable quality and reliability. Perhaps, all organisations are destined to continually struggle for the balance in production and protection, and accidents represent a map of that struggle to succeed and survive.

References

American Bureau of Shipping (1998). The application of ergonomics to marine systems, Guidance Notes, Houston, Texas.

Apostolakis, G. E., Mancini, G., van Otterloo, R. W., and Farmer, F. R. (Eds.) (1990). Reliability engineering and system safety, Elsevier, London.

Aven, T. and Porn, K. (1998). Expressing and interpreting the results of quantitative risk analysis: review and discussion, Reliability Engineering and System Safety, Vol. 61, Elsevier Science Limited, London, UK, 1998.

Avigutero, T. and Bea, R. G. (1998). "Effects of damage and repairs on the lateral load capacity of a typical template-type offshore platform", *Proceedings International Offshore and Polar Engineering Conference*, International Society of Offshore and Polar Engineering, Golden, CO.

Bea, R. G. (1974). Selection of environmental criteria for offshore platform design, *Journal of Petroleum Technology*, Society of Petroleum Engineers, Richardson, Texas, pp. 1206–1214.

Bea, R. G. (1975). Development of safe environmental criteria for offshore structures, *Proceedings Oceanology International Conference*, Brighton, UK.

Bea, R. G. (1992). Marine structural integrity programs (MSIP), Ship Structure Committee, SSC-365, Washington, DC.

Bea, R. G. (1996a). Human and organisation errors in reliability of offshore structures, *Journal of Offshore Mechanics and Arctic Engineering*, American Society of Mechanical Engineers, New York, Nov.–Dec. 1996.

Bea, R. G. (1996b). Quantitative and qualitative risk analyses – the safety of offshore platforms, *Proceedings of the Offshore Technology Conference*, OTC 8037, Society of Petroleum Engineers, Richardson, Texas.

Bea, R. G. (2000a). Achieving step change in risk assessment and management (RAM), Centre for Oil and Gas Engineering, http://www.oil-gas.uwa.edu.au, University of Western Australia, Nedlands, WA.

Bea, R. G. (2000b). Performance shaping factors in reliability analysis of design of offshore structures, *Journal of Offshore Mechanics and Arctic Engineering*, Vol. 122, American Society of Mechanical Engineers, New York, NY.

Bea, R. G. and Lawson, R. B. (1997). Stage-II analysis of human and organisational factors, *Report to JIP on Comparative Evaluation of Minimum Structures and Jackets*, Marine Technology and Development Group, University of California at Berkeley.

Bea, R. G., Brandtzaeg, A., and Craig, M. J. K. (1998). Life-cycle reliability characteristics of minimum structures, *Journal of Offshore Mechanics and Arctic Engineering*, Vol. 120, American Society of Mechanical Engineers, New York, NY.

Bea, R. G., Holdsworth, R. D., and Smith, C. (Eds.) (1996). *Proceedings 1996 International Workshop on Human Factors in Offshore Operations*, American Bureau of Shipping, Houston, Texas.

Bier, V. M. (1999). Challenges to the Acceptance of Probabilistic Risk Analysis, Risk Analysis, Vol. 19, No. 4.

Bucknell, J. (2000). Defect assessment for existing marine pipelines and platforms, *Proceedings APEC (Asia-Pacific Economic Cooperation) Workshop on Assessing and Maintaining the Integrity of Existing Offshore Oil and Gas Facilities*, Beijing, China, U.S. Minerals Management Service, Herndon, VA.

Center for Chemical Process Safety (1989). Guidelines for technical management of chemical process safety, American Institute of Chemical Engineers, New York.

Center for Chemical Process Safety (1992). Guidelines for investigating chemical process incidents, American Institute of Chemical Engineers, New York.

Center for Chemical Process Safety (1993). Guidelines for auditing process safety management systems, American Institute of Chemical Engineers, New York.

Center for Chemical Process Safety (1994). Guidelines for preventing human error in process safety, American Institute of Chemical Engineers, New York.

De Leon, D. and Heredia-Zavoni, E. (2001). Probability distributions of mechanical damage for offshore marine platforms, *Proceedings of the Eleventh (2001) International Offshore and Polar Engineering Conference*, International Society of Offshore and Polar Engineers.

Demming, W. E. (1982). Out of the crisis, Massachusetts Institute of Technology Center for Advanced Engineering Study, Cambridge, MA.

Dougherty, E. M., Jr. and Fragola, J. R. (1986). Human reliability analysis, John Wiley and Sons, New York.

Faber, M. H. (1997). Risk Based Structural Maintenance Planning, Probabilistic Methods for Structural Design, Soares, C. G. (Ed.), Kluwer Academic Publishers, Amsterdam, The Netherlands.

Fischhoff, B. (1975). Hindsight does not equal foresight: the effect of outcome knowledge on judgment under uncertainty, *Journal of Experimental Psychology, Human Perception, and Performance*, Vol. 1, New York.

Gertman, D. I. and Blackman, H. S. (1994). Human reliability and safety analysis data handbook, John Wiley andSons, New York.

Gierlinski, J. T. and Rozmarynowski, B. (1999). Task I.2 and II.13: system reliability of intact and damaged structures under extreme environment and fatigue conditions, Report no. WSA/AM3681/Task I.2, *Report to JIP on Comparative Evaluation of Minimum Structures and Jackets*, WS Atkins Consultants Ltd, Leatherhead, UK.

Groenewg, J. (1994). Controlling the controllable, the management of safety, DSWO Press, Leiden University, The Netherlands.

Haber, S. B., O'Brien, J. N., Metlay, D. S., and Crouch, D. A. (1991). Influence of Organisational Factors on Performance Reliability – Overview and Detailed Methodological Development, *U. S. Nuclear Regulatory Commission*, NUREG/CR-5538, Washington, DC.

Hale, A., Wilpert, B., and Freitag, M. (1997). After the event, from accident to organisational learning, Pergamon Press, Elsevier Sciences Ltd., Oxford, UK.

Harris, D. and Chaney, F. (1969). Human factors in quality assurance, John Wiley and Sons, New York.

Hessami, A. G. (1999). Risk management: a systems paradigm, Systems Engineering, John Wiley and Sons, London, UK, 1999.

International Standards Organisation (1994a). *ISO 9000 Series*, quality management and quality assurance standards, British Standards Inst. Publication, London, UK.

International Standards Organisation (1994b). Quality systems – model for quality assurance in design/development, production, installation, and servicing, ISO 9001, London, UK.

International Standards Organisation (1994c). Health, safety, and environmental management systems, Technical Committee ISO/TC 67, Materials, Equipment and Offshore Structures for Petroleum and Natural Gas Industries, Sub-Committee SC 6, Processing Equipment and Systems, London, UK.

Jones, R. B. (1995). Risk-based management – a reliability centered approach, Gulf Publishing Co., Houston, Texas.

Kirwan, B. (1994). A guide to practical human reliability assessment, Taylor and Francis, London, UK.

Klein, G. (1999). Sources of power, MIT Press, Cambridge, Massachusetts, 1999.

Kletz, T. (1991). An engineer's view of human error, Institution of Chemical Engineers, Rugby, UK.

Knoll, F. (1986). Checking techniques, Modeling Human Error in Structural Design and Construction, Nowak, A. S. (Ed.) American Society of Civil Engineers, Herndon, Virginia.

Kontogiannis, T. and Lucas, D. (1990). Operator performance under high stress: an evaluation of cognitive modes, case studies and counter measures, Report No. R90/03, Nuclear Power Engineering Test Center, Tokyo, Japan, Human Reliability Associates, Dalton, Wigan, Lancashire, UK.

Libuser, C. (1994). Managing organisations to achieve risk mitigation, PhD Dissertation, Andersen School of Business, University of California, Los Angeles.

Marine Technology Directorate (1989). Underwater inspection of steel offshore installations: implementation of a new approach, Report 89/104, London, UK, 1989.

Marine Technology Directorate (1992). Probability-based fatigue inspection planning, Report 92/100, London.

Marine Technology Directorate (1994). Review of repairs to offshore structures and pipelines, Report 94/102, London, UK.

Marshall, P. W. and Bea, R. G. (1976). Failure modes of offshore platforms, *Proceedings of Behaviour of Offshore Structures Conference, BOSS '76*, Vol. II, Trondheim, Norway.

Matousek, M. (1990). Quality assurance, *Engineering Safety*, Blockley, D. (Ed.) McGraw-Hill Book Co., London, UK.

Melchers, R. E. (1993). Society, tolerable risk and the alarp principle, *Proceedings of the Conference on Probabilistic Risk and Hazard Assessment*, Melchers, R. E. and Stewart, M. G. (Eds.), The University of Newcastle, N.S.W., Australia.

Moan, T. (1997). Current trends in the safety of offshore structures, *Proceedings International Offshore and Polar Engineering Conference*, International Society of Offshore and Polar Engineers, Golden, Colorado.

Molak, V. (Ed.) (1997). Fundamentals of risk analysis and risk management, CRC Lewis Publishers, New York, 1997.

MSL Engineering Ltd. (1999). Effect of vessel impact on intact and damaged structures, Doc. Ref. C209R007-Rev 1, *Report to JIP on Comparative Evaluation of Minimum Structures and Jackets*, London, UK.

RAMBØLL (1999). Conceptual design summary report, Job No. 978503, Doc. No. 340-005, Rev. 1, 1999-01-05, *Report to JIP on Comparative Evaluation of Minimum Structures and Jackets*, Copenhagen, Denmark.

Rasmussen, J. (1996). Risk management, adaptation, and design for safety, *Future Risks and Risk Management*, Sahlin, N. E. and Brehemer, B. (Eds.), Kluwer Publishers, Dordrecht.

Rasmussen, J., Duncan, K., and Leplat, J. (Eds.) (1987). New Technology and human error, John Wiley and Sons, New York.

Reason, J. (1990). Human error, Cambridge University Press, London, UK.

Reason, J. (1997). Managing the risks of organisational accidents, Ashgate Publishers, Aldershot, UK.

Roberts, K. H. (1989). New Challenges in Organisational Research: High Reliability Organisations, *Industrial Crisis Quarterly*, Vol. 3, Elsevier Science Publishers, Amsterdam, the Netherlands.

Roberts, K. H. (Ed.) (1993). New challenges to understanding organisations, McMillan Publishing Co., New York.

Rochlin, G. I. (1997). Trapped in the net: the unanticipated consequences of computerisation, Princeton University Press, Princeton, New Jersey.

Shetty, N. (Ed.) (2000). Comparative evaluation of minimum structures and jackets, Synthesis Report, *Report to Joint Industry Project Sponsors*, W. S., Atkins Consultants Ltd., Report No. AM3681, Leatherhead, UK.

Soares, C. G. (Ed.) (1998). Risk and reliability in marine technology, A. A. Balkema, Rotterdam, The Netherlands.

Spouge, J. (1999). A guide to quantitative risk assessment for offshore installations, CMPT Publication 99/100, ISBN I 870553 365, London, UK.

Stewart, M. G. and Melchers, R. E. (1988). Checking models in structural design, *Journal of Structural Engineering*, Vol. 115, No. 17, American Society of Civil Engineers, Herndon, Virginia.

Swain, A. D. and Guttman, H. E. (1983). *Handbook of Human Reliability Analysis with Emphasis on Nuclear Power Plant Applications*, NUREG/CR-1278, U.S. Nuclear Regulatory Commission, Washington, DC.

Vinnem, J. E. (1998). Evaluation of methodology for QRA in offshore operations, Reliability Engineering and System Safety, Vol. 61, Elsevier Science Limited, London, UK.

Weick, K. E. (1995). Sensemaking in organisations, Sage Publishers, Thousand Oaks, CA.

Weick, K. E. (1999). Organizing for high reliability: processes of collective mindfulness, *Research in Organisational Behaviour*, Vol. 21, JAI Press Inc.

Weick, K. E. (2000). The neglected context of risk assessment – a mindset for method choice, Risk Management in the Marine Transportation System, Transportation Research Board, National Research Council, Washington, DC.

Weick, K. E. and Quinn, R. E. (1999). Organisational change and development, Annual Review of Psychology, New York.

Weick, K. E., Sutcliffe, K. M., and Obstfeld, D. (1999). Organizing for high reliability: processes of collective mindfulness, Research in Organisational Behaviour, Staw and Sutton (Eds.), Research in Organisational Behaviour, JAI Press, Vol. 21, Greenwich, CT.

Weick, K. E. and Sutcliffe, K. M. (2001). Managing the unexpected, Jossey-Bass, San Francisco, CA.

Wenk, E. Jr, (1986). Tradeoffs, imperatives of choice in a high-tech world, The Johns Hopkins University Press, Baltimore, MD.

Winkworth, W. J. and Fisher, P. J. (1992). Inspection and repair of fixed platforms in the North Sea, *Proceedings Offshore Technology Conference*, OTC 6937, Society of Petroleum Engineers, Richardson, Texas.

Woods, D. D. (1990). Risk and human performance: measuring the potential for disaster, Reliability Engineering and System Safety, Vol. 29, Elsevier Science Publishers Ltd., UK.

Wum, J. S., Apostolakis, G. E., and Okrent, D. (1989). On the inclusion of organisational and management influences in probabilistic safety assessments of nuclear power plants, *Proceedings of the Society for Risk Analysis*, New York, 1989.

Xu, T., Bea, R. G., Ramos, R., Valle, O., and Valdes, V. (1999). Uncertainties in the fatigue lives of tubular joints, *Proceedings Offshore Technology Conference*, OTC 10849, Society of Petroleum Engineers, Richardson, Texas.

Handbook of Offshore Engineering
S. Chakrabarti (Ed.)
© 2005 Elsevier Ltd. All rights reserved.

Chapter 13

Physical Modelling of Offshore Structures

Subrata K. Chakrabarti
Offshore Structure Analysis, Inc., Plainfield, IL, USA

13.1 Introduction

This chapter will describe the need, the modelling background and the method of physical testing of offshore structures in a small-scale model. The physical modelling involves design and construction of scale model, generation of environment in an appropriate facility, measuring responses of the model subjected to the scaled environment and scaling up of the measured responses to the design values. The purpose of duplicating the environment experienced by an offshore structure in a small scale is to be able to reproduce the responses that the structure will experience when placed in operation in the offshore site. This enables the designers to verify their design methods and take any necessary corrective actions for the final design of the structure before it is released for construction. The physical model also allows the proof of the concept for a new and innovative design for a particular application as well as verifies the operational aspects of a designed structure.

For a successful physical modelling, the following areas should be known and will form the basis for this chapter:

- Needs for model tests
- Similarity laws for modelling
- Froude number and related scaling
- Reynolds number and its effect
- Towing resistance and drag effect
- Scaling of a hydroelastic model
- Offshore model testing facilities and their qualifications and limitations
- Important components of a wave basin
- Modelling of environment
- Instrumentation requirements and measurement accuracy
- Modelling difficulties and distortion in scaling

- Challenges in testing of deepwater and ultra-deepwater structures
- Data analysis and reporting

While a general discussion on the physical modelling problem will be made, emphasis has been placed on modelling the present-day offshore structures in a small scale. Particular attention has been given to the testing of deep-water offshore structures. The chapter is laid out in such a way that it may be used in developing a request for testing proposal for a forthcoming model test.

13.1.1 History of Model Testing

Model testing has been an integral part of the development of offshore structures starting with the shallow water structures in the early fifties to the present day. The operational elements for an offshore structure are routinely examined through model testing. Many of the design parameters are verified through model tests. As the water depth for offshore structures is getting deeper, the technique for small-scale testing is becoming increasingly more challenging. Model testing for today's deep-water structures, however, is essential for a better understanding of the stability behaviour and survival characteristics of deep-water structures. A description of the role of model testing has been given in Dyer and Ahilan (2000). Many detailed aspects of physical modelling and testing may be found in Chakrabarti (1994).

Experimental testing of physical scale models in a wave basin, in which the critical response parameters are determined by direct measurement, has been the traditional way of investigating the behaviour of offshore vessels [ITTC, 1999]. It is recognized as the most reliable tool for reproducing realistic and extreme situations an offshore structure is expected to experience in its lifetime. In particular, it may be important for complex systems, where various kinds of static and dynamic coupling effects among the system components may occur. Additionally, the physical models have the advantage over the numerical models that unknown phenomena and effects, not described by theoretical models, can be discovered.

Typically, model scales in the range 1 : 50–1 : 70 are used for such testing. The complete floater system with moorings and risers are modelled. Often a simplified modelling is employed including a truncated mooring system and a reduced number of "equivalent" risers. These will be discussed further in the subsequent sections.

Because of the limitation in the available basin depths, such truncation is a common occurrence. As the water depth goes deeper, question may be raised as to the suitability of such testing of distorted models. It is probably safe to state that deep-water development will proceed with or without model testing. The designers will use whatever design tools are considered appropriate for their design without direct verification through model testing. However, in practice, such will seldom be the case. Decisions for placing deep-water structures will be made at a level different from designers and experimenters. All such developments will include plans and budget for model testing. In fact, no new structure will probably be built and installed without some scale model testing. Therefore, intelligent decisions should be made in planning such testing when the full model scaling is not possible. This, however, is not new.

Offshore structure development has taken a similar course in the past. We should not overlook the fact that many of the theories were developed only after a phenomenon was discovered during a model (and prototype) observation. One example is the high tendon loads experienced by the tension leg platform (TLP). Model tests revealed very high vertical loads in the TLP tendon even though heave natural period was very low. Theory has evolved since its discovery, which mathematically describes the source of such loads commonly known as "springing". Since then, impact-type loading has also been discovered in full-scale TLP measurements and theory for this "ringing" load has been developed subsequently. Other areas of measurements of model response include slow drift oscillation of a soft moored floating structure, green water impact on decks, air gaps, stability of floating structures, etc. Many of these tests used truncated models. Some of these physical phenomena still do not have adequate analysis tools. Structures have been developed and installed successfully in spite of these deficiencies.

Let us illustrate by a simple example. In 1965, Chicago Bridge & Iron Co. (CB&I) built its first fixed offshore structure in the Persian Gulf, named the Khazzan Dubai oil storage tank. This structure was an ideal candidate for the 3-D diffraction design tool for a general shape, which did not exist at the time. In fact, in 1970, the traditional 3-D linear diffraction theory programme was first developed and commercially used in offshore structure development. This programme was not used in the design of Khazzan tank, which used simply the Froude–Krylov theory along with a large safety factor. The first model test of the Khazzan tank was very crude with substandard measurements compared to today's technology. Later, verification of model test with the diffraction theory showed that the design was conservative. Today after 30 years, Conoco is still successfully operating the three CB&I-built Khazzan storage tank complex in the Persian Gulf.

A scale model test in a wave basin is carefully controlled, and has far superior accuracy and sophistication today. For ultra-deepwater structures, scale distortion and truncation are here to stay no matter where the testing is performed. On the other hand, truncation is nothing new in model testing. Coastal people have been running successful tests with distorted models for many many years. Most of the recent deep-water model tests for the new generation semi-submersibles and SPARs needed truncation in the area of, among others, the mooring lines and risers in the system. They have been considered successful and meaningful results were obtained for motions, sectional structure loads and mooring line loads, air gaps, slamming loads, green water effects, etc.

Moreover, there has hardly been a model test where the experimenters not learnt something new, no matter how trivial the tests were. Admittedly, the ultra-deepwater testing poses additional challenges. This means that such testing should be planned more carefully. In fact, as much time should be spent in the planning of these tests as the actual testing time in the basin. The test goal should remain focused in these cases, rather than all-inclusive. It is possible that different types of testing should be designed for different goals for the same system. Some testing of the structure component may be included in the overall plan along with the complete (distorted) testing. Sometimes, the multiple model scales of the system may be warranted.

New phenomena will continue to be discovered for these ultra-deepwater structures yet undefined by theory. In particular, the stability as well as non-linearity issue is an open question. Additionally, model testing is a simple and efficient technique in improving and optimising a system or a concept, which is more difficult to achieve alternatively.

Therefore, physical model test of the overall system that takes care of distortion in a systematic way should continue. What modelling technique will work, of course, will depend on the particular system in question. In fact, certain specific ground rules may be laid out for deep water testing on the basis of what is available to us today. This may be an analysis of dos and donts and pitfalls to avoid. Some of these areas will be covered in the latter sections.

13.1.2 Purpose of Physical Modelling

One of the principal benefits of model testing is that valuable information is provided which can be used to predict the potential success of the prototype at relatively small investment. The physical model provides qualitative insight into a physical phenomenon, which may not be fully understood currently. The use of models is particularly advantageous when the analysis of the prototype structure is very complicated. In other situations, models are often used to verify simplified assumptions, which are inherent in most analytical solutions, including non-linear effects. An example of this is the discovery of slow drift oscillation of a moored floating tanker through model testing before a theory describing the second-order oscillating drift force and the associated motion was derived. Model test results are also employed in deriving empirical coefficients that may be directly used in a design of the prototype.

Therefore, the following list gives the principal benefits to be gained from a model test:

- Validate design values
- Problem difficult to handle analytically
- Obtain empirical coefficients
- Substantiate analytical technique
- Verify offshore operation, such as, a specific installation procedure
- Evaluate higher order effects normally ignored in the analysis
- Investigate unpredicted or unexpected phenomena

13.2 Modelling and Similarity Laws

It is important to have a clear understanding of the scaling laws before the model measurements may become meaningful. Why do we need scaling laws? We can cite the following reasons for scaling laws:

- Testing is generally done in a small scale,
- The scaling laws allow scaling up of the measured data to full scale.

Modelling laws relate the behaviour of a prototype to that of a scaled model in a prescribed manner. The problem in scaling is to derive an appropriate scaling law that accurately describes this similarity. In modelling a prototype structure in a small scale, there are, at least, three areas where attention must be given so that the model truly represents the prototype behaviour – structure geometry, fluid flow and the interaction of the two. Therefore, we shall seek similarity in the structure geometry, similitude in the fluid kinematics and the similitude in the dynamics of the structure subjected to the fluid flow around it.

13.2.1 Geometric Similitude

Geometrically similar structures have different dimensions, but have the same shape. In other words, a model built for testing in a small scale must resemble the prototype in shape, specially the submerged sections. At least, the important submerged elements must be modelled accurately. This can be easily achieved if we assume that a constant scale ratio exists between their linear dimensions

$$\frac{\ell_p}{\ell_m} = a \tag{13.1}$$

where ℓ_p and ℓ_m are any two corresponding homologous dimensions of the two structures namely, prototype and model, respectively and a is the scale ratio between them. In this case, we say that the two structures are geometrically similar. The ratio of the two similar dimensions (e.g. diameter and length of a particular member) will, therefore, remain constant and establishes the scale factor for a model. This factor will be defined as λ throughout this book.

13.2.2 Kinematic Similitude

The kinematic similitude is achieved in the model if the ratio of the fluid velocity and fluid acceleration are preserved. Thus, the ratio of the prototype velocity to the corresponding model velocity will be a prescribed constant. This applies to all velocities including fluid particle, wind speed, towing speed, model velocity in a particular direction, etc. Similarly, the ratio of the acceleration will be a different constant. Their relationships will be determined from the scaling laws. When these laws are satisfied for velocity and acceleration, the model is considered kinematically similar to the full-scale structure.

13.2.3 Hydrodynamic Similitude

Consider the masses of two similar structures in similar motions. Noting that the induced force may be written using the Newton's law as the product of mass and acceleration, all corresponding impressed forces must be in a constant ratio and similar direction. Therefore, geometrically similar structures in similar motions having similar mass systems are similarly forced. When the model is forced similar to the prototype, the model is considered dynamically similar to the prototype. The chosen scaling laws establish this scaling relationship for the model.

In order for a model to truly represent the full-scale structure, all three conditions, namely the geometric, kinematic and dynamic similarities must be maintained. Then only the model test data may be scaled up to the full scale without any distortion. Hydrodynamic scaling laws are determined from the ratio of forces. Table 13.1 gives the most common scaling laws from the fluid structure interaction problem. Several ratios may be involved in a particular scaling. One of these may be more predominant than others. The dynamic similitude between the model and the prototype is achieved from the satisfaction of these scaling laws. In most cases, only one of these scaling laws is satisfied by the model structure. Therefore, it is important to understand the physical process experienced by the structure and to choose the most important scaling law which governs this process.

Table 13.1 Common dimensionless quantities in offshore engineering

Symbol	Dimensionless number	Force ratio	Definition
Fr	Froude Number	Inertia/Gravity	u^2/gD
Re	Reynolds Number	Inertia/Viscous	uD/ν
Eu	Euler Number	Inertia/Pressure	$p/\rho u^2$
Ch	Cauchy Number	Inertia/Elastic	$\rho u^2/E$
KC	Keulegan–Carpenter Number	Drag/Inertia	uT/D
St	Strouhal Number	Eddy/Inertia	$f_e\ D/u$

In table 13.1, D = member diameter, T = wave period, g = gravity, ν = kinematic viscosity, p = pressure, E = modulus of elasticity and f_e = vortex (eddy) shedding frequency. The Froude number applies to gravity waves. The Reynolds number is related to the drag force in the structure. The Euler number is not as important as these quantities, except for vertically loaded long, slender structures. The Cauchy number plays an important role for an elastic structure, such as, compliant towers, risers and tendons. The Keulegan–Carpenter number is very important for small structural members where forces are computed based on hydrodynamic inertia and drag coefficients (see Chapter 4). The Strouhal number is the non-dimensional vortex shedding frequency and should be considered for a moving structure when the flow past a structural member separates and produces vortices past the structure.

The typical current or wave–structure interaction problem involves Froude number, Reynolds number and Keulegan–Carpenter number. For structures that are subject to deformation, Cauchy number should additionally be considered. For structures vibrating in fluid medium, the Strouhal number is also included.

The frequency of vortex shedding, f_e, from a stationary circular cylinder of diameter D in a fluid stream of velocity u has been shown to be a linear function of Reynolds number *Re* over a wide range. A relationship between the Strouhal number *St* and Reynolds number *Re* exists in steady flows. It is generally accepted that $St \approx 0.2$ in the range $2.5 \times 10^2 < Re < 2.5 \times 10^5$. Beyond this range, *St* increases up to about 0.3 and then, with further increase in *Re*, the regular periodic behaviour of u in the wake behind the cylinder disappears. Some variation in this trend has been observed in experiments by several investigators, particularly, outside the constant range of *St*.

In the offshore structure problem, the most common among the dimensionless scaling laws presented in table 13.1 is Froude's law. The Reynolds number is also equally important in many cases. However, Reynolds similarity is quite difficult, if not impossible, to achieve in a small-scale model. Simultaneous satisfaction of *Fr* and *Re* is even more difficult. The Froude law is the accepted method of modelling in hydrodynamics.

13.2.4 Froude Model

The Froude number has a dimension corresponding to the ratio of $u^2/(gD)$ as shown in table 13.1. Defining *Fr* as

$$Fr = \frac{u^2}{gD} \tag{13.2}$$

the Froude model must satisfy the relationship:

$$\frac{u_p^2}{gD_p} = \frac{u_m^2}{gD_m} \tag{13.3}$$

Assuming a scale factor of λ and geometric similarity, the relationship between the model and full-scale structure for various parameters may be established. Table 13.2 shows the scale factor of the common variables that the Froude model satisfies. The variables chosen are the most common ones that are encountered in the offshore structure testing. For a scale factor of 1 : 50 and 1 : 100 for a model, the multiplying factors for these variables are also shown in the table. For scale factors other than these, this table may be easily converted to yield the multiplying factor for the desired responses. Thus, for a Froude model, the scaling of the model response to the prototype values is straightforward. There are instances, however, where this scaling may not be achieved simply. A few examples will be cited later where this table is not directly applicable and possible remedies or corrections that may be adopted for the above method will be discussed.

One should note that fluid density and viscosity are different between the model and prototype, even though the difference is generally small. This difference is often ignored due to small corrections. However, the scaled up values may be corrected by the ratio of these quantities if desired. Chapter 3 lists the values of these quantities for the temperature difference. Examples of a few prototype quantities and environmental parameters are given in table 13.3. A few structure responses are also included in the table. Their values at different scale factors are listed. This is an exercise to illustrate what scale factor for a particular test requirement may be appropriately chosen and what becomes of the magnitudes of quantities in the model scale. Such a table will guide the user to choose the most appropriate scale given the limitation of a chosen testing basin and measurements.

13.2.5 Reynolds Model

If a Reynolds model is built, it will require that the Reynolds number between the prototype and the model be the same. Assuming that the same fluid is used in the model system (viscosity ratio = 1), this means that:

$$u_p D_p = u_m D_m \tag{13.4}$$

Table 13.2 Scaling of variables using Froude law

Variable	Symbol	Scale factor	λ=50	λ=100
All linear dimensions	D	λ	50	100
Fluid or structure velocity	u	$\lambda^{1/2}$	7.07	10
Fluid or structure acceleration	$\dot{u}$	1	1	1
Time or period	t	$\lambda^{1/2}$	7.07	10
Structure mass	m	λ^3	1.25E3	1.0E6
Structure moment of inertia	I	λ^5	3.125E8	1.0E10
Section moment of inertia	I	λ^4	6.25E4	1.0E8
Structure displacement volume	V	λ^3	1.25E3	1.0E6
Structure restoring moment	C	λ^4	6.25E4	1.0E8
Force	F	λ^3	1.25E3	1.0E6
Moment	M	λ^4	6.25E4	1.0E8
Stress	σ	λ	50	100
Spring constant	K	λ^2	2500	1.0E4
Wave period	T	$\lambda^{1/2}$	7.07	10
Wave length	L	λ	50	100
Pressure	p	λ	50	100
Gravity	g	1	1	1
Fluid density	ρ	1	1	1
Fluid kinematic viscosity	ν	1	1	1
Reynolds number	Re	$\lambda^{3/2}$	353.6	1000
Keulegan–Carpenter number	KC	1	1	1

if a scale factor of λ is used in the model, then this equality is satisfied if

$$u_m = \lambda u_p \tag{13.5}$$

In other words, the model fluid velocity must be λ times the prototype fluid velocity. In general, this is difficult to achieve, especially if a small-scale experiment is planned. It also points out the difficulty of satisfying both the Reynolds and the Froude number simultaneously.

On the other hand, if the Froude's law is used in modelling, the distortion in the Reynolds number is large. As noted earlier, for a Froude model, the Reynolds number scales as:

$$Re_p = \lambda^{3/2} Re_m \tag{13.6}$$

Table 13.3 Scaling of typical prototype parameters for various scale factors

Parameter	Unit	Prototype	1 : 25	1 : 50	1 : 100	1 : 200
Length	m	500	20	10	5	2.5
Draft	m	100	4	2	1	0.5
Column diameter	m	50	2	1	0.5	0.25
Structure mass	kg	1E6	64	8	1	0.125
Max wave height	m	30	1.2	0.6	0.3	0.15
Min wave height	m	2	0.08	0.04	0.02	0.01
Max wave period	sec	20	4	2.8	2	1.4
Min wave period	sec	5	1	0.7	0.5	0.35
Current speed	m/s	1	0.2	0.14	0.1	0.07
Load	N	1E6	64	8	1	0.125
Displacement	m	2	0.08	0.04	0.02	0.01

Therefore, the larger the scale factor, the larger is the distortion in the Reynolds scaling. In fact, it is possible that the model flow will be laminar, while the prototype flow falls in the turbulent region. Experiments have shown that the flow characteristics in the boundary layer are most likely to be laminar at $Re < 10^5$, whereas the boundary layer is turbulent for $Re > 10^6$. In this case, two different scaling laws apply, (namely, both Froude and Reynolds), which cannot be satisfied simultaneously. (Use of different fluids to match the Reynolds number may not be practical.) In this case, it is most convenient to employ Froude scaling and to account for the Reynolds disparity by other means. There are several methods that may be used to account for the distortion in the Reynolds scaling for a Froude model:

- Maximise scale of the model to simulate the prototype effect closer
- Correct Reynolds effect in scaling up of data to full scale
- Trip the incoming flow by roughing the model surface in the forward area
- Induce turbulence in the flow by external means ahead of the model

The larger the model, the closer is the flow simulation. This is, however, difficult to achieve for offshore structures. Sometimes, fluid of lower viscosity than water is used to increase the value of Reynolds number in the model. For equality of both Froude and Reynolds number, a fluid whose kinematic viscosity is about $1/\lambda^{3/2}$ of that of water should be used. When λ is large, such as for offshore structures, this is impossible to achieve.

One method of achieving a proper Reynolds number effect at the boundary layer is to deliberately trip the laminar flow in the model by introducing roughness on the surface of the forward part of the model. Then the model in most part will see turbulent flow. This works for a long model, because once the flow regime is turbulent, the drag effect is only weakly dependent on the Reynolds number. In testing tanker models, external means, such as studs, pins or sand-strips attached near the bow, are often used to induce turbulence.

Flow can also be tripped ahead of the model by introducing a mesh barrier submerged from the surface.

In towing tests, with horizontally long structures, such as, ships or barges, the skin friction resistance is comparable to the wave-making resistance and is dependent on Reynolds number. Thus, towing resistance depends on both Froude and Reynolds law. Corrections are made in the friction factor (which is known as a function of Reynolds number) based on the respective Reynolds number before the data on model towing resistance is scaled up to the prototype value. If this difference is ignored in scaling, the (scaled up) prototype data will generally be non-conservative.

13.2.5.1 Towing Resistance of a Ship Model in Wave Basin

For a ship/barge model, the scaling is done with Froude scale and corrective measures are taken to scale up the measured values in model scale. The following steps are adopted routinely as a standard procedure:

- Measure model resistance R_m at model speed u_m by towing the model in water with carriage
- Compute total resistance coefficient C_{tm} by dividing R_m by the factor $(0.5\rho A_m u_m^2)$ where A_m is the model submerged surface area
- Compute model friction coefficient C_{fm} by the Schoenherr formula based on model Re_m number

$$C_f = \frac{0.075}{(\log_{10} Re - 2)^2} \tag{13.7}$$

- Compute residual coefficient $C_{rm} = C_{tm} - C_{fm}$ which corresponds to wave making resistance and is Froude-scaled
- Add to C_{rm} a ship appendage correlation allowance Ca of 0.0004 (based on ITTC recommendation)
- Compute prototype ship frictional resistance coefficient C_{fp} using equation (13.7) for the prototype Reynolds number Re_p
- Add friction coefficient to the residual C_{rm} to obtain total prototype ship resistance coefficient C_{tp}
- Multiply by the normalisation factor $0.5\rho A_p u_p^2$ (where A_p is the prototype submerged surface area) to obtain the full-scale ship resistance
- Make correction in the density between the sea water and the fresh water used in the model test by multiplying by the ratio of the two
- Compute the prototype horsepower requirement.

The above procedure is illustrated by an example calculation of total resistance of a barge from the model test results in table 13.4. The model represents a 1 : 55 scale model of a barge. The model was towed at the scaled speed (column 2) and towing loads (column 3) were measured at these speeds. The subsequent columns follow the steps outlined above until the total resistance of barge in full scale (in kips) is obtained on the last column. Since no appendages were present, no allowance for the appendages was considered in this example. This table illustrates how the correction for the Reynolds number distortion is accounted

Table 13.4 Scaling of measured model resistance to prototype resistance

Scale factor = 55 Model wetted area = 29.61 ft^2 Model length = 10.39 ft					Prototype water mass density = 1.98 Model water mass density = 1.94 Model kinematic viscosity of water = 1.17E-05 ft^2/s					
Towing Speed		Model					Proto			
Proto-type	Model	Resis-tance	*Ctm*	*Re*	*Cfm*	*Crm* =*Crp*	*Re*	*Cf*	*Ct*	Resis-tance
knot	ft/s	(lb)	$\times10^{-2}$	$\times10^{5}$	$\times10^{-4}$	$\times10^{-3}$	$\times10^{8}$	$\times10^{-4}$	$\times10^{-3}$	(kips)
3.95	0.9	0.147	0.632	7.97	3.65	5.95	3.25	5.22	6.48	25.6
4.83	1.1	0.218	0.627	9.74	3.49	5.92	3.97	5.18	6.44	38.0
6.15	1.4	0.289	0.513	12.4	3.32	4.80	5.06	5.14	5.32	50.8
7.03	1.6	0.419	0.570	14.2	3.22	5.38	5.78	5.12	5.89	73.5
7.91	1.8	0.5	0.537	15.9	3.14	5.06	6.50	5.10	5.57	88.0
8.79	2	0.6	0.522	17.7	3.07	4.92	7.22	5.08	5.42	105.8
10.11	2.3	0.791	0.521	20.4	2.99	4.91	8.31	5.06	5.41	139.7
10.99	2.5	0.92	0.513	22.1	2.94	4.83	9.03	5.04	5.34	162.6
11.86	2.7	1.06	0.506	23.9	2.89	4.77	9.75	5.03	5.28	187.6
13.18	3.0	1.334	0.516	26.6	2.83	4.88	10.8	5.02	5.38	236.1
14.06	3.2	1.492	0.507	28.3	2.79	4.79	11.6	5.01	5.29	264.4
14.94	3.4	1.68	0.506	30.1	2.76	4.78	12.3	5.00	5.28	297.9
15.82	3.6	1.832	0.492	31.9	2.73	4.65	13.0	4.99	5.15	325.4
17.14	3.9	2.312	0.529	34.5	2.68	5.02	14.1	4.98	5.52	409.6
18.02	4.1	2.872	0.595	36.3	2.66	5.68	14.8	4.97	6.18	506.7

for in the resistance calculations. The scaled-up prototype resistance is shown in fig. 13.1. For comparison, the model values are directly scaled up by Froude scale (see table 13.2 for the factor) without regard to Reynolds number distortion and are also plotted in the figure. In this case, the friction force was small compared to the inertia force so that the difference in magnitude between the two is quite small.

13.2.5.2 Drag Resistance of an Offshore Structure

Unlike the ship-shape, many offshore structures are not elongated and have high forward speed and skin friction force is not a concern. However, many members of an offshore structure are subject to drag forces, which experience a similar problem of Reynolds distortion from the Froude scaling. This drag force is called form drag and has been

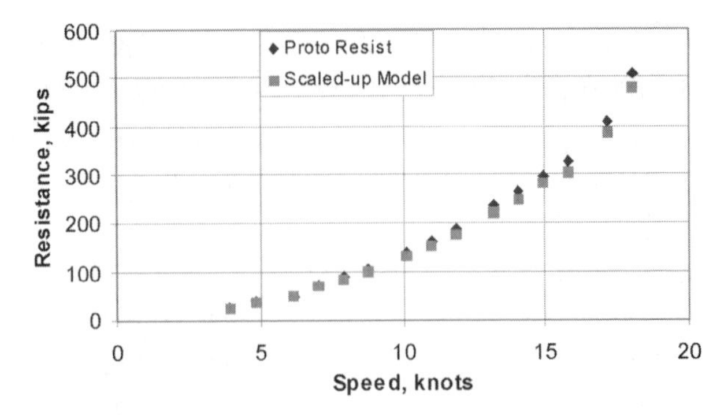

Figure 13.1 Scaled prototype towing load

described in Chapter 4. The towing resistance for these offshore structures is expected to include both inertia and drag forces.

For an offshore structure, where form drag is important, a similar procedure as in Section 13.2.5.1 may be adopted:

- Measure towing resistance of an offshore structure in a towing tank
- Obtain C_D for drag members from available chart on published model test data
- Compute drag force on drag members at model speed
- Correct C_D due to any shielding effect from members using literature data
- Subtract model drag force from the measured total load
- Scale up the difference (residual force) by Froude scale
- Compute prototype drag force on drag members using design guide
- Account for any shielding from design guide (e.g. API guidelines)
- Add prototype drag force to scaled up data
- Correct for surface roughness, fluid density, etc., as needed.

It is possible to derive the drag coefficients for the model and prototype members from Hoerner, (1965) and Sarpkaya (1976) as well as the certifying agency guidelines [e.g. API, 1979] for offshore structures. The above correction is needed because of the difference in the flow regime between the model and the prototype.

The laminar flow in the model may be equivalently compensated by artificial stimulation. In this case, no corrections are necessary in the process of scaling the model data. This is illustrated in figs. 13.2 and 13.3 for a semi-submersible production model. In fig.13.2 a semi-submersible rig model is seen being towed in a wave basin with the help of an overhead carriage. The load cells to measure the towing resistance may be seen between the model and the carriage. A submerged grid screen may be seen in the foreground mounted on the same carriage about 3 m (10 ft) ahead of the model (to simulate turbulence). The grid consists of taut strings about 6.4 mm in diameter spaced about 30 mm apart covering the frontal area of the semi-submersible.

The test was done with and without the grid to illustrate its effect. The results are provided in fig. 13.3. It is clear that the presence of the grid disturb the flow (as was observed during

Figure 13.2 Towing of a semi-submersible on a carriage (courtesy Offshore Model Basin, Escondido, CA)

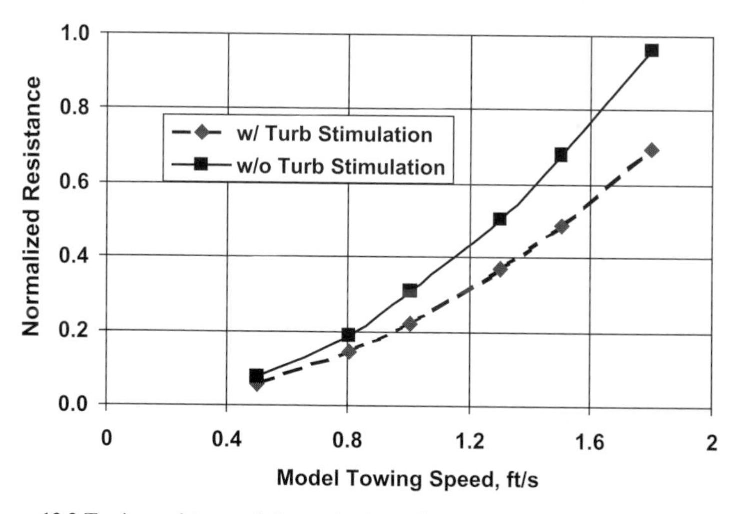

Figure 13.3 Towing resistance of the semi-submersible with and without turbulence screen

the model test) seen by the model and the net effect is a reduction in the measured resistance. Since the prototype drag coefficient (in turbulence flow) is expected to be lower than the corresponding model drag coefficient (near laminar flow), the net load is expected to be lower. Thus, while the degree of turbulence compared to the prototype is unknown, the presence of the screen appears to duplicate the prototype flow. Therefore, the measured resistance will be close to the scaled prototype resistance and no additional corrections due to Reynolds number distortion are necessary in this case.

13.2.6 Cauchy Model

Let us now consider another type of structure where the flexibility of its members becomes important. In this case, the member is expected to undergo deformation due to the interaction with waves and this effect should be accounted for in the model for proper simulation.

It is often desired to test structures to determine stresses generated in its members due to external forces, for example, from waves. It is well known that for long slender structures, the stiffness of the structure is important in measuring the response of the structure model in waves. In this case, the elasticity of the prototype should be maintained in the model. Hydroelasticity deals with the problems of fluid flow past a submerged structure in which the fluid dynamic forces depend on both the inertial and elastic forces on the structure. Therefore, in addition to the Froude similitude, the Cauchy similitude is desired.

The Cauchy similitude requires that stiffness, such as in bending, of a model must be related to that of the prototype by the relation:

$$(EI)_p = \lambda^5 (EI)_m \tag{13.8}$$

where E = modulus of elasticity and I = moment of inertia. This provides the deflection in the model which is $1/\lambda$ times the deflection in the prototype (Froude's law); also, stress must be similarly related, such that, $\sigma_p = \lambda\sigma_m$ (table 13.5).

Let us consider the example of a cantilever beam. The maximum deflection at the end of the beam is given by $\delta_{\max} = Fl^3/(3EI)$ where F is the load at the end of the cantilever

Table 13.5 Scaling of structure stiffness parameters for combined Froude/Cauchy scale

Item	Formula	Prototype/Model
Bending moment	$M_{yp} = \lambda^5 M_{ym}$	λ^5
Bending stiffness	$(EI)_p = \lambda^5 (EI)_m$	λ^5
Axial stiffness	$(EA)_p = \lambda^3 (EA)_m$	λ^3
Section modulus	$I_p = \lambda^4 I_m$	λ^4
Young's modulus	$E_p = \lambda E_m$	λ
Stress	$\sigma_p = \lambda\sigma_m$	λ
Torsional rigidity	$(GI)_p = \lambda^5 (GI)_m$	λ^5
Torsional modulus	$G_p = \lambda G_m$	λ

of length l. Equation (13.8) satisfies Froude's law for this relationship. Since the section moment of inertia satisfies

$$I_p = \lambda^4 I_m \tag{13.9}$$

we have:

$$E_p = \lambda E_m \tag{13.10}$$

Thus, the Young's modulus of the model material should be $1/\lambda$ times that of the prototype. Scaling parameters for different important variables are given as the ratio of full scale to model values in terms of scale factor λ in table 13.5. Assuming steel for the prototype material ($E_p = 2.07 \times 10^8$ kPa or 30×10^6 psi) and $\lambda = 36$, the model E_m should be 5.7×10^6 kPa (83,300 psi). Therefore, a suitable material should be chosen with this value to build the model that will be elastically similar.

13.3 Model Test Facilities

Since scale models of an offshore structure are used to measure their responses accurately so that they may be applied in a design with confidence, it is important to choose a model testing facility that will fulfill the requirements of a testing programme. The primary purpose of wave tank study is to obtain reliable data by minimising scale effects and measurement error.

The model testing facility for offshore structures should consist of the following capabilities – model building, instrumentation, simulation of environment and the software to record and analyse data. The physical facility should consist of a basin with the capability of generating waves, wind and current. An efficient wave absorption system is also essential in a basin. The simultaneous generation of waves and current allows the study of their combined interaction with the model. The wind effect is simulated on the superstructure of the model (the portion above the water) and is often accomplished using a series of blowers located just above the water surface near the model.

A testing facility should have the following optimum requirements so that a variety of structures may be tested in the facility:

- Tests at a reasonable scale (1 : 50–1 : 100 preferred)
- Capability of generating regular and random waves over a wide frequency range
- Wave spreading for certain structures may be preferred
- Period of waves from 0.5 s to 4 s
- Height of waves from a few centimetres (inches) to about 0.6 m (24 in.)
- Towing carriage with a steady speed range of 0.15–3 m/s (0.5–10 ft/s) and capable of carrying a large displacement structure without appreciable structure deformation
- Wind generating capability with a movable bank of fans
- Current generating capability with a return flow system
- Non-contacting motion measurement
- Underwater video documentation capability

The preferred requirements for instrumentation and measurements are described later.

13.3.1 Physical Dimensions

In today's development of the deepwater fields, it is desirable to have a deep model basin. In fact, for ultra-deepwater, an extremely deepwater facility is required, which is not available. Even the deepest basin in the world is not adequate for the practical simulation of the deepwater depth of today and the full mooring system simulation. Therefore, it is recognised that some distortion in scaling all the important parameters in a model test is inevitable.

The choice of scale for a model test is often limited by the experimental facilities available. However, within this constraint, optimum scale should be determined by comparing the economics of the scale model with that of the experiment. It should be kept in mind that too small a scale may result in scale effects and error. Too large a model is often very expensive and may introduce problems from physical handling of the model.

When Reynolds effect (such as, presence of drag force) are important, a large scale is recommended to minimise the problem of scale effects. However, the adverse effects of the tank walls must also be considered and avoided in this case. As a rule of thumb, for circular cylindrical structures, the overall transverse dimension should not exceed 1/5th the width of the tank. When larger three-dimensional structures are tested in a wave tank, undesirable transverse reflections may generate in the tank from its sidewalls. This effect may be minimised by introducing lateral wave absorbers along the basin wall.

For offshore structure modelling, a two-dimensional wave basin with a mechanical wavemaker is often utilised. There are two main classes of mechanical type wavemakers. One of them moves horizontally in the direction of wave propagation and has the shape of a flat plate driven as a flapper or a piston. The other type moves vertically at the water surface and has the shape of a wedge. In deeper water, a double flapper is often used. A double flapper wavemaker consists of two pivoted flappers, an actuation system driven hydraulically and a control system. For a flapper type wavemaker, the backside (outside the basin) may be wet or dry. Both have advantages and disadvantages, which are taken into account in the design of such a system. The dry-back system appears to be more popular.

The wave basin sometimes has a false bottom, which is adjusted to obtain the scaled water depth. In this way a facility may be made suitable for both deep and shallow water testing. Several facilities also have a deeper pit near their middle suitable for slender deep-water structures. A list of representative larger facilities of the world that perform commercially on a contract basis is given in table 13.6. Some overall particulars of these facilities including gross dimensions and capabilities are included in the table. These should be useful to a user for initial screenings. The website addresses of these facilities are also included, which should be consulted for further details about the capabilities of the facilities. They may help a user to obtain additional information to check against their needs for a suitable match.

The earlier wave tanks built prior to 1980 only produce waves that travel in one direction. These are suitable for reproducing long-period ocean waves that are unidirectional. Wind-generated multi-directional waves require facilities that can generate multidirectional waves. These facilities have widths comparable to their lengths. Many modern facilities have this capability. These facilities are identified in table 13.6.

Table 13.6 Selected database on available test basins suitable for deepwater testing

No.	Facility	Size	Depth	Centre hole	Type waves	Periods	Wave height	Tow speed	Current	Wind
		(m)	(m)	(m)		(s)	(m)	(m/s)	(m/s)	(m/s)
1.	Bassin d'Essais des Carenes, France *www.iahr.org/hydralab*	545 × 15	7	none	long	0.3–5.0	1.0	12		
2.	CEHIPAR, Madrid, Spain *www.cehipar.es/English/*	152 × 30	5	5	long and short	1.7 m–15 m length	0.9	5.0		
3.	Danish Hydraulic Institute, Denmark *www.dhi.dk*	30 × 20	3	12	long and short	0.5–4.0			none	fans
4.	Danish Maritime Institute, Lyngby, Denmark *www.danmar.dk*	240 × 12	5.5	none		0.5–7.0				fans
5.	DTMB (MASK), CDNSWC, MD *www.dt.navy.mil*	79.3 × 73.2	6.1	none	long and short	0.5–3.0			none	fans
6.	DTMB, CDNSWC, MD (Deep Basin) *www.dt.navy.mil*	846 × 15.5	6.7	none	long	1.0–3.0				
7.	IMD, NRC, Newfoundland *www.nrc.ca/imd/*	200 × 12	7	none	long and short	0.5–10				fans
8.	KRISO, Korea *www.kriso.re.kr*	56 × 30	4.5	12	long and short	0.5–5.0	0.8	–	0.5	20.0
9.	MARIN, The Netherlands (Seakeeping and Manoeuvering) *www.marin.nl*	170 × 40	5		long and short		0.45	6.0		

(Continued)

Table 13.6 Continued

No.	Facility	Size	Depth	Centre hole	Type waves	Periods	Wave height	Tow speed	Current	Wind
		(m)	(m)	(m)		(s)	(m)	(m/s)	(m/s)	(m/s)
10.	MARIN, The Netherlands (Offshore) *www.marin.nl*	45 × 36	10.5	20	long and short	0.3–3.0	0.4	3.2	0.1–0.4	
11.	MARINTEK, Norway *www.marintek.sintef.no*	80 × 50	10	none	long	0.8–10	0.9	5	0.2	fans
12.	Shanghai JT Univ., China *www.sjtu.edu.cn*	50 × 30	6	–	long	0.5–3.5	0.5	1.0	0.2	10.0
13.	OMB, Escondido, CA *www.modelbasin.com*	90 × 14.6	4.6	9	long	0.7–4.0				fans
14.	COPPE, Rio, Brazil *www.laboceano.coppe.ufrj.br*	50×30	10		long and short		1.0			
15.	OTRC, Texas A&M *otrc.tamu.edu*	47.5 × 30.5	5.8	16.7	long and short	0.5–4.0	0.9	0.6	0.6	12.0

13.3.2 Generation of Waves, Wind and Current

In model testing, the environment experienced by the structure should be properly simulated in the laboratory. Two of the major environmental parameters required in offshore testing are waves and wind. The following capabilities are generally requested in model tests:

- Regular unidirectional wave
- Random unidirectional wave (including white noise)
- Wave group
- Multi-directional wave
- Wind generation
- Current generation

The following sections will describe the generation of these environments in a testing basin.

The important frequency band applicable to offshore structures lies in the range of 5–25 s. The maximum energy of the ocean waves of design importance falls in the area of 10–16 s depending on the severity of the storm. Therefore, the model basin should have the capability of generating these waves with maximum heights (based on the prevalence of wind in the area) at a suitable scale. The magnitudes of these wave parameters are very important in the selection of a suitable scale for the model test.

Generation of high frequency wave components at a small scale is difficult in a wave tank. For example, at a scale of 1 : 200, a 0.5 s model wave represents a 7.0 s prototype wave. The wave generators seldom have quality wave generation capability much below 1 Hz. On the other hand, ocean waves at 5–7 s may have significant effect on the dynamic response of floating structures.

13.4 Modelling of Environment

The model testing facility should have the capability of simulating the wave, wind and currents commonly found at the offshore sites. The generation of model waves is essential for offshore structure testing. Many deep-water structures experience large current, which may also be an important environment, needed in a model scale. Often, towing of the entire structure and its components with the help of a carriage is used to simulate the uniform current speed on the structure. Wind is simulated by various means as well.

13.4.1 Modelling of Waves

Modelling of regular waves is straightforward. The regular waves are given in terms of a wave height and a wave period. These quantities are appropriately reduced to the model scale by the selected scale factor. The waves of the given height are generated by the harmonic oscillation of the wavemaker at the required amplitude. For an acceptable regular wave, the waves should be of near permanent form and the height of the waves from one cycle to the next within the test duration should have minimum prescribed fluctuations.

Random waves are generated in the model basin to simulate one of the many energy spectrum models proposed to represent sea waves (see Chapter 3). For the generation of random waves, a digital input signal is computed from the target spectrum taking into account the transfer function for the wavemaker. The transfer function generally accounts for the relationship between the mechanical displacement of the wavemaker to the water displacement, and the hydraulic servo control system.

13.4.2 Unidirectional Random Waves

Two of the most common methods of wave generation [Chakrabarti, 1994] in the basin are the random phase method and the random coefficient method. The former is spectrally deterministic while the latter is non-deterministic. The former is straightforward and approaches non-deterministic form for a large number of wave components. This is the most common method of wave generation in a basin and is described here.

The random sea surface is simulated with the summation of a finite number of Fourier components as a function of time. Thus, the generated surface profile $\eta(t)$ having the energy density of a specified (or chosen) spectral model has the form:

$$\eta(t) = \sum_{n=1}^{N} a_n \cos(2\pi f_n t + \varepsilon_n) \tag{13.11}$$

There are three quantities on the right hand-side that should be calculated from the specified spectral model. The quantities a_n, f_n and ε_n are the amplitude, frequency and phase of the wave components and are obtained as discussed in Chapter 3 (Section 3.6.3).

The number of wave components ($N=200$ minimum, preferably, 1000) is chosen. The spectral model is subdivided into N equal frequency increments as shown in fig. 3.17 having width Δf over the range of frequencies between the lower and upper ends of the frequency spectrum, f_1 and f_2 (cut-off frequencies based on the basin limitations). For each of these frequency bands, the Fourier amplitude $a(n\Delta f)$ is obtained from the spectrum density value $S(n\Delta f)$ as

$$a_n = a(n\Delta f) = \sqrt{2S(n\Delta f)\Delta f}, \quad n = 1, 2, \ldots N \tag{13.12}$$

The frequency f_n is chosen as the centre frequency of the nth bandwidth in the spectral model. The corresponding phase ε_n is created from a random number generator with a uniform probability distribution between $-\pi$ and $+\pi$. The quantity f_n is, sometimes, chosen arbitrarily within the nth bandwidth to provide further randomness.

13.4.3 Multi-directional Random Waves

In the case of a directional sea, the directional spectrum is obtained as

$$S(f, \theta) = S(f)D(f, \theta) \tag{13.13}$$

where the spectral energy density $S(f)$ is the same as used for the unidirectional sea. The directional spreading function usually has the same form for all the frequencies in the spectrum. The directional spectrum has been discussed in Chapter 3 (Section 3.6.4).

For directional seas, the simulation of the surface profile is shown in Chapter 3 (Section 3.6.5). It is important to calibrate the wave a priori so that the appropriate waveform, which satisfactorily matches the spectral shape, may be duplicated.

Waves in the basin should be calibrated without the presence of the model. The random waves should be generated from the digital time history signal computed from the desired spectral model and the generated spectra should be matched with the theoretical (e.g. P–M or JONSWAP) spectral models. Once an acceptable match is found (e.g. fig. 3.19 or 3.21), the setting will be saved for later use. This will assure repeatability of the wave spectrum from one test run to the next. This repeatability in the wave generation is very important for the success of the test programme and should be satisfactorily demonstrated by the model basin. Sufficient duration for the random waves and white noise will be given during the test runs so that reliable estimates of the spectra and transfer functions as well as the short-term statistics may be made. Random sea state records should allow test duration of 30 min (prototype) for test data. For slow drift tests, a duration of 120 minutes (prototype) is recommended.

The acceptable tolerances for the wave parameters are as follows:

Regular Waves

- Average wave height H, of a wave train consisting of at least 10 cycles: Tolerances: $\pm 5\%$
- Average zero-up-crossing wave period T: Tolerances: ± 0.2 s full scale.

Irregular Waves

- Significant wave height H_s: Tolerances: $\pm 5\%$
- Spectral peak period T_p: Tolerances: ± 0.5 s full scale
- Significant part of the measured spectrum shape: maximum offset $\pm 10\%$

13.4.4 White Noise Seas

White noise spectra denote wave spectra with near uniform energy over the full range of wave frequency of interest. Sufficient energy content is required from say 5–20 s prototype to obtain significant motion response of the structure. The white noise is difficult to experience in a physical system at sea. However, the advantage of white noise spectrum is that it allows to spectrally analyse the response signal and develop response transfer functions, phases and coherence over the given wave period range in one single run. The method of data reduction is straightforward with the help of cross-spectral technique [Bendat and Piersol, 1980].

The generation of white noise with significant amount of energy over a wide band of frequencies is a difficult task at any model basin. The energy level, particularly at the two ends, tapers down and hence, it is, sometimes, referred to as pink noise. The overall energy level is necessarily low and the response will be expected to be small as well. What may make the data reduction difficult is that most floating structures will respond to the white noise with a slow drift oscillation of significant amount. This will require possible digital filtering of data at the low frequencies. The cross-spectral analysis does not need filtering and the reliable range of transfer function is determined from the high value of coherence. The areas of low coherence are eliminated from the RAO.

Generally, this method provides reasonable accuracy and has the advantage of obtaining the transfer function from one single test run. However, it is recommended that at least limited number of regular waves may be tested to verify these values of transfer function from the white (pink) noise run as well as any anomaly observed during the random wave tests.

13.4.5 Wave Grouping

The motions of floating structures are found to be sensitive to wave groups. The wave group is defined by the envelope wave of the square of the wave elevation. The slow drift oscillation, which depends on the difference frequency, is shown to differ significantly by the groupiness present in the irregular wave. In other words, two spectral realisations, having same energy contents, but different group spectra, will yield two different low-frequency responses. Therefore, it is important to model the groupiness function in the generated wave in addition to the spectral shape. The groupiness function is computed from the squared integral of the spectral density and represents the relationship among the difference frequencies within the energy density spectrum.

This function should be computed for both the spectral model and the simulated wave in the basin during its calibration. The two should be matched as closely as possible to insure proper grouping of waves for the slow drift oscillation tests. An example of the comparison of the groupiness function for a JONSWAP wave is shown in fig. 13.4.

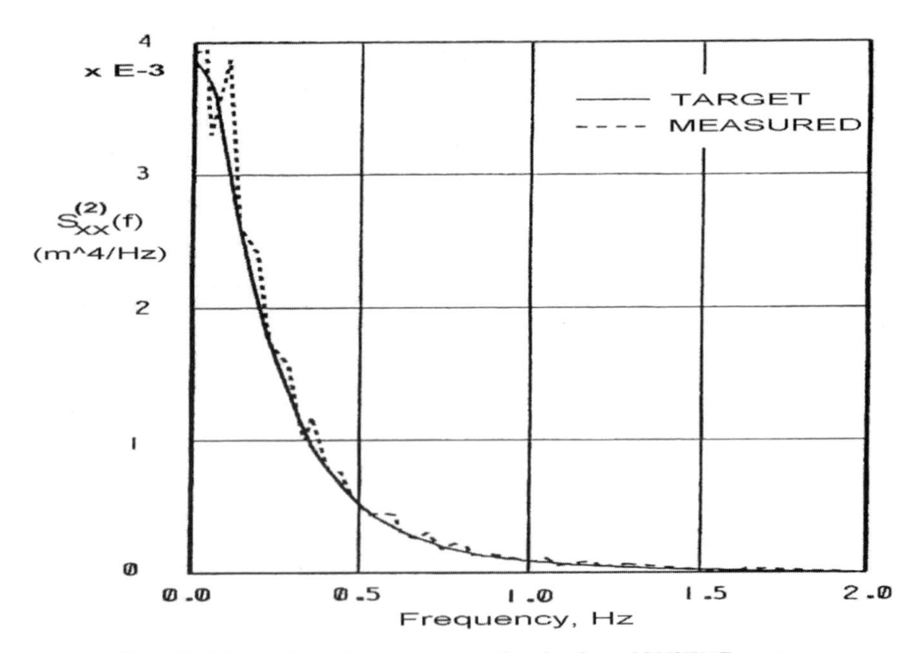

Figure 13.4 Comparison of wave groupiness function for a JONSWAP spectrum

13.4.6 Modelling of Wind

The wind loads on the structure may be particularly important in the design of such structures as a floating moored structure. However, it should be emphasised that these loads in the model system are limited by the associated scaling problems. Wind loads are functions of Reynolds number and *Re* is (an order of magnitude) smaller in the model compared to the prototype *Re*. Therefore, it is possible that the prototype wind effect falls in the turbulent region while the corresponding model wind effect is in a laminar region. In this case, the model test results may be considered conservative. This is why oftentimes, it is the properly scaled mean wind load that is simulated rather than the wind speed. In other words, wind speed in the model is adjusted so that the mean model wind load is matched.

Wind may be generated with blowers positioned strategically in front facing the model. In this case, the model superstructure must be accurately modelled. While wind velocity is often taken as a steady value, the wind spectrum may be important in some applications. The frequency range of a wind spectrum is quite broad-banded, often covering a range from 0.005 to 1 Hz (see Chapter 3 for the description of wind spectrum model).

While a bank of fans is most commonly used in generating wind over a model superstructure, there are several other methods of simulating wind effect on offshore structures. The earliest method of simulating a steady wind load on an offshore structure is still in use. It is achieved with a weight hanging in the direction of wind from the model with the help of pulleys and strings. A fan mounted on the deck of a floating model has also been applied to simulate steady and oscillating wind load. The pros and cons of these three wind simulation methods in a model scale are compared in table 13.7. Each method has its place in model testing.

To demonstrate the capabilities of the model mounted fan, we present an example of a wind condition that was considered in a model test of a floating platform. Table 13.8 summarises the parameters that were used in wind spectral simulation.

There are two separate data files needed for the test. The first file is the time history of the signal that is used to control the pitch angle of the generator blades. The second one is a time history of the load computed from the wind drag formula [equation (4.13)]. This load is considered appropriate for the generation of wind load on the model. The measured load during the test is compared with the computed one. Adjustments are made in the control signal to obtain a satisfactory match. For the above example, the desired and computed wind load spectra are shown in fig. 13.5. The comparison between the two in the load spectra is considered acceptable in this case.

13.4.7 Modelling of Current

In many deepwater locations considerable current prevails. In the design of structures in these locations, the effect of current may be extremely important. For such structures, simulation of current in the wave basin is essential. While current may be simulated well with towing the model (already discussed), this method may not be adequate for many offshore components.

There are several options in generating local current in the general area of the model in the test basin [Chakrabarti, 1994]. The current generator should, in general, have the

Table 13.7 Comparison of wind generation methods

Different Simulation Methods	*Pros*	*Cons*
Weights hung from the model superstructure	• Simple to implement • No detailed superstructure model needed	• Simulates only the mean wind speed • Can add some inertia to the model
Fixed bank of fan	• Generates the steady load as well as the spectrum	• Difficult to generate accurately over a large area • Need a precise model of the superstructure
Fan mounted on the model	• Generates the steady load as well as the spectrum • More accurate generation of the wind possible • No modelling of the superstructure • Easy to change wind heading	• Fan becomes part of the model and its inertia should be included as part of model • Causes change of wind heading with model yaw • Large pitch angle changes the scaled wind load

Table 13.8 Wind parameters used in the example

Variable (see Chapter 3)	Wind condition
Reference elevation (m)	10
Mean velocity (m/s)	29
Area (m^2)	3884
Air density (kg/m^3)	1.21
Peak frequency coeff.	0.025
RMS velocity	0.15*V (1h, z_R)
Starting frequency (Hz)	0
Ending frequency (Hz)	0.1
Frequency components	4096
Sampling period (ms)	54

following capabilities:

- The hardware to generate current is reasonably transparent and has minimum influence on the waves generated simultaneously in the basin.
- The current in the region of the model is reasonably steady.

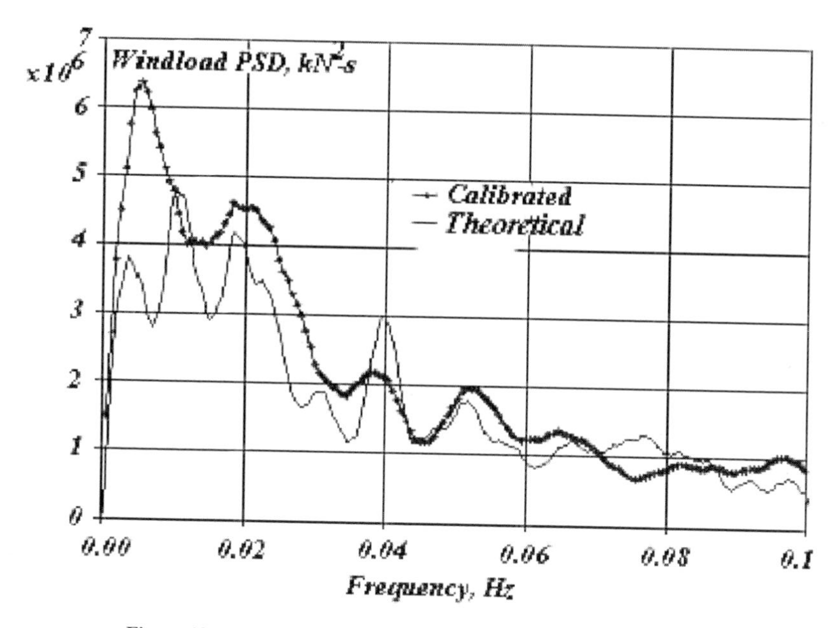

Figure 13.5 Comparison of target and measured wind load spectra

- In any physical generation of current in a basin, some turbulence is present. Small turbulence may be acceptable, as it will better simulate the prototype situation and minimise the effect of distortion in the model Reynolds number.
- The current profile is extended to a desired depth over the width of the model region.
- Some vertical shear in the current profile is possible by selectively throttling the flow.

The modelling of current in a laboratory test with or without waves is an important consideration in offshore structure modelling. This capability in a wave basin allows studying the wave–current interaction on the model. In current modelling in a facility, the uniformity and distribution of current should be carefully investigated. The generation of current is simplified if a closed loop is placed in the facility. It is one of the most desirable methods and is achieved by pumping water into and out of the two ends of the tank by a piping system. If a false bottom exists in the facility, underwater pumps can circulate the water in a loop above and below the false floor. Counter-current is generated by reversing the flow.

If an installed current generation is not available, local currents are often generated by placing portable current generators in the basin. These may take the form of series of hoses with outside water source or portable electric outboard motors. Uniformity of flow is achieved by proper control of the velocity.

For a local current generation, a manifold may be created over the area covering the width and depth of the model in the basin. The manifold may consist of small-diameter PVC pipes of adequate size and number through which flow can be generated. The manifold is

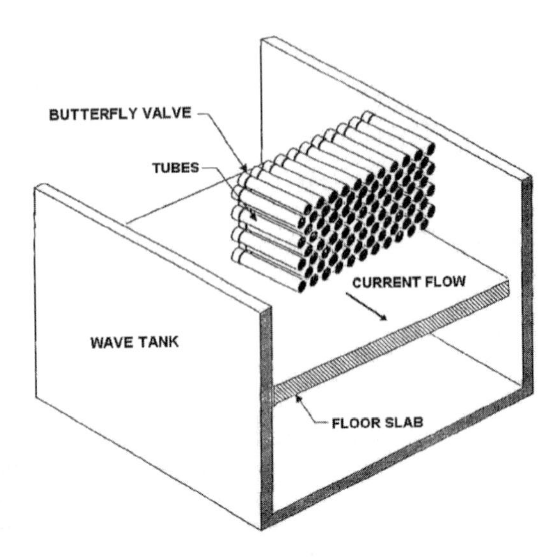

Figure 13.6 Technique for generation of shear current in test basin

supported on a structure and hung from the carriage above and placed, say, about 3 m ahead of the model. The flow is created and controlled by a controllable pump. The water is circulated from an intake pipe from the wave basin. Flow straighteners, such as a tube bundle, may be accommodated in the basin to stabilise the flow as long as they do not interfere with the waves. Individual controls are provided in manifold at each elevation with valves so that the flow through them may be individually controlled. It is possible to generate some vertical shear in the current profile by selectively throttling the flow. This is illustrated in fig. 13.6 with the butterfly valves on flow strengtheners on a false bottom of a testing basin.

If current is inadequate or unavailable, it is sometimes simulated by attaching the model to a towing carriage and towing the model at steady speeds down the tank with or without waves. While towing does not duplicate the current effect exactly, it is generally considered acceptable for steady currents.

A thorough calibration of the current generation should be performed before the model is placed in the basin. The uniformity and distribution with depth of the current profile is established at the test site by a series of current probes placed vertically. The temporal variation of current should be limited to 10% or less for a steady current test. The current velocity required for the test will depend on the scale factor and is generally scaled with Froude scaling. If current is an important consideration in the testing, the scale factor should be chosen such that the available current can simulate the desired environment.

13.5 Model Calibration

While the calibration of the environment is being carried out in the basin, the following calibration procedures may be simultaneously undertaken with the model itself for a floating structure. The completed model is weighed without ballast. The centre of gravity as well as the natural period in pitch/roll of the model in air are determined. Calculations are performed to determine the amount and location of the ballast to achieve the necessary properties of the model. These ballast weights are placed in the model and the location of CG and the natural period of the model in air is verified on the calibration table.

For a non-rigid model, the actual stiffness of the model should be carefully determined and compared to the computed stiffness. For verification of numerical modelling software, it may not be necessary to match the computed stiffness very closely as long as the model stiffness is established well.

13.5.1 Measurement of Mass Properties

The mass properties of the structure model are measured using a specially built calibration table (fig. 13.7). The table is designed to accommodate the largest structure expected to be tested in the basin. We describe here one such table being used at the offshore model basin (OMB). The tilt table has a large bed to hold the structure. The table is set on a pair of knife-edge fulcrums at its central axis such that it is free to swing in a vertical plane. The position of the fulcrum is adjustable in the vertical direction. The use of counter weights allows the tilting platform to be balanced at any fulcrum adjustment. At each fulcrum adjustment, counter weights are moved to align the CG of the table with the tilt axis. The table is attached to coil springs at its edges (see fig. 13.7) with known spring constant.

The weight, centres of gravity, and pitch and roll radii of gyration of the model structure are measured with the help of the calibration table. Before the model is placed on the tilting

Figure 13.7 Setup of the model on calibration table (Courtesy of Offshore Model Basin)

table, the proper model displacement is achieved by ballasting the model to the desired draft with specified weights. Next, the table height is adjusted such that the desired KG of the model is measured between the knife edge fulcrum and the top of the table. Then, the ballasted model is centred on the platform table. Ballast is arranged vertically to arrive at the model CG on the tilt axis of the table so that the model and table have the same KG. The radius of gyration (k_{yy}) of a mass (m) is defined by

$$k_{yy} = \sqrt{I/m} \tag{13.14}$$

where I is the moment of inertia of the mass about an axis of interest. The moment of inertia of the table is defined as

$$I_t = \frac{T_n^2(t)K_r}{4\pi^2} \tag{13.15}$$

where K_r is the rotational spring constant and $T_n(t)$ is the period of oscillation of table for small angles.

The pitch radius of gyration is set using the tilting table restrained by the spring system. The natural period of oscillation of the tilt table alone is measured first about the tilt axis. The moment of inertia of the table without the model is then obtained using the relation in equation (13.15). The moment of inertia of the platform with the model is measured by observing the natural period of oscillation of the system with the model. The moment of inertia, and consequently the radius of gyration of the model structure are computed by subtracting the moment of inertia of the table from the combined inertia of the table – structure system and then using equation (13.14). The inertia of the model alone is then defined as

$$I(m) = I(t+m) - I(t) \tag{13.16}$$

where the local variables m and t stand for model and table respectively.

For a long model, a compound pendulum called a bipolar system, may be used to define the roll radius of gyration of the model. The compound pendulum is made out of two single pendulums, one supporting the bow and the other supporting the stern of the model. The moment of inertia of the model about the pin axis is measured by observing the natural period of oscillation. The moment of inertia of the model about the pin axis is defined as:

$$I_{pin} = \frac{T_n^2 mgL}{4\pi^2} \tag{13.17}$$

where g is the gravitation constant, and L is the distance from the pin to the CG. The moment of inertia of the model about its centre of gravity is obtained from the parallel axis theorem defined as

$$I_{CG} = I_{pin} - mL^2 \tag{13.18}$$

The roll radius of gyration (k_{xx}) is obtained from the relation:

$$k_{xx} = \sqrt{I_{CG}/m} = \sqrt{\frac{T_n^2 gL}{4\pi^2} - L^2} \tag{13.19}$$

Similar expressions may be obtained for the pitch direction.

Table 13.9 Model ballasting in pitch and roll on tilt table

m (model) = 1446 lb = 44.99 slugs
KG (model) = 1.70 ft
d = centre of table to centre of placed weight = 2.83 ft
x = centre of table to deflection measurement point = 3.25 ft

(a) Calibration of spring					
Load (lb)	Reading (in.)	Defl. (ft)	Mom. = Load*d	θ = Defl./x	Mom./θ
0	0.89				K_r
20	1.05	0.0133	56.67	0.0041	13821.95
40	1.21	0.0133	56.67	0.0041	13821.95
60	1.38	0.0133	56.67	0.0041	13821.95
80	1.54	0.0133	56.67	0.0041	13821.95
100	1.70	0.0133	56.67	0.0041	13821.95

(b) Calibration of model				
Item	Units	Table	Roll	Pitch
Measured natural period*	s	0.825	1.19	1.303
Computed moment of Inertia	Slug.ft^2	233.8	250.5	350.2
Desired moment of Inertia	Slug.ft^2		221.7	353.2
Measured radius of gyration	ft		2.36	2.79
Desired radius of gyration	ft		2.22	2.802

* Average over 10 cycles; measured roll and pitch periods include the table

An illustrative example of a model calibration on a table is given here showing the details of the calibration of the table and model inertia. The properties of the model and the table springs are given on the top section (a) of the table. The pitch and roll properties (as found from the calibration table) are included in the bottom portion (b) of table 13.9.

After the dry properties are known and verified for accuracy, the model is placed in the water and ballasted to the proper draft with ballast weights. The static tests are carried out by adjusting the location of the ballast adjusted to achieve the scaled GM values of the model. The positions of the weights are chosen such that the moment of inertia of the model is relatively unchanged. The natural period in heave, pitch and roll of the model are determined by displacing the model from its equilibrium position and recording its motion with the help of a rotational transducer or an accelerometer.

The mooring lines may be calibrated by choosing a short section of each type of material making up the line and determining its elastic properties by a tension test. Care should be taken in choosing the springs for the non-linear mooring system. These springs are calibrated to establish the scaled stiffness of all the individual mooring lines.

13.6 Field and Laboratory Instrumentation

In model testing, the simulated environment and the model response to that environment are measured. Usually, the test environment is intended to scale a specified ocean environment. In order to verify that the sea state has been properly modelled in the laboratory test, measurements are made with the wave height gauge (e.g. resistance or capacitance wave probe) and current meters. These instruments are commercially available. The instruments are placed near the model to measure the wave profiles experienced by the model. One probe is often placed across from the model in line with its centre to determine the phase relationship between the model response and the corresponding environment.

The instruments that are necessary for the successful measurement of the model environment and the responses of the model are described here.

13.6.1 Type of Measurements

Structure responses of interest might include environmental loads on a fixed structure, motions of a floating or moored structure and stresses on individual members or components of a structure. The interaction effect of waves with a structure may also be of importance in a design. For example, wave reflection or the run-up of waves on the face of a structure can be an important consideration in the design of an offshore platform. The instruments in these measurements are often specially designed to meet the requirements of the model. For example, load cells are designed to fit between the model and its mounting system in the wave tank in the range of expected loads. Strain gauges are mounted directly on the model surface to measure stresses.

The standard instruments that are required during a typical fixed/floating structure test and their applications are listed in table 13.10. Typically acceptable in-place accuracy of these instruments on the model is noted in column 3.

13.6.2 Calibration of Instruments

Transducers receive a physical input from the model such as displacement, acceleration, force, etc. subjected to a model environment and produces an equivalent electrical output. The transducer is designed so that this transformation from the measured response to volts is in the linear range for the level of response expected. This allows a single-scale factor for conversion of the output to the required engineering unit. A few common means of measuring an input signal include a bonded strain gauge, a linear variable differential transformer (LVDT) and a capacitance gauge. These components are placed in a transducer stock, which is designed to measure an expected response in a model test.

For example, the strain gauge is glued strategically on a tension/compression member of a load cell designed for the desired load range. The load cell is attached between the model and the mounting system. As the model is subjected to waves, the load imposed by the wave on the model is recorded by the load cell. Before this placement, these instruments are

Table 13.10 Typical instruments for offshore model testing

Instrument	Application	Accuracy	Comments
Wave probes	Measures incident and phase waves	1/16 in.	Mounted from the air above the water surface
Air gap probes	Measures air gap between the free surface and the deck of the model	1/16 in.	Splashing of water introduces inaccuracy in measurement
Non-contact optical tracking system	Six degrees of freedom (DOF) motion of floating structures	1/8 in. and 1/2 deg. after application of software	Cameras are mounted on the side wall or the carriage
Motion sensing transducer (MST)	Light specially built mechanical system having linear and angular potentiometers to measure six DOF	1/8 in. linear and 1/2 deg. angular	Inertia and damping effect of the mechanical system on the structure should be known
Load cells	Measures loads between the model parts where attached, e.g. towing loads, and wave loads on member	0.1 lb or less depending on load range; 5% cross talk	Attaches between model and fixtures or two members of a model
Ring gauges	Measures line tensions of the mooring lines at the fairlead	0.1 lb	Mounts on model at the mooring line fairlead
Strain gauges	Measures the stresses on the mounting point on the model	0.01 μ	May be mounted directly on elastic members
Accelerometers	Measures (XYZ) accelerations at the point of attachment, e.g. CG of the model	0.01g	Mounted on the model where acceleration/displacement is desired
Towing speed indicator	Records towing speed of the towing carriage	0.1 ft/s	Part of the towing carriage
Towing dynamometer	Two component (XY) load cell capable of measuring the towing load on the structure	0.1 lb	A hinge provided between the staff and dynamometer to allow freedom in pitch

placed on a specially designed calibration stand and calibrated over the range of expected values. For example, the load cell is fixed on the calibration stand and known standard weights are hung in the direction of measurement from the load cell in increments and the associated voltages are recorded. In case of a capacitance wave probe, the calibration is achieved by placing it submerged at the water surface and moving it up and down in water. The linearity of the instrument is verified and a scale factor in terms of the response unit per volt is generated. This factor is used to multiply the output voltage during the testing.

Each instrument should be checked for waterproofing and calibration prior to setting up the test in the wave tank. The wave probes are calibrated by lowering and raising the probe in still water over the range of water level covering the height of the generated waves for the test programme. The ring load cells used in the mooring lines are calibrated in tension by hanging the load cells vertically and using standard weights over the range of mooring line loads expected. For sectional loads on a structure component, the XYZ load cells are calibrated in each direction on a calibration stand. Cross talks between two orthogonal axes (i.e. reading on one due to loading on another) are recorded. If the cross talk is high, the instrument should be rejected or re-assembled. In each case, the linearity of the gauges is assured by least square technique and checking the correlation coefficient and standard deviation.

The MST (six degrees of freedom motion sensing transducer) is calibrated in each direction, and a calibration curve is developed for each transducer. For pitch and roll angles, the angular potentiometer is turned in steps. For heave, the model mounting plate in MST is raised in steps. For surge and sway, the MST mounting plate is moved forward or sideways in steps. For yaw, the mounting plate is rotated about its vertical axis in steps. Additional calibration checks are performed to demonstrate that the calibrations, polarities and uncoupling software result in measured data, which corresponds to actual displacements by displacing the MST in several directions at the same time.

If a non-contacting position tracking system is used, then a complete dry calibration is required for the system on a calibration stand before mounting it on the basin. In-place calibration should also be performed to verify the set-up and the software used for the data reduction for the camera system.

Instruments are electrically connected to an automatic data acquisition system (DAS) so that the transducer signal may be automatically recorded. A simple schematic of a data acquisition system is shown in fig. 13.8. The typical transducer signal is such that its output is given in microvolts. It is first amplified by a gain factor to yield a voltage in the limit

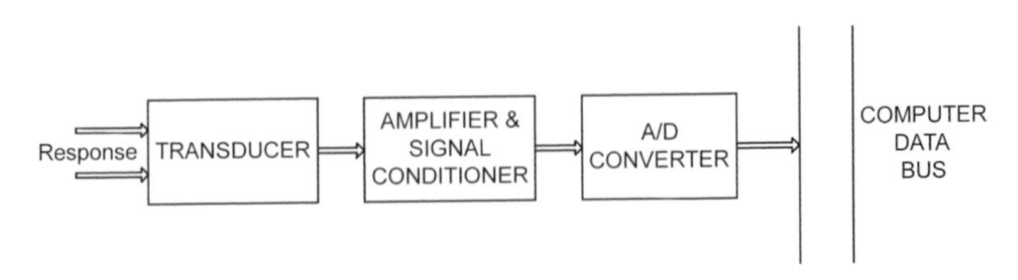

Figure 13.8 Schematic of data acquisition system

of 0–5 or 0–10 V. The signal is conditioned for recording, which may include analogue filtering of noise and other unwanted signals and then converted from the analogue to digital form through an A/D converter. Unlike analogue signal, digital signals are non-continuous and stored into a computer memory at the specified sampling rate. Today these operations are accomplished efficiently on a desktop personal computer.

Instruments for the measurement of responses at a small scale may pose a problem due to its size compared to the model. What creates the inaccuracy in the system is the introduction of superfluous physical phenomenon not present in a larger scale model or prototype, for example, effect of the instrumentation cables, and physical size of instruments. However, many small precise and reliable instruments are available today. The measurement accuracy or instrument sensitivity at a small scale, say 1 : 100, is not a serious problem. The generally accepted overall in-place measurement accuracy is about 5%. At a much smaller scale, this accuracy may drop down to as much as 20%. For a small-scale testing (smaller than 1 : 100), this measurement error must be recognized and considered in the correlation and extrapolation of data.

Regular checks of the instrumentation should be performed during testing to confirm that the instrumentation has not undergone any significant changes during the test programme. Checks should be performed each day prior to commencement of data acquisition and whenever the test set-up is changed. Typically, this will include cleaning of the wave probes, re-adjustment of load cell zeroes to correct for drift errors, and simple static tests.

13.7 Pre-Tests with Model

Before the test set-up begins in the wave tank, it should be assured that the properties of the model and associated parts are properly modelled. The following tests, at a minimum, should be performed on a floating structure model.

13.7.1 Static Draft, Trim and Heel

Purpose of Test: Record draft, trim and heel.

Test Procedure: The floating model is placed in water and the draft, trim and heel are recorded and compared with the scaled values. If there is a discrepancy on the draft, then it is rectified before proceeding.

13.7.2 Inclining Test

Purpose of Test: To determine the metacentric height (GM) of the model.

Test Procedure: Weights in increments are set at accurately measured distances from the floating model centreline, and the inclinations measured. From these measurements, the metacentric height is evaluated, and compared with that calculated for the model with the specified KG, corrected for the inclining mass. Inclining tests are performed in the transverse and longitudinal directions at increments of the heel and trim angles and the righting moments are determined and verified. Any adjustments are made in the model properties to match the calculated value within 5%.

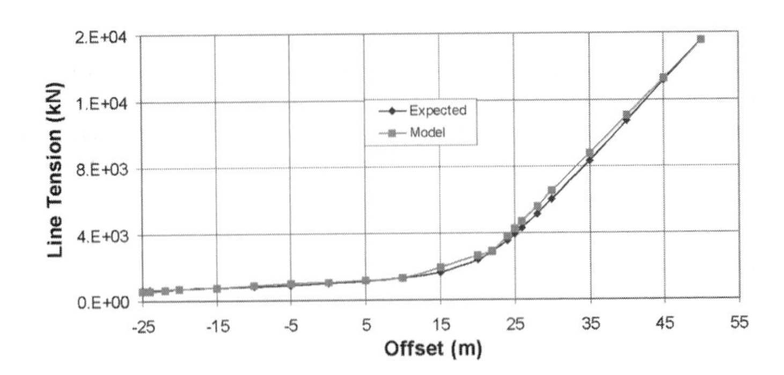

Figure 13.9 Mooring line offset test in model

13.7.3 Mooring Stiffness Test

Purpose of Test: The aim is to measure the restoring force characteristic of the moored model, and to demonstrate that this characteristic is representative of the full-scale mooring system.

Test Procedure: With the model moored at the specified pre-tensions, the draft is measured, and compared with the expected value. A set-up should be provided (e.g. with a line and pulley system) for applying a known steady horizontal force to the model above the water line. Forces in equal increments should be applied, and the resulting offsets, vessel trim and mooring line tensions are to be measured. The offsets are to include horizontal and vertical components and the inclination. The force should be applied in two directions, the first along the longitudinal direction, and the second in the transverse direction. The expected load range in the positive and negative direction should be covered. An example of the measured restoring force of mooring lines compared to the computed model line forces is shown in fig. 13.9. The data represents scaled-up prototype values. The offset shown is the expected range of offset during the model test.

Regarding the measurement system, particularly the motion measurement system, the carriage with the instrumentation system is positioned after the pretension displacement of the model has taken place. This will allow the measuring system to stay within the limits of motion of the model from the wave and slow drift oscillations.

13.7.4 Free Oscillation Test

Purpose of Test: To determine the natural periods and damping coefficients of the moored model in free oscillatory modes in six DOF including surge, sway, heave, roll, yaw and pitch.

Test Procedure: The model is located reasonably well away from the edges of the tank, to avoid reflections of radiated waves. The model is given an initial displacement one at a time in the selected mode of motion and is released. Time histories of the resulting motions in all six DOF are recorded by MST or motion sensors. The tests are conducted in calm sea conditions. Care should be taken to achieve a nearly pure single degree of oscillation. If the oscillation in any other direction is significant, indicating coupling effect, then the test should be repeated. This motion time history will provide the natural period of oscillation and damping of the system in the degree of oscillation in question. Note that it is not necessary to measure the displacement of the model in particular. An adequately measurable response in the oscillation mode from any model-mounted instrument (such as wave gauge, accelerometer, etc.) will provide the desired results. The idea is to record a decaying oscillation from the instrument mounted on the model as the model moves.

13.7.5 Towing Resistance Test

Purpose of Test: To evaluate the towing resistance/current drag of the complete vessel.

Test Procedure: The towing carriage tows the model from one end of the basin to the other. The steady part of the towing speed is used to record the test run. Towing may be performed in waves, as well. Note that the encounter frequency of wave by the model will be different from the generated frequency by the Doppler shift (see Chapter 3), the magnitude of which will depend on the towing speed. For towing in an irregular wave, multiple test runs may be necessary so that the total run length is sufficient for the RAO and other statistical calculations. In this case, the subsequent waves should start where the last one was left off. During towing tests, the quantities measured are towing speed, towing load, and centre of resistance. Towing may also provide the values of drag coefficient for the model.

13.8 Moored Model Tests in Waves and Current

In this series of tests the floating model is moored in its permanent in-place condition. The tests are performed in wind, waves and current. Tests are carried out with the vessel and mooring intact. For each test, the environmental conditions are generated, and the behaviour of the model is recorded with the installed instruments. The following measurements are usually made during such tests:

- Wave elevation at several positions in the tank to measure wave profile and phasing
- Six DOF motion response of the floater about a fixed coordinate system
- Surge, sway and heave accelerations measured at the desired deck level
- Stresses at several locations on the model if it is flexible
- Tension in each mooring line
- Air gap at several model location under deck measured by capacitance probe or similar

13.8.1 Regular Wave Tests

Purpose of Tests: To establish transfer functions for all measured responses in regular monochromatic wave conditions. To observe any non-linearity in the response transfer function by varying wave height at a few selected wave periods. To define steady state drift force for each regular wave of given period and height.

Test Procedure: The model is subjected to a series of regular waves. The data sampling rate and test duration are chosen such that the steady state values of the responses may be obtained accurately. Typically, about 10 cycles of steady state data are recorded. The offset from zero value gives the steady drift force. The second-order drift force on a floating vessel due to regular waves is proportional to the square of the wave height.

13.8.2 White Noise Test

Purpose of Tests: To define the transfer functions for the model motion, and mooring loads over the expected range of wave periods.

Test Procedure: The model is subjected to a wide band spectrum having nearly equal spectral energy level. The data sampling rate and test duration are chosen such that the transfer function may be obtained spectrally, using a cross-spectral approach. Typically, about 10 min of model scale data will be required for reliable spectral results.

13.8.3 Irregular Wave Tests

Purpose of Tests: To establish the behaviour of the complete moored vessel in an irregular sea state with and without the influence of current and wind. Generally, the sea states experienced at the offshore field are simulated in these tests to study the operational and survival characteristics of the system.

Test Procedure: The irregular wave runs correspond to the random waves calibrated without the model in the basin. The model is moored with a specified mooring system and pre-tensioned the specified amount. In the absence of physical current or towing, the steady load due to current may be simulated with a line and a force transducer attached to the model. The transducer is pre-tensioned a specified amount representing the scaled steady load at the desired point of application. Alternately, the wind and current are physically generated in the basin where such a facility is available. The length of the test run should be sufficient such that a reliable spectrum may be estimated from the measured channels. Irregular wave tests are performed for a period of 120 min full scale to better define the spectral and second-order response characteristics. The transfer functions are computed for the responses from these test runs. They also allow the statistical analysis for the short-term extreme responses.

13.8.4 Second-Order Slow Drift Tests

Purpose of Tests: To establish the quadratic transfer functions for the second-order motions of the moored model.

Test Procedure: In addition to the steady drift force, a slowly oscillating drift force is generated on a moored floating structure due to an irregular wave. This drift force is excited around the long natural period of the system from the difference frequencies in the irregular wave components. Thus, a low-frequency response of the vessel is expected covering the frequency band around the natural frequency of the system. This response spectrum due to a random wave is related to the wave spectrum through an integral in terms of a quadratic transfer function.

It is often difficult to establish the values of this quadratic transfer function through the spectral approach from irregular waves. Therefore, an alternate technique is recommended to develop the quadratic transfer function for the slow drift motion. Since the slow drift motion appears as the difference in the frequencies in the irregular wave and has a bandwidth around the natural frequency of the system, this bandwidth can be established from the irregular wave runs. Then, frequency pairs can be chosen from the input wave spectra such that they produce a difference frequency in this band. This will give rise to a symmetric matrix based on the pairs of wave frequencies, which form a wave group.

Wave groups from the frequency pair of equal amplitude are generated in the basin with these frequency pairs and the responses are measured. The low-frequency components are filtered through fast Fourier transform (FFT) and the response amplitudes are derived. These response amplitudes are normalised with respect to the square of the wave group amplitudes to give the quadratic transfer function as a matrix.

The number of test runs will depend on the width of the response spectrum. This method is quite accurate, since it will directly measure the slow drift response for one difference frequency pair.

13.9 Distorted Model Testing

Distorted models are often used in offshore structure testing. The distortion appears in several areas, one of which is model scale. In shallow water coastal engineering, it is quite common to use two different scale factors – one for the vertical direction and one for the horizontal direction. Because of the limited water depth in the testing facility, the vertical components of a deepwater offshore structure, e.g. mooring lines and tendons are truncated in the model. This distortion should be carefully designed so that the goal of the model test is achieved and the information for the full scale may be derived from the test. This section describes the common distortions found in the model of an offshore structure and the usual remedies taken to correct the problems.

13.9.1 Density Effects

In a wave tank, almost invariably fresh water is used to represent the seawater found in a prototype application. This creates a small difference in the density, which is about 3%. This difference reflects a similar change in the measured forces, which need to be corrected.

All model weights should be corrected for the difference in water density between that at the test facility and sea water (1025 kg/m^3). This is achieved by the ratio of the two water densities.

13.9.2 Cable Modelling

In modelling very long cables in a laboratory facility, experimentally realistic diameters should be maintained. This is achieved by combining the proper choice of the elastic material, the role of drag coefficient in conjunction with buoyant devices, and increased kinematic viscosity of the test fluid. Elasticity of a cable/wire (tensile stiffness) is an important property that should be scaled with a suitable material at a small scale. This involves the Cauchy similarity as well as the Froude similarity. The Reynolds number based on cable diameter is involved indirectly with the drag coefficient, C_D.

The requirements for scaling a large cable structure in a laboratory are governed by its length, $L_p = \lambda L_m$ Once the length scale is chosen, the flow velocity is determined from the

Froude number. In addition, the density ratio is fixed, which determines the modulus of elasticity for the model cable, namely, $E_p = \lambda E_m$. Material, such as plasticised polyvinylchloride (PVC), can be used in the model cable to provide the required modulus of elasticity. The proper density for the material may be achieved by impregnation of powdered lead. The diameter of the model may be determined by making proper adjustment of the drag coefficient based on the Reynolds number.

13.9.3 Modelling of Mooring Lines, Risers and Tendons

The following are the properties for a mooring line or a steel catenary riser (SCR) listed in order of their importance. These should be modelled as accurately as possible.

- Vertical and horizontal pretension components
- Vertical stiffness
- Line mass and drag characteristics
- Horizontal stiffness over the range of anticipated offsets.

There are several alternates we can use to model mooring chains and catenary risers:

- Model the stiffness curve with multiple springs that is pre-tensioned – drag damping on the mooring line or SCR is not considered here.
- Use model chain that has the correct submerged weight – the geometry is complex and calculation of drag on the model chain is difficult.
- Use an outer flexible tube e.g. a thin-walled tygon tubing of a diameter representative of the model diameter and a weighted cable inside – this method provides the scaled drag effect, but is time-consuming in modelling and may have large bending stiffness.
- Use a plastic rod of suitable material of correct submerged weight per foot. Segment the rod in about 1–2 ft length connected by eye hooks. This will provide a uniform diameter for the lines and risers (except for the small area of the hooks). It is easier to build and still provide a reasonable estimate for the drag coefficient.

The most important property of a chain is its weight (per unit length). The material and size of the model chain can be chosen such that the weight can be achieved at a small scale. The elasticity of the material should be verified to ensure the order of magnitude. The geometry of the chain is difficult to scale, which introduces inaccuracy in simulating hydrodynamic damping of the chain generated from its own motion as well as from the wave and current action in the upper part of the ocean. Hydrodynamic damping of the mooring line has been shown to have a significant effect on damping for the low-frequency response of the floating structure.

Distortion due to truncation in length is provided by additional springs. Means of correcting for the damping effect from the truncated chain may be introduced in the model chain. This area of truncation is discussed further later.

13.9.3.1 Truncated Mooring Line Simulation

The (taut and catenary) mooring system needs careful attention since the dimensions, and especially, the depth of the tank, often do not allow a direct scaling of the geometry of a typical deepwater floating structure. This is particularly true today as the exploration and production of minerals are going into deeper water.

A truncated mooring line is a common occurrence in testing models of structures placed in deep water. The stiffness of the missing line segment is modelled by additional springs at the bottom of the truncated line. The truncation appears at the basin floor. The tension and initial angle of the mooring line are matched at the fairlead to the prototype design condition. However, the tension and the bottom angle at the line truncation point rarely duplicate the prototype situation. Moreover, the line angle changes as the floater moves in waves and even the fairlead angle cannot be maintained at the prototype values.

The difficulty of modelling and set-up of a floating moored system in a basin arises from the following considerations:

- The mooring stiffness is often non-linear
- The fairlead angle changes with time and loading for a given environment
- The initial line angle requires change as different environmental conditions and model orientations are simulated.

The mooring lines are usually modelled such that the correct non-linear stiffness behaviour is achieved at the fairlead connection points. A truncated mooring spread is often considered acceptable as long as the stiffness properties at the vessel are correctly represented. It is often important to model all mooring lines individually.

For the success of a small-scale testing, it is important that the mooring system simulation is kept simple and the mooring arrangement does not change with every environment. If the pre-tensioned line force is roughly linear with the line extension, the mooring line may be modelled with a set of linear springs arranged in a straight line. The spring set is attached to a cable to achieve the required length of the mooring line. When required, the mooring line model can be non-linear. The springs are chosen such that the stiffness may be easily adjusted to match the linear slope by adding or removing a set of springs. One end of the cable is attached to the fairlead at the model through a load cell. The other end of the cable is attached to an anchor plate at the bottom of the basin in order to maintain the initial fairlead angle for the particular environment. This bottom attachment point is sometimes brought to the carriage with a pulley system so that the initial angle can be adjusted from the surface.

The initial fairlead angle is adjusted in order to match the calculated values. It is understood that this angle will change with loading from the environment. Since there is a large pre-tension in most cases, the error in the angle with load will be small. The initial tensions at the fairleads, which are monitored with the help of the load cell located at the end of the mooring lines at the fairleads are adjusted and maintained for different wave headings.

The procedure during testing is as follows: the model is moored with the taut mooring system with the initial fairlead angles and the ballasted anchor plates set at pre-marked locations at the basin floor. For the test runs where steady loads are needed, the load is applied with the line and pulley arrangement. The model displaces in the aft direction under this load. The model is pulled back by the anchor lines to its initial position (marked on the carriage). This will maintain the initial position while properly pre-tensioning the mooring lines.

In order to simplify set-up changes, the anchor plates remain at the same locations between test runs with different drafts. Under this arrangement with draft, the initial fairlead angles will be different, as expected in the real case. The anchor plates will be re-located for different wave heading.

Another possible set-up used in several model basins is a Simple Mooring and Riser Truncation (SMART coined by the Offshore Model Basin). SMART consists of a combination of lines running from the model's fairlead to ring gauges downward at the elevation angle to a specified weight fastened on the line and then continuing up to a fixed point on a vertical pole. This arrangement is represented in fig. 13.10. SMART is geometry dependent. The restoring force of the SMART system follows the desired stiffness characteristics of the non-linear mooring line. There are four variables. Adjusting the distances A, B and C, together with the magnitude of the suspended weight, it is possible to model the desired mooring line characteristics. The mooring line loads can be decoupled for drag estimates.

The calibration and installation of a SMART system is simple and fast. The stiffness characteristics are set by the location of the fairlead and weight. Additionally, the weight contributes a realistic inertia load of the hanging chain, which is, generally, absent in the simulation with springs. The weight introduces some hydrodynamic damping as well. The pretensions are set automatically, which allows dynamic load readings. It is quickly adjustable for model draft, easily rotated to change model heading, and readily towed for the simultaneous simulation of current load by towing.

This system has been used by several model basins in various projects, including truss Spars and semi-submersibles. A static offset test was conducted on the model of this system shown in fig. 13.11(a). The measured horizontal forces due to horizontal displacements are shown in fig. 13.11(b). The system static offset characteristics are checked in

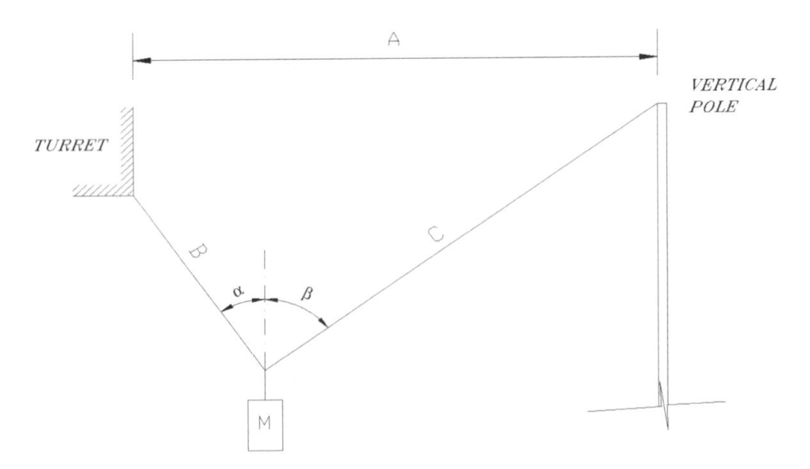

Figure 13.10 Mooring line arrangement in model

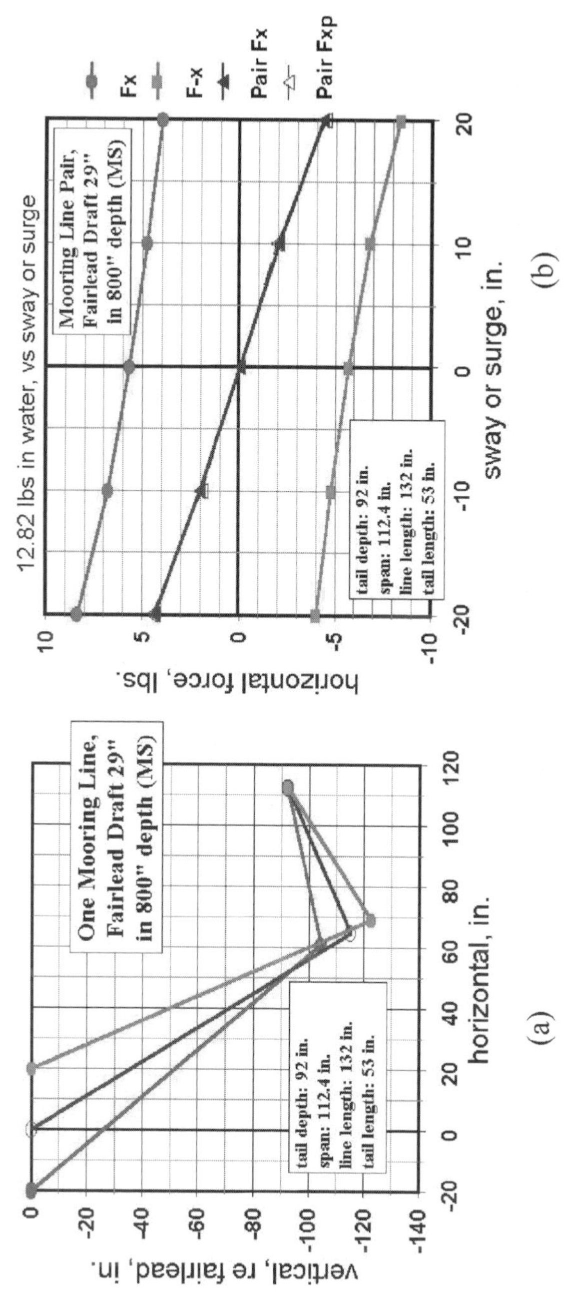

Figure 13.11 SMART mooring line configuration and static offset test results

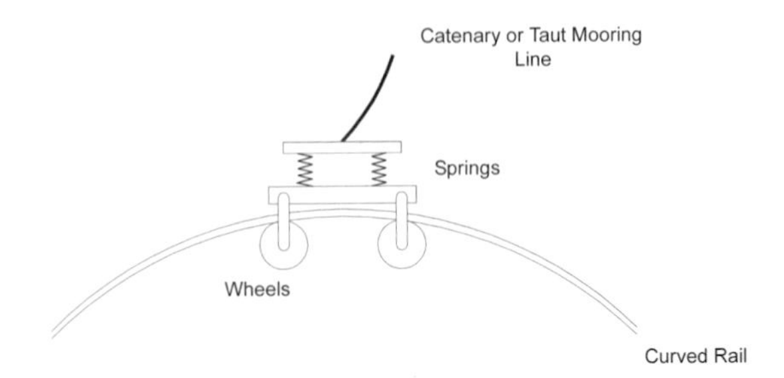

Figure 13.12 Simulation of the bottom end of a truncated mooring line

the basin by applying a series of horizontal loads (with weights over pulleys) to the model while measuring offset distances and line tensions.

Similar to the mooring lines, steel catenary risers can be simulated to match the forces induced from the risers on the model. Risers can either be modelled individually or combined into one system for a group to match the non-linear (at large offsets) restoring force exerted at their attachment point on the model. Decoupling the loads on the individual SMART lines also provides useful information of the mooring/riser-induced moments on the vessel.

An alternative arrangement to the above-mentioned truncated mooring line systems is to use a moving system at the bottom joint of the truncated line. The purpose is to allow the scaling of the fairlead angle during the model motion. This is accomplished by using a curved rail of a pre-selected curvature and the line is attached to a set of wheels travelling on the rail (fig. 13.12). The spring simulating the stiffness of the balance of the line length is added as described before. The curved wheel is designed such that the angle at the fairlead changes to the scaled value with the motion of the floater. The size of the mooring line may be set to incorporate the approximate load (and associated damping) experienced by the entire line. This will provide a closer scaling of the coupled motion of the floater and line. One difficulty of this arrangement is the possible additional damping introduced by the wheels moving on the rail. On the other hand, the mooring line on the ocean bottom produces frictional damping. The friction in the wheels may be designed to approximate this effect.

13.10 Ultra-deepwater Model Testing

Traditionally, model testing verifies the hydrodynamic response of new systems for oil and gas production systems. It is preferred to perform tests in laboratory basins, which can accommodate the full depth of moorings and risers. For ultra-deep waters, however, the modelling of full-depth system becomes difficult, since no test facility is sufficiently large or deep to perform the testing of a complete floating system with compliant mooring in 1500–3000 m depth, at a reasonable model scale. In this case, the validity of truncation described in the earlier section may be questioned. Various procedures have been proposed and developed to meet this challenge in ultra-deepwater testing. Some of these are:

- Physical model tests of complete system – Ultra small-scale testing ($\lambda \gg 100$)
- Passive Equivalent Mooring Systems [see Buchner 1999]

- Active Equivalent Mooring Systems – e.g. Active Truncated Line Anchoring Simulator (ATLAS) [see Buchner, et al 1999]
- Outdoor large-scale model tests at sea or in lakes
- Field tests in full scale
- Numerical wave tank, e.g. computational fluid dynamics
- Combination of model tests and computations

The actual choice may depend on several factors, such as the type of structure to be modelled, most important parameters to be studied, and the environmental conditions (depending on the location, etc.). The last procedure in the list above combines model test at a reduced depth coupled with computer simulation. This is termed hybrid method. Some of these alternatives are briefly discussed here.

13.10.1 Ultra Small-scale Testing

As discussed earlier, the first alternative (i.e. complete system modelling) is considered to be the most direct, independent method for determining model response. Considering the size of the existing model testing basins, a very small-scale model is needed for testing a complete system in deep water. This scenario is illustrated in table 13.11 where, as an example, the available depth of the basin versus equivalent model scale depth is shown for a few available ocean basins. As can be seen, a depth of 3000 m will require a scale factor of 1 : 200 for a complete model in the deepest available basin of the world. This scale should be considered limiting for a full model test. At a scale of 1 : 300, the modelled depth goes up to 4500 m.

For use of ultra-small scales, one has to assure that the uncertainty of results is within specified acceptable levels, and there is a need for quantification of these uncertainties. Some of the practical restrictions are the reduced repeatability of waves, currents and wind modelled at very small scales (1 : 150 and smaller). This may be improved if small portable generators are used closer to the models. On the other hand, their presence may have a direct influence on the model response. Several additional areas of concern may be stated as follows:

Table 13.11 Available prototype depth in different ocean basins

Model Basin	Available depth (m)	1 : 100	1 : 200	1 : 300
DTMB MASK	10	1000	2000	3000
MARIN	10	1000	2000	3000
MARINTEK	10	1000	2000	3000
OMB	5	500	1000	1500
OTRC	5.8	580	1160	1740
COPPE	15	1500	3000	4500

- Accuracy related to model construction
- Scaling of geometry and mass properties, and response levels
- Accuracy of instrumentation
- Possible influence of instrument probes and cables on model response
- Generation of environmental condition, and capillary effects
- Viscous scale effects
- Increasing importance of current loads
- Damping and inertia effect of the mooring and riser systems

A few tests were performed in multiple scales in the same basin so that the scale effects may be studied. An FPSO was tested in scales of 1 : 170 and 1 : 55 and comparisons of results were made [Moxnes and Larsen 1998]. A similar study was made with a semi-submersible [Stansberg, et al 2000], where tests in scales of 1 : 55, 1 : 100 and 1 : 150 were compared. Particular care was needed during the planning, preparation and execution of these model tests, since the required accuracy is at a level considerably higher than for conventional scales. The experience from these studies shows that model testing in ultra small-scales down to 1 : 150–1 : 170 is, in fact, possible, at least for motions and mooring line forces of FPSOs and Semis in severe weather conditions. For floating systems, not requiring a large "footprint" area on the bottom, such as TLPs, tests in deep pit section of the wave basin may be an alternative [Buchner, et al 1999]. It is, however, difficult to generate a specified current over the entire depth in that case.

13.10.2 Field Testing

A field test of large models or prototypes, of course, is one method for the verification of design tools. Note, however, field experiments are very expensive and complex, are not guaranteed for success and are at the mercy of the environment. Brazilian Oil Company, Petrobras is a pioneer in deepwater development using many first-of-a-kind technologies. Their philosophy has been that the field experience will prove these technical firsts.

Testing in fjords or lakes is another alternative to basin tests, and presently may be the only one, without having to compromise on scale and system simplifications. For research projects, and for use as reference check (benchmark test) of the numerical computations of specific details, testing in fjord is a very attractive alternative. Examples are reported in Huse et al (1998) and in Grant et al (1999). The main problem of Fjord-testing is, of course, that the environmental conditions are not controllable and, therefore, cannot be used on a routine basis as a design tool. In conjunction with an installed technical facility at sea (e.g. a floating dock, a wavemaker, a top-end actuator, etc.) it may be possible to bring in some control of the environment, even though they will be expensive to install.

An at-sea test of the small deepwater semi-submersible, called Motion Measurement Experiment was performed by the US Navy at a site with 900 m (2910 ft) water depth off the coast of Port Hueneme, California. The submersible was proposed as an unmanned Navy facility to support offshore aircrew combat training programme. A three-point mooring system was used in which each line comprised of chain platform pendant, polyester line, anchor chain and anchor. The reason for the full-scale testing for this system is obvious because of the deep water and small structure size. It was expected that the dynamics of the mooring lines themselves would have a substantial coupling effect on

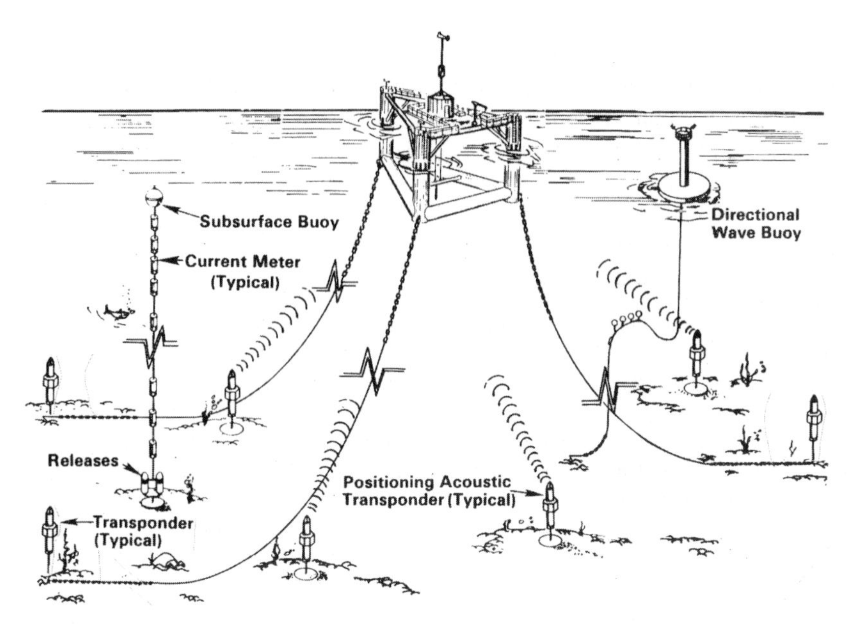

Figure 13.13 At-sea test of a small semi-submersible [Shields, et al. (1987)]

the motions of the semi-submersible. The test set-up is shown as a schematic in fig. 13.13. The environment was measured by a directional discus buoy, wave staffs and electromagnetic current meters. The platform responses were measured by a motion sensor package including accelerometers and rate gyroscopes. Shackle load cells measured mooring line loads. The platform experienced a storm with significant height as high as 8 m, which was close to the design wave height for the system.

The advantage of a full-scale testing is that generally minimal scaling effect is involved in the measurements. However, full-scale testing has its own drawbacks:

- Such testing can only provide feedback on the design after the structure has been built, but fails to provide information at the design stage.
- There is little control on the environment, so that the structure seldom can be tested in survival conditions such as a design storm.
- The environment on wind, waves and current are not well defined and are at the mercy of nature.
- The interference effect of the offshore structure with the instrument and the measurement accuracy is not known.
- From the point of view of cost and practical aspects, the testing is often limited and only a small number of measurements are possible.
- The reliability and accuracy of full-scale measurements are influenced by large loads, vessel-mounted instruments and vessel motions, the difficulty with the reference values (zero values, position reference) and external aspects (wave directionality, turbulence, temperature, etc.).

13.10.3 Truncated Model Testing

We have already discussed the truncated system in Section 13.9.3 in which mooring lines and risers are truncated. In designing truncated systems, one needs to apply an efficient methodology in choosing the right system. For example, one should apply an optimisation technique to establish a truncated system with the required properties. The method has to consider at least the following items:

- Uncertainties in model scale versus uncertainties introduced by the gap between full-depth system and truncated system
- The importance of interaction effects between the mooring/riser system and the floater motions
- More important loading effect, e.g. wind, waves, current, VIV etc.
- Room to explore unknown effects in the test setup.

There is also a need for general guidelines to help set the criteria for the requirements for the properties of the truncated system. These requirements are dependent on the system (and site) in hand and have to be evaluated on a case by case basis.

13.10.4 Hybrid Testing

A realistic alternative is the use of a hybrid form of testing. In this case, the challenge for the design verification of a deepwater system is to apply model tests and numerical computations in such a manner that the reliability is ensured and the critical system parameters are verified at an "acceptable" level of accuracy. Reliability analysis will quantify the effect of the uncertainties. Ultimately, the accuracy of the design verification must be reflected in the selection of the level of the safety factors in the design of the deepwater system. Of course, for a cost-effective design, these safety factors should be optimised.

Another important issue is the very long natural surge/sway periods of deepwater systems and their impact on the procedures used in statistical analysis for the verification. For hybrid verification, the complete modelling is replaced by a hybrid modelling, which introduces an uncertainty gap. The question is how to know that the final simulations give the same results as would have been obtained from a complete model test. Proper model scale and proper truncated set-up should be chosen to reduce these uncertainties. A schematic illustration of how the uncertainty of the verification process depends on the model scale and the degree of truncation is given in fig. 13.14. It qualitatively shows that the uncertainty increases in physical modelling as the scale factor increases, while the uncertainty in the hybrid system increases with smaller value of the scale factor. Therefore, an optimum scale factor shown by a range in the middle of the intersected curves should be arrived at for the model test. Possible hybrid approaches are discussed in more detail in the following sections.

13.10.4.1 Truncated Systems with Mechanical Corrections

The simplest approach with a truncated system is the one without computer assistance at all. This has been discussed already. In this case, all connections to the full depth system is incorporated passively in the model test set-up itself, by means of springs, masses and

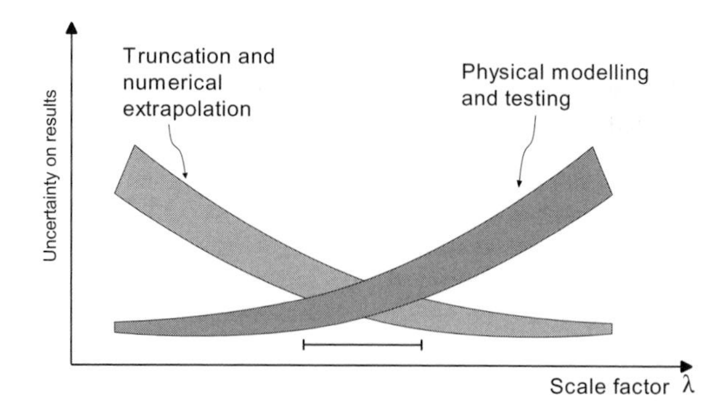

Figure 13.14 The balance between uncertainties related to truncation and to small scales [MARINTEK, 1999]

mechanisms connected to the floater. Although static characteristics can be modelled quite well by this method [Clauss and Vannahme, 1999], it has been found [Dercksen and Wichers, 1992; Oritsland, 1996; Chen, et al 2000] that it is difficult to combine a proper line dynamics that reproduces floater damping. When such issues are of less significance, this procedure may be considered as an alternative.

A passive system involves model tests with truncated system (equivalent mooring/riser system) and subsequent extrapolation to full depth by use of numerical simulations. The main motivation to perform model test with truncated system is to validate and/or calibrate the numerical tool for a system similar to the actual full-depth case. Various procedures have been described for combining a "passive" truncated test set-up with a subsequent off-line computer analysis. For examples, see Dercksen and Wichers (1992), Kim et al (1999), Chen et al (2000) and Stansberg et al (2000).

13.10.4.2 Hybrid Passive Systems

In order to reduce the uncertainties related to an off-line extrapolation of test results from a truncated to the full-depth systems, one should strive at obtaining the same motion responses of the floater as would result from the full-depth mooring. The truncated mooring system should preferably have a similarity to the physical properties of the full-depth system. In practice, the design of the test set-up should follow the following rules, in order of their priority:

- Model the correct net, horizontal restoring force characteristic
- Model the correct quasi-static coupling between vessel responses (for example, between surge and pitch for a moored semi-submersible)
- Model a "representative" level of mooring and riser system damping, and current force
- Model "representative" single line (at least, quasi-static) tension characteristics.

To the extent that these requirements may not be fully realised, the philosophy of the procedure is that the numerical simulations shall take care of the effect of the deviations between the full-depth and the truncated system.

The purpose of the model test will dictate the actual procedure proposed. Thus, if the purpose of the experiment is to study only a specific effect, the main focus of the physical modelling is placed on that particular detail, while other details are simulated on the computer. For example, tests can be run with a single mooring line for a study on line dynamics, or with the vessel moored in a very simple spring system to study only the vessel hydrodynamics. On the other hand, if the aim is to observe the behaviour of the total system, one will try to model the physical model as much as possible, including, for example, individual mooring line models, albeit truncated. In the latter case, the purpose of the tests is to check and calibrate the numerical programme on the whole system, including the vessel and the lines and risers, on the reduced depth system. Subsequently, the full system is executed along with the numerical model with the relevant information in an extrapolated version. There may also be an "intermediate" case, where lines and riser systems are modelled in a realistic way, but where the main focus is still on the floater. The more advanced the available computer programmes, the more "new" information can be expected from the computations. But they will have to be extensively verified a priori against a range of experiments.

A particular two-step (passive) hybrid verification procedure was developed by Stansberg, et al (2000) for numerical reconstruction. Similar ideas have been suggested in Dercksen and Wichers (1992). The principle is illustrated in fig. 13.15, and can be summarised as follows:

1. Design truncated set-up (according to above guidelines)
2. Select and run a proper test programme with representative tests for the actual problem
3. Reconstruct the truncated test (coupled analysis) numerically for calibration and check of the computer code
4. Extrapolate to full depth numerically. For the computer simulations, coupled analysis is generally recommended.

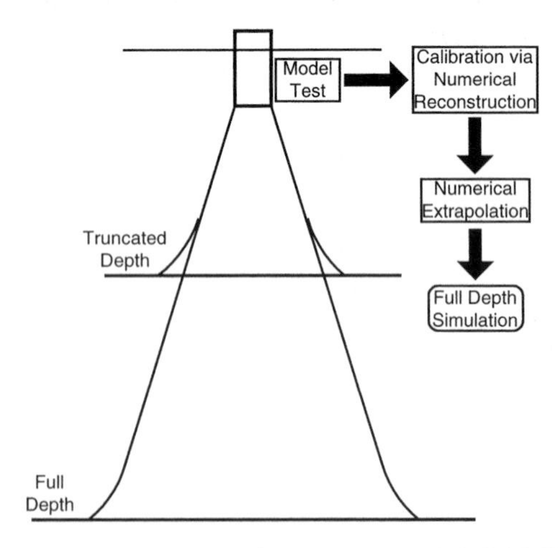

Figure 13.15 Two-step hybrid verification procedure [MARINTEK, 1999]

13.10.4.3 Hybrid Active Systems

Active hybrid model testing systems make use of real-time computer-controlled actuators replacing the truncated parts of moorings and risers. The system must be capable of working in real time in model-scale, based on feedback input from the floater motions. Thus, the mooring line dynamics and damping effects are artificially simulated in real time, based on a computer-based model of the system. System identification from model tests of a single mooring line can be used as input to the computer model. A feasibility study with such a system used on a 1 : 80 scaled FPSO model moored in a relatively shallow water basin has been described in Watts (1999, 2000).

Buchner et al (1999) described another system, which might be used in a deep-water basin. In place of a passive system, an active system is installed at the truncated end of the lines. The main features for such a system may include a robot arm on the basin floor (e.g. the MARIN ATLAS system) which will be driven from the surface via an analogue control. The system is designed in such a way that it actively simulates the behaviour of the truncated portion of the mooring lines or risers. The virtual mooring lines (and risers) below the basin floor are coupled to the real mooring lines in the basin. It requires a rigorous computational effort on a real-time basis that simulates the behaviour of the complete mooring (and riser) system. The system can accommodate the soil mechanical aspects of the problem as well. The method can simulate the interaction effect of the mooring/riser system on the low-frequency vessel motions.

However, such testing procedure is highly dependent on the accurate performance of sensitive electronic equipment at the basin floor controlled by the numerical simulations. Moreover, the robot arms can induce hydrodynamic effects themselves, which can interact with the mooring and riser system.

A complete model test verification system based on these ideas is a challenging, but interesting task for future considerations. It requires powerful computers, as well as well advanced and accurate control systems. The motion range required in 6 degrees of freedom for actuators simulating very deep systems may be another limiting factor. One should also ask: how "intelligent" does the computer model have to be for hydrodynamic verification purposes? It is expected that significant developments will take place in this field in the future.

Advantages and disadvantages of this system are:

- No numerical representation of the floater force model exists. Scaling is taken care of by real-time tests visually resembling "real" model tests.
- It is difficult to validate/verify correct performance of numerical simulations that control the actuators.
- Advanced (intelligent) software is needed, requiring rigorous computations.

13.10.4.4 Challenges in Numerical Simulation

Whether a passive or an active system is applied, a numerical tool is essential. For an active system, the numerical tool has additional requirements. The computational tool should have the following attributes:

- Faster and more efficient computers. Real-time feedback requires ultra-fast data computation.
- Faster and more efficient algorithms in general.
- Efficient algorithms for time-domain wave kinematics (viscous drift forces and local wave loading on individual mooring lines and risers).
- Utilisation of multiprocessor hardware.
- Coupled vs. uncoupled analysis (uncoupled approach needs verification with coupled analysis).
- Improved mathematical formulation for the floater force model.
- Formulation of non-linear material properties.
- Hysteresis effects/energy dissipation for taut mooring made of synthetic ropes.

13.11 Data Acquisition and Analysis

So far, we have discussed the modelling technique, scaling methods and measurements. In this section, we briefly comment on the data collection in the test and the analysis procedure that is adopted. The purpose is to obtain technically meaningful results that can be used by the structural engineer in the design of the full-scale structure.

13.11.1 Data Acquisition System

The data acquisition system should be automatic using an A/D system to convert the analogue signal from the instruments to digital form. The signal conditioners should consist of amplifiers, switchable filters and bridge sensors. There should be ample data channels available for accommodating all the instruments required for a test. The data throughput capability of all channels should be high of the order of 50–100 kHz. The data collection/reduction system should be such that the data after each test run may be examined within a short time after the run is completed.

13.11.2 Quality Assurance

Several steps should be taken to assure that the data collected in the basin during the tests are accurate and that all instruments are working properly. The signal conditioners should be checked every morning to check any drifting of instrumentation. Suspected instruments should be check-calibrated and any problems should be fixed before testing continues. The wave probes should be cleaned periodically to avoid erroneous reading. In-place calibration is performed of all installed instruments to ensure proper measurement and their accuracy.

Proper verification of the data acquisition and software routines required for the processing of the recorded data should be made prior to the testing. Several verification problems on the software for the wave generation and data acquisition should be run and the programmes verified. The hardware including the amplifier system should be checked for accuracy using standard calibration technique.

A known calibration wave should be run daily and checked against specifications. If the resulting calibration signal is outside the specified region, a logical procedure should be instigated to verify the component parts of the wave making process i.e. input control signal, wave maker motion, wave probe, logging system and pre-processing. The problem should be fixed before the testing resumes. The collected data should be compared with the standard run made at the commencement of the test in order to make sure all channels are giving similar results within acceptable tolerance.

13.11.3 Data Analysis

Data analysis consists of several steps. All data are collected in the time domain using a suitable high pass filter that removes the high-frequency electrical noise inherently present in the system.

All data are normally presented in prototype units using scale factors discussed earlier (table 13.2). Preliminary results of testing should be made available to the client after each test. This should include:

1. Force vs. offset results after the offset tests
2. Natural period and damping estimates after pluck tests
3. For regular wave tests and white noise tests, motion RAOs plotted for the structure
4. For regular and irregular wave tests, statistics of each channel should be calculated (including mean, maximum, minimum and standard deviation of all responses)
5. Selected time history plots of the data channels as necessary to examine the data quality and trend.

The regular wave test data are reduced to obtain the transfer functions (RAO) and plots are presented showing the RAO results for the various motions, sectional loads, stresses on the hull and the mooring line loads. Any problem related to the natural period response of model should be discussed.

The offset due to wave drift force is measured and accounted for in the determination of the transfer function at the wave frequencies. The magnitude of the wave drift of the vessel is reported.

For irregular waves, spectral energy densities are calculated and compared with the theoretical values. The spectral calculation of the responses is given. The RAO of the low- and high-frequency responses is computed by a cross-spectral method.

For channels, which are subject to statistical analysis, the following parameters, at a minimum, should be determined.

- Mean, minimum, maximum, standard deviation
- Significant values
- Mean periods

It is recommended that the design software be executed prior to testing once the model and test conditions are known. These results should be available during the test runs. This allows a direct comparison with the test data during the data reduction while the test is being executed. This permits uncovering and rectifying possible problems encountered in the test. It also allows redesigning the test to investigate and understand a particular discrepancy between the model tests and design tool results.

In addition to the analysis listed above, the following data analysis should be included at a minimum in the final report:

1. Estimate of damping factor vs. response amplitude for 6-DOF motions from pluck tests
2. Comparison of RAOs from white noise and irregular wave tests
3. RAO for response and airgap for each gauge location for regular wave and white noise tests
4. Tension RAOs for mooring lines and risers for regular and irregular waves and white noise tests
5. Plots showing coherence and phase along with all RAOs
6. Time history plots and spectral density plots of all channels for irregular wave tests
7. Extremal analysis of responses for the irregular wave tests

The final report should include:

- Model test set-up, coordinate system and sign conventions
- Detailed drawings and pictures for the model as used in the test
- List of instruments and their functions
- Measured mass properties and distributions compared with the computed
- Wave, current, wind and instrument calibration
- Test matrix
- Test results including transfer functions as noted above
- Any significant visual observation during the test of significance.

References

American Petroleum Institute (1979). "Recommended Practice for Planning, Designing and Constructing Fixed Offshore Platforms", API-RP2A, March, Washington, DC.

Bendat, J. S. and Piersol, A. G. (1980). Engineering Applications of Correlation and Spectral Analysis, John Wiley and Sons, New York.

Buchner, B. (1999). "Numerical simulation and model test requirements for deep water developments," *Deep and Ultra Deep Water Offshore Technology Conference*, March, Newcastle.

Buchner, B., Wichers, J. E. W., and De Wilde, J. J. (May 1999). "Features of the state-of-the-art Deepwater Offshore Basin", *Proceedings on Offshore Technology Conference*, OTC 10841.

Chakrabarti, S. K. (1994). Offshore Structure Modelling, World Scientific Publishing, Singapore.

Chen, X., Zhang, J., Johnson, P., and Irani, M. (2000). "Studies on the dynamics of truncated mooring line", *Proceedings on the 10th ISOPE Conference*, Vol. II, Seattle, WA, USA, pp. 94–101.

Clauss, G. F. and Vannahme, M. (1999). "An experimental study of the nonlinear dynamics of floating cranes", *Proceedings on the 9th ISOPE Conference*, Brest, France.

Dercksen, A. and Wichers, J. E. W. (1992). "A discrete element method on a chain turret tanker exposed to survival conditions", *Proceedings on the BOSS'92 Conference,* Vol. 1, London, UK, pp. 238–250.

Dyer, R. C. and Ahilan, R. V. (2000). "The place of physical and hydrodynamic models in concept design, analysis and system validation of moored floating structures", *Proceedings of Offshore Mechanics and Arctic Engineering Conference*, Paper No. OMAE2000/ OFT-4192, New Orleans, LA, USA.

Grant, R. G., Litton, R. W., and Mamidipudi, P. (1999). "Highly compliant (HCR) riser model tests and analysis", *Proceedings on Offshore Technology Conference*, OTC Paper No. 10973, Houston, TX, USA.

Hoerner, S. F. (1965). Fluid Dynamic Drag, Published by the author, Midland Park, New Jersey.

Huse, E., Kleiven, G., and Nielsen, F. G. (1998). "Large scale model testing of deep sea risers", *Proceedings on Offshore Technology Conference*, OTC Paper No. 8701, Houston, TX.

ITTC (1999). Environmental Modelling, Final Report and Recommendations to the 22nd ITTC. *Proceedings of 22nd ITTC Conference*, Seoul, Korea.

Kim, M. H., Ran, Z., Zheng, W., Bhat, S., and Beynet, P. (1999). "Hull/mooring coupled dynamic analysis of a truss spar in time-domain", *Proceedings on the 9th ISOPE Conference*, Vol. I, Brest, France, pp. 301–308.

MARINTEK (1999). "Deep Water Model Test Methods: Recommendations and Guidelines on Hybrid Model Testing", Report No. 513137.15.01, Trondheim, Norway. (Restricted).

Moxnes, S. and Larsen, K. (1998). "Ultra small scale model testing of a FPSO ship," *Proceedings of Offshore Mechanics and Arctic Engineering Conference,* OMAE-98-381, June, Lisbon.

Oritsland, O. (1996). "VERIDEEP. Act. 2.4, Simplified Testing Techniques – Type I", MARINTEK Report No. 513090.45.01, Trondheim, Norway (Restricted).

Sarpkaya, T. (1976). "In-line and transverse forces on cylinder in oscillating flow at high reynolds number", *Proceedings on Offshore Technology Conference*, OTC 2533, TX, USA, Houston, pp. 95–108.

Shields, D. R., Zueck, R. F., and Nordell, W. J. (1987). "Ocean model testing of a small semisubmersible", *Proceedings on Offshore Technology Conference*, Houston, pp. 285–296.

Stansberg, C. T. (2001). "Data interpretation and system identification in hydrodynamic model testing", *Proceedings on the 11th ISOPE Conference*, Stavanger, Norway.

Stansberg, C. T., Yttervik, R., Oritsland, O., and Kleiven, G. (2000). "Hydrodynamic model test verification of a floating platform system in 3000 m water depth", *Proceedings of Offshore Mechanics and Arctic Engineering Conference*, Paper No. OMAE00-4145, New Orleans, LA.

Stansberg, C. T., Ormberg, H., and Oritsland, O. (2001). "Challenges in deep water experiments – hybrid approach", *Proceedings of Offshore Mechanics and Arctic Engineering Conference*, OFT-1352, June, Rio de Janeiro, R. J., Brazil.

Watts, S. (1999). "Hybrid hydrodynamic modelling", *Journal of Offshore Technology*, The Institute of Marine Engineers, London, UK, pp. 13–17.

Watts, S. (2000). "Simulation of metocean dynamics: extension of the hybrid modelling technique to include additional environmental factors", *SUT Workshop: Deepwater and Open Oceans, The Design Basis for Floaters*, February, Houston, TX, USA.

Wichers, J. E. W. and Dercksen, A. (1994). "Investigation into scale effects on motions and mooring forces of a turret moored tanker", *Proceedings on Offshore Technology Conference*, OTC paper 7444, Houston.

Handbook of Offshore Engineering
S. Chakrabarti (Ed.)
© 2005 Elsevier Ltd. All rights reserved.

Chapter 14

Offshore Installation

Bader Diab and Naji Tahan
Noble Denton Consultants, Inc., Houston, Texas

14.1 Introduction

While civil engineering structures are normally built at their installation site, offshore structures are built onshore and transported to the offshore installation site. The process of moving a structure to the installation site involves three distinct operations referred to as the loadout, transportation and installation operations. Collectively, these operations are also known as the "temporary phases" and the engineering work associated with them as "Installation Engineering".

During the temporary phases, the structure is subjected to loads that are different in magnitude and direction from the in-place loads. The shape, the weight and the cost of offshore structures are, therefore, influenced by these temporary phases. The temporary phases also affect the choice of the fabrication yard and the cost and schedule of the overall project.

Given a large number of the temporary phase concepts and the numerous types of offshore structures, it would be difficult to present a comprehensive study of installation in a single chapter. The objective of this chapter is to provide the reader with a basic understanding of the most common concepts together with their advantages, disadvantages and limitations. While some design guidance is offered within, the chapter is not meant to provide a comprehensive design guidance on all the installation concepts. For such guidance, the reader is referred to the volumes of technical literature such as research papers, codes or recommended practice, regulatory authority publications and the rules of the classification societies and the marine warranty surveyors.

Different types of structures require different methods of transportation and installation. Different installation methods can also be used for the same type of offshore structure. The work presented in this chapter is arranged along the types of structure and the installation concepts.

14.2 Fixed Platform Substructures

14.2.1 Types of Fixed Platform Substructures

A fixed substructure is that part of an offshore platform which sits on the seabed and is rigidly connected to it by means of foundation piles (e.g. jackets) or under the effect of its weight (e.g. gravity base structure). The installation methods of the following substructures are covered in this section:

- Jackets
- Compliant towers
- Gravity base structures.

14.2.2 Jackets

The jacket is a space frame structure made of tubular steel members. The jacket legs and braces transmit environmental and topsides loads into the piles and subsequently into the seabed. Jackets typically have three, four, six or eight legs. Jackets with three legs are known as tripods. Jackets with a single caisson type leg also exist. These are also known as monopods.

Piles made of tubular steel are installed through the legs of the jacket or through the pile sleeves connected to the jacket legs at its base. The piles installed inside the jacket legs normally extend to the top of the legs. Through leg piles are connected to the jacket legs at the top using shim plates, known as "crown shims", that are installed in the annulus between the leg and the pile and are welded to both. In some structures, the annulus between the jacket and the pile is grouted, although this is no longer a common practice. Piles installed through sleeves on the outside of the leg structure are connected to the sleeve by grouting the pile-sleeve annulus.

Regardless of the size or the type of jacket installation, once the jacket is on the seabed, its weight is temporarily supported by mudmats. Mudmats are added to the bottom of the jacket legs to provide the required bearing area to support the jacket weight and resist environmental loading during installation and until the strength of the piles has sufficiently developed. This phase is known as the "unpiled stability" phase. They are flat panels that are made of stiffened steel plate or, to reduce weight, from glass reinforced plastics. Mudmats are sized so as to support the combined loads of the jacket weight and buoyancy, weight of piles that have to be supported on the jacket and environmental loads associated with the installation window. Section 14.9.4 lists the typical unpiled stability requirements.

The method of installation depends on the weight and the physical dimensions of the jacket and on the capacity of the installation equipment. The following methods are the most common for a jacket installation.

14.2.2.1 Lift and Lower in Water

This method is used for small jackets, in very shallow water, which are transported on barges in the upright position already pre-rigged for offshore lift and installation by a crane vessel. Once offshore, the jacket is lifted off the deck of the barge and lowered down to the seabed. Jackets installed in such a configuration are typically less than 50 m tall.

The foundation piles for this size of jacket structure are typically transported together with the jacket on the same cargo barge. Once the jacket is set on the seabed, the piles are installed using the same crane vessel and a pile hammer of an adequate size.

14.2.2.2. Lift and Upend

As the size of a jacket structure increases, it is built and transported on a cargo barge in the horizontal position. The jacket is lifted off the cargo barge using one or two cranes. Following pick-up, the cargo barge is withdrawn and the jacket is upended. Single cranes with two blocks can be used for upending smaller jackets with the jacket length aligned with the plane of the crane boom. There are several methods of upending jackets:

- Two-block upending – upending in air or partially in water using two crane blocks. In this method, the jacket does not have sufficient buoyancy to float without crane assistance. Instead, the upending is achieved by hoisting down the block of one of the two cranes while the other is hoisted up. Figure 14.1 shows a two-block upend operation. The size of the jackets that can be upended with a single crane is limited.
- Single-block upending. A jacket installed using this method needs to have sufficient buoyancy to float in the horizontal position by itself. In this method, the jacket is pre-rigged with two sets of four slings. The first set of slings – the lifting slings – are attached somewhere along the top jacket frame, while in the horizontal position. The second set of slings – the upending slings – are attached to padeyes at the top of the legs when the jacket is in the upright position. The jacket is lifted off the cargo barge with the lifting slings and lowered into the water until its buoyancy balances its weight. The lifting slings are then disconnected from the crane hook and the upending slings are connected

Figure 14.1 Two block upending (Marathon East Brae jacket)

Figure 14.2 Single block upend

to the hook. The jacket is then ballasted in a controlled manner until it is upended a few meters above the seabed. Further ballasting is then carried out until the jacket is positioned on the seabed. This method only requires one crane albeit it has to be capable of lifting the full jacket weight without assistance. The jacket legs need to be made buoyant by installing rubber diaphragms at the bottom of the legs and steel caps at the tops. Additional equipment such as flooding valves, umbilicals and pumps are also needed. The jacket buoyancy has to be designed so as to allow easy access for rigging the upending slings, while the jacket floats horizontally. Sufficient buoyancy and subdivision is also required to ensure that flooding of one compartment does not lead to the jacket sinking or making the installation operation impossible to complete. Some consideration should be given to provide remotely operated valves with manual back-up. Figure 14.2 shows a single block upend operation.

14.2.2.3 Launching

Jacket structures that are too heavy to be lifted can be launched into the sea off a launch barge. A launch barge is a flat top cargo barge equipped with skid beams, a rocker arm, launch winches and a suitable ballasting system. Jackets are designed to be either self-upending or upended with the assistance of a crane vessel. Launched jackets need to have sufficient reserve buoyancy in order to ensure they float at the end of the launch sequence. The jacket legs are made buoyant by the use of rubber diaphragms at their bottom ends and steel caps at the top. Additional buoyancy located appropriately is sometimes required to achieve the required level of reserve buoyancy or to ensure the jacket will upend itself at the end of the launch sequence.

Launching operations require the jacket to be fitted with a launch truss. The launch truss is an integral part of the jacket structure and serves to transfer the weight of the jacket into the skid beams and the rocker arm during the launching operation. The weight of the launch truss normally constitutes a significant part of the jacket weight.

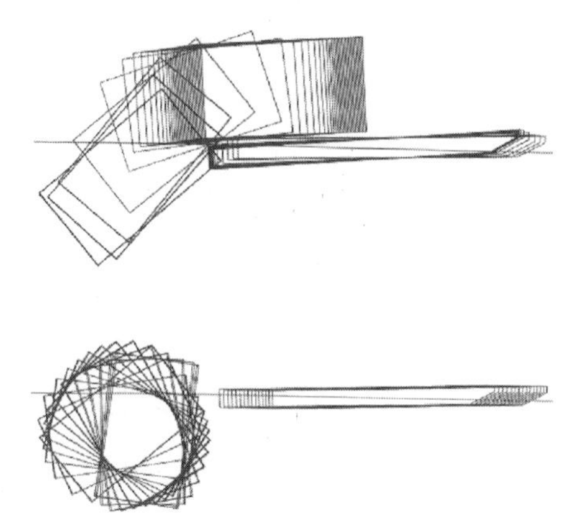

Figure 14.3 Launch simulation of a self-upending jacket

The rocker arms are two beams that are installed at the stern of the barge in line with the skid beams. They are connected to the stern through hinges. The rocker arms serve to support the jacket weight as it rotates over the barge stern and dives into the water. As such, the rocker arms and the supporting hinges can be substantial structures. Figure 14.3 shows a typical launching sequence of a jacket that was designed to be self-upending. Sections 14.8.2 and 14.9.3 include more information on launching.

14.2.3 Compliant Towers

Compliant towers are made of several rigid steel sections joined together by hinges such that the tower can sway under environmental loads. A compliant tower's mass and stiffness characteristics are tuned such that its natural period would be much greater than the period of waves in the extreme design environment. This reduces their dynamic response to such environment and extends the applicability of fixed platform to deeper water such as 1000 m.

A compliant tower structure can be divided into four basic structural components:

- The foundation piles,
- The base section,
- The tower section(s). Depending on the water depth and the means of transport, the tower can be made in one or more sections,
- The deck.

The base and the tower sections are lattice space structures fabricated from tubular steel members and thus termed the jacket base and the jacket tower sections. Normally the tower section is much larger than the base section.

A typical installation sequence of a single tower section is described next. The jacket base section is transported on and launched off the deck of a launch cargo barge at site. The top of the jacket would be connected to a derrick barge and the bottom to its assisting tugs. Once in water, the jacket base section would be upended by the derrick barge assisted by

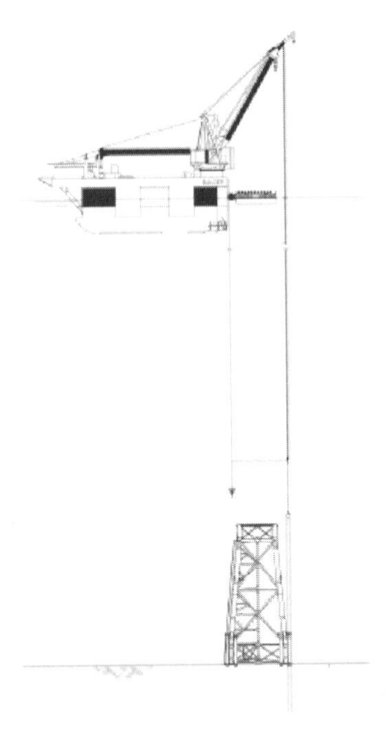

Figure 14.4 Installation of the Baldpate piles [De Koeijer, et al 1999]

the jacket buoyancy. Once vertical, the jacket will be lowered and manoeuvred into position often with the guidance of a pre-installed docking pile.

Piles are transported to the site on cargo barges, lifted off and upended, using the cranes of the derrick barge, lowered, stabbed through the jacket base pile sleeves and driven to target penetration as shown in Fig. 14.4. Pile driving is addressed in Section 14.4.2 where a more detailed description is provided.

After, the verticality and orientation of the jacket base are achieved, piles are grouted to the pile sleeves. The base structure would now be safely secured to the seabed and ready to receive the next tower section.

Then, the tower section is transported on the deck of a launch barge and launched into water. Due to the large weight and height of the tower section, it is designed such that it is self-upending after separating from the launch barge and going into the water. Once vertical, the tower section, is ballasted to the required float-over draft. The tower section is then towed and positioned over the pre-installed jacket base section as shown in Fig. 14.5. With assistance from the attending derrick barge, and position-holding by tugs, ballasting continues until the pins at the base of the tower section engage one by one in their respective receiving buckets at the top of the pre-installed base section. Grout is then injected into the gap between the pin and bucket, which provides the structural continuity and the integrity of the entire subsurface structure (base and tower sections). The tower is now ready to receive the topsides deck. The topsides deck can then be lifted by the derrick barge and set onto the tower structure.

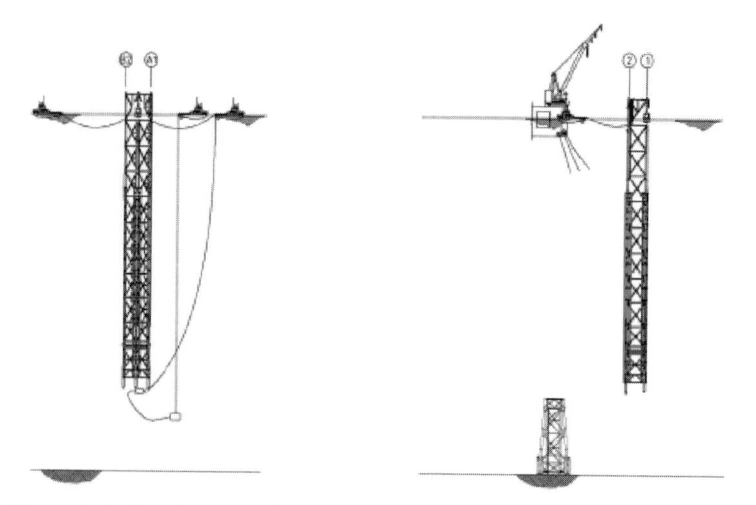

Figure 14.5 Upend and transfer of Baldpate tower section [De Koeijer, et al 1999]

14.2.4 Gravity Base Structures

Gravity base structures (GBS) are very large structures that sit on the seabed and resist sliding and overturning loads by friction and soil bearing capacity. The hull of a GBS is made of several tanks that are used to store oil and ballast. The lightship displacement of the gravity based structures can be of the order of several hundred thousand tonnes. GBS have been installed in water-depths of up to 300 m. Most gravity based structures are made from concrete although one steel gravity base platform, Maureen, was installed in the North Sea.

Concrete platforms are built and installed in a different way from steel jackets. The construction commences in a dry dock adjacent to the sea. The structure is built vertically, from the bottom up, in a similar manner to onshore buildings. When the structure is complete, the dock is flooded and the structure floats under its own buoyancy. The topside structures are normally installed at an inshore location by deck mating or any other suitable method. Multiple tugs are used to tow the structure to its offshore location. Once on location, the structure's tanks are filled with sea water to a predetermined ballasting plan and the structure is sunk down to its final position on the seabed. The GBSs are typically trial-ballasted prior to tow to site.

GBS are towed at a large draft and their towing requires very detailed analyses and marine procedures including the following aspects:

- Available water depth, underkeel and horizontal clearances in the tow route.
- Stability and freeboard.
- Required number of tugs, bollard pull and design of the towing attachments.

Given its size, several tugs tow the GBS at a very slow speed of 2 knots or less. Table 14.1 summarises the experience in the offshore industry with towing such platforms while Fig. 14.6 shows the tug towing arrangement of one of the early GBS.

Table 14.1 Previous GBS towing distances and duration

Platform	Installed	Water depth	Towed distance and duration
Ekofisk	1973	70 m	216 nm 7 days
Frigg CDP-1	1975	104 m	120 nm 5 days
Beryl A	1975	120 m	170 nm 6 days
Brent B	1975	140 m	170 nm 6 days
Brent D	1976	140 m	160 nm 6 days
Frigg TCP-2	1977	104 m	80 nm 4 days
Statfjord A	1977	146 m	220 nm 7 days
Statfjord B	1981	146 m	220 nm 7 days
Statfjord C	1984	146 m	230 nm 7.5 days
Gullfaks A	1986	135 m	160 nm 6 days
Gullfaks B	1987	142 m	160 nm 6 days
Oseberg A	1988	109 m	130 nm 5 days
Gullfaks C	1989	216 m	160 nm 6 days
Snorre A	1990	309 m	180 nm 6.5 days
Draugen	1993	251 m	333 nm 8.9 days
Sleipner A	1993	82 m	156 nm 7 days
Troll	1995	303 m	80 nm 6.5 days
Hibernia	1997	80 m	260 nm 9 days

Figure 14.6 Tow of Beryl A GBS

14.3 Floating Structures

14.3.1 Types of Floating Structures

The most common floating production storage and offload (FPSO) vessels are converted tankers. While most of the new-build FPSO retain the aspect ratios of tankers, their bow and stern hull shapes tend to be more square than the ship-shaped tankers. An FPSO, as the name suggests, supports production and storage operations with some of the largest ones being built today capable of storing 2 million barrels of oil. Given their length-to-width aspect ratio, environmental loading on the beam of the vessel is much higher than that on the bow or stern. Turret mooring systems that allow the FPSO to weather-vane so as to minimise the environmental load are a common choice for station keeping particularly where there is very little directionality in the design environment such as in the areas exposed to hurricanes or typhoons. In environments where the weather is directional, such as in West Africa, or semi-directional as in Brazil, there is scope for using the spread-moored systems for station keeping.

Semi-submersible vessels are also referred to as the column-stabilised units. Their most common hull form consists of four columns supported by two pontoons. The pontoons are submerged under normal operations and the only water-piercing part of the hull are the columns. These vessels are commonly used for drilling operations in water depths in excess of 100 m. Several semi-submersibles have also been used as production rigs in deep water. Some of these vessels support the combined drilling and production operations but have no storage capabilities. While mooring remains the most common type of station keeping, deepwater semis are equipped with thrusters that maintain station with a dynamic positioning (DP) system. Semi-submersibles are sensitive to additional weight and increases in water depth as their operating water plane area is small. With a length-to-width aspect ratio close to 1.0, a spread mooring pattern is usually adopted for the semi-submersibles.

Conventional tension leg platforms (TLP) have similar hull forms to semi-submersible vessels with water-piercing columns and pontoons. TLPs are anchored to the seabed via vertical tendons that are made of high strength steel pipes commonly joined by mechanical connectors or, less frequently, by welding. Tendon tensions at the operating draft are balanced by hull buoyancy. This system is self-restoring since any offsets from the mean position caused by environmental loads results in a gain in hull buoyancy and tensions and generating a restoring force that pulls the TLP back to its mean position. TLPs have been installed in water depths ranging between 148 m and 1432 m. A conventional TLP can support drilling and production operations while the smaller mini TLPs can only support production operations. Because of the high stiffness of the tendons, the TLP motions are much smaller than the semi-submersibles and FPSO. Figure 14.7 shows different TLP configurations. The TLP foundations are typically driven piles although other pile types are feasible. Sometimes, foundation templates are used.

Deep draft caisson vessels (DDCV), also known as spars, are an alternative to the TLP in deep water. A conventional spar hull form consists of a vertical cylinder made of a combination of voids and ballast tanks. Truss spars are a variation on the theme with the lower part of the length of the cylinder substituted by a truss structure. Spars are inherently stable as their centre of gravity is located below their centre of buoyancy. They

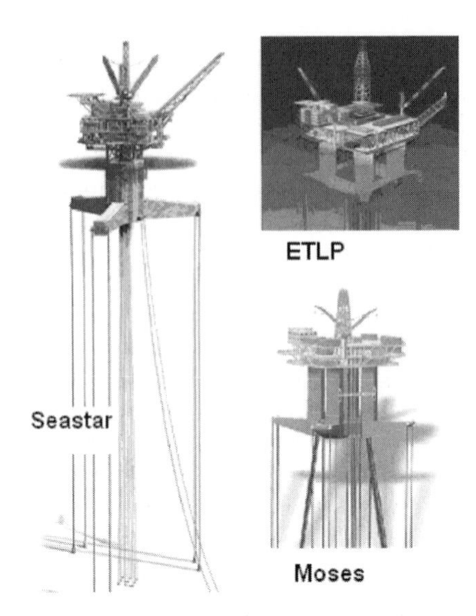

Figure 14.7 Schematic of Seastar, ETLP and Moses TLP designs

are normally moored by a semi-taut spread mooring system although at least one spar is currently being designed with a taut leg polyester mooring system. Figure 14.8 shows a schematic of a spar. The spar dimensions vary with the largest built to date being of the order of 150 ft in diameter and 750 ft long.

The immersed part of the spar hull consists of a hard tank (usually the mid-section) which provides the buoyancy, and a soft tank at the bottom where the fixed ballast is stored.

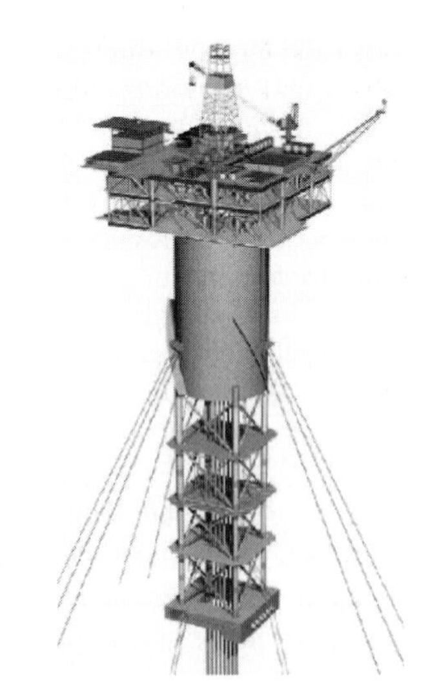

Figure 14.8 Schematic of a truss spar

The mooring line fairleads are positioned close to the pitch centre which is well below the water line. This minimises the fairlead excursions in rotational movements allowing the mooring system to be reasonably taut which, in turn, minimises lateral excursions. The heave motions are also low because of the spar's low water plane area compared with its hydrodynamic mass giving low motion characteristics overall.

14.3.2 Installation of FPSOs

Although the installation of the topsides onto the hull of the FPSO is considered to be part of the construction phase, the topside integration lifts are often carried out by floating crane vessels making the operation akin to an installation operation. Since the integration lifts are carried out along the quay, or in sheltered conditions, the criteria that are applied to the lift are those for an inshore lifting. The availability and the size of the lifting equipment in the vicinity of the yard is normally a significant consideration when selecting a construction yard for the integration of the FPSO topsides. If the capacity of the available lifting equipment is low, the topsides would have to be split into a greater number of modules of a manageable size.

The installation of the topside is followed by a period of a few months during which the installed modules are hooked up to the ship systems or to each other. During this phase, the FPSO has to remain moored along the quay.

Whether FPSOs are converted tankers or purpose-built vessels, they are unlikely to have any propulsion, since it is not required during the service life. They are therefore towed to site using at least one tug and, more likely two or three tugs.

The mooring system of the FPSO is installed prior to the arrival of the FPSO and laid on the seabed or, in the case of polyester mooring lines, suspended at mid-depth using buoyancy cans.

The FPSO is towed over the mooring pattern and the tow switches from the towing configuration to the station-keeping configuration. While the tugs hold the FPSO in position, other tugs pick up the ends of the pre-laid mooring lines and bring them towards the FPSO fairleads where they are connected to winches or chain jacks that are installed on the FPSO. The tugs are released when a sufficient number of lines are connected. The mooring line hook-up operation continues until all the lines are connected and tensioned.

14.3.3 Installation of Semi-Submersibles

Drilling semis normally carry their mooring legs and anchors on board. The mooring system consists of a chain, wire or a combination of both. When the vessel arrives on location, its anchors are handed over to anchor-handling (A/H) vessels. The A/H vessel then moves towards the designated anchor installation position while the mooring line is paid out from the semi's on the winches. The anchor is lowered to the seabed at the designated location. Preloading the anchors and tensioning of the mooring lines is carried out using the anchor handling vessel and the on-board winches.

With the introduction of the taut leg polyester mooring systems for semis, particularly in the case of the production vessel, the mooring system can be pre-laid ahead of the semi arriving on site. The semi is then "hooked-up" to the mooring legs one by one, using temporary or permanent winches or chain jacks installed on board.

14.3.4 Installation of Tension Leg Platforms

The main components of the TLP are the hull, the deck, the piles and the tendons. Pile installation is discussed in Section 14.4. The tendons can be installed ahead of the hull or installed at the same time as the hull. Similarly the deck can be integrated with the hull at the fabrication yard or installed after the hull. This section describes the various stages involved in the installation process.

14.3.4.1 Wet Tow of Hull and Deck

Once installed, TLPs derive their stability from the tendons. Free floating stability is deemed to be an issue only during the temporary phases including wet tow and installation. This issue determines at what stage the deck is installed. There are two installation philosophies of TLPs:

a. Installation of Complete Platform

The deck is installed inshore at or near the integration site and the completed platform is transported to site. This saves the cost of expensive derrick barges and hook-up and commissioning work offshore. The transportation operation involves a wet tow for at least a part of the voyage. The platform is therefore required to have adequate free floating stability. The hull is designed to provide sufficient buoyancy and the water plane area to meet the stability requirements during wet tow and installation. More recently TLP designers added temporary stability tanks to the hull in order to meet the stability requirements. These tanks are removed once the TLP installation is complete, thus leaving the hull with only the necessary structure to meet the in-place conditions. ABB's Extend Pontoon TLP (ETLP) is an example of such a concept. This is thought to reduce the cost of the platform as the temporary stability tanks can be re-used and their cost can be spread over several projects.

b. Installation of the Hull and Deck separately

This installation philosophy was adopted on several mini-TLPs such as the Seastar and the Moses. With weights of less than 6000 tonnes, the topsides can be installed in a single lift and offshore integration time is not perceived to be a significant handicap. A crane vessel is required on-site during the installation of the hull, hook-up to the tendons and deck installation. During installation of the hull, an additional hull stability is often required during hull ballasting for installation. This is achieved by applying an upward load on the hull by the crane hook.

14.3.4.2 Tendon Assembly

This section describes the means of delivering the tendons to site and assembly.

a. Dry Tow of Tendon Sections and Assembly Offshore

A typical tendon string is made up of a bottom section, several main body sections and a top section made of a length adjustment joint (LAJ). The bottom connects through a mechanical connector to the pile or foundation template. The individual sections are joined together with a mechanical connector such as the Merlin connector. The main body sections are typically fabricated in sections of 240–270 ft lengths, shipped on a cargo barge to the installation site, where they are lifted and upended by a crane barge. During the tendon assembly process, the weight of the tendon string that has already been assembled is supported on a tendon assembly frame (TAF) which is a purpose built structure that is installed over the side of the derrick barge. The maximum length of individual sections is determined by the available hook height of the derrick barge. Tendon strings with longer sections require fewer mechanical connectors but a larger installation crane boom.

b. Wet Tow of Complete Tendon

As an alternative to using tendon connectors, tendon strings are assembled by welding individual sections together. The tendons are subsequently launched and wet towed to the site in the same way as the pipe bundles. Buoyancy modules may be strapped onto the tendons to provide additional buoyancy and control stresses during the wet tow as shown in Fig. 14.9. Once at site, the tendons can be upended with the help of winches or cranes and controlled removal of the buoyancy modules. This method saves the cost of mechanical connectors. The tow operation has to be designed carefully to ensure that failure of any component during the wet tow does not lead to the total loss of the tendon string.

14.3.4.3 Tendon Hook-up

a. Pre-installed Tendons

Tendons can be installed prior to the hull arrival to site. To ensure that the tendon and its components remain taut, upright and to keep the stresses within design allowables, temporary buoyancy modules are provided in the form of steel cans which connect to

Figure 14.9 Buoyancy modules fitted to the Heidrun TLP tendons

the tendon at its top as described in Section 14.3.4.4. This temporary buoyancy module (TBM) is clamped to the tendon after the tendon assembly is complete. The tendon and the TBM assembly is lifted from the side of the derrick barge and manoeuvred until the tendon bottom connector is stabbed into the foundation and latched in position. The TBM is then deballasted such that it applies sufficient tension to the tendon until it is hooked up to the TLP. Figure 14.10 shows a schematic of a pre-installed tendon.

When the platform arrives on site, it is ballasted until the tendon connector engages the LAJ teeth at which point the connector is locked off. Once the connector is locked off the connector allows the downward movement of the platform under wave action but prevents any upward movement. This is known as "ratcheting". The ballasting operation continues in parallel with the ratcheting motions, until the desired draft is reached. At that point the ballast water is pumped out causing the tension in the tendons to increase while the hull draft only reduces marginally by the amount of tendon-stretch. The de-ballasting operations are considered complete when the desired pre-tension is reached in the tendons. Figure 14.11 shows a typical time history of the tendon loads during the ratcheting operation.

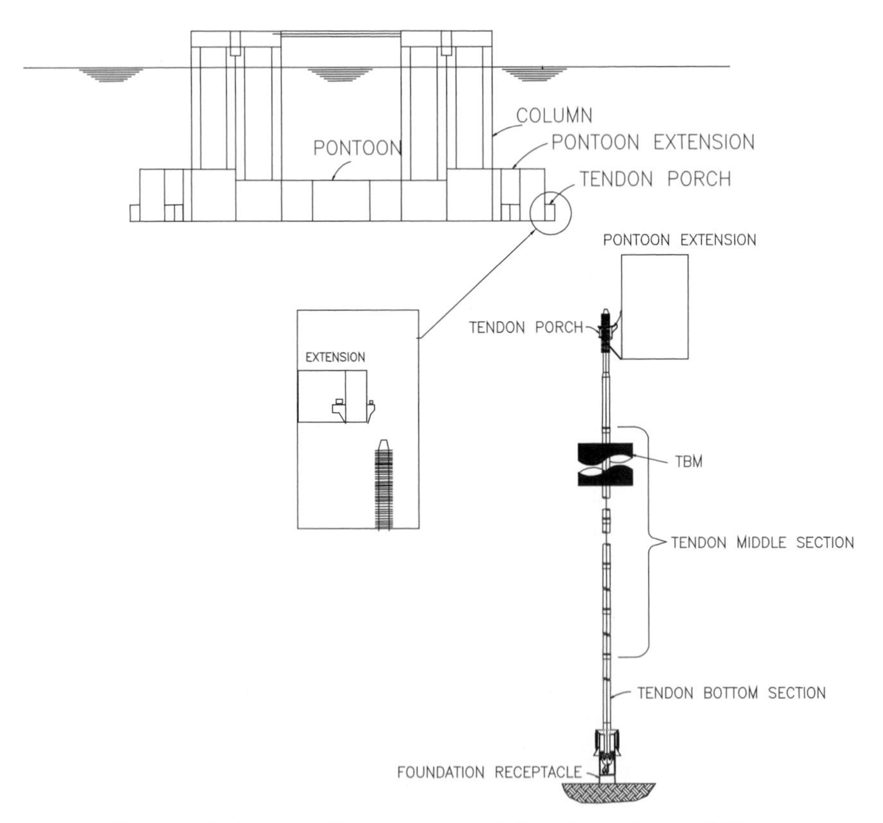

Figure 14.10 A pre-installed tendon with a TBM before hook-up to ETLP

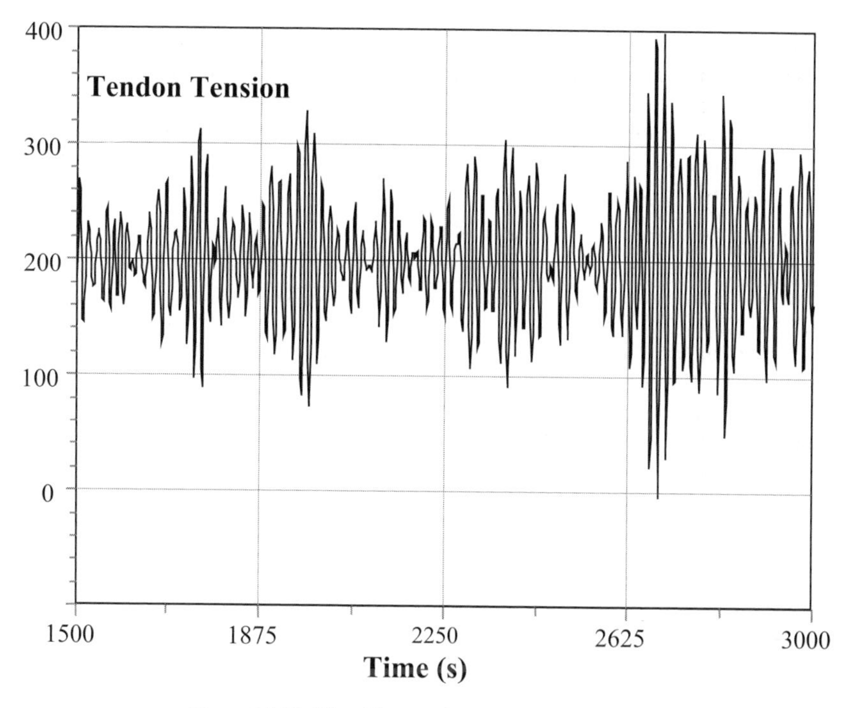

Figure 14.11 Time history of tendon ratcheting loads

b. Hull and Tendons Installed Concurrently

Once the tendon is assembled on site, the derrick barge hands it over to the platform where it is hung from the tendon porches. Once all the tendons are hung from their respective porches, their bottom connectors are stabbed into their piled foundations. Tendon pre-tensioning is achieved using mechanical tensioners similar to chain jacks. The pre-tensioning operation can proceed in several stages with only one group of tendons being tensioned during each stage in order to limit the number of the tensioning devices required.

The tendon porches in this type of installation have to be open on one side to allow the tendon to be inserted laterally. This restriction does not apply to the pre-installed tendons. Figure 14.12 shows a schematic of the tendon stabbing operation.

14.3.4.4 Temporary Buoyancy Tanks

Where the tendons are pre-installed, temporary buoyancy modules (TBMs) are used to maintain tension in the tendon until the hull arrives. Each TBM is subdivided into several chambers or a cluster of tanks such that the tension in the tendon is not lost with any one compartment getting accidentally flooded. The TBM is normally clamped to the tendon after the tendon is assembled, while it is still hanging over the side of the derrick barge. The TBMs are located below the LAJ such that they do not interfere with the operation of hook-up to the platform. The TBMs are flooded when the tendon is stabbed into the

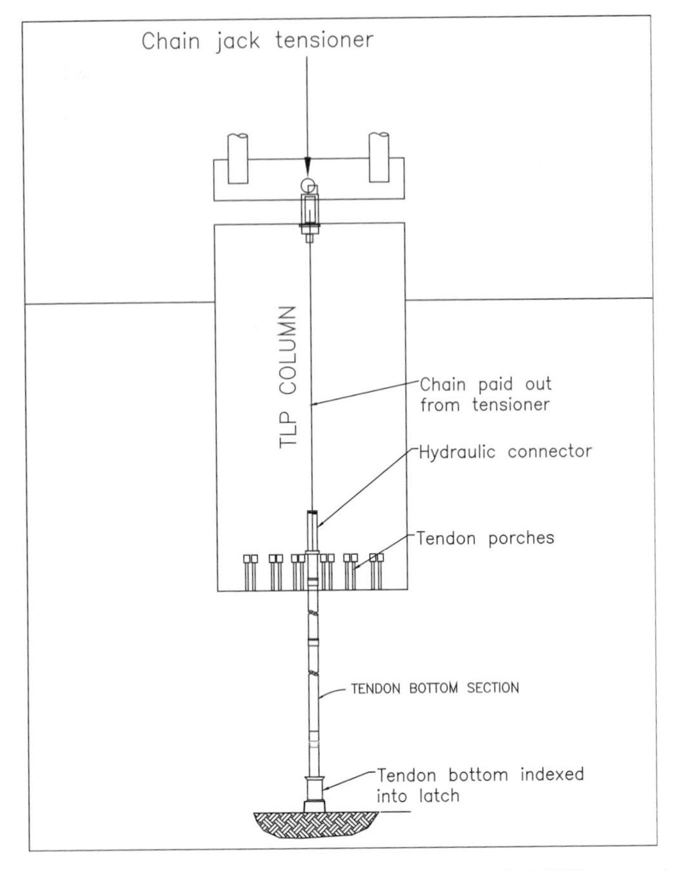

Figure 14.12 Stabbing of tendons hung-off from the Auger TLP (Offshore engineer)

foundation piles. While the top of the tendon is still supported by the crane hook, the TBM is de-watered by pumping compressed air into it, until the desired buoyancy is achieved. The top of the tendon is subsequently released from the crane hook.

TBMs can have a closed bottom or an open bottom. One of the critical areas of design is to ensure that the TBMs have sufficiently large openings in their top to allow air to escape while they are lowered through the wave zone. Once the TLP is hooked up to its tendons and sufficient tension is achieved in the tendons, the TBMs can be flooded so that they become neutrally buoyant and removed by the attending installation vessel.

14.3.5 Spar Installation

14.3.5.1 Wet Tow and Upending

Spar hulls and decks are normally installed separately. The hull is normally wet-towed to site and upended by ballasting.

Figure 14.13 Upending of Nansen spar [Beattie, et al 2002]

In conventional spars, fixed ballast is added to the soft tank at the bottom end of the hull followed by variable ballast added to the hard tank at the top. A fixed ballast in the soft tank could either be water or hematite.

The truss spars are made of a hard tank at the top, a truss section which substitutes the soft tank in a conventional spar, and a fixed ballast tank at the bottom.

The ballasting operations can be done by free-flooding the tanks. In this case, large "rip-out" plugs are removed from the tanks to facilitate free flooding. The vent size has to be carefully designed so as to allow the escape of large volumes of air in a very short period of time. Where tanks are ballasted by pumping water into the tank, the pumping rates need to be maximised to ensure a rapid operation that can be completed within a reasonable window. Figure 14.13 shows a typical spar upending operation.

14.3.5.2 Mooring Line Hook-up

Mooring lines are installed prior to the arrival of the spar on site and laid on the seabed until the spar arrives or, in the case of the polyester mooring lines, kept above the seabed using buoyancy devices. The spar end of the mooring line is normally made of a chain. This segment could either be pre-installed with the rest of the mooring line or, alternatively, installed during the hook-up operation.

Recovery of the mooring lines is normally performed using a crane vessel. The weight of the mooring line dictates the size of the crane vessel required. The connection of the mooring line to the spar is performed using the pull-in winches installed on the spar. A messenger line is deployed from the spar through the fairleads and connected to the end of the mooring line which is supported by the crane vessel as shown in Fig. 14.14. Once the messenger line is connected to the mooring line, the winch pulls the messenger/mooring line assembly back. Once the correct pretension level is achieved, the chain stoppers are locked off.

The hook-up operation described above often requires a substantial pull-in system. The size of the pull-in system can be reduced by supporting the weight of the mooring line at an intermediate point close to its top end on a clamp installed on the crane vessel [Dijkhuizen,

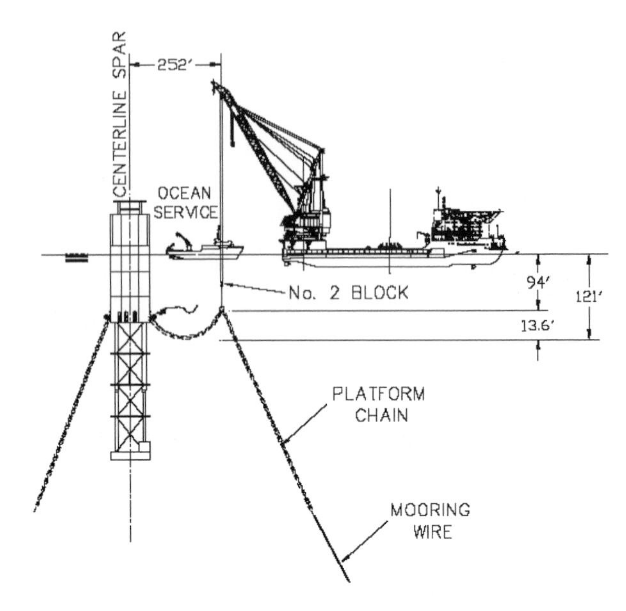

Figure 14.14 Handover of mooring line to Nansen spar messenger line [Beattie, et al 2002]

2003]. An equaliser system is rigged up so as to allow the crane vessel and the spar to be winched closer together. The short length of chain beyond the clamp is handed over to the spar for connection. This system is shown in Fig. 14.15.

During the hook-up operation, the spar is held in position using tugs connected to the spar hull. These tugs serve to keep the spar on location during the hook-up operation. The tugs have to be sized to resist environmental loads and loads from the mooring lines already connected to the spar hull.

14.4 Foundations

14.4.1 Types

There are four main types of foundations:

- Driven piles
- Drilled and grouted piles
- Suction embedded anchors
- Drag embedded anchors

Each type requires a different method of installation. Gravity structures may be regarded as a type of foundation, but are considered in this chapter as a "fixed platform". Refer to Section 14.2.4.

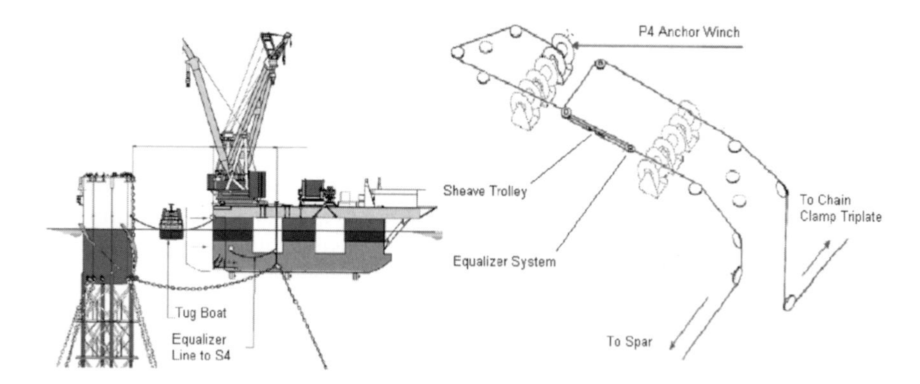

Figure 14.15 Pulling Horn Mountain spar and crane vessel with equaliser system [Dijkhuizen, 2003]

14.4.2 Driven Piles

Driven offshore piles are steel tubular members which consist of a driving head, the main body of the pile and a driving shoe. The pile length, diameter and the wall thickness depend on the soil characteristics and the magnitude of design loads. Pile lengths to over 500 ft and diameters greater than 96 in. have been installed.

14.4.2.1 Transportation and Installation

Piles are normally transported on cargo barges to the offshore location. They may be lifted off the deck of the cargo barge and transferred onto the deck of the installation vessel before the commencement of the installation activities. Alternatively, they can be upended immediately after lifting from the cargo barge.

Piles are lifted and upended using two crane blocks or a single block with an internal lifting tool (ILT). The ILT is a specially designed tool, which consists of a mechanical device inserted into the inside of the tubular pile head, with hydraulic pistons which push a set of grippers against the inner walls of the pile driving head and support the weight of the pile through the friction generated between the ILT grippers and the inside wall of the pile head.

Other lifting options have been used such as the padeyes welded to the exterior of the pile, at some distance below the top of the pile so as to avoid any interferences with the pile driving hammer.

Once in the vertical position, each pile is lowered through the water and stabbed into the seabed or the template structure.

For steel jackets piles can be driven through the jacket legs or through the pile sleeves connected to the jacket legs at its base. The jacket leg and the pile sleeve both act as a guide for the positioning and the directionality of the pile.

14.4.2.2 Hammer Types and Sizes

The most common types of offshore pile driving hammers are steam and hydraulic hammers. The steam driven hammers can be used when driving piles through jacket legs or in shallow water where pile followers may be used which ensure that the hammer remains out of the water. However, with offshore developments moving into deeper waters, hydraulic hammers have been used to drive piles both below and above water. Hammers vary in size, weight and capacity depending on the characteristics of the pile to be driven and the soil properties to be driven into. They can be classified in terms of the maximum energy they can deliver. Existing hammers can drive piles up to 120 in. in diameter. Hydraulic hammers are more efficient than the steam hammers in terms of the energy delivered to the pile and, as such, their energy output needs to be carefully controlled and monitored.

The hammer has to be sufficiently large to drive the pile to design penetration in the given soil conditions. Typically the soil conditions considered correspond to both the lower and upper bounds. Also, the pile is assumed to be either "plugged" or "unplugged". The plugged condition refers to the case where the soil plug inside the pile is assumed to have become an integral part of the pile and moves with it as the pile closes at the bottom. The unplugged refers to the case where the soil plug is assumed to remain in level with the soil outside the pile and that resistance from skin friction continues to develop both inside and outside the pile. The combinations of soil upper and lower bounds and plugged and unplugged behaviour give rise to four cases of analysis, which need to be considered in the pile design.

A pile "driveability" analysis is normally carried out to establish the following:

- Whether the pile can be driven to the required depth with the proposed hammer size/energy in the four analysis cases.
- Whether the dynamic stresses in the pile exceed allowable stresses.

The driveability analysis is based on the wave equation method, first proposed by Smith (1960). In the absence of specific driveability analyses being carried out, guidance is available in the industry on required pile wall thickness and diameter combination for a given hammer size [API RP2A]. Almost all offshore pile installation projects, however, are now-a-days based on the pile driveability analyses.

Pile driving criteria are summarised in Section 14.9.5. Figure 14.16 shows a typical pile driveability analysis plot of blows per foot for the expected penetration depth.

14.4.3 Drilled and Grouted Piles

The drilled and grouted steel pile concept has been used successfully in offshore applications. Typically, a hole is drilled to a given depth into the sea floor through the leg of a jacket structure. A pile is then fed through the jacket leg and lowered into the drilled hole. Cement is then pumped down from the top through and around the pile to fill the gap between the pile and the sides of the hole in the seabed. Pumping is continued until the annulus between the jacket leg and the pile is filled with grout cement. In this way, structural continuity and load transfer is achieved from the jacket to the pile through the grout annulus between the pile and the inside wall of the jacket leg.

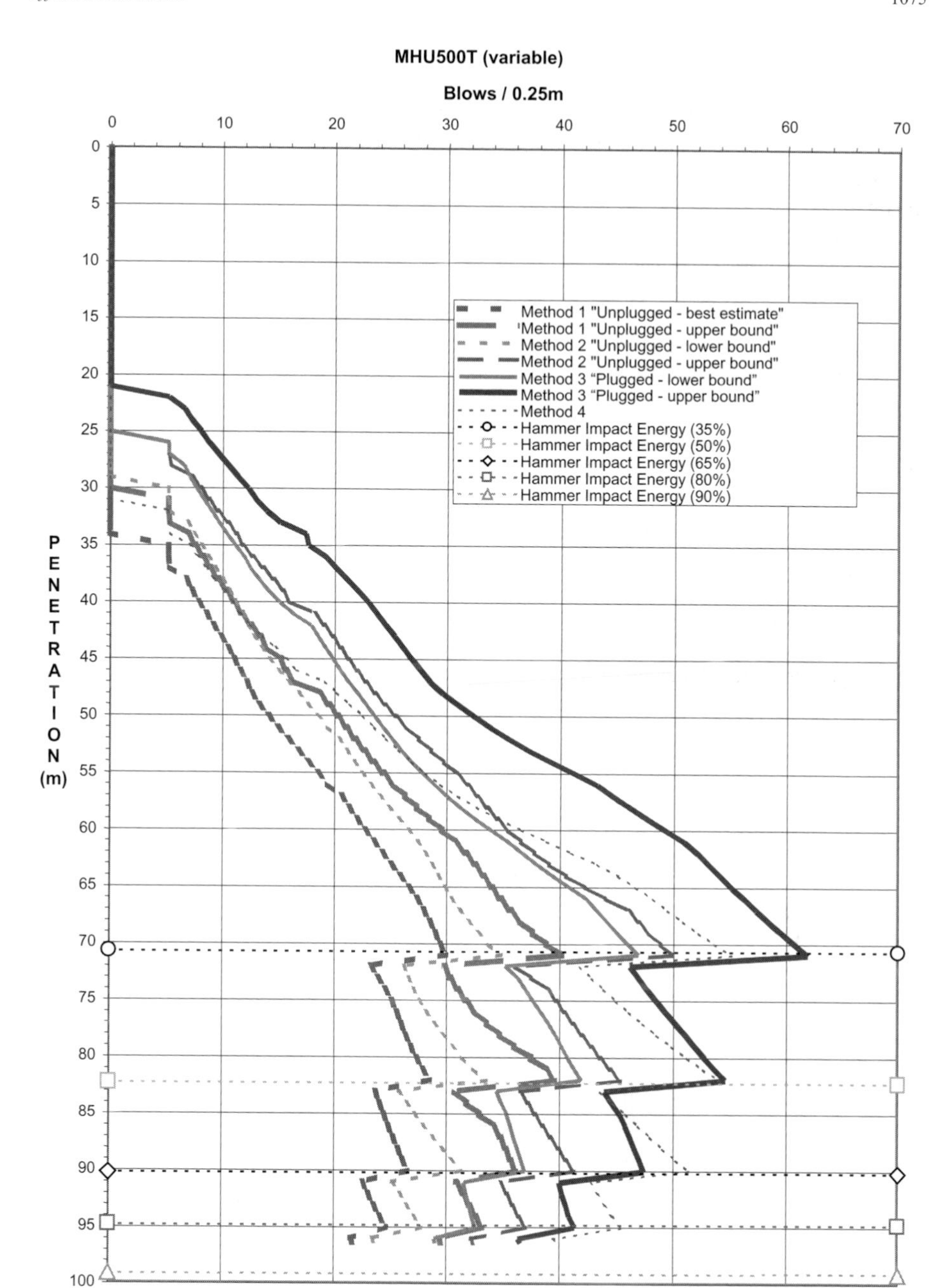

Figure 14.16 Pile blow count/penetration analysis plot

Drilling operations should be done carefully to minimise the possibility of hole collapse. Steel pipe casings are used when a hole instability is expected.

It is worth noting that the drilled foundations have a distinct advantage over the other types where holes can be drilled through the rock while pile driving may not be considered as an option.

14.4.4 Suction Embedded Anchors

Recently, suction embedded anchors have been used to anchor floating exploration and production platforms particularly in a soft cohesive seabed soil. They have been introduced in deepwater applications where alternative foundation concept may prove more costly and most probably require the use of a large derrick barge.

The suction piles are made of an open bottommed cylinder with a hole somewhere near the top through which water is pumped out to "suck" the pile into the seabed as shown in Fig. 14.17.

The suction anchors have been installed in water depths from as shallow as 40 m to as deep as 2500 m. Diameters ranging from 3.5 to 7 m have been used, with a penetration up to 20 m.

Unlike the drag embedment anchors, the location of the suction piles can be determined with great accuracy. This provides a distinct advantage in fields with congested subsea facilities. An added benefit of the use of suction anchors is that they do not need to be

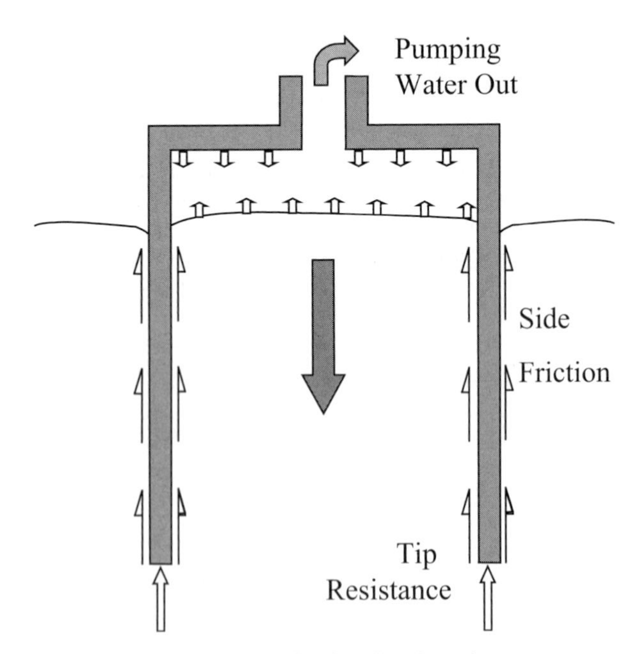

Figure 14.17 Suction pile schematic

dragged in order to be proof loaded. The choice of the installation vessel depends on the size of anchor and other operations that are taking place during the same installation campaign. For deep water mooring installations, the suction anchor is often installed at the same time as the mooring line, thus avoiding the need to connect those two components under water. There are also connectors which can be used to connect mooring lines to a pre-installed suction anchor.

The suction anchors can be lifted or skidded onto the deck of an anchor handling tug (AHT) which transports it directly to its offshore location ready for installation. The installation process consists of the following stages:

- Over-boarding
- Lowering to the seabed
- Penetration into the ground.

Deploying the pile over-board the installation vessel can be carried out using a crane or an A-frame depending on the size of the pile. Other low cost installation options are also available.

Once in water, the pile is lowered to the seabed using the vessel crane or deck-mounted winches. The most critical phase of the lowering process is the "hovering" stage where the suction pile is suspended several meters above the seabed. During this phase, successive heave cycles can cause the pile to partially penetrate and then retract from the seabed. As the pile approaches the seabed, the entrained water escapes below the lower rim and through the hole on top, thus creating a damping force on the pile motions. It is important to ensure at this stage that the damping loads and the seabed resistance to penetration do not cause slackening of the slings leading to subsequent snatch loads. Heave compensators fitted to the crane or the winch help make this stage much more controlled. Once the soil resistance to penetration exceeds the self-weight of the pile, the crane wires are slackened.

A survey package is normally attached to the suction pile at its top to give verticality and orientation information, but more usually this function is provided by an attending ROV which attaches itself to the anchor. The horizontal positioning of the anchor may be assisted by using pre-measured and installed guide-ropes which are tied back to an existing structure e.g. wellhead frame. Alternatively, a set of small buoys can be pre-installed to mark the target position of the pile.

Once the self penetration ceases, the attending ROV, which is equipped with suction pumps connects to the suction valves and pumping of water from the inside of the anchor can commence. The anchor penetrates the soil as a result of the water being pumped out of the hole at the top thus creating an under-pressure that drives the pile into the ground. Water is pumped out at a pre-determined and controlled rate so as not to implode the anchor.

The total soil resistance to penetration, R_{TOT} is the sum of the resistance from the side friction, R_{side} and the resistance from the tip including any stiffeners that may be present, R_{tip}:

$$R_{TOT} = R_{side} + R_{tip} \tag{14.1}$$

The amount of under-pressure, Δu, needed to penetrate the pile into the soil is:

$$\Delta u = \frac{R_{\mathrm{TOT}} - W}{A} \qquad (14.2)$$

where,

W is the submerged weight of the pile,
A is the projected horizontal area inside the pile.

The required under-pressure is inversely proportional to the pile projected area, A. Since A is proportional to the square of the pile diameter while R_{TOT} is proportional to the diameter, the required suction pressure reduces as the pile diameter increases.

If the suction pressure exceeds the soil capacity, the soil fails by upheaval in the soil plug inside the pile. The suction pressure is also limited by the structural integrity of the pile. It is important to verify that the soil capacity is not exceeded either during the lowering stage when the pile is accelerating while suspended from the crane hook, or during the penetration stage.

Since the required suction pressure is inversely proportional to the pile diameter, piles with larger diameter can achieve higher penetrations before reaching refusal. Refusal is defined by the suction pressure being equal to the limiting soil-failure loads.

Care has to be taken into designing of the suction anchor to ensure that the anchor does not rotate during penetration. The provision of the vertical cross walls inside the anchor, in the lower part, can stabilise the anchor during the penetration phase.

14.4.5 Drag Embedded Anchors

Drag anchors have been in use by ships for a very long time and have been used in the offshore industry, since its early days for mooring semi-submersible drilling vessels, single point moorings (SPMs) and installation vessels.

The drag anchors generate their holding capacity by self-embedding in the seafloor when pulled horizontally mobilising the shear strength of the seabed soil to resist the pulling force. The ultimate holding capacity of the drag embedment anchors is several multiples of its weight, depending on its type and on the soil conditions.

Some anchors embed themselves in the soil, irrespective of their orientation on contact with the seabed, for example the Bruce anchor. Other types of anchors, for example the Delta anchor, will only embed if it arrives at the seabed in the correct orientation. The installation of such anchors will involve the use of a second line, in addition to the anchor line, for correct orientation.

Anchors are normally installed by anchor handling tugs (AHT). When mooring a vessel the AHT approaches the vessel stern until it is in close proximity to the fairlead. The anchor is handed over to the AHT winch and the AHT heads towards the designated anchor location while the mooring vessel's winch pays out the mooring line. The anchor is then lowered to the seabed by either a wire attached to a ring chaser or a pendant wire. In the case of a ring chaser, the AHT pulls the ring back along the mooring line by the line attached to the ring and offers it to the offshore vessel. In the case of a pendant wire, the free end of the wire is attached to a buoy and left in position.

Once all anchors are in place, the mooring lines are subjected to tension test loads by pulling diagonally the opposite mooring lines against each other (cross-tensioning). The anchors can also be dragged by the AHT some distance (up to some 200 ft) along the seabed to achieve the required holding capacity. Piggyback anchors may also be added, if additional holding capacity is required.

14.5 Subsea Templates

Subsea templates are fabricated from steel members; they vary in size and weight depending on their functional requirements. Template weights typically range from a one hundred-tonne skid frame to several hundred tonnes.

14.5.1 Template Installation

Subsea templates can be installed using the same mobile offshore rig (MODU) used for well drilling or a heavy lift crane vessel.

Figure 14.18 shows a procedure for installing a subsea template by keelhauling it below the MODU. This installation method involves pre-installing the piles through a temporary pile guide frame, keelhauling the template below the rig and lowering the template to the sea floor using the drill pipe. In this type of installation, the weight of the template is restricted

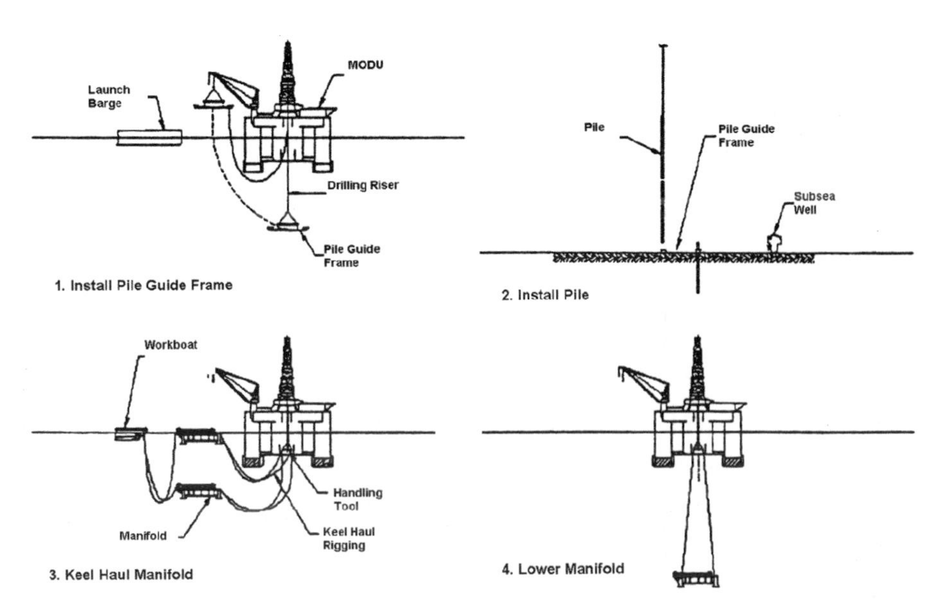

Figure 14.18 Installation of a subsea template by a MODU [Homer, 1993]

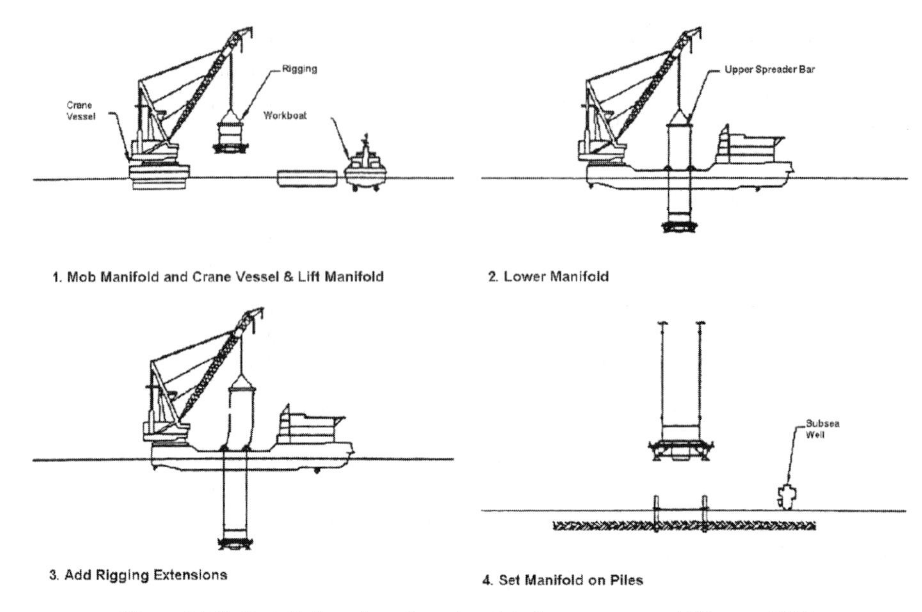

Figure 14.19 Installation of a subsea template by a crane vessel [Homer, 1993]

to the lift capacity of the rig's draw-works. The keelhauling phase can be simplified by the use of templates which are buoyant. Buoyancy may be obtained by using steel tubulars in the construction of the template structure. However, as the water depth increases, wall thickness to diameter ratio increases rapidly negating any perceived benefit from using the MODU for installation.

Figure 14.19 shows a procedure for installing a subsea template using a heavy lift crane vessel. The heavy lift vessel directly lifts the template off the deck of a transport barge and lowers it into water. The template is then further lowered to the sea floor on the crane hook using the crane's underwater block.

If an underwater block is not available or that its capacity is insufficient, the template can be transferred to a deep-water-lowering winch system. The transfer from the crane hook to the lowering system typically occurs after the template is in water. Particular attention has to be paid to the method of transfer as it imparts additional risk to the overall installation operation.

14.5.2 Positioning and Monitoring

Templates are installed within tight tolerances in terms of position, direction and level. The position and the orientation of the template are achieved through the use of the pre-drilled wellheads as guides. The docking piles, installed either before or after the drilling activities, are also used as guides.

The template is lowered to some height, typically a few feet, above the seabed, at which point the position and orientation of the template are verified and corrected, if necessary.

Inclinometers are mounted on and used to monitor the levelling of the template on the seabed. The inclinometers can be linked to a control room on the installation vessel through an umbilical line. Levelling is achieved by using hydraulic jacks, which act by pushing against the pre-installed piles and are remotely operated.

14.5.3 Rigging Requirements

Whether a MODU or a heavy lift is used, as the water depth increases, the lift capacity is diminished by dynamic load margins and by the weight of the lowering string.

Subsea templates are typically lifted using a single point lift with four wire slings connected to a single hook at the top and attached to four lift padeyes at the corners of the template structure. Figure 14.20 shows a subsea template ready for installation.

The rigging of the template is designed for the three phases of the installation operation:

- The lifting in air of the dry weight of the template from the transport barge. Dynamic factors apply, which account for the lifting by a floating structure from another floating structure in an offshore environment. The dynamic amplification factors for this phase are typically less than 1.25.
- The lowering of the template into water through the wave zone. The drag and the inertia due to the direct wave load impart additional loads on the template and the supporting riggings. Slam and slap loads can also be significant.
- The lowering into water of the template to the seabed. The drag and inertia loads in this phase result from the template's dynamic motions that are caused by the motions of the installation vessel. The combination of the template mass and hoisting wire stiffness can give rise to natural heave period for the template that are in the same range as the installation wave periods. The resulting resonant response of the template induces

Figure 14.20 Subsea template lifting

dynamic tensions that equal, or exceed, static tension in the rigging. A dynamic analysis can be carried out to calculate the motions and line tensions in either time or frequency domains. Cranes or winches with heave compensation are often used in deepwater installation to avoid such resonant response.

14.5.4 Existing Subsea Facilities

The design and the installation method of a subsea template should take into account any existing installations on the seabed. Wellheads are generally pre-installed; flow lines may also be installed before the template. Other cluster well systems, jumpers, may exist in close proximity of the intended location of the subsea template. All these factors need to be taken into consideration in the design and installation methodology of the subsea template.

14.5.5 Seabed Preparation

The surface of the seabed is rarely horizontal or even. The horizontality of the template is crucial to successful drilling operations. For these and other reasons, piles are installed prior to the installation of subsea templates. The piles can be driven, jetted or drilled and grouted. For foundation types and installation methods, refer to Section 13.4.

14.6 Loadout

The phase of transferring the completed structure from the quay onto the deck of a cargo vessel is referred to as the loadout operation. Most loadout operations take one of four forms:

a. Trailer loadout where mutliwheel hydraulic trailers are brought underneath the structure, in order to lift it and wheel it onto the deck of a barge which is placed right up against the quay;
b. Skidded loadout where the structure rests on steel rails and winches are used to push or pull the structure onto the deck of a barge which would have to be equipped with skid beams to take the structure onto its final location on the barge;
c. Lifted loadout where the modules are lifted onto the deck of the barge using shore-based cranes or floating crane barges;
d. Float-away loadout where a structure is built in a dry dock facility, such as semi-submersible hulls, TLP hulls, FPSO hulls, etc. Upon completion, the dry dock is flooded, or ballasted down in the case of floating dry docks and the structure which floats under its own buoyancy is towed away by tug boats.

The decision on the type of loadout should be made as early as possible in the design process as, it has direct consequence not only on the configuration and size of structural members but also on the economy of the project.

14.6.1 Loadout Methods

The choice of a loadout method depends on a multitude of factors such as the geometry of the structure, its weight and the availability of trailers close to the fabrication site.

The experience of the designer and the fabricator also influences the choice of the loadout method. Like any other project phase, pure commercial factors are quite often the reason behind a certain loadout method to be adopted.

14.6.1.1 Trailer Loadout

For a trailer loadout, the module is supported on multi-wheel trailers for the movement onto the cargo vessel. The trailers may be self-propelled or may be pushed or pulled onto the vessel. Trailers accommodate uneven ground surfaces and small movements between the barge and the quay.

The support configuration over the trailers is likely to be different from the in-place configuration leading to different load path and set of stresses being imposed on the structure. A separate analysis is normally carried out to verify the structural integrity in this mode.

Trailers are normally arranged in three hydraulic groups such that the load on each group can be calculated by simple statics. The reactions from the axles in each group is applied as a uniformly distributed load acting upwards against the weight of the structure. They can be regarded as a series of linear springs, if necessary.

In this type of loadout, it is important that the loadout barge maintains elevation against the quay within a specified tolerance, which is typically a few inches. An adequate ballasting system with sufficient redundancy is essential for the success of the loadout operation. The ballast system compensates not only for the transfer of load but also for the effects of tide. The global and the local strengths of the loadout barge, in addition to the stability of the system, are important considerations in determining a ballast plan for the loadout operation. A typical trailer loadout is shown in Fig. 14.21.

Figure 14.21 Topside module loadout on trailer

14.6.1.2 Skidded Loadout

In a skidded loadout, the structure is pushed (by jacks) or pulled (by winches) onto the cargo vessel. Skidding may also be achieved by utilising skid units which travel together with the skidded structure. The moving is effected by a combination of push/pull hydraulic jacks and clamps.

The structure is usually supported on skid shoes that are guided over the skid beams.

The force required to move the loadout object along the skid rails depends on the friction between the skid shoes and the skid rails. The initial load required to move the structure from the static, typically the erection location, is referred to as the breakout or the static load. As the structure moves forward, the force required to keep it moving is less than that at breakout.

Table 14.2 shows typical values for the static and the dynamic coefficients of friction.

Teflon pads are sometimes mounted on top of the skid rails, with grease applied to them in advance of the moving structure to minimise frictional resistance. The speed of the loadout operation is dictated by several factors such as the stroke of the jacks, the number of parts on the pulling winch or the speed of the ballasting system.

For planning and design purposes, the capacity of the skidding equipment such as jacks, wires and anchor points should exceed the breakout force described.

14.6.1.3 Lifted Loadout

When considering a lifted loadout, the designer should take into account the lift capacities of available cranes. These may consist of land cranes with lift capacities measured in hundreds of tonnes, or of floating sheer-leg cranes with capacities reaching thousands of tonnes. The same rigging arrangement as for the offshore lift can sometimes be used. A visual inspection of the lift points is required upon completion of the lifted loadout. If the rigging arrangement is different from the installation one, a separate loadout lift analysis is required. The stability of the land cranes or the floating sheer leg also requires checking.

Table 14.2 Coefficient of friction used in skidded loadout operations

Contact Surfaces	Static	Dynamic
Steel/Steel	0.15	0.12
Steel/Teflon	0.12	0.05
Stainless Steel/Teflon	0.10	0.05
Teflon/Wood	0.14	0.06
Steel/Waxed Wood	0.10	0.06

14.6.1.4 Float-away Loadout

The weight of some offshore structures increases to levels where it is not feasible or economical to load them out from the quay directly onto the transport vessel. In this case, these structures are designed to be launched from a dry dock and wet-towed either to their offshore location or to an awaiting transport vessel.

Examples of floated-away structures include the hulls of semi-submersibles, Tension Leg Platforms, FPSOs and FSOs.

14.6.2 Constraints

The type of constraints that need to be considered depends on the loadout method chosen. For the skidded or trailer loadouts, the following parameters need to be considered:

- Quay and barge local and global strength.
- Barge freeboard in relation to quay height.
- Tidal range and rate of variation with time.
- Pump capacity and redundancy.

Pressures imposed by trailer wheels on the ground or transport barges are often less than 10 ton/m^2. Typically, the local quay strength and the barge deck strength are adequate for this level of loading. Skid beams, on the other hand, impose concentrated loads on the quay and the barge deck. The skid beams are often supported on piled foundations on the quay. Where possible they are also aligned with the barge's longitudinal bulkheads to minimise stress on the barge transverse frames.

Transfer of the ballast water, either between the barge tanks, or between the barge and the sea is carried out to compensate for the weight of the structure or the effects of tide. The pumps should have adequate capacity to keep the barge level with the quay within a specified tolerance that depends on the loadout equipment. Pumps should have some redundancy and spare capacity, typically 50%, to allow for individual pump failures.

Timing is also another factor where, in order to take advantage of the tidal conditions, a loadout operation may commence at low tide so that unnecessary de-ballasting of the barge is avoided.

Water depth and quay height above the water line at the loadout quay represent additional constraints. A minimum underkeel clearance of around one metre shall be maintained at all stages of the operation. In some instances where the water depth is not sufficient, a grounded loadout is considered where the barge sits on a well leveled and prepared seabed. If a barge needs to be grounded for a certain loadout operation, due to limitations on the water depth and quay height, the condition, levelness and bearing capacity of the seabed at the quay are some important considerations. In this case, the ballast plan has to be developed so as to ensure that only a proportion of the barge and cargo weight is resisted by the seabed. This proportion is limited to the bearing capacity of the soil.

The constraints associated with the lifted loadouts are similar to those considered in any lifting operation. Mooring large crane vessels in the vicinity of loadout facilities is sometimes difficult. When land-based cranes are used, the strength of the quay and the load-sharing between the cranes, if more than one is used, are important considerations.

Other constraints relate to the weather and include the swell, the current, the wind speed, visibility and the general weather conditions.

14.6.3 Structural Analysis

The loadout procedure provides a detailed description of all the stages of the loadout operation. A representative structural model is normally set up to incorporate the support configuration during the various phases of the operation. For the trailer and lifted loadouts, the support configuration does not change significantly during the loadout and a single analysis would be adequate to model all the loadout stages. Care should be taken to ensure that the hydraulic connectivity of the trailers and the potential for variations in the load sharing between the different trailer groups is understood and accounted for in the analysis.

In the case of a structure supported on four skid shoes or more, the structure needs to be analysed and checked for settlement or loss of support due to the barge movement or ballasting inaccuracies. The amount of mis-alignment that needs to be considered depends on the loadout procedure. It is usually difficult, and too restrictive, to keep the mis-alignment below 25 mm.

For analysis purposes, a loadout is regarded to be a static condition. If the loadout analysis is carried out on the basis of the working stress design code (e.g. AISC or API RP2A-WSD) no increase in the allowable stresses can be taken into account.

14.7 Transportation

14.7.1 Configuration

Structures can either be wet or dry-transported. In a wet transport the structure floats on its own hull and is towed by one, or more tugs to the offshore site. In the case of dry transport, the structure is loaded onto a flat top cargo barge, on a general purpose cargo carrier or on a purpose built submersible ship often referred to as a heavy lift vessel (HLV). Topsides, jackets, piles and subsea units have no or little buoyancy and are normally transported "dry". Structures such as semi-submersible vessels, gravity base platforms, tension leg platforms, spars and jack-up rigs can be either wet or dry transported. The decision to transport these structures dry or wet depends on:

- Dimensions, weight and centre of gravity height of the structure: The current cargo weight record on submersible ships is 60,000 ton.
- Transport route design environment: If the direct environmental loads or motions associated with a wet tow are too onerous on the structure, it needs to be dry transported.
- Distance and schedule constraints: Heavy lift vessels and general purpose ships are the fastest mode of transportation and are therefore the most common modes for long

distance transportations. Typically, heavy lift vessels can achieve calm weather speeds of 12–16 knots while the wet tow speeds are in the range of 4–8 knots. Large structures as the gravity base platforms, spars and large TLPs are towed at speeds of less than 4 knots.

- Cost: Heavy lift vessels are competitive for long and medium distance transports, while towing is more cost effective for short tows.
- Ability to avoid bad weather: In areas with tropical revolving storms or generally harsh environment, tows can only be undertaken at certain times of the year. Tows are generally too slow to change course to avoid forecast bad weather or seek shelter.

14.7.2 Barges and Heavy Lift Ships

Transport barges vary in size and capacity. Their availability also depends upon the geographical location. There are mainly two types of transport vessels:

- Towed barges and
- Self-propelled ships including submersible heavy lift vessels.

Cargo Barges

These are barges which are towed or pushed by tug boats to transport from one location to another. These, in the majority, are flat top and bottom and are simply equipped with navigational lights, fairleads and towing points. A small proportion of these barges are designed to be submerged so as to pick up floating cargoes. These are equipped with a forecastle and a deck structure at the bow and have their own ballast system. Large steel boxes, stability casings, are added at the stern to provide additional water plane area necessary for the stability of the barge and its cargo as the deck goes through the water line. These stability casings are removable and can be stowed away on the deck of the barge or stored onshore when not required.

Towed barges are classified not only by their length and width and also by their mode of utilisation (e.g. Launch barges, submersible barges). The typical barge sizes and their uses are:

- Barges less than 200 ft in length and 50 ft wide. These are small pontoons used for carrying small structures in sheltered inshore waters.
- 250 ft × 70 ft barges. These are relatively small pontoon barges with no ballast systems of their own. They are used to transport small offshore modules, small jacket and piles, tendon sections for TLPs, containers for pile driving hammers, modules of drill rigs, etc.
- 300 ft barges. These can be 90 or 100 ft wide barges. They represent standard cargo barges used quite extensively in the offshore industry. Most of these barges are not equipped with a ballast system of their own. Medium size structures, in the region of 3000 Te have been transported on barges of this type.
- 400 ft × 100 ft barges. These barges are often equipped with a ballast system of their own. Due to the deck space available on the barge, more than one structure can be transported onboard these barges. These barges are ideal for transporting piles and

bridges as they avoid the risk of immersion in water and wave slamming. Some of these barges are used for launching shallow water jackets.

- Barges of 450 ft and longer have been used for jacket launching. These barges are equipped with ballasting systems in addition to skid beams and rocker arms at the stern to enable the launching of jackets. Heerema's H851 barge, which is nominally 850 ft long by 200 ft wide, is the largest barge available in the industry.
- Submersible barges. These are towed barges equipped with stability casings aft and a ship-like bow structure and a bridge, sufficient to enable the submerging of the barge above its main deck. The Boa barges (nominal dimensions 400 ft × 100 ft), the AMT barges (nominal dimensions 470 ft × 120 ft) and the recently built Hyundai barges (nominal dimensions 460 ft × 120 ft) are examples of these submersible barges. These barges can submerge up to 6–8 m above their decks.

Vessel owners and operators publish data of their vessels in terms of deadweight which provides a broad indication of their carrying capacity. Additional requirements need to be met in terms of their global strength, local deck and frame strengths and height of the cargo's centre of gravity. While a cargo barge may be able to transport a 10,000 ton structure with low vertical centre of gravity and supported on a large number of points on the deck, it may only be able to transport a 6000 ton topsides module which has a relatively high centre of gravity and supported on fewer support points.

A typical tow arrangement with a towing bridle is shown in Figs. 14.22 and 14.23. Two lines run from tow brackets through fairleads on the barge and connect to a triplate through towing shackles. These two lines are referred to as the towing bridle. A third line connects the triplate to the winch of towing tug. An emergency wire is installed along the length of the barge. The line is attached to a synthetic rope that terminates with a buoy which trails behind the barge during tow and forms part of the towing arrangement.

The size of a tug is determined on the force required to hold the tow in a given environment. The Noble Denton Guidelines require the tow to hold station in a Beaufort Force 8 with a corresponding significant wave height H_s of 5.0 m, a wind speed of 20 m/s (at 10 m above mean sea level) together with a current speed of 1 knot with the barge heading into the wind. The resulting load is multiplied by an efficiency factor, which accounts for the difference in the tug-pulling capacity between calm weather and storm conditions. Further reduction in the efficiency applies when multiple tugs are used.

The towline pull required (TPR) is usually calculated by adding the wind, the current and the wave forces. The wind force (in tonnes) is calculated as:

$$F_w = 0.0625 \sum (AC_h.C_s) V_w^2 \tag{14.3}$$

where,

A_w is the projected wind area (in m^2)
C_h is the height factor (from MODU)
C_s is the shape factor (from MODU)
V_w is the wind speed in m/s.

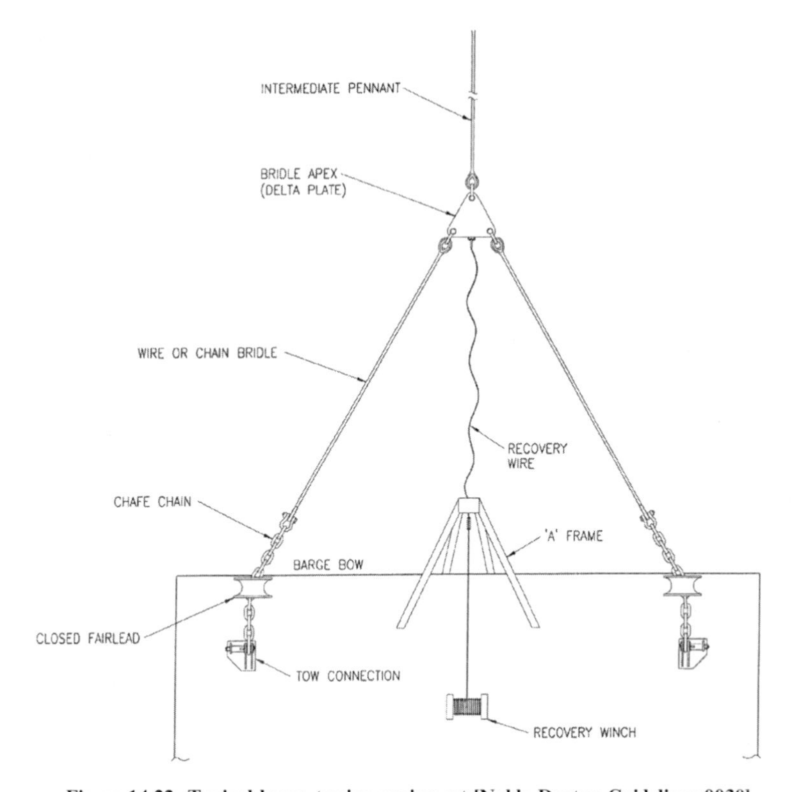

Figure 14.22 Typical barge towing equipment [Noble Denton Guidelines 0030]

Figure 14.23 Typical arrangement of tow line and bridle

The current force (in tonnes) is calculated as:

$$F_c = 0.5\rho \sum (A_c C_d V_c^2)/g \qquad (14.4)$$

where,

A_c is the projected current area (in m^2)
C_d is the drag coefficient which varies between 1.0 for a barge with a flat bow and 0.3 for the spoon bows
V_c is the current speed.

The wave force (in tonnes) is normally calculated from a diffraction analysis. In the absence of any specific data, wave forces can be conservatively calculated using:

$$F_v = \rho B H_S^2/16 \qquad (14.5)$$

where,

B is the beam of the barge (in m)
H_s is the significant wave height (in m)

Submersible Heavy Lift Ships

These ships often have two propulsion systems that are independent of each other and provide an adequate margin of safety against the ship being completely incapacitated. Some heavy lift ships also have a dynamic positioning system. The ship's ballast system enables it to submerge its deck, allowing the floating cargo to be floated on or off. The speed of these ships makes them attractive for long haul transportation operations. Their speed also gives a greater ability to avoid forecast storms. This is considered to be a distinct advantage in places and seasons that are prone to severe weather conditions such as tropical revolving storms.

Table 14.3 lists the largest self-propelled transport vessels with some of their characteristics.

Topside decks as well as semi-submersible and TLP hulls have been carried on the decks of heavy lift ships. For structures that float on their own hulls, the heavy lift ship submerges such that its deck and cribbing clear the keel of the floating cargo by a safe margin of about 3 ft. The cargo is then floated over the ship's deck and positioned against pre-installed guides by means of wires and winches. The ship is then de-ballasted back to transportation draft. Figure 14.24 shows the dry transport of a semi-submersible vessel with a displacement in excess of 40,000 ton. The topside structures are loaded on the heavy lift ship using one of the conventional methods such as skidding, using trailers or using cranes.

14.7.3 Design Criteria and Meteorological Data

Stability and strength are the main aspects of a transportation operation that need to be verified. The following engineering studies are normally undertaken when planning a

Table 14.3 List of largest self-propelled dry transport vessels

Vessel Name	Vessel length	Vessel beam	Deck depth	Submergence depth above main deck	Dead weight
	(m)	(m)	(m)	(m)	(ton)
Swan Class (Tern, Swift and Teal)	180.5	32.26	13.3	7.3	32,650
Tai An Kou and Kang Sheng Kou	156.0	32.2	10.0	9.0	18,000
Transshelf	173.0	40.0	12.0	9.0	34,000
Mighty Servant 3	181.2	40.0	12.0	10.0	27,700
Black Marlin	217.8	42.0	13.3	10.1	57,000
Mighty Servant 1	190.0	50.0	12.0	14.0	41,000
Blue Marlin	217.8	63.0	13.3	10.1	78,000

Figure 14.24 Dry transport of P-40 on self-propelled vessel mighty servant 1

transportation operation:

- A route study to evaluate the design environmental criteria. This is normally carried out when a voyage-specific motion analysis has to be carried out.
- A stability study to demonstrate that the carrier vessel, in the case of a dry transport, or the hull of the transported structure, in the case of a wet tow, meet the requirements of the IMO or the classification society. The analyses are normally carried out using the generic wind speeds of 100 knots for intact stability assessment and 70 knots for

damaged stability. Lower wind speeds are sometimes considered on a case-by-case basis for restricted tows in sheltered waters. The stability requirements are covered later in this section.

- Motions and accelerations study. Typically, motion analyses are carried out with the voyage specific environmental criteria using diffraction or strip theories. In the absence of such meteorological data, deterministic motions are often used.
- A structural assessment taking into account the loads associated with the motions and accelerations.
- Seafastening design.
- A local and global strength assessment of the carrier vessel in the case of a dry transport.

The most widely used deterministic motions criteria are those introduced by Noble Denton for flat bottom cargo barges and other types of carrier vessels. The criteria are:

20° roll angle in 10 s period $\pm$ 0.2 g heave acceleration.

12.5° pitch angle in 10 s period $\pm$ 0.2 g heave acceleration.

When deriving the voyage-specific environmental data for the transportation route, the 10-yr return environment is normally considered. Given the temporary nature of the transportation phase, the data is normally derived specifically for the departure month so as to take advantage of seasonal variations. The transportation route is normally split into several sectors within which the environment is assumed to be uniform as shown in fig. 14.25. The duration of exposure within each of those sectors is calculated based on the

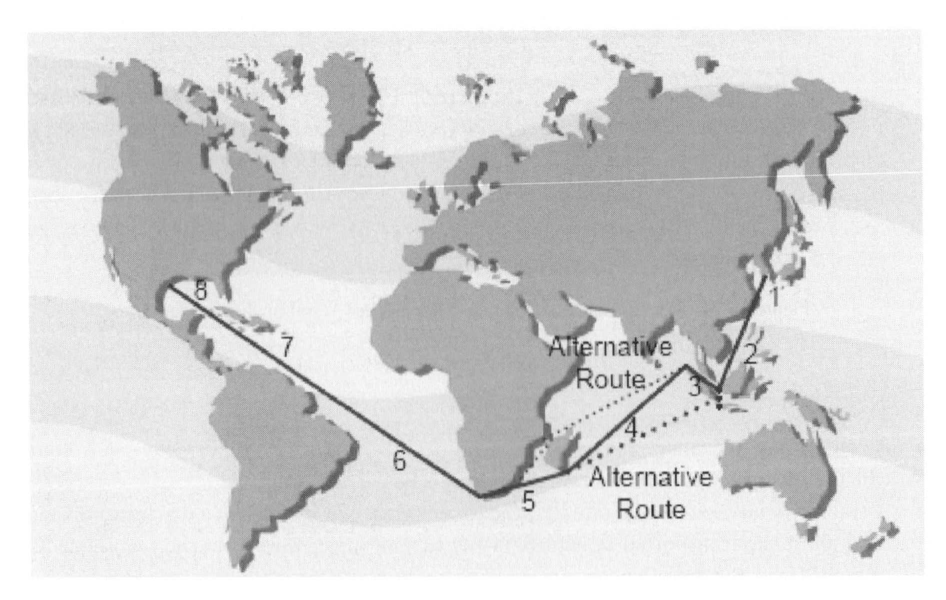

Figure 14.25 Typical transport route sectors between Korea and the North Sea

vessel speed. Given that the exposure periods are normally less than 1 month, the environmental data may be reduced to allow for the shorter exposure periods.

The monthly wave-height scatter for each sector are normally used to define the Weibull distributions using the method of moments. Additionally, the Fisher–Tippet Type 1 distributions can also be fitted to the wave data. An average month in a 10-yr period will have approximately 2435 periods of 3 hour storms. The probability of non-exceedence associated with the 10-yr return monthly storm is therefore 0.9996.

Meteorological data sources include the satellite databases and the voluntary observatory fleet (VOF) data sets. The most comprehensive satellite data set available is a satellite radar altimeter data for which 15 years of data is now available. Each altimeter measures the significant wave height over a 5–10 km footprint every second (corresponding to 7 km steps) giving an accuracy comparable with estimates of wave height from a 20 min buoy record.

Synthetic Aperture Radar (SAR) data allows the computation of the directional wave spectra from the satellite-measured data so that all the wave parameters are available for analysis. This type of information has only become available recently and may not be as accurate as the satellite altimeter data for wave height, but nevertheless it provides very useful descriptions of the sea surface. The most comprehensive databases are the CLIOsat database and the ARGOSS internet-based wave climate database.

VOF data sets include visual observations of wind speed, wind direction, wave height and direction, wave period and swell height, period and direction, among other parameters, provided voluntarily by ships officers of many nations.

14.7.4 Transport Route

Transportation routes are selected based on the economic, environmental and safety considerations. The following factors are considered:

- The environmental conditions along the transport route affect the motions of the vessel and the voyage speed. The weather conditions after the commencement of the transport operation often dictate local deviations from the planned route.
- The existence of safe havens. As part of a contingency planning, particularly for long transports, safe havens have to be identified in case the conditions require the vessel to seek refuge in a port.
- Vessel or cargo dimensions and hull draft which restrict passage below certain obstructions, such as bridges, or in shallow water or through locks and waterways.
- Costs of the passage through canals, such as the Suez Canal.

14.7.5 Motions and Stability

Motion analyses are carried out to estimate the motions and accelerations of a vessel during transport normally using the frequency domain analysis techniques. These analyses are often carried out in two phases. In the first phase, the motion response to regular waves for a range of wave periods is derived in the form of vessel response amplitude operators (RAOs) for all six degrees of freedom. In the second phase, the response to irregular waves is derived using the significant wave height and wave period.

The following parameters are needed for the motion analysis:

- Significant wave height representative of the tow route
- A range of peak wave periods
- Wind speed and
- Vessel heading relative to the waves.

The design wave height, H_s, can be based on the 10-yr return adjusted for the periods of exposure. The range of peak wave periods, T_p, is used to account for the different wave steepnesses and can be obtained from the following expression:

$$\sqrt{13 \cdot H_s} \leq T_p \leq \sqrt{30 \cdot H_s} \tag{14.6}$$

where H_s is expressed in metres and T_p in seconds. If the peak roll period of the barge falls outside the T_p range for the design wave, smaller waves with periods similar to that of the barge roll period are also analysed.

In the absence of a motions analysis, the loads can be combined deterministically as follows:

$$F_{HR} = W\left[\pm\frac{4\pi^2\theta z}{T_R^2 g} + (1 \pm 0.2)\sin\theta\right] \tag{14.7}$$

The vertical force is given by:

$$F_{VR} = W\left[\pm\frac{4\pi^2\theta y}{T_R^2 g} + (1 \pm 0.2)\cos\theta\right] \tag{14.8}$$

The rotational moment of inertia is given by:

$$M_R = I_{yy}\left[\pm\frac{4\pi^2\theta}{T_R^2 g}\right] \tag{14.9}$$

where,

T_R Roll, or pitch, period (in seconds)
θ Roll, or pitch, angle (in radians)
z Height above the centre of rotation (assumed to be at the waterline)
y Horizontal distance from the centreline of the barge
g Gravitational acceleration (m/s^2)
I_{yy} Moment of inertia of the cargo about its longitudinal axis
F_{HR} Inertia force parallel to the vessel's deck
F_{VR} Inertia force normal to the vessel's deck.

Intact/Static Stability

Stability requirements are stipulated by classification societies or, in some cases, marine warranty surveyors. Stability requirements such as the range of positive stability, the required area ratio and the damaged stability scenarios to be considered depend on the shape of the hull. The following are extracts from Noble Denton's stability requirements for ships and barges:

"The range of intact stability about any axis shall not be less than 36° for large barges and 40° for small cargo barges (less than 23 m in beam or 76 m in length). Alternatively, if model tests or motion analyses are carried out, the minimum range of static stability shall not be less than $(20+0.8^*\theta)^\circ$, where θ is the maximum amplitude of motion plus the static angle of inclination from the design wind. The buoyancy of a watertight cargo may be considered in the computation of the stability characteristics.

Any opening giving an angle of down flooding less than $(\theta+5)^\circ$ shall be closed and watertight when at sea, where $\theta=20^\circ$ for large barges and 25° for small barges. A cargo overhang shall not immerse as a result of heeling in a 15 m/s wind in still water conditions.

The area under the righting moment curve to second intercept of the righting and wind overturning moment curves or the down flooding angle shall not be less than 40% in excess of the area under the wind overturning moment curve to the same limiting angle".

14.7.6 Seafastenings/Tie downs

Where possible, the strong points on the cargo, such as the legs of topside decks, are located over the strong points of the carrier vessel, such as the bulkheads. Where this is not possible, the weight of the structure is supported on steel grillages that distribute the static and the dynamic loads into the carrier vessel's strong points.

When the dry transport cargo is a plated structure, such as the hulls of mobile offshore drilling units or tension leg platforms, the weight of the cargo is distributed into the deck of the carrier vessel through wood cribbing. Cribbing could be aligned with the cargo's frames to avoid overstressing it or arranged in a random fashion. Where random cribbing is used, the dynamic stresses in the cribbing are normally limited to 1 N/mm^2. Otherwise stresses are limited to 4 N/mm^2. Other limiting stresses are considered depending on the type of wood used.

Seafastenings are structural members that are made of steel members or steel wire that are used to restrain the structures on board a vessel against movement due to the vessel motions. Steel wires lashings are normally used for smaller cargoes that are transported on board cargo ships. For large offshore structures, seafastening can consist of a system of steel tubular members which are welded to the cargo and to the deck of the transport vessel.

The design of the grillage and seafastening is usually carried out to the requirements of the AISC and the API RP2A. For the design of the seafastening members, the allowable stress may be increased by a third to reflect the transient and extreme nature of the transport load. The third increase in the allowable stresses does not however apply to the local strength of the deck of the carrier vessel.

14.7.7 Structural Analysis

When seafastening members are modelled, the structural analysis is carried out in two phases:

- Still water static condition and
- Transport dynamic condition.

The first phase consists of a static analysis in still water condition, where the full structure is modelled and the seafastening members omitted from the analysis. Only the gravity loads are considered at this stage.

In the second phase, the seafastening members are added, or the boundary conditions are modified to reflect their addition. The analysis is carried out against the dynamic loads such as:

- Vessel motions and accelerations listed above. Standard motions are combined as follows:
  $\pm$Roll $\pm$ heave
  $\pm$Pitch $\pm$ heave
- Deflections of carrier vessel, if they are significant
- Direct wave loads due to the inundation of cargo overhang, if they are present.

Where the voyage-specific motion analyses are carried out, direct wind loads on the cargo are also considered. However, to allow for the coinciding of the low probability of extreme motions and extreme wind loads, the combined loads are reduced by 10%. The design load is then the highest of:

- Motion-induced loads alone
- Wind-induced load alone
- $0.9 \times$ combined motion and wind loads.

Static and dynamic loads are combined to calculate the highest stresses in the cargo and the highest and lowest reactions to the carrier vessel.

14.7.8 Inundation/Slamming

Parts of large cargoes that overhang the transport vessel may be subject to direct wave loading such as slamming, drag and buoyancy. Slamming loads result when the structure makes the first contact with the water surface. These are impulsive loads with durations in the order of several milliseconds. As such, their effect is generally localised to the area of the slamming with little or no loads transmitted globally to the rest of the structure. The drag and the buoyancy loads result from the subsequent submergence of the structure in the water. The duration of these loads is of the same order of magnitude as the vessel motions and wave periods. They may therefore be combined with other dynamic loads. However, inundation loads are often not in phase with the global inertia loads and may act to reduce these loads. It is recommended to carry out several analyses that include and exclude the inundation loads and design for the worst case.

Cargo inundation changes the hydrodynamic stiffness and the motion characteristics of the carrier vessel. If significant inundation is expected, its impact on motions should be considered. While simplified methods are available for this purpose, the non-linear behaviour introduced by cargo inundation is best predicted by model testing or time domain analysis.

Slamming is dependent on the encounter relative velocity between the structure and the instantaneous water surface. It also depends on the encounter angle and the shape of the structural member and the amount of entrained air in the water. Most research carried out on slamming corresponds to idealised conditions such as the slamming of wedges or flat plates on flat calm water surfaces. This makes their results conservative.

The wave slamming forces can be evaluated on the basis that the impact slamming force is equal to the rate of change of momentum of the water, given by equation (14.10):

$$F = \frac{d(mV_s)}{dt} \tag{14.10}$$

where m is the mass amount of water and V is the velocity of an equivalent circular cylinder. There are difficulties in estimating the impacting mass of water and the velocity of the equivalent cylinder which varies with time.

The slamming force is given below:

$$F = 0.5C_s\rho V_s^2 \, \ell d/g \tag{14.11}$$

where,

- C_s is the slamming coefficient, typically taken as equal to π for a tubular member. Generally this coefficient has to be agreed upon for shapes other than tubular members
- d is the diameter of the cylinder
- ℓ is the length of the cylinder
- V_s is the encounter slamming velocity between the member and the wave
- ρ is the density of water.

14.8 Platform Installation Methods

14.8.1 Heavy Lift

This is the most common method of installation of offshore structures. In this method, the structure is lifted off the transportation vessel by a crane vessel and lowered into position. The lifted structure is equipped with lifting lugs and slings that are connected to the lugs and the hook of the crane vessel. Figure 14.26 shows a typical lifting arrangement.

The most common method of attachment of slings to the lifted structure is by the use of shackles connected to padeyes that are welded to the structure. Shackles of up to 1000 tonnes Safe Working Load (SWL) size are available. The slings can be alternatively wrapped around trunnions or cast padears that are tailor designed for the lift. Cast padears are normally used in larger lifts.

Given that the structure is supported in the lifting mode in a different configuration from the in-place condition, its integrity in the lift condition needs to be verified. Also, the slings are often attached to the structure at an angle to the vertical, thus imparting horizontal loads in the area of the lifting lugs. Additional bracing is normally required to resist these loads. Where this is not possible, the slings are kept vertical by the use of spreader bars or

Figure 14.26 Lifting arrangement

spreader frames. For statically indeterminate lifting configurations, the structural integrity also has to be verified against any sling length mis-match that cause the redistribution of loads to individual slings.

The main constraint associated with the lifted installation is the capacity and the availability of crane vessels. There are only a few crane vessels with a lifting capacity in excess of 5000 ton. Furthermore, the availability and cost of such vessels is normally a major consideration when planning an offshore installation operation.

14.8.2 Launch

This method is typically used for installing jackets with weights that exceed the lifting capacity of available cranes. Launching operations are performed over the stern of the launch barge. The launch barge arrives on site with the launch rigging already attached to the jacket and with the jacket overhanging the barge stern. The launch operation starts by trimming the barge typically by about 4–5° by the stern. In order to initiate the sliding of the jacket over the skid beams, the launch winches pull the jacket towards the stern. As the jacket travels towards the stern the barge trim increases and the sliding process is accelerated till the centre of gravity of the jacket passes over the rocker arm hinges. At this point, the jacket starts to rotate and enters the water. The barge accelerates in the opposite direction of the jacket and a complete separation between the two is achieved. This operation normally lasts several minutes.

Figure 14.27 Launching of a 4-Leg Jacket

The trajectory of the jacket during the launch should be such that the jacket clears the seabed by a sufficient margin. Launch trajectory is predicted by a launch analysis. In the launch analysis, the equations of motion of the barge and the jacket are solved at small discrete intervals during the launch sequence. The jacket-entrance into the water introduces drag and inertia loads onto the jacket members that resist the jacket motion. The launch trajectory is dictated by the relative magnitude of weight and buoyancy of the jacket, the relative positions of their respective centres and by damping introduced through the drag loads on the jacket members. Figure 14.27 shows a launched jacket entering the water.

14.8.3 Mating

Also known as "deck mating" and "floatover", this method is used for a deck installation when the weight of the deck exceeds the available crane capacity. The mating operation is executed using the transporting vessel which may be a flat top cargo barge or a heavy lift ship.

The most common deck mating method in offshore environments is the internal floatover where the transportation vessel is maneuvered between the legs of a fixed platform jacket. Deck mating is also used for installation of decks of the semi-submersible vessels over their hulls. The weight of the deck is transferred to the jacket, or floating substructure, solely or largely by ballasting the barge down until contact is made between the jacket and the deck and the load is transferred completely from the barge to the jacket. Figure 14.28 shows an internal floatover operation.

Figure 14.28 Floatover of the Malampaya topsides (Heerema)

Deck mating requires the deck to be supported during transportation and installation at locations other than its normal in-place supports. Additional steel trusses are therefore required to transfer the weight of the deck into these temporary supports. The jacket also has to be designed for an internal floatover. The distance between the jacket legs has to be greater than the width of the barge by a sufficient margin to allow safe entrance and withdrawal. In addition to increasing the size and weight of the jacket to accommodate the barge entrance, this has a knock-on effect on the deck design since the deck now has to span over a larger distance between supports in its in-place condition. Furthermore, no jacket braces can be installed in the area of barge operations. The leg structures have to be made stronger to compensate for diagonal bracing. This can be a significant factor in locations subject to seismic loads.

While the main mechanism for a load transfer is through changing the draft of the transportation vessel, there are several proprietary systems that speed the load transfer operation or increase its operating envelope. The most common systems use:

- Sand jacks. In this system, the top of the jacket legs are turned into enclosures that are filled with fine sand. The weight of the deck is first transferred to the sand "jacks" by ballasting the barge down. Once the deck separates from the barge, the sand is allowed to drain from the bottom of its compartments and the deck settles into its final position. This method requires strict quality control of the sand, moisture-content and drain valves.
- Hydraulic Jacks. The deck weight is transferred to the jacket by a combination of ballasting and hydraulic jacks. Given their ability to change the elevation of the deck rapidly the jacks help shorten the installation period. They also allow the operation to be carried out using shallower barges which helps reduce the impact on the jacket design.

External floatovers are less common than internal ones and are more sensitive to environmental loading than internal floatovers. Historically they have been used in the installation of modules over gravity based platforms in the North Sea. In this type of installation, two barges support the extremeties of the deck while the area under the middle section of the deck is kept free to avoid clashing with the platform structure during the floatover operation. The deck structure has to be capable of being supported at the extreme ends. The most critical type of loads are the racking loads that could result from relative motions, particularly the pitch motions, of the two barges. Such loads have historically limited the application of this method to inshore sheltered locations, lakes and fjords.

14.8.4 Hook-up to Pre-Installed Mooring Lines

Moored platforms are installed on station by hooking up the pre-installed mooring lines. This operation requires the use of vessels with sufficient winching or lifting capacity to handle the weight of pre-installed mooring lines. Equipment is also required on board the vessels such as winches or chain jacks.

14.8.5 Heavy Lift Vessels

14.8.5.1 Types

The lifting capacities of the floating crane vessels have increased over the years in parallel with the increase in platform sizes. Lifting topsides in larger modules reduces the cost of offshore hook-up and commissioning. The current offshore lifting record stands at 12,000 ton.

Heavy lift vessels can be categorised as follows:

- Semi-submersible crane vessels (SSCVs) with dual cranes such as the two largest lift vessels in the world, Saipem's S7000 and Heerema's Thialf.
- Ship-shaped monohull lift vessels with slewing cranes. Seaway's Stanislav Yudin is an example of this type of heavy lift vessel.
- Flat bottom monohull lift vessels with slewing cranes. Saipem's S3000 is an example of such a vessel.
- Sheer leg crane barges. These are flat bottom barges with an A-frame type boom that can boom up and down. Often, the position of the boom can be adjusted along tracks on the deck of the barge for given lift configurations. Smit's Taklift 4 is an example of this type of lift vessels.

Some heavy lift vessels have dual lifting and pipelay capabilities and are referred to as derrick lay barges.

14.8.5.2 Lift Capacities

Table 14.4 shows a listing of heavy lift crane vessels with capacities exceeding 300 ton. The capacity referenced in the table show the maximum lift capacity of the main block at the minimum radius and with the booms tied back, where applicable.

Table 14.4 Capacities of heavy lift vessels

Vessel name	Max. capacity (ton)	Vessel name	Max. capacity (ton)	Vessel name	Max. capacity (ton)
Nan Tian Ma	300	DLB 750	680	Semco L-1501	1500
Cairo	318	HD-423	680	Asian Hercules	1600
Mohawk	318	Yamato	700	Yamashiro	1626
Thor	350	DB 3	725	Toltika	1814
Q4000	359	Kuroshio II	725	DLB 1601	1814
Southern Hercules	362	Arapaho	726	Hercules	1815
Asian Helping Hand III	400	DLB-KP1	730	Taklift 1	1900
Smit Typhoon	400	Teknik Perdana (DLB 332)	750	Huasteco	2032
Taklift 3	400	Ocean Builder	755	Kongo	2050
Taklift 5	400	DB 16	780	Castoro Otto	2177
Nan Tian Long	450	DB Raeford	780	Suruga	2200
Illuminator	465	DB16	780	HD2500	2267
DB Sara Maria	476	Comanche	784	Kuroshio	2272
Sara Maria	476	Eide lift 2	800	DB 30	2300
Atlantic horizon	499	Pacific Horizon	800	Master Mind	2400

L M Balder	500	Mixteco	812	Stanislav Yudin	2500
Mexica	500	DLB 801	817	Taklift 6	2800
Nanyang	500	DB 26	820	Taklift 7	2800
Crawler	540	Cherokee	839	S3000	3000
Lili Bisso	544	Enak	863	Taklift 8	3000
DB 17	562	HD-1000	907	DB101	3175
BOS 355	590	Shawnee	909	Asian Hercules II	3200
OFSI DB-1	599	Castoro 2	998	Musahi	3600
Avon Senior	600	Chesapeake	1000	DB 50	4000
Koeigo	600	Mnicoperi 30	1000	Rambiz	4000
Uglen	600	Roland	1000	Taklift 4	4000
Field development ship	605	Smit Cyclone	1000	SLC 5000	4536
DB 1	615	Teknik Padu (DLB 264)	1000	Columbia	4600
Courageous	620	Nan Tian Long-900	1200	Balder	7200
Seminole	632	DB 27	1260	Hermod	9000
DB General	635	DLB 1000	1290	Thialf	12,000
Cappy Bisso	635	Nagato	1300	S 7000	14,000
OFSI DB RAREFORD	635	DLB Polaris	1500		

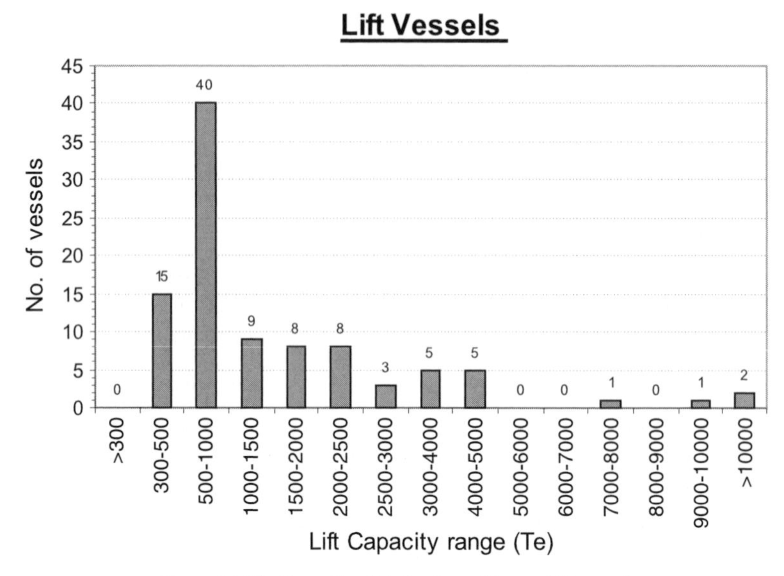

Figure 14.29 Capacity distribution of heavy lift vessels

Figure 14.29 presents the distribution of heavy lift installation vessels in terms of their maximum lifting capacities. The plot is presented in such a way that vessels are grouped in the ranges of their lift capacities.

14.8.5.3 Station Keeping of Heavy Lift Vessels

The station keeping of a heavy lift vessel consists of a conventional spread mooring or a dynamic positioning (DP) system. On some vessels, the dynamic positioning system is used to assist the conventional spread mooring system.

Spread mooring systems are used to hold the crane vessel on location, as well as moving the vessel slowly during installation operations, between the load pick-up and the load set down locations. The mooring lines are typically made of wire rope and the anchors are conventional drag-embedded anchors. The mooring system has to be designed for typical operating conditions such as the 1-yr return storm and should prudently be checked for the single damaged line case. Anchors are normally pre-loaded to ensure no further anchor slippage takes place during installation.

In deepwater installation, the time required for the deployment of the mooring lines becomes impractically long in relation to the duration of the overall installation campaign. Some heavy lift vessels have been fitted with DP systems to meet the requirements of deepwater installation. Vessels equipped with DP systems can set up on location and depart in short time periods. The dynamically positioned vessels can also change headings quickly to reduce the environmental loads and motions.

A DP system consists of a control, a sensor, a thruster and a power system. Its positioning may be accomplished through the use of an acoustic, mechanical, satellite or a radar positioning system. Vessels are given the notation of DP1, DP2 or DP3 depending of the levels of redundancy and segregation of their DP systems.

14.9 Platform Installation Criteria

14.9.1 Environmental Criteria

Defining the limiting environmental criteria is an essential part of the installation planning. The limits set for a particular operation are a function of the duration of the operation, the ability to forecast weather and/or change the course of the operation once it starts.

The following summarises the industry's approach to the environmental criteria:

- Low environmental limits are set for operations that can be completed within a short period of time and can be carried out in a relatively sheltered environment and under controlled conditions. Most loadout operations, for example can be completed in less than one day and are started upon receipt of a good weather forecast. Typically loadout operations are designed for wind speeds of 20 knots. A lifting operation is another such example.
- For operations that require several days to complete, such as a jacket lifting and piling operation, the jacket is designed to meet a design storm of 1 yr or greater.
- Operations that require longer than three days, but less than 30 days to complete, such as most of the towage operations, are designed for the 10-yr return storm. Adjustments for limited exposure to certain sectors of the tow route are possible such that the effective return period for the storm is reduced to somewhere between 1 and 10 yr.
- Operations that require longer than 30 days, such as mooring of FPSO and TLP at the quayside during hook-up and commissioning work, are designed for storms with return periods that range from 20 to 50 yr.

When determining the duration of the operations, it is important to take into consideration the possibility of equipment-failure, slowing down the progress of the operation.

While a particular operation may be designed for a storm of a certain return period, the operation does not normally commence in weather of that magnitude. All the installation-related operations should be started in good weather. It is also important to note that the operability of the installation equipment is often the most limiting factor. Lifting operations are often limited to seas of less than 5 ft while the piling operations that require crane assistance are limited to seas of less than 8 ft.

Where it is impractical, or uneconomical, to design against a storm of a particular return period, it is possible to plan the operation for a lesser limit provided:

- Good weather forecasting is available on a frequent basis.
- It is possible to avoid the inclement weather, given sufficient notice.

A typical example of such a case is the weather routing of transportation operations with sensitive cargoes carried on heavy lift ships with speeds of 12 knots or greater. Regular forecasts of at least once daily are relayed to the ship. When an inclement weather is forecast, the ship changes course to avoid the weather.

14.9.2 Heavy Lift

Hook Load

The heavy lift operation consists of three distinct systems, the crane vessel, the lifting slings and the structure being lifted.

The load experienced by the crane vessel is referred to as the hook load. This consists of the weight of the structure, the weight of the rigging and any dynamic factors caused by the dynamic motions of the crane vessel and or the transportation barge. Weight contingencies are a function of the uncertainty in the weight estimate. For example, a structure at preliminary stage design could have a weight factor of 1.10–1.15 while the factor drops to 1.03 for a structure that has been weighed.

Dynamic amplification factors (DAF) can be derived analytically from motion studies of the crane vessel where the boom tip accelerations are calculated using an appropriate hydrodynamic theory. Coupled body dynamic studies can also be used to calculate the DAF where the dynamics behaviour of the transportation barge is significant or in deep water installations where the lifted structure is under water. Standard DAFs are often used in lieu of such studies. These are covered by the guidelines of the marine warranty surveyors (MWS) or the classification societies. Typically the DAFs are larger for the smaller structures and for offshore lifts, as shown in Table 14.5.

Additional load factors are applied for structures lifted by two or more cranes. These account for the inaccuracies in the location of the centre of gravity (COG), the tilt of the lifted structure and the relative movement of the crane vessels.

The hook load should not exceed the capacity of the crane at the installation radius as given by the crane chart. Where the crane chart includes an allowance for the dynamics, the hook load need not include a dynamic amplification factor unless the dynamic response of the lift is expected to be excessive.

For subsea installation work in deepwater, the DAF may be in excess of the values listed in table 14.5 and needs to be calculated explicitly. In order to carry out a dynamic lift analysis for deepwater installation, the following information is required:

- Response amplitude operators of the installation vessel.
- Mass and stiffness of crane hoisting wires.
- Installation wave height and range of periods.
- Mass of the lifted structure, added mass and the drag coefficient.

The equation of the dynamic equilibrium is:

$$\left(M + M_a + \frac{m_{\text{wire}}}{2.718}\right)\frac{d^2Z(t)}{dt} + C\frac{dZ(t)}{dt} + kZ(t) = kZ_v(t) \tag{14.12}$$

where,

$Z(t)$ is the heave of the lifted structure.

Table 14.5 Dynamic amplification factors

Weight (ton)	DAF	
	Offshore	Inshore
< 100	1.30	1.15
100–1000	1.20	1.10
1000–2500	1.15	1.05
> 2500	1.10	1.05

$Z_v(t)$ is the vessel heave.
M is the structure dry weight.
M_a is the structure added mass.
m_{wire} is the dry weight of the hoisting wires.
C is the vertical drag coefficient of the structure.
K is the stiffness of the hoisting wires.

The dynamic amplification factor due to the heave motions is given by:

$$DAF = \frac{K}{\left[((M + M_a + (m_{\text{wire}}/2.718))\omega^2 - K)^2 + (C\omega)^2\right]^{0.5}} \tag{14.13}$$

where ω is the natural frequency of the spring (rigging) and the mass (structure) system.

Load in Rigging

For lifts that are statically indeterminate, such as a deck lifted by four slings connected to a single hook, the load distribution of the individual sling is affected by variations in the lengths of the slings as well as the relative distance of their attachment points from the centre of gravity (COG). This mis-match causes a skew load which is accounted for using one of the several methods available in the industry:

- Assume that one diagonally opposite pair of slings carries 75% of the weight, while the other carries 25%. A similar approach is to apply a skew load factor equal to 1.25 to the worst loaded sling. For a symmetrical lifting arrangement with a COG located in the centre, this approach is equivalent to a 62.5–37.5% distribution, which is less onerous than the first one [Noble Denton Guidelines 0027].
- Provided the mis-match does not exceed 1.5 in. or 0.25% of the length of any one sling or 3 in. or 0.5% difference between the shortest and the longest slings, it can be ignored as recommended by the API RP2A. Mis-matches of greater magnitude need to be accounted for analytically.
- The skew load factor can be calculated analytically as recommended by the Det Norske Veritas Rules for Marine Operations (1996). For a 4-point lift with double symmetrical sling arrangement, the skew load factor, *SKF*, is given by equation (14.14).

$$SKF = 1 + \frac{\varepsilon_0}{\varepsilon} \tag{14.14}$$

where,

ε, is the average strain in the sling at a load 30% greater than the dynamic hook load.

ε_0, is the strain associated with the total sling and the padeye fabrication tolerance.

Shackles are normally designed to a factor of safety of 4.0, if the loading includes dynamics and 5, if it does not. The factor of safety is defined as the ratio of the minimum break load (MBL) to the safe working load (SWL). Slings are typically designed to a minimum factor of safety of 3 which applies to the hand-spliced terminations. If the ratio of the pin (or trunnion) diameter to the sling diameter is less than 4, sling bending efficiency may become critical and a higher safety factor would apply. The required sling minimum break load is given by equation (14.15) [Noble Denton Guidelines 0027].

$$MBL = \frac{F_{\text{sling}} \times 2.25}{\eta} \tag{14.15}$$

where,

F_{sling}, is the dynamic force in the sling.

η, is the lesser of the sling termination factor, *STF*, or the bending efficiency factor, *BEF*.

$STF =$ 0.75 for hand-spliced slings.
1.0 for resin sockets or swage fittings.

$BEF =$ is given by equation (14.16).

$$BEF = 1 - 0.5\sqrt{\frac{\phi_{\text{sling}}}{\phi_{\text{pin}}}} \tag{14.16}$$

where, ϕ_{pin} and ϕ_{sling} are the diameters of the pin and sling respectively.

Load in Lifted Structure

The global integrity of the lifted structure has to be assessed against the following loads:

- Self-weight including any weight contingencies.
- Dynamic loads typically applied as a DAF.
- Skew loads due to sling mis-match.

Given their criticality, the structural members that frame into the lifting points are assessed against higher loads using a "consequence factor".

The API RP2A lumps the three factors listed above into a single factor of 1.35 while for the members framing into the lift point, the factor is increased to 2.0. A typical lift analysis carried out to the requirements of the API RP2A would consist of three load cases as follows:

- Static load cases to calibrate the model weight against the weight report.
- Static load case $\times$ 2.0 to investigate the integrity of members framing into the lift point. A factor of 1.5 can be used instead for lifts in sheltered conditions.

- Static load case × 1.35 to investigate the integrity of other members. A factor of 1.15 can be used instead for lifts in sheltered conditions.

The approach described earlier for the development of the skew load in rigging components can also be used. In this case, the effects of dynamics and weight contingencies are applied as a load factor while the skew load effects (if the lift is statically indeterminate) are accounted for by forcing the diagonally opposite slings to carry 75% (or 67.5%) of the load.

Loads in Lifting Points

Lifting points are normally designed using a load factor of 2.0 on the statically resolved rigging load [API RP2A]. To allow for the effects of uncertainty in the alignment of the rigging, a load equal to 5% of the lifting point design load is applied so as to cause bending in the weak axis of the lifting attachment. The load factor of 2.0 is consistent with the API RP2A's load factor for members framing into the lift point.

Where the skew load factors are used explicitly, such as in the 67.5–32.5% distribution, an additional consequence factor of 1.35 is applied to the load in the lift point. The lift point load factor is therefore built up from:

Lift point load factor = 1.1 (weight contingency) × 1.1 (DAF) × 1.25 (skew load factor) × 1.35 (consequence factor) = 2.05. This is close to the load factor stipulated by the API RP2A of 2.0.

Structural Design Requirements

In assessing the global integrity of the lifted structure and the strength of lifting points, the steel components have to meet the requirements of the API RP2A for tubular members or the American Society for Steel Construction, the Allowable Stress Design (AISC-ASD) for non-tubular members. The 1.33 increase on allowable stress is not permitted in the design and analysis of steel for lifting operations.

The load factors listed above are consistent with the working stress design (WSD). Lifting operations can also be designed to Load Resistance Factor Design (LRFD) codes such as the API RP2A LRFD where different load factors are stipulated.

In the design of lifting points, it is normally preferable to rely on load transmission to the primary steel through shear rather than tension. Full penetration welds and plates made of steel with through thickness properties (Z-quality) are also preferred. Pin holes in the padeyes are normally line bored after the cheek plates are installed in order to ensure an even bearing surface against the pin.

Lifting Point Inspection and Re-Use

A suitable scope of non-destructive testing is normally specified including an ultrasonic testing (UT) of full penetration welds. Where attachments are used for more than one lift, the critical welds and the inside of the pin hole should be inspected for cracks using a suitable technique, such as magnetic particle inspection (MPI). A visual inspection for mechanical damage should also be carried out.

14.9.3 Launching

Jacket Buoyancy and Trajectory

The reserve buoyancy of a launched jacket should typically be 15% in the intact condition and 5% in the damaged condition [Noble Denton Guidelines 0028]. The damage scenarios should include the largest compartment which is typically a jacket leg. Further subdivision of the jacket structures is sometimes required to ensure the damaged stability requirements are met. The trajectory of the jacket during the launch should be such that the lowest point would clear the seabed in the intact condition and by a safe margin (typically 10% of water depth) in the damaged condition.

Different reserve buoyancy and clearance levels are sometimes specified by owners depending on their own experiences. When assessing the jacket buoyancy and the launch trajectory, it is important to consider variations in the magnitude of weight and buoyancy and the locations of their respective centres.

The jacket legs are made buoyant by the installation of rubber diaphragms at the base and steel caps at the top. The legs are kept under a small net positive nominal pressure. The diaphragms are designed and tested against the highest hydrostatic pressure that they are likely to encounter plus an additional margin. The effects of slamming should also be considered. The quality of the diaphragm material and the means of securing them to the jacket legs are important considerations.

Jacket Structure

The launch analysis produces a set of consistent forces for each stage of the launch operation which are used to assess the integrity of the structure. These include:

- Jacket weight and instantaneous buoyancy.
- Inclination.
- Reactions from the barge.
- Drag and inertia loads.
- Hydrostatic pressure.

The requirements of the API RP2A for the structural design of the jacket members apply. The 1.33 increase in the allowable stress is permitted. The effects of local loads on the jacket members such as slamming should also be investigated.

Other Requirements

The jacket and barge combination need to meet the stability criteria prior to and during the launch operation. Between the initiation of sliding and the rotation of the jacket, the barge has to meet reduced stability criteria [Noble Denton Guidelines 0028]. All the winches and the associated launch rigging should be adequately sized against friction during the breakout and continuous movement.

14.9.4 Unpiled Stability

Once the jacket is upended, its submerged weight along with the environmental loads and the weights of any piles have to be supported by the jacket mudmats.

Typically, environmental loads in this phase are based on a reasonable return period such as the one-year return storm, thus allowing sufficient time to install sufficient piles to resist a greater storm. The jacket at this stage is analysed as a gravity structure and the design criteria are:

- Ensure no uplift.
- Ensure the soil-bearing capacity is not exceeded. Typically a factor of safety against soil bearing failure of 2.0 is required against static loads (self-weight and buoyancy) and 1.5 against the combined static and environmental loads. Similarly a factor of safety of 1.5 is expected against sliding.
- Mudmat structure is designed against the allowable values of AISC or an appropriate code for the material used.

14.9.5 Pile Installation

The pile lifting and upending analyses are normally carried out to establish that the lifted pile sections are not overstressed at the various angles of inclinations. The pile structures are assessed against the combined bending and compressive forces using the beam-column formulation in the API RP2A.

A pile "stick-up" analysis is also carried out to establish that the length of pile that is laterally unsupported does not buckle. The unsupported length of pile in this case is typically analysed as a cantilever with an effective length factor of 2.1. In addition to the piles self-weight and the hammer weight, a lateral load equal to 2% of both the weights is assumed to be applied at the pile head. Where piles are battered the additional lateral forces associated with the batter are also included. Stresses due to the pile stick-up are static stresses and the 1/3 increase in allowable stress is not permitted in this type of analysis.

Pile driveability analyses typically need to ensure that the dynamic driving stresses do not exceed 90% of the specified minimum yield strength and that the combined static (stick-up) and dynamic stresses do not exceed yield.

Pile fatigue due to driving stresses needs to be assessed in some cases particularly when the pile in-place fatigue life is marginal. This is done using Miner's rule or the rainflow technique.

The piling operation is terminated when the pile reaches "refusal". A refusal is when the pile does not advance significantly under the successive blows from the driving hammer. Refusal criteria are often agreed between the installation contractor and the owner and are influenced by earlier experience with the specific soil conditions, the hammer and the pile dimensions. In the absence of a specific criteria, the API RP2A suggests that refusal at 250 blows per one foot for five consecutive feet or 800 blows per foot for any single foot of penetration. These criteria should be considered as no more than a guidance and more specific criteria should be agreed upon at the onset of the project.

The contractor also needs to demonstrate that he can overcome the effects of soil set-up due to short stoppages associated with welding additional pile sections or any hammer mechanical problems.

14.9.6 Deck Mating

Deck mating operations involve the following stages:

Barge Entrance

During this stage, the barge motions should be so as to allow sufficient clearance between the deck structure and the top of the jacket under the design environmental and tide conditions. Typically a minimum clearance of 0.3–0.5 m is stipulated between the deck cones and the top of the jacket structure after allowing for the barge and deck motions.

Sufficient clearance needs to be kept between the barge and the jacket legs after allowing for any fenders that may be installed. Operations with clearances as small as 0.3 m between the barge and the jacket fender have been carried out.

Fenders need to be designed to resist potential impact from the barge. Impact energies can be derived as stipulated in the HSE Guidance Notes (1990). The integrity of the jacket structure should be assessed against the impact loads.

The barge stability, the local and global strength needs to be assessed for this phase.

Ballasting for Clearance Reduction

In addition to the requirements of the previous stage the effects of occasional impact between the deck legs and the jacket should be assessed. Fenders are typically installed to absorb the energy of such impact. Both the jacket and the deck need to be designed for the impact loads acting in conjunction with self weight, motions, direct hydrodynamic loads as well as barge deflections.

The requirement to maintain clearance between the jacket and the deck does not arise here as the objective of this stage is to reduce that clearance to start load transfer.

If the barge is moored to the jacket, the loads in the mooring system need to be investigated. The mooring system may be designed against the requirements of the API [API RP2SK] for both the intact and the one-line damaged conditions.

Load Transfer Stage

While the potential for separation between the deck and jacket remains, impact loads are not likely to be as significant as they were during the previous stage. In addition to the loads applied earlier, the motions of the barge will now be significantly influenced by the stiff connection between the jacket and deck. Furthermore, the hydrodynamic and wind loads applied to the barge and deck are transferred directly to the jacket structure. The barge, deck, jacket and any other structures connecting the three components need to be assessed against the combination of:

- Direct environmental loads on all three structures.
- Self-weight of the deck and jacket. Several intermediate stages of load transfer, for example 25, 50, 75 and 100%, need to be investigated.
- Barge deflections.

Barge Withdrawal Stage

During this stage, the potential impact of the barge and any deck supporting structures that remain attached to the barge against the deck need to be investigated.

Prior to starting the withdrawal of the barge, sufficient clearance has to be maintained taking into account clearances between the barge and the deck above and between the barge and jacket bracing below after allowing for barge motions.

Structural components are normally designed to the requirements of the API RP2A and the AISC with the 1.33 increase on allowable stress.

14.9.7 Tension Leg Platforms

14.9.7.1 Tendon String

The stresses in the tendon sections have to meet the requirements of the API RP2A during the lifting and upending operation.

Once installed in their tendon receptacles, the net buoyancy in the TBM has to be sufficient to maintain tension in the tendon and ensure that the tendon offsets remain within the allowable design limits of the tendon connector. The strength of the tendon in this configuration has to be assessed against loads caused by an environment of a suitable return period such as the 1–10-yr storm. Current loads are likely to be the predominant loads given that the top of the tendon is well below the wave zone. In addition to direct current loads the effects of vortex induced vibrations should also be investigated with different current profiles. Fatigue damage in the tendons in the free-standing phase particularly due to vortex shedding should be investigated.

14.9.7.2 Platform Floating Over the Tendon

The design of the floatover operation has to ensure that:

- There is adequate clearance between the platform and the top of the tendons, taking into account platform motions and tidal effects.
- The platform has adequate static and dynamic floating stability.
- The platform structure can support the motions during this operation.

14.9.7.3 Platform Ballasting for Tendon Connection

The stability criteria tend to vary depending on the type of TLP and client specifications. Mini-TLPs such as the Seastar are be installed with the assistance of a crane vessel. The hull is connected to the hook of a crane and a hook load is applied to ensure a minimum GM is maintained. The larger TLPs typically meet the static and dynamic stability requirements that are similar to those required for towage operations.

14.9.7.4 Lock-Off and Ratcheting Stage

During this transient stage, the stability of the TLP is enhanced by the tendon stiffness. The loads on the teeth of the LAJ are checked against their respective allowable loads.

14.9.7.5 De-ballasting to Achieve Desired Tendon Tensions

Once the final draft is achieved and the tendon connectors are closed, the loads in the tendons due to the motions of the TLP, should remain tensile. Tendons are effectively tensioned to a set pretension by pumping out ballast water from the hull. A "storm safe" condition may be defined where there is sufficient tension in the tendons to resist the 10-yr return storm.

The installation operation should be designed such that the operations starting with floatover till a storm-safe condition is achieved can be completed within a reliable weather forecast of, typically, 72 h.

14.9.8 Spar

Spars are installed by wet towing so site, controlled upending and hook-up to mooring lines.

- For the wet towing operation, the spar needs to meet the strength and stability requirements associated with the design environment for the tow route.
- Intact and damaged stability requirements need to be met at the various upending stages. The spar global strength is also assessed at these stages.
- The mooring system has to have adequate factors of safety during its incremental hook-up to the spar. Depending on the expected duration of the hook-up operation, the design environment could be between the 1-yr return and 10-yr return.

14.9.9 FPSO

Floating Production Storage and Offloads (FPSOs) are often built or converted a long distance away from their final installation site. Whether the FPSO is towed or is self-propelled, the transportation to site is a temporary phase for which the hull and the topsides need to be checked. This is particularly important in the case of FPSOs installed in benign environments such as West Africa where the environmental conditions for the tow to site are often more severe than those at the installation site. The topside structures and the hull design could therefore be governed by the transportation condition.

The design criteria for the tow is typically the 10-yr return storm for the worst part of the tow route adjusted for the exposure period. The bending and torsional moments in the hull structure and the motions and accelerations are normally evaluated using the diffraction or the strip theories. The topside strength is investigated using structural analysis programs employing the finite element theory. The integrity of the topside is investigated under the effect of the combined motions, accelerations, hull deflections, wind load and gravity.

Stability requirements are normally those of the Class society and are agreed with the Marine Warranty Surveyor. Seakeeping characteristics are often investigated using model testing. SOLAS requirements have to be met as well if the FPSO is manned during the tow to site.

Since the FPSO is typically moored at the quay for several months while the topsides are added, the quayside mooring system has to withstand a design storm with a return period of 50 yr. Shorter return period storms may apply for short quayside integration campaigns. The design of the mooring system is normally carried out to the requirements of the Oil Companies International Marine Forum [OCIMF, 1978] which uses a static approach. For exposed locations with the possibility of significant seas developing the vessel motions may become important and need to be taken into account.

The following factors are considered when designing the tow to site operation:

- Hull girder strength.
- Motions and accelerations and their effect on the strength of the topside structure.
- Vessel deflections and their impact on the strength of the topside structure.
- Vessel stability.
- Bollard pull required.
- Strength of tow line connections.
- Seakeeping issues such as yaw motion while under the tow known as "fishtailing".

In addition, marine considerations have to be taken into account such as the tow route, ports of shelter and life saving equipment.

When they arrive at the installation site, FPSO vessels are normally positioned over pre-installed mooring lines. The lines are individually "hooked-up" to the FPSO while the vessel is held in position by tugs. Typically, the FPSO is considered to be storm-safe when a sufficient number of mooring lines are installed to resist the 10-yr return storm. Mooring analyses are carried out to verify the storm safe condition. The mooring line components are expected to meet the requirements of API RP2SK for both intact and damaged conditions. Where a synthetic mooring system is used, the requirements of the API RP2SM need to be met.

The pre-installation of the mooring lines is normally carried out using a vessel equipped with a suitable crane and winches that can support the weight of the mooring line. The weight of the anchor may have to be supported too if it is installed at the same time as the mooring line. The loads seen by the mooring lines are normally a fraction of the loads they experience in-service. The limitations on their installation are normally related to the installation technique and the equipment used. For example, there is a minimum bend radius for steel wires or polyester lines installed from reels. These are normally specified by the manufacturer. Typical ratios of drum diameter to rope diameter are [Noble Denton, 1999]:

- 24 for a steel spiral strand.
- 14 for steel wire rope.
- 6–15 for braided jacket fibre rope.
- 20–30 for extruded jacket fibre rope.

The equipment handling the mooring lines present another potential limitation as in deep water, the weight and the size of the mooring line requires large lifting or winch capacities. Similar considerations apply to the hook-up equipment whether linear winches or chain jacks. Installation rigging should include features such as swivels or torque balanced lowering wires to avoid locking in torsional loads in the mooring line.

14.10 Installation of Pipelines and Risers

14.10.1 Types of Subsea Pipelines

Offshore pipelines are predominantly made of carbon steel although pipelines made of other steels are used in the industry. The same installation methods and constraints apply to all steel pipelines. The steel pipe is normally supplied for the offshore installation in single joints of standard 40 ft length or, in lengths of double joints (80 ft) or triple joints (120 ft). The steel pipes are invariably coated against corrosion using a fibre bonded epoxy coating of typically ¼ in. thickness. Concrete weight coating is often used to provide on-bottom stability for the pipeline. This coating is installed at a coating mill before the pipe joints are shipped offshore.

A pipe-in-pipe is a relatively recent development in the industry. It consists, as the name suggests, of a pipe installed inside another pipe. The outside pipe provides a thermal insulation for the inside pipeline in cold deepwater, thus preventing wax and hydrate formation.

14.10.2 Methods of Pipeline Installation

The method of pipeline installation depends on a combination of factors such as the water depth, the diameter and the weight of the pipe. The installation methods are divided into the following categories:

- S-Lay
- J-Lay
- Reeled
- Towing

The relative merits of these methods of installation are discussed in more detail in Chapter 11. A brief explanation of the methods is included in this section.

14.10.2.1 S-Lay

In this shallow water installation method, the pipeline is configured in the shape of the letter "S" where the pipeline bends in a hogging mode as it exits the lay barge (referred to as overbend) and in a sagging mode as it touches the seabed (referred to as sagbend). In order to keep stresses in the pipe in the overbend within acceptable levels, the pipe is kept under tension during the laying operation.

The pipeline is laid using a lay barge. Figure 14.34 shows a typical S-lay lay barge. The lay barge is equipped with pipe storage space and handling equipment, welding stations, NDT stations, hydraulic tensioners, winches, ramps, rollers and, often, a "stinger".

The pipe joints are lined up in the pipe tunnel where they are welded together – large lay barges are equipped with several welding stations to speed the laying rate. Once the pipe joints are welded, the welds are tested for defects using a suitable non-destructive testing (NDT) technique, such as X-ray. The welded pipe is then fed through a tensioner that maintains the necessary tension in the pipeline to keep stresses within acceptable levels. In addition to maintaining adequate tension on the pipe, the stresses in the pipe are

Figure 14.30 Shore pull operation

manipulated by changing its radius of curvature as it exits the lay barge. This is achieved by changing the heights of the rollers in the ramp. Additionally, a "stinger" can be installed to the aft end of the barge to provide additional support while the radius of curvature is changed gradually. Stingers are made of steel tubulars and buoyancy tanks that are framed together to provide structural strength and buoyancy. Stingers can be in one or more articulated sections. The depth of the stinger is changed by manipulating its buoyancy.

In addition to the normal pipelaying operation described above, pipelaying involves a start-up, a laydown and, in some cases, emergency abandonment and recovery. The startups and the laydowns are normally adjacent to a beach (shore pull operation) or an offshore platform while emergency abandonment and recovery could be anywhere. In all these operations, a fixed point has to be provided on the seabed or onshore to provide reaction against the tension applied to the pipe by the tensioner on board the pipelay vessel. This anchor point is often referred to as a "deadman anchor". Figure 14.30 depicts a shore pull operation.

14.10.2.2 J-Lay

This type of operation was developed to cater to deepwater pipeline installation. In the J-lay configuration, the pipeline is assembled in a near-vertical plane which allows it to exit the vessel with little bending curvatures, thus eliminating the overbend that characterises the S-lay configuration. The tension requirements are much smaller than those in the S-lay and are limited to the pipe's own weight. The pipe departure angle from the vessel can be manipulated by changing the angle of the J-lay tower. Because of the lower laying tensions, the potential for the development of free spans on the seabed are lower. Free spans are lengths of unsupported pipe which have to be controlled to avoid strength problems and vortex-induced-vibration (VIV) problems.

Most J-lay towers are equipped with a single welding station although a few have two. In order to speed the installation rate double, triple or even quadruple length, joints are used thus saving valuable offshore welding time. Figure 14.34 further down shows a typical J-lay barge.

14.10.2.3 Reeled Installation

In this method of installation, the pipe is spooled on a reel that is installed on the pipelay vessel. At offshore installation site, the pipe is "unspooled" and deployed to the seabed. This method of installation is faster than both S-lay and J-lay since the pipe does not need to be welded on board the vessel.

The pipe is subjected to a plastic strain when it is spooled and unspooled. For a specific amount of allowable plastic strain the larger the pipe diameter, the greater the required radius of the reel. This sets a practical limitation on the size of pipe that can be laid by reeling. Reels can be installed in the horizontal or the vertical planes with the vertical reels generally being smaller in diameter. The Deep Blue has a vertical reel with a diameter of 20 m and a pipe weight capacity of 5000 tonnes. The Hercules' reel is a horizontal one with a diameter of 35 m and a weight capacity of 6300 tonnes giving it the ability to lay up to 10 miles of 18 in. diameter pipeline. Figure 14.31 shows the Deep Blue reel pipelay vessel.

Since spooling and unspooling a pipeline induces plastic strains in the pipeline material, a reduction in the fatigue life due to the installation process is a concern. Typically the

Figure 14.31 Deep Blue reel pipelay vessel

concern is over the growth of welding defects, acceleration of crack growth and residual strain issues. Research projects including full scale testing of pipes of up to 16 in. in size concluded that the plastic deformation had little effect on the fatigue performance of girth welds provided good quality control and fabrication tolerances are applied. [Bell, 2000].

To address concerns about fatigue, it is also possible to lay pipelines by reeling along the majority of their routes and to transfer the laying methods to the J-lay mode in the fatigue sensitive section such as the steel caternary riser (SCR). The pipelay vessel would have to be equipped with both the reels and a J-lay tower.

14.10.2.4 Towed Pipelines

Short pipelines or pipe bundles can be fabricated on land and launched into the water to be towed to the installation site using tugs. The pipelines are kept buoyant during the operation and additional buoyancy aids are often used at selected locations to reduce stresses. The pipeline in this mode is subject to hydrodynamic loads as well as its own weight and buoyancy. The pipeline is either towed near the surface or well below the wave action zone to keep the hydrodynamic loads to a minimum. Once at the installation site, the pipeline is lowered to the seabed in a controlled manner by flooding and by removing the buoyancy tanks.

14.10.3 Types of Risers

The discussion in this section will focus on three different types of risers:

- Rigid risers in shallow water.
- Rigid risers in deep water.
- Flexible risers.

Rigid risers in fixed platforms are typically straight sections of steel pile that extend vertically between the seabed and the topsides of the fixed platform. These risers are supported laterally by the fixed platform substructure.

Deepwater rigid steel risers can be either top tensioned risers (TTRs) or steel catenary risers (SCRs). The TTR are kept under tension by a mechanical system installed on the platform such as in the case of the TLP, or by the use of air cans such as in the case with spar platforms. The SCRs are suspended from the production platform in the shape of a catenary that meets the platform at an angle. To accommodate changes in this departure angle, the SCR is typically connected to the platform piping through a flex joint.

Flexible risers are constructed of multiple layers of steel-textile reinforcement with an inner layer of elastomer or polymer material for product containment. The profile of flexible risers can take several shapes such as a free hanging catenary, a lazy-S, a lazy wave and a steep wave. The flexible riser adopts the shape of an S or a wave with the help of the buoyancy elements.

14.10.4 Methods of Riser Installation

The installation of a rigid riser on a shallow water fixed platform is typically done separately from the installation of the pipelines. The subsea pipeline is normally terminated

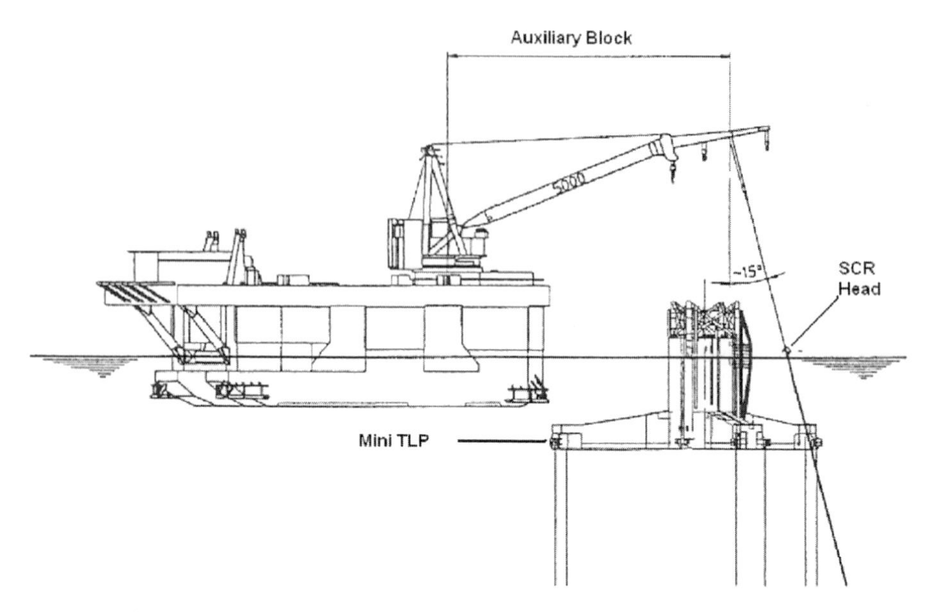

Figure 14.32 Hook-up of SCR to Matterhorn mini TLP by Hermod (Heerema)

close to the base of the platform. A derrick lay barge lifts the end of the pipe from the seabed using davits while the riser is lifted in the near vertical configuration by the vessel crane. The connection between the pipeline and the riser is made above air and the riser/pipeline combination is laid down onto the seabed using the crane and davits. The pipeline can alternatively be pulled to the surface through rigid bends called the J-tubes, that are installed at the base of the jacket.

The SCRs are a continuation of the subsea pipeline and get installed at the same time. SCRs have been installed using the S-lay or J-lay configurations and by reeling. Once the installation is complete, the SCR is laid on the seabed till the production platform is installed. The SCR is then picked up by the installation vessel and hooked up to the production platform as shown in fig. 14.32.

Top tensioned risers are normally installed from the TLP or the spar platform. Where air cans are used to provide tension to the risers, the riser sections can be installed through the air cans. In this case, the air cans hang off the spar hull. The riser sections are then inserted through the air cans as shown in Fig. 14.33.

14.10.5 Vessel and Equipment Requirements

The limiting water depth for the S-lay operations depends largely on the size of tensioner on board the pipelay vessel and the availability of vessel buoyancy to support the weight of suspended pipe. Since pipelaying operations require the vessel to move continuously along a designated route, station keeping is an important consideration for choice of pipelay vessel. Pipelay vessels that are conventionally moored (3rd Generation vessels) are limited

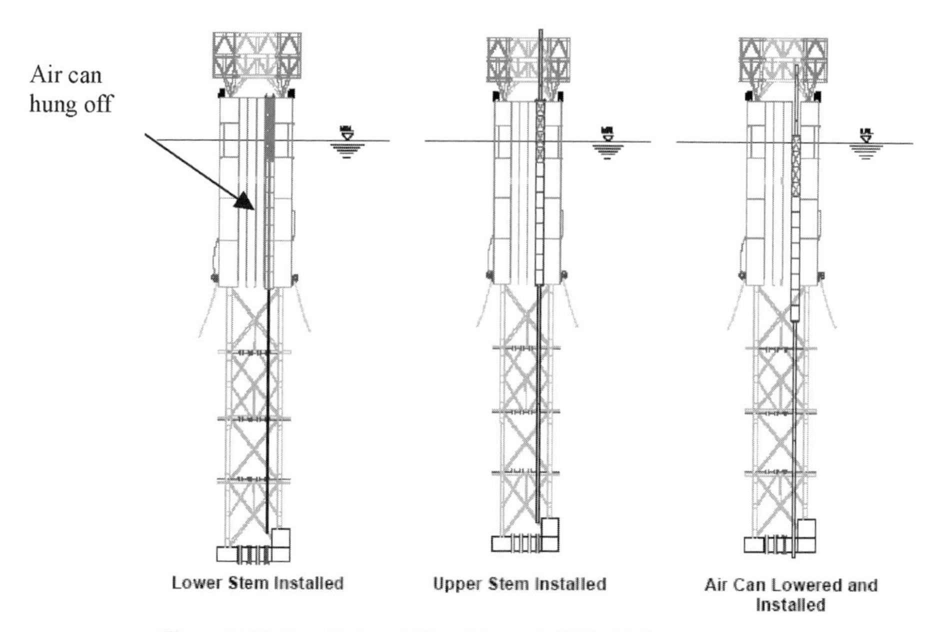

Figure 14.33 Installation of Horn Mountain TTRs [O'Sullivan, 2003]

to water depths in the region of 500 m. Beyond this limit, the anchor moving operations become critical path activities and tend to slow down the laying operation. Vessels equipped with DP systems (4th Generation monohulls) and tensioners in excess of 500 tonne capacity have laid pipe in just under 2000 m water depth using the conventional S-lay technology. Table 14.6 lists the number of pipelay vessels available along with their laying configuration and the station keeping characteristics.

The size of the tensioner available on the laying vessel therefore determines the vessel's pipelaying limitations. The pipelay vessels are also equipped with an abandonment and recovery (A & R) winch that is used for the startup and laydown operations as well as emergency abandonment and subsequent recovery. The winch capacity is normally similar to that of the tensioner as it has to apply tension loads to the pipe of similar magnitude. Table 14.7 lists some of the largest pipelaying vessels available with their respective tensioner capacities and their available laying configurations.

14.10.6 Analyses Required

A single pipelaying operation often straddles a range of water depths, pipe wall thicknesses and concrete coating thicknesses. The installation contractor analyses all reasonable combinations of these parameters. The aim of the analyses is to demonstrate that the subsea pipeline can be installed safely using the proposed equipment and procedures.

When designing a pipelay operation, it is important to set reasonable weather limitations for the operation. Such a limitation can be derived analytically or based on earlier experiences with similar operations with the vessel. The installation contractor provides

Table 14.6 Types and numbers of pipelay vessels

Installation method	Vessel type	Feature	No. of vessels
S-Lay	2nd Generation	Flat bottom barge	> 40
	3rd Generation	Semi-submersible	4
	4th Generation	DP Vessel	2
J-Lay	All DP, either semi-submersible or ship shaped	All DP with either tracked or fixed position tensioners	6
Reeled	Flat bottom barge, semi-submersible or ship shaped (DP or anchored)	Horizontal reel	16
		Vertical reel	10

Table 14.7 Capacities of the largest pipelay vessels

Vessel	Tensioner capacity (ton)	Maximum pipe diameter (in.)	Reel	J-Lay	Station keeping
Solitaire	523	60	None	No	DP
Deep Blue	551	26	Dual	Yes	DP
Skandi Navica	386	19	Dual	Yes	DP
Polaris	240	60	None	Yes	DP
Hercules	545	60	Rigid	No	DP
DCV Balder	568	30	None	Yes	DP/moored
DB 50	352	20	Dual	Yes	DP
FDS	401	22	Dual	Yes	DP
S 7000	525	32	None	Yes	DP
S3000 (ex Maxita)	291	20	Dual	No	DP
Catsoro Sei	330	60	None	No	Moored
LB 200	341	60	None	No	Moored

procedures for laying the pipe down if the weather is forecast to exceed the agreed limits and for subsequent recovery when the inclement weather subsides.

14.10.6.1 S-Lay and J-Lay

A static stress analysis is carried out to demonstrate that the combination of pipe curvatures, self weight, buoyancy and applied tensions do not overstress the pipe.

Figure 14.34 J-Lay (FDS) and S-Lay (Chickasaw) vessel configurations

The configuration of the roller, the stinger or the J-lay tower, as shown in Fig. 14.34, are modelled into this analysis. The pipe is discretized into small beam column finite elements, which can undergo large displacements.

A dynamic analysis may also be carried out in addition, or instead, of the static analysis. This analysis takes into account the dynamic motion characteristics of the pipelay vessels and is very useful for defining the limiting weather criteria for the installation operation. In addition to the data required in the static analysis, this analysis requires the response amplitude operator (RAO) curves for the vessel to be input as well as the environmental data.

In addition to the normal laying analyses, the following analyses are carried out:

- Startup and laydown, abandonment and recovery stress analyses to demonstrate the feasibility of these operations.
- Contingency procedures and analyses to describe what engineering may be required in case of dry and wet buckles. These include de-watering requirements in case of wet buckles or, if de-watering is not an option, a wet buckle lift/recovery analysis.
- Weld repair calculations to determine the permissible length, depth and location of weld gouging while the pipe is under tension.
- Crossing analysis to demonstrate that the pipe would not be overstressed when laid over crossings.
- Mooring analyses in the case of moored vessels. This aims to demonstrate that the proposed anchor patterns are adequate capacity to resist the proposed installation weather combined with pipelay tensions.
- In deepwater, the direct environmental loads on the suspended pipe string, such as current loads, can be important and should be taken into account.

14.10.6.2 Reel Lay

The following analyses are typically carried out:

- Packing analysis. This analysis is carried out to derive the strains in the pipe while they are being spooled over the installation reel and to derive the required spooling tensions.
- Static initiation, laydown and normal pipelay stress calculations. These are similar to the calculations performed for the S-lay operations. The dynamic analyses can also be carried out where the vessel motions are taken into account. The significance of direct environmental loads on the pipe string should also be assessed.
- Contingency analyses such as stress analyses of a flooded pipe.
- Abandonment and recovery stress analyses.
- Fatigue analyses. The fatigue analysis takes the dynamic motion characteristics of the reel lay vessel into account. Various scenarios are analysed with the welds assumed at different locations with respect to the vessel. The aim of these calculations is to determine how long the vessel can remain in a given pipelay configuration without exceeding a specified fatigue damage limit.

14.10.6.3 Acceptance Criteria

Acceptance criteria for the pipelay stress analysis can be based on limiting stress or on limiting strain. If a static analysis is carried out the allowable stress is typically 72% of the specified minimum yield stress (SMYS) in the sagbend and 85% SMYS in the overbend. If a dynamic analysis is carried out, the allowable stress can be increased to 96%.

The limiting strain criteria are typically 0.15% for the sagbend and 0.2% for the overbend. The approach takes into account the non-linear relationship between stress and strain such as the representation by Ramberg-Osgood (1943):

$$\varepsilon_m = \frac{\sigma}{E}\left[1 + c\left(\frac{\sigma}{\sigma_{0.7}}\right)^{n-1}\right] \tag{14.17}$$

where,

σ is stress.

$\sigma_{0.7}$ is the stress corresponding to the intersection of the experimental stress strain curve with a secant having a slope equal to 0.7E.

ε is strain.

E is the elastic modulus.

c and n are Ramberg-Osgood coefficients and exponent respectively. These are determined by fitting the above formula to the steel stress–strain curve.

The Ramberg-Osgood equation can be written in the form of a moment (M)–curvature (K) relationship to allow the bending strain, ε_m, to be calculated:

$$\frac{K}{K_y} = \frac{M}{M_y} + A\left(\frac{M}{M_y}\right)^B \tag{14.18}$$

$$\varepsilon_m = \frac{\sigma_y}{E}\left[\frac{M}{M_y} + A\left(\frac{M}{M_y}\right)^B\right] \tag{14.19}$$

where,

σ_y is the SMYS.
K and M are the applied moment and curvature respectively.
K_y and M_y are the moment and the curvature when the extreme fibre of the pipe is at yield.
A and B are coefficients that are determined such that the moment–curvature equation above stands as a best representation of the pipe behaviour.

In addition to limiting the stress and strains, buckling and ovalisation limit states have to be checked against the Det Norske Veritas' Offshore Standard DNV OS-F101 (2000).

Generally, where installation plastic strains exceed 0.3%, an engineering criticality assessment is carried out to determine the required material fracture toughness required to tolerate the welding flaw sizes allowed based on the NDT used. Alternatively, the assessment can be used to determine acceptable flaw sizes. For installation strains exceeding 2%, additional testing is required to ensure against unstable fracture. The material characteristic strain resistance is verified by realistic testing of the girth welded pipe such as full scale testing. The material is also expected to meet additional ductility requirements in terms of ratios of yield strength to tensile strength and minimum elongation [DnV OS-F101, 2000].

References

American Petroleum Institute – RP 2SK (March, 1997). "Recommended practice for design and analysis of stationkeeping systems for floating structures", (2nd ed.).

American Petroleum Institute – RP 2SM (March, 2001). "Recommended practice for design, manufacture, installation, and maintenance of synthetic fiber ropes for offshore mooring", (1st ed.).

American Petroleum Institute – RP 2A WSD (December, 2002). "Recommended practice for planning, designing, and constructing fixed offshore platforms", (21st ed.).

Beattie, S. M., Pyles, S. R., McCandless, C. R., and Kuuri, J. (May 2002). "Nansen/Boomvang field development – construction and installation", *Offshore Technology Conference*, Paper # 14092, Houston, Texas.

Bell, M. (2000). "Fatigue performance of steel catenary risers installed by reel ship", *Deepwater Pipeline and Riser Technology Conference*, Houston, March 7–9.

De Koeijer, D. M., Renkema, D., Edel, C. J. C., Willis, C. H., and Payne, D. (May 1999). "Installation of the baldpate compliant tower", *Offshore Technology Conference*, Paper # 10919, Houston, Texas.

Dijkhuizen, C., Coppens, T., and van der Graaf, P. (May 2003). "Installation of the horn mountain spar using the enhanced DCV balder", *Offshore Technology Conference*, Paper # 15367, Houston, Texas.

Det Norske Veritas (1996). "Rules for planning and executing of marine operations", Part 2, Chapter 5.

Det Norske Veritas Offshore Standard OS-F101 (January 2000). "Submarine pipeline systems".

Homer, S. T., Chitwood, J. E., Childers, T. W., and Verret, A. J. (May 1993). "Deepwater templates and cluster well manifolds: is there a single correct approach?", *Offshore Technology Conference*, Paper # 7268, Houston, Texas.

HSE (Health and Safety Executive) (1990). "Offshore installations: guidance on design, construction and certification", (4th ed.), and subsequent amendments.

Noble Denton (1999). "Deepwater fiber moorings – an engineers' design guide", (1st ed.), OPL Publications.

Noble Denton International (May 2002). "General guidelines for the transportation and installation of steel jackets", Guideline 0028/NDI, 1st Revision.

Noble Denton International (October 2002). "Guidelines for lifting operations by floating crane vessels", Guideline 0027/NDI, 3rd Revision.

Noble Denton International (May 2004). "General guidelines for marine transportations", Guideline 0030/NDI, Rev. 0.

Oil Companies International Marine Forum (OCIMF) (1978). "Guidelines and recommendations for the safe mooring of large ships at piers and sea islands", Witherby and Co., London.

O'Sullivan, E. J., Shilling, R. B., Connaire, A. D., and Smith, F. W. A. (May 2003). "Horn mountain spar risers – evaluation of tension and installation Requirements for deepwater dry tree risers", *Offshore Technology Conference*, Paper # 15385, Houston, Texas.

Ramberg Osgood (July, 1943). "Description of stress strain curves by three parameters", National Advisory Committee for Aeronautics, Technical Note 902.

Smith, E. A. L. (1960). "Pile driving analysis by the wave equation". *Journal of Soil Mechanics and Foundaitons Engineering Division*, ASCE, Vol. 86, No. SM4.

Handbook of Offshore Engineering
S. Chakrabarti (Ed.)
© 2004 Elsevier Ltd. All rights reserved.

Chapter 15

Materials for Offshore Applications

Mamdouh M. Salama
ConocoPhillips Inc., Houston, TX, USA

15.1 Introduction

Cost, safety and reliability of offshore developments depend largely on the cost-effective and proper selection of materials for the different components. This chapter reviews the important materials and corrosion issues and discusses the key factors that affect materials selection and design. The chapter includes several sections that provide performance data and specifications for materials commonly used for offshore developments. In addition, the chapter discusses key design issues such as fracture, fatigue, corrosion control and welding.

15.1.1 Factors Affecting Materials Selection

Structural, production and process components are fabricated using different materials including carbon steels, corrosion-resistant alloys, concrete, ceramics, elastomers, plastics and composites. Proper materials selection requires a clear definition of the following operating conditions, as well as consideration of the electrochemical, mechanical and processing compatibility amongst the different materials:

1. Operating loads and environment
2. Possible extreme and upset conditions
3. Special operating practices
4. Operating temperature
5. Corrosivity of production fluids and external environment
6. Corrosion control strategy
7. Service life
8. Maintenance flexibility
9. Environmental restrictions
10. Regulations

The materials selection process must take into account the influence of these operating conditions on materials. While it is obvious that the lowest cost–acceptable material option is to be selected, other factors such as strength level, fracture toughness, availability, weldability and machinability may make the selection of a more expensive material more economical.

15.1.2 Classification of Materials

Materials are characterised based on several parameters including type, strength, fracture control, corrosion resistance, chemistry, microstructure, weldability, etc. The following are the different types of materials used for offshore structural and production applications.

1. Structural steel: These are carbon and low alloy steels used for structures and pipelines.
2. Production equipment steel: These are carbon, low alloy and alloy steels used for tubulars, pipes, fittings and production/process equipment.
3. Corrosion resistance alloys: These materials are used for production and process equipments that are subjected to corrosive environments containing CO_2 and H_2S. They involve stainless steels, nickel base alloys, cobalt base alloys, nickel–copper alloys and titanium alloys.
4. Non-metals: These involve elastomers, coatings, plastics and composites.

15.2 Structural Steel

Structural steels are generally specified based on the appropriate national or industry standards such as ASTM, API, BSI, ISO, etc. In most cases, standards provide mainly the basic requirements such as limits on chemical composition and tensile properties. During the mid-1960s several in-service and structural fabrication problems were encountered illustrating that the common pipes such as API 5L B and structural steels such as ASTM A7 and ASTM A36 do not always meet the design/service need for the offshore industry. Failure analysis studies on several salvaged structures have shown that low notch toughness, laminations, lamellar tearing and poor weldability were major contributors to the failures [Peterson, 1969; Carter, et al 1969]. This made offshore operators and certifying authorities conscious of the need for more restrictive standards to ensure that the steel is of high quality and satisfies strict fracture toughness and weldability requirements [Peterson, 1975; Salama, et al 1988]. Therefore, standards such as API 2H, 2Y and 2W were developed. The structural steels addressed in these standards include: killed fine grain normalised, controlled rolled, quenched and tempered, and controlled rolled and accelerated cooled (referred to as TMCP) [Salama, et al 1988; Peterson, 1987; Masubuchi and Katoh, 1987]. In addition to the above API grades, special grades from general standards such as ASTM and BSI are also used in specifying steels for offshore structures. Table 15.1 provides a summary of the chemical composition and the mechanical properties of some offshore structural steels.

To meet the demand for high-quality offshore structural steels with higher strength, improved weldability and higher fracture toughness as well as lower costs, significant advances in steel making processes were made by steel companies. These advances included

Table 15.1 Specifications and typical chemical composition of TMCP Grade 60 Pipe

Element	Specifications	Composition
Carbon	0.10 max.	0.065
Manganese	1.15–1.40	1.35
Silicon	0.15–0.30	0.18
Sulphur	0.005 max.	0.0025
Phosphorus	0.018 max.	0.007
Aluminium, total	0.02–0.05	0.03
Titanium	0.003–0.020	0.015
Niobium	0.01–0.03	0.02
Nickel	0.25 max.	0.21
Copper	0.25 max.	0.21
Chromium	0.10 max.	0.035

the close control of the blast furnace to the supply of desulphurised iron, the wide spread use of continuous casting of thick slab for rolling to plate, the introduction of vacuum arc degassing, vacuum degassing, argon stirring and injection techniques, and the almost exclusive use of basic oxygen steel making. These improvements resulted in significant control on alloying elements (e.g. C, Mn, Nb, V, Al), major reduction of impurities (e.g. S, P, N_2) and improved uniformity of composition and properties. Also, advances in computer control and rolling capacity led to the development of a new class of HSLA steels, namely TMCP (Thermo-Mechanical Control Process) steels. The TMCP involves both controlled rolling and controlled (accelerated) cooling to produce a fine ferrite grain steel (ASTM 10–12). The main aim of TMCP is to increase the strength and fracture toughness and improve weldability by the reduction of carbon equivalent and appropriate control of chemical composition. Due to the steel making process and the low carbon content, TMCP steels have higher residual stresses than conventional normalised steels and are more sensitive to HAZ softening due to high heat input welding. The API 2W specification covers TMCP steel plates whose minimum yield strength is between 290 and 414 MPa (42–60 ksi). TMCP steels have been successfully used in many applications such as offshore structures, pipelines, vessels and TLP tendons. Table 15.1 provides the specifications and the typical chemical composition of TMCP steel. A capacity that needs to be assessed when using TMCP steel is the potential of softening of the heat-affected zone (HAZ) combined with the presence of local brittle zones (LBZ). While LBZ is not unique to TMCP steels, the potential of HAZ softening is generally associated mainly with TMCP steels [De Koning, et al 1988; Denys and Dhooge, 1988]. However, the presence of LBZ requires special attention for TMCP steels because unlike normalised steel in which the HAZ yield strength is higher than the base plate, the HAZ yield strength of TMCP steels tends to be lower than both the weld metal and the base plate. The combination of lower structural redundancy, higher stresses and the location of lower strength HAZ normal to the loading directions can result in situations where fatigue cracks can sample more LBZ regions thus increasing the possibility of brittle fracture. Recognising the industry concern regarding LBZ, some steel companies have developed LBZ free steels by alloy modifications to promote the austenite to ferrite transformation and to prevent the bainite transformation [Ohnishi, et al 1988; Suziki, et al 1989].

15.3 Topside Materials

During the last two decades, several key advances have occurred regarding material selection on oil- and gas-producing platforms. However, the general approach for the materials selection approach remains the same. Corrosion assessment, including corrosion calculations, is generally carried out for all process and utility units – particularly where the process fluid is associated with wet CO_2 and H_2S. While the main concern with CO_2 containing environment is corrosion, the main concern with H_2S and chloride-containing environments is stress corrosion cracking. The CO_2 corrosion rate of carbon steel is predicted using one of the industry accepted corrosion prediction model such as deWaard and Milliams model for CO_2 corrosion [deWaard and Milliams, 1976; deWaard, et al 1991].

The corrosivity of the production environments can be broadly categorised as follows:

1. Non-corrosive production: This includes conditions when corrosion is not expected to occur. This includes conditions when the CO_2 partial pressure is less than 5 psia and conditions when the pH value is higher than 5.2 even if the CO_2 partial pressure is 30 psia. The pH value depends mainly on alkalinity of water, acetate, temperature and CO_2 partial pressure. This could also include extremely corrosive conditions where the corrosion is mitigated by the use of corrosion inhibitors.
2. Production containing CO_2: This corresponds to conditions where the in situ pH value is lower than 5.2 or the partial pressure of CO_2 exceeds 30 psia. Under this environmental category, H_2S can be present as long as its partial pressure does not exceed 0.1 psia.
3. Production containing H_2S, or CO_2 and H_2S: This corresponds to production containing H_2S with a partial pressure that exceeds 0.1 psia. Under this environmental category, unlimited levels of CO_2 and chlorides can also be present.
4. Water injection: This is associated with seawater injection. In such a case, the oxygen and chlorine contents are the important factors governing the corrosivity. If produced water is reinjected, then the materials selection option must also take into account the CO_2 and H_2S levels in the produced water.

While different oil companies may have different guidelines for material selections, the following is the general basis of these guidelines, with some variations in the corrosion rate limits and corrosion allowances:

1. For non-corrosive fluids, carbon steel with 1.5 mm corrosion allowance is used. The corrosion allowance is specified to account for offset conditions.

2. For fluids resulting in a corrosion rate of less than 0.13 mm/yr (5 mpy), carbon steel with 3–6 mm corrosion allowance is selected. This is the case if the corrosion is controlled using corrosion inhibitors.
3. In cases when the corrosion rate is higher than 0.13 mm/yr (5 mpy), corrosion resistance alloy (CRA) is specified. While the general corrosion for CRAs is generally negligible, compatibility with the environment regarding stress corrosion cracking and localised forms of corrosion requires special attention.

15.3.1 Materials Applications

In the past, plain carbon steel was the material of choice for seawater, firewater, process piping and equipment. Corrosion control was mainly addressed by painting, galvanising, or concrete lining. Because of excessive corrosion and often-needed replacements, stainless steels AISI 303, 304, 316 and 321 were used in piping of chemicals and hydraulic oil, and even in seawater systems [Haven, et al 1999]. External corrosion attack soon appeared, and only 316 proved to be resistant to the offshore atmosphere. The Cu–Ni alloys were also used for their resistance to seawater corrosion. However, the Cu alloys are found to be sensitive to seawater velocities because they can suffer erosion–corrosion at high velocities (V > 3m/s, 10 ft/s) and some pitting corrosion may occur at stagnant conditions. If, however, there is a possibility of H_2S contamination, Cu–Ni is not recommended. As a follow-up, exotic materials such as 22 Cr and 25 Cr duplex and 6 Mo stainless steels, Ni base austenitic alloys and titanium alloys began to be introduced. These new alloys offered excellent corrosion resistance and are weldable.

While the 300 series, the austenitic and the duplex stainless steels offered excellent internal corrosion resistance, some failures occurred due to external stress corrosion cracking (SCC) caused by the chloride-rich atmosphere offshore. The Ni content plays an important role and the lowest resistance is experienced at about 8% Ni. While 316 stainless steels are not recommended for temperatures above about 140°F (60°C), duplex stainless steels have a much better resistance against SCC due to 50% ferrite content of the structure. Exposed to offshore atmosphere, the 22 Cr duplex shows resistance to 230°F (110°C). Above 230°F (110°C), 6 Mo or 25 Cr is generally recommended.

However, chloride stress corrosion cracking temperature limits for the alloys are always subject to debate. While some operators use 140°F (60°C) as specified above, others use lower limits and some use higher limits. As an example, the EEMUA 194 publications states that austenitic steel grades such as 316/316L are susceptible to chloride stress corrosion cracking where the material temperature exceeds about 50°C (122°F) and oxygen and chloride containing water are present. To a lesser extent duplex stainless steels are also affected, though the threshold temperature increases to about 120°C (248°F) for the 22% Cr and to about 150°C (302°F) for 25% Cr grades, dependent upon fluid chloride content, temperature, pH and oxygen level. Oxygen levels in produced hydrocarbons are usually too low to give rise to this problem. Where oxygen may be introduced locally as a result of raw seawater ingress or chemical injection, or where brines are very concentrated, consideration should be given to the use of nickel alloys resistant to chloride stress cracking. The results from over 700 stainless steel shell and tube heat exchangers in heating/cooling water service including boiler water feeds showed that on the water side, chloride stress corrosion cracking will not occur in austenitic stainless steels exchangers when the system water temperature is less than 80°C (176°F), regardless of the chloride content. At a chloride content of <7 mg/L, SCC will not occur regardless of water temperature. Chitwood and Skogsberg (2004) have reported that 316 stainless steel can be safely used in deaerated production environments containing upto 0.5 psi (0.003 MPs) H_2S and 50,0000 ppm Cl^- at a minimum pH of 3.5 and a maximum temperature of 175°F. At 0.5 psi H_2S, the maximum temperature can be raised to 225°F if the maximum chloride content is reduced to 10,000 ppm. For non-sour environment, 316 can be used to a maximum temperature of 350°F when the chloride level is less than 150,000 ppm.

Where a process fluid contains wet H_2S and CO_2, Incoloy 825 or Inconel 625 are selected. For components operating at low pressure (<20 bar), internal lining with an organic/inorganic coatings can be used as an alternative to cladding with 825 or 625. Where H_2S is present, all materials must meet the requirements of NACE MR0175 and ISO 15156-1/2/DIS. Low-temperature carbon steel (LTCS) is used for service down to −40°C and austenitic stainless steel, type 316L/304L, for design temperatures below this, unless corrosive conditions dictate higher alloy grades.

15.3.2 Materials for Seawater Systems

The recommended materials for seawater service below 60°F (15°C) are 6 Mo and 25 Cr duplex stainless steels, and for service greater than 60°F (15°C) Ti or fibre-reinforced polymer (FRP). Some problems were encountered with early application of FRP that were attributed to improper design, handling and installation. For stainless steels, a minimum PRE value (PRE = % Cr + 3.3 × % Mo + 16 × % N) of 40 must be specified. The 6 Mo or 25 Cr duplex stainless steel piping can be used in higher temperature service up to 95°F (35°C), if crevices are avoided or they are overlaid by Inconel 625. Flanges and threaded connections must also be manufactured from crevice corrosion-resistant alloys such as the superaustenitic material (654 SMO or UNS S34565). A small amount of oxygen in the water causes pitting and crevice corrosion in 22 Cr, while 25 Cr and 6 Mo will stand higher oxygen content. Figure 15.1 provides a comparison between different allowable stress and cost for these grades [Haven, et al 1999]. While materials selection for firewater systems can be similar to the seawater systems, the small nozzles have no tolerance to any corrosion products, otherwise they will plug. Therefore, the preferred material for firewater piping is Ti or FRP.

15.3.3 Materials for Process Piping and Equipment

Produced fluids are generally corrosive due to the presence of water, CO_2, H_2S and chlorides. Assessment of the corrosivity of the produced fluids is established using prediction models. The most widely used models are for predicting CO_2 corrosion. The common practice is to use carbon steel if the predicted corrosion rates can be accounted for by the addition of less than 0.25″ (6 mm) as a corrosion allowance. If a higher corrosion allowance is required, 316 stainless steel is used and also 22 Cr duplex stainless steel is often used when higher strength is required. The 25 Cr duplex stainless steel or the superaustenitic materials are sometimes specified if an even higher strength is required.

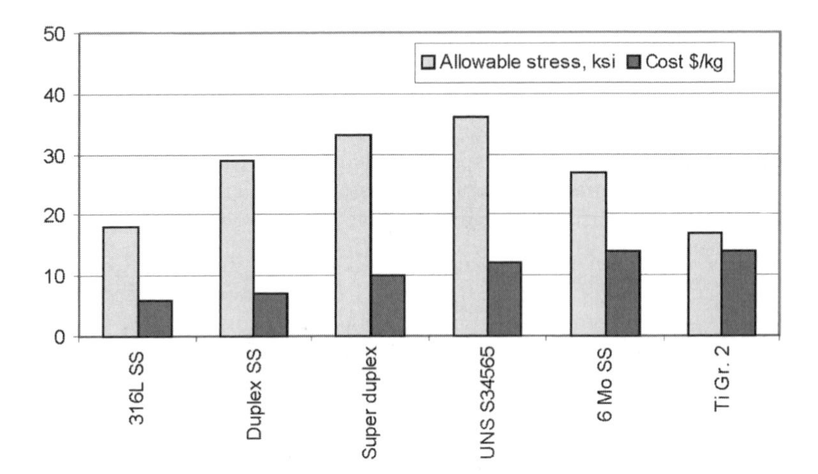

Figure 15.1 Cost and allowable stress for different alloys

Often, the piping in made of 25 Cr or superaustenite, and valves are manufactured using 22 Cr because of its better availability.

The material selected for equipment such as vessels and pumps must be compatible with the attached piping. For low-pressure vessels, internally coated carbon steel and anodes are used. For high pressures, carbon steel with a corrosion-resistant alloy cladding is the preferred option. Some vessel manufacturers prefer applying internal Inconel alloy 625 weld overlay of 0.125–0.2″ (3–5 mm) thickness.

15.4 Material for HPHT Applications

High pressure, high temperature (HPHT) offshore fields are characterised by pressures in the order of 15,000 psi and temperature in the order of 180°C (350°F). In addition, the produced fluids contain CO_2 of about 4% and H_2S of about 50 ppm. These conditions present several challenges that include the need to use higher strength materials, larger thickness and heavier components. The use of high strength and thick materials makes it necessary to address issues such as stress corrosion cracking, welding, brittle fracture, heat treatment requirements, handling, etc. The high temperature adds an additional challenge because the material's allowable strength is reduced with temperature. At 180°C (350°F), the strength is reduced by 5% for steel and up to 20% for cold worked alloys such as duplex stainless steel. At high temperature, design issues associated with buckling due to thermal expansions also become important. The combination of corrosive fluids and high temperature increases the corrosion rate of carbon steel and reduces the effectiveness of inhibitors. Therefore, expensive corrosion-resistant alloys are used.

15.4.1 Limitations of Materials for HPHT Application

Under the operating environments of HPHT fields, carbon steels suffer high corrosion rate that can reach 250 mpy at a high temperature (350°F) and high strength steels suffer H_2S

cracking at a low temperature (< 170°F). To avoid cracking, sour service grade steels with controlled chemistry and restricted hardness are required. The typical strength limit for sour grade steels is generally equivalent to the 95 grade steel (95 ksi yield strength). However, sour grade P110 steel for use as casings in mildly sour service is also available. High strength steels (80–100 ksi) are also considered for flowlines. Technical issues that should be carefully considered include sulphide stress cracking, welding, defects assessment, potential need for PWHT and possible use of mechanical joints.

The premise of using carbon steel components is based on the ability to implement an effective corrosion inhibition program. The use of inhibitors for temperature up to 200°F is considered state-of-practice and for temperatures in the range of 200–250°F is considered state-of-the-art. For temperatures that reach 350°F, non-environmental friendly inhibitors are available, but they impose an environmental challenge because they require a very high concentration, thus making it necessary to install a costly recovery system.

Due to the above challenges, the use of solid or clad corrosion-resistant alloys for development of corrosive HPHT fields is becoming common. The conventional 13 Cr and 22 Cr stainless steel tubing are not suitable because of potential cracking, pitting and high corrosion rates. Typical materials recommendation for HPHT development include the use of 825 or 28 Cr alloys for production tubing, 625 (20 Cr, 8 Mo, Ni) clad for trees and 825 (20 Cr, 3 Mo, 40 Ni, Fe) clad for manifold, subsea safety valve, flowlines and risers. While these high corrosion-resistant alloys are available, their cost is high. Solid alloys cost 10–20 times carbon steel cost, and 625 and 825 clad on carbon steel will cost 5–10 times carbon steel.

15.5 Advanced Composite Materials

Composite materials offer several advantages for marine construction because of their low density, corrosion resistance, high thermal insulation, high structural damping and excellent fatigue performance. In addition, the use of composites allows for greater design flexibility by tailoring the properties to meet specific design requirements, thus promoting better system-oriented solutions. On a one-to-one replacement basis, composite components are often more expensive than their steel counterpart. However, on a performance-equated basis, the economic incentive to use composite components can often be demonstrated based on their capability to reduce system and life cycle costs.

Fibreglass composites, and to a lesser extent carbon composites, have been used by the offshore oil industry in a variety of applications. Some specific examples for the offshore application of fibreglass composites include firewater piping, seawater piping, storage vessels, grating, fire and blast walls, cable trays, mud mats and subsea wellhead. Almost all Tension Leg Platforms (TLPs) in the Gulf of Mexico have used fibreglass pipe for the firewater ring main and gratings. While in the past many operators had very disappointing experiences with leaks of fibreglass pipe joints, recent applications of fibreglass pipes on several offshore facilities have shown that the leak rate of several thousands of field joints was less than 0.1%. Composites have also been used for many high-pressure vessel applications. These vessels are manufactured from glass and carbon fibre composites with a thermoplastic liner. Composite pressure vessels have been developed and qualified for use as mud gas separators. Composite accumulator vessels have been used for

Figure 15.2 Composite drilling and production risers and spoolable carbon fiber composite tether

production-riser tensioning systems. High pressure spoolable composite pipes have been used for on-shore and they are now being considered for offshore applications. The future applications of advanced composites for the offshore industry are for risers and tethers (Salama, et al, 1999, 2001, 2002). Figure 15.2 shows composite tether and drilling risers that are being qualified for deepwater applications.

Since the fire resistance of composites presents a significant technical issue that has limited the use of composites in many surface applications. The fire performance of materials is characterised by their fire growth (ignitability, flame spread, heat release and flashover), habitability (smoke and toxicity), and residual strength (structural integrity). A survey of various commercially available thermoset and thermoplastic composites showed that the phenolic-based composite materials offer the best cost-effective fire performance.

The long-term durability of composites in seawater depends on the type of resin, fibre, fibre sizing and laminate construction. Studies on the effect of seawater on composites showed that the interfacial shear strength of seawater-saturated E-glass epoxy composites with good sizing was decreased by less than 25%. When improper sizing was used, the interfacial shear strength was decreased by about 50%. Test results have also showed that the impact of moisture on carbon fibre/epoxy composites is far less than that on fibreglass composites.

Joining of composites to other composite or metal components is always challenging. The goal is to select a joining concept that achieves high load transfer efficiency, high reliability and durability, minimum joint thickness, simple manufacturing, minimum cost and minimum weight. The two basic joining approaches are adhesive bonding and mechanical interlocking. For low-pressure piping, adhesive bonding has been proven successful. For high-pressure piping, such as risers, joining between the metal connectors and the composite tubes has been successfully demonstrated by using mechanical traps.

15.6 Elastomers

Elastomers are used in many offshore applications such as seals, corrosion resistance liners and flexible joints. Table 15.2 provides a list of the main elastomers that are used in oilfield applications and a qualitative comparison between their performances. Since the common

Table 15.2 Qualitative comparison between high performance oilfield elastomers

Resistant to	Kalrez	Chemraz	Aflas	Viton GF	Viton A	Camlast	HNBR	NBR
Gen. chemicals	5	5	3	3	2	2	2	1
Corr. inhibitors	5	5	5	3	2	2	2	1
Scale inhibitors	5	5	5	2	1	3	3	2
Methanol	5	5	5	5	2	3	3	3
Crude oil	5	5	3	5	4	2	2	2
H_2S	5	5	5	4	3	2	2	1
Toluene	5	5	3	5	4	2	2	2
Acidic brines	5	5	5	5	5	2	2	1
Acids	5	5	5	5	4	2	2	1
Hydraulic fluids:								
Oil	5	5	5	5	5	4	4	3
Water/Glycol	5	5	5	5	3	4	4	2
Solvents/Diesel	5	5	2	5	4	2	2	2
High temp. (350°F)	5	5	4	3	3	2	2	1
Low temp. (0°F)	2	4	2	3	4	3	3	5
Extrusion resist.	2	2	2	3	3	4	4	5
Compression set	2	2	2	3	4	3	3	4
Abrasion resist.	2	2	3	3	3	5	5	5
Cost (A is highest and F is lowest)	A	B	C	D	E	D	E	F

Note: Scale of 1 to 5:
1: Not recommended
2: Recommended only for short exposure
3: Moderate effect (can be used for static application)
4: Minor effect
5: No effect

practice is to specify elastomers on the basis of their generic performance and tensile properties that are traditionally established between the equipment supplier and the rubber manufacturer, it is important to consider two important precautions. The first is that for the same elastomer, most of the physical properties including important ones such as resistance to explosive decompression and extrusion resistance can be greatly influenced by compounding. Therefore, not all elastomer types with the same hardness will have the same properties and thus qualification of the actual material must be established. The second precaution is that some trademarks, such as Viton, represent a family of elastomers with very different capabilities. An example, standard Viton (Viton A or E60C) has poor methanol resistance but Viton B or GF has good resistance. On the other hand, Viton B or GF has poor low-temperature performance while Viton A and E60C have good performance. There are several important factors that affect the selection of elastomers that include:

1. Contact fluids: Elastomers must be selected to be compatible with the various fluids that come in contact with them such as production fluid, workover fluid, completion fluid, kill fluid, acidising fluid, hydraulic control fluid, corrosion inhibitors, scale inhibitors, solvents, gas hydrate control fluid, etc. For example, chemicals such as methanol or glycols are often used for gas hydrate control or as part of a corrosion inhibitor delivery system for gas wells, gas lift operations and pipelines. These solvents can cause excessive swelling or softening of some of the commonly used materials like Viton A- or E60C-type elastomers, and thus special alcohol-resistant grades such as Viton GF need to be selected.

2. Seawater temperature: In northern climates, the temperature at the sea bottom is about 40°F (4°C). Some of the more chemically resistant elastomers like Kalrez and Aflas cannot function as a dynamic seal at this temperature. The limiting service temperature will be influenced by the seal design and function. The lower limiting service temperature of an elastomer can be improved by the inclusion of a high molecular liquid, a plasticiser. If this plasticiser can be extracted out or is lost when the seal is subjected to excessive heat, the beneficial effect is lost. On the other hand, if a seal with poor low temperature properties is slightly swollen by the contact fluid such as crude oil, the "plasticised" seal shows improved low-temperature performance.

3. Service life: The selection of elastomers can be influenced by service life. Under similar environmental conditions, NBR can be acceptable for short service (< 5 yr) while fluoroelastomers like Viton, Aflas and Kalrez are required for long service.

15.7 Corrosion Control

Corrosion control strategy involves two facets, namely corrosion control against production fluids and corrosion control against seawater. The use of chemical inhibitors for corrosion control against production fluids is widespread in the oil industry. But, the reliability of chemical injection valves and mandrels limit the applicability of this option for downhole corrosion control. However, the use of corrosion inhibitors is viable for protecting topside equipment and pipelines. When deciding whether to use carbon steel option with corrosion inhibitors or corrosion resistance alloys, three issues need to be considered:

1. Capital and operating costs
2. Reliability of inhibitor delivery systems
3. Possible future restrictions based on environmental considerations.

The capital costs involve the cost of control, injection and distribution hardware, corrosion monitoring systems, storage weight and space. The operating costs involve the cost of chemicals, maintenance, monitoring, manpower, additional selection programs and technical service. Corrosion control against seawater is achieved using cathodic protection or coatings. When coating is used, proper surface preparation is crucial to achieving the desired performance. Cathodic protection is also used in combination of coating to provide the added protection in areas of coating damage.

15.8 Material Reliability and Monitoring

Although failure of materials may appear to occur instantaneously and at random, failures often result from a gradual degradation. Condition monitoring offers an approach for monitoring deterioration, changes in processing conditions and specific events that precede equipment failures. Unlike inspection, condition monitoring provides evidence about the condition of a component or system and monitors deteriorations that precede failures without requiring shutdowns and dismantling for inspection.

Each condition monitoring system relies on the measurement of specific parameters, either continuously or semi-continuously, as a function of time. It is always desirable to monitor parameters that can be directly related to degradation such as wall thickness. Frequently, the monitored parameter is indirectly related to deterioration and, therefore, the data is interpreted in terms of a model of the degradation process. Corrosion, wear, leak, vibration and cracking are the main parameters that are monitored. Table 15.3 describes the most relevant and widely used systems for monitoring corrosion.

15.9 Fracture Control

Materials are designed to withstand the combination of the different operating loads without exceeding their ultimate strength, instability condition or fracture limits. Safety factors are included in the design to account for uncertainties associated with loading, analysis and material performance. The traditional approach is to use the working stress design (WSD) approach in which a safety factor is applied to the maximum allowed stress. The WSD approach does not allow separating the uncertainties in the load from those of the material. Therefore, the approach that is currently being used for offshore structural design involves the use of partial safety factor (PSF) method and is also known as the load-resistance factor design (LRFD) method. In this approach, calibrated safety factors using reliability methods are applied to both loads and strength. Since different design codes apply the safety factors in slightly different ways, it is inappropriate to mix values from different codes.

The strength parameter will vary according to the expected failure mode, i.e. yielding, buckling, instability, brittle fracture or fatigue. Therefore, several material parameters are generally required to perform the proper material selection and design. These parameters include:

1. Yield strength
2. Ultimate strength
3. Elongation to failure
4. Reduction in area
5. Elastic modulus
6. Fracture toughness
7. Crack growth rate
8. Fatigue S–N curve

Offshore structures generally include complex welded joints that have large local stress concentrations and are subject to fatigue loadings induced by environmental forces.

Table 15.3 Corrosion monitoring systems

Technique	Monitored parameter	Comments
Linear polarisation	Corrosion rate Corrosivity of fluids	Involves passing a small dc or a fixed low frequency ac (5 or 10 Hz) current between two probes and monitoring the potential change, which is related to corrosion rate. Flush mounting and careful choice of probe material are necessary to minimise differences between actual and probe corrosion.
Electrochemical impedance	Corrosion rate Corrosivity of fluids	More accurate than linear polarisation particularly for high-resistance electrolytes and inhibited oils. It involves passing a variable frequency ac current between probes and monitoring both amplitude and phase of the ac potential with frequency.
Electrical resistance	Corrosion rate	Corrosion rate is related to the wire resistance, which increases as the cross-sectional area decreases due to corrosion. Accuracy of results depends on the wire geometry, material and cross-sectional area. The presence of conductive sulphide scales lead to erroneous results.
Iron analysis	Corrosion rate	The method is applicable in long flow lines containing sweet fluids. The method cannot distinguish between general corrosion and localised corrosion. It also cannot be used for monitoring H_2S corrosion because iron sulphide precipitates in solution.
Coupon testing	Corrosion rate	Coupon testing by spool pieces is simple, but may be limited in the context of a fully instrumented corrosion monitoring system.
Hydrogen probe	Corrosion rate	Most corrosion reactions produce molecular hydrogen and some atomic hydrogen in case of sour environment. Atomic hydrogen diffuses through the steel and its amount can be measured by a probe installed on the external surface.
Remote visual	Surface condition	Boroscopes and other fibre optic devices are used for internal inspection. In general, this is considered an off-stream technique.

This, in addition to fabrication defects that are often present in welded structures, will result in the early initiation of fatigue cracks. Since it is not practical or economical to fabricate defect-free structures, premature failures are avoided by the use of appropriate inspection and quality control procedures to limit the defect size and by the proper account of these defects in the design. To minimise the probability of failures, the design of offshore structures is based on the combined use of classical design and structural integrity design. Structural integrity design or as often called engineering critical assessment (ECA) is the basis for precluding structural failure due to brittle fracture or premature fatigue cracking. Integrity design provides a tool to assess fracture resistance by integrating stress analysis, fabrication quality and mechanical properties of the steel. Mechanical properties that are required include fatigue crack growth curves, fracture toughness and the basic tensile properties (e.g. yield strength and tensile strength). Currently, all design guidelines, codes or standards for critical applications emphasise fracture control procedures and provide requirements for fracture toughness, weldability, inspection, etc. The commonly used guideline for ECA is the British Standard BS7910: 1999 that includes detailed procedures for assessing both fracture and fatigue.

The most common approach for fatigue design involves the use of the S–N curves. The S–N curve provides a relationship between the cyclic stress range (ΔS) and the number of cycles to failure (N) as presented by the following form:

$$N = \frac{A}{\Delta S^m} \tag{15.1}$$

The values of A and m are constants determined experimentally and depend on the material, joint details and the operating environment. The value of m for steel is in the range of 3–5 and can be higher for other materials. This makes fatigue life predictions very sensitive to the assumptions upon which the stress range at a specific location is based. This sensitivity is the cause of one of the major shortcomings of the S–N method in estimating the fatigue life. To partially account for this shortcoming, design codes specify different curves according to the geometry of the joint as shown in fig. 15.3.

Also, the standard S–N curves are based on constant amplitude cyclic load which is not representative of the actual loading that is variable. Therefore cyclic fatigue calculation requires the use of an accumulative damage rule; the most common one is the Miner's rule that has the following form:

$$D = \frac{n_1}{N_1} + \frac{n_2}{N_2} + \frac{n_3}{N_3} + \cdots = \sum \frac{n_i}{N_i} = 1 \tag{15.2}$$

where n_i is the number of cycles for which the stress is subjected to a stress range ΔS_i. Failure occurs at ΔS_i when the number of cycles reaches N_i. Assuming that the value of D equals 1 is not realistic because it ignores the effect of loading sequence. Therefore, almost all codes specify lower values that vary between 0.1 and 0.5 depending on the criticality of the component and the difficulty of the in-service inspection and repair. In order to apply the Miner's rule, a break down of the cyclic load spectrum into blocks of ΔS_i and n_i is required. The most common method for decomposing the stress spectrum where the stress ranges are ill defined is the rainflow method. Fatigue analysis is performed using either a deterministic approach by simply applying the Miner's rule or the spectral

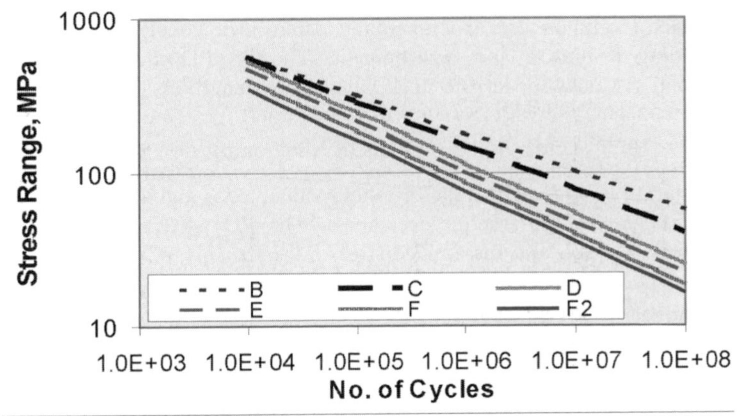

S–N curve	Welding condition
B	Machined surface and welds normal to main loading direction.
C	Two-side weld with overfill and dressed flush.
E	Two-side or one-side weld with temporary backing strip.
F	One-side weld with permanent backing strip.
F2	One-side weld with no backing strip.

Figure 15.3 Typical fatigue curves for different steel joints

(frequency domain) approach. The spectral analysis is applicable for structures that are subjected to random loading, but which respond linearly with wave height at any wave period.

Since the *S–N* approach does not directly account for existing crack-type defects in the component, fatigue life predictions based on crack growth analysis is often used. In this case, the fatigue life is estimated by the integration of an appropriate crack growth equation such as the Paris law between the allowable initial defect (a_i) and the final defect (a_f) at which failure occurs. The size of the final defect depends on the fracture toughness of the material and the applied stress. The Paris crack growth equation provides a relationship between the crack growth rate (da/dN) and the stress intensity factor range (ΔK) in the following form:

$$\frac{da}{dN} = C\Delta K^m \tag{15.3}$$

The stress intensity factor range, ΔK, is defined by:

$$\Delta K = \Delta S Y \sqrt{\pi a} \tag{15.4}$$

where ΔS = cyclic stress range, Y = a correction factor dependent on component and crack geometries, a = half length of through thickness rectilinear crack.

The parameters C and m are experimentally determined constants which depend on material, loading condition and environment. The BS PD6493 [1980] provides the following C and m values for ferritic steels with yield strength up to 600 N/mm^2:

$$C = 3 \times 10^{-13} \quad \text{and} \quad m = 3 \text{ (in units of N and mm)} \tag{15.5}$$

The Paris equation is bounded by the threshold value, ΔK_o, and the critical value, K_{max}, which is a measure of the fracture toughness. The PD 6493 provides the following relationship between ΔK_o and the applied stress ratio R:

$$\Delta K_o = 190 - 144\,R \text{ N/mm}^{3/2} \tag{15.6}$$

This relationship provides the lower bound to all published threshold data for grade 50 D steel, in air and seawater. But other data for similar steels and for austenitic steels lie below the PD6493 line. As a result BSI 7910:1999 proposed the following relationship based on 97.7% probability of survival (Eq. 15.7).

$$\begin{aligned} \Delta K_o &= 170 - 214\,R \text{ N/mm}^{3/2} && \text{for } 0 \le R < 0.5 \\ &= 63 \text{ N/mm}^{3/2} && \text{for } R \ge 0.5 \end{aligned} \tag{15.7}$$

The offshore industry has used several advanced fracture mechanics methodologies and testing to establish allowable final defect (a_f). These methodologies include crack tip opening displacement (CTOD) and J_R and failure assessment diagram (FAD) methods. Ensuring against brittle fracture by specifying a blanket CTOD value is difficult without performing a detailed fatigue life calculations. Toughness specifications in terms of CTOD values are valuable, because when used with fatigue crack growth rate data in the framework of fracture mechanics analysis, they can provide valuable information on tolerable defects, remaining life and allowable loading conditions. Because of the complexity of the CTOD testing, most design codes still rely mainly on Charpy energy and transition temperature concepts as the main fracture toughness acceptance criteria.

References

Carter, R. M., Marshall, P. W., Swanson, T. M., and Thomas, P. D. (1969). "Materials problems in offshore platforms", *Proc. of 1st Annual Offshore Technology Conference*, Vol. 2. Paper OTC 1043.

Chitwood, G. and Skogsberg, L. (2004). "The SCC resistance of 316L expandable pipe in production environments containing H_2S and chloride", NACE paper 04138.

De Koning, A. C., Harston, J. D., Nylar, K. D., and Ohm, R. K. (1988). "Feeling free despite LBZ", *Proc. of 7th Int. Conf. On Offshore Mechanics and Arctic Engineering*, Vol. 3, pp. 161–181, ASME.

Denys, R. M. and Dhooge, A. (1988). "Mechanical tensile properties of weld simulated HAZ microstructures in normalised, quenched-tempered and TMCP low carbon manganese steels", *Proc. of 7th Int. Conf. on Offshore Mechanics and Arctic Engineering*, Vol. 3, pp. 207–219, ASME.

de Waard, C. and Milliams, D. E. (1976). "Prediction of carbonic acid corrosion in natural gas pipeline", BHRA Conference, University of Durham, The Netherlands, Vol. 28, p. 24.

de Waard, C., Lotz, U., and Milliams, D. E. (1991). "Predictive model for CO_2 corrosion engineering in wet natural gas pipelines", Corrosion 91, Paper 577, NACE.

Haven, T., Kolts, J., Salama, M. M., and Tuttle, R. N. (1999). "Materials status and developments for oil and gas industry offshore, topside", *Proc. of Int. Workshop on Corrosion Control for Marine Structures and Pipelines*, February 6–11, Galveston, Texas, USA, pp. 465–490, American Bureau of Shipping.

Masubuchi, K. and Katoh, K. (1987). "Uses of thermo-mechanical control process (TMCP) steels for ships and offshore structures and welding considerations," *Proc. of 6th Int. Conf. On Offshore Mechanics and Arctic Engineering*, Vol. 3, pp. 137–144, ASME.

Ohnishi, K., Suziki, S., Inami, A., Someya, R., Sugisawa, S., and Furusawa, J. (1988). "Advanced TMCP steel plates for offshore structures", *Proc. of Microalloying 88*, pp. 215–224, ASM International.

Peterson, M. L. (1969). "Evaluation and selection of steel for welded offshore drilling and production structures", *Proc. of 1st Annual Offshore Technology Conference*, Vol. 2, Paper OTC 1075.

Peterson, M. L. (1975). "Steel selection for offshore structures". *Journal of Petroleum Technology*, Vol. 27, pp. 274–282.

Peterson, M. L. (1987). "TMCP steels for offshore structures", *Proc. of 19th Annual Offshore Technology Conference*, Vol. 4, Paper OTC 5552.

Salama, M. M., Peterson, M. L., and Williams, T. H. (1988). "Experiences with structural steel in the offshore industry", *Proc. of Microallying '88*, ASM International, pp. 131–142.

Suziki, S., Arimochi, K, Furusawa, J., Bessyo, K., and Someya, (1989). "Development of LBZ free low Al-B-treated steel plates", *Proc. of 8th Int. Conf. On Offshore Mechanics and Arctic Engineering*, Vol. 3, pp. 657–663, ASME.

Handbook of Offshore Engineering
S. Chakrabarti (Ed.)
© 2005 Elsevier Ltd. All rights reserved.

Chapter 16

Geophysical and Geotechnical Design

Jean M. Audibert and Jun Huang
Fugro-McClelland Marine Geosciences, Inc., Houston, TX 77081, USA

16.1 Preface

For a professional geotechnical engineer, geologist and geophysicist, there are many excellent textbooks, articles and papers, as well as numerous international, national and industry codes of practice and guidance notes on the subject of seabed investigation. However, there are very few informative handbooks that cater to the non-specialist Project Manager and other professionals requiring a working knowledge of the subject to better facilitate meaningful dialogues with their specialist advisors and with their contractors.

This chapter is based, to a large degree, on a handbook that was developed by the FUGRO Group, and is the result of consultations with some of the leading specialists in the fields of geophysics and geotechnical investigations. These discussions have been transcribed into plain language without a recourse to complex science, mathematics, or lengthy descriptions of complicated procedures.

The objective of this chapter is to provide an overview of the geophysical and geotechnical techniques and solutions available for investigating the soils and rocks that lie beneath the seabed. Every project and every situation is different; the subject itself is highly technical. A project's successful outcome depends on securing the services of highly competent contractors and technical advisors. It should also be noted that any reference in this document to achievable soil/rock penetration, production rates or weather limitations and the like, are provided for general guidance only. What is achievable will always be governed by a combination of factors, such as geology, water depth, environment and vessel capabilities.

It is hoped that this chapter will fill a knowledge-gap and provide a useful guide to science, its application and technology.

Kind permission[1] to use this originally copyrighted material was granted by the FUGRO Group to the authors, who are particularly grateful to Mr. Eugene Toolan, Chief Operating Officer, for granting such permission.

16.2 Introduction

In the infancy of the offshore industry, the soil exploration program was performed simultaneously with construction. The soil boring served as a construction guide, rather than a design tool. Today, soil investigations are done months to perhaps years ahead of construction. The information is used to evaluate the type of structure best suited for the site and to complete the sophisticated designs.

Offshore investigations involve both direct and indirect methods. Direct methods are those which provide actual physical evidence of the materials, such as soil borings, drop cores and in situ testing. The indirect methods are those which sense remotely, such as electromechanical and geophysical profiling.

The scope of an investigation should be considered carefully. The investigative methods are influenced by the following factors:

- Water depth
- Type of structure
- Environmental loading
- Soil conditions
- Local experience
- Geologic hazards
- Potential foundation savings

The collective consideration of these factors will lead to the selection of the scope and the type of site investigation that will provide the appropriate technical information at a reasonable cost to effect an economical design.

Geophysical surveys and geotechnical investigations are seldom performed without an end objective in mind. In general, the objective is the engineering design, construction and the installation of some sort of seabed structure.

The environments in which these operations take place vary greatly and can have a major influence on the choice of surveying and geotechnical system(s) used and

[1]**Disclaimer:** Please note that the specifications of equipment described in this handbook are continuously evolving. The authors accept no liability for the accuracy of the information herein provided or the use to which it is put.

have an impact on the field operations, not the least of which is safety. To better categorise these environments, the geo-industry has developed an empirical operating scale:

Category	Description
Harsh	An environment such as the North Sea and the North Atlantic seaboards where there is a high frequency of sudden storms.
Tropical seas	Normally benign and swell-free regions but which lie within tropical storms paths. Such storms are invariably announced by weather warning notices.
Bounded seas	Enclosed seas such as the Caspian, Mediterranean and Black Sea that are free of oceanic swells but where storms can be sudden.
Benign tropics	Areas, such as the west coast of Africa, with continental shelves open to the ocean where storms are infrequent but which suffer from prolonged intervals of long-period swells.
Arctic	In general, the high latitudes bounded by the limits of summer sea ice. These areas are subject to sudden storms and, beyond their equinoctial circles, provide limited working opportunities.

Water depths also affect geophysical and geotechnical activities and dictate the sort of techniques and instrument systems required and their operational effectiveness. Generalising, water depth limitations of geophysical remote sensing systems differ from those that constrain the geotechnical systems. While this is not a practical difficulty, it is worth considering as it can influence the mode of operations, especially where the geophysical and the geotechnical activities are combined.

Geophysical depth ranges		Geotechnical depth ranges	
Inshore, ports and harbours	< 25 m	Shallow water/near-shore	< 20 m
Shallow water	25–250 m	Offshore	20–500 m
Medium depth	250–1500 m	Deepwater	500–1500 m
Deepwater	1500–3000 m	Ultra-deepwater	> 1500 m
Ultra-deepwater	> 3000 m		

16.2.1 Regulations, Standards and Permits

All marine activities are subject to international and/or national regulations and industry operating standards. A number of the international regulations such as those of the International Maritime Organization (IMO) have not necessarily been ratified by all participating nations although they may, in whole or part, have been adopted by, or have become accepted practice of, individual nation states.

Many of the operational and technical facets of geophysical surveying and geotechnical investigations are included within the various standards and codes of practice; a list of some of these is included in Section 16.12. Invariably, a program of offshore work will require permits from the maritime authorities and from the various departments having jurisdiction over the operating areas such as offshore oil and gas fields and their associated infrastructure of pipelines and work zones. Likewise, cable surveying and installation operations will require permits that will include beach landfalls and site access.

While preparing a specification for an operation, there is sometimes the temptation to assume that the standards and regulations, requirements, procedures and permit arrangements of an earlier job can be applied. Unless there are substantial grounds for believing this, such a practice should be avoided, as there is the greatest risk of oversight that can have serious safety, legal and financial consequences.

16.2.2 Desk Studies and Planning

The chances for a successful outcome to a seabed investigation are significantly improved when the work program commences with a properly structured desk study. Time and again this sensible precaution has demonstrated savings in time and cost, and has always led to an improved end-product while providing the engineer with an early overview of site conditions and expectations upon which to base preliminary designs.

Desk studies must not only focus on the requirements of the end product such as a platform, pipeline, cable, or anchor installation, but must also consider the environmental impact of the proposed engineering and the wider consequences of the work. Desk studies comprise the collection of information from public, in-house and commercial sources that can be evaluated to develop overviews on:

- The regional quaternary geology, surface sediments and seabed morphology,
- Probable geotechnical conditions, nature of seabed soils and rocks, etc.,
- The local topography (bathymetry),
- The meteorological and ocean environment, e.g. tides, currents, weather patterns and sea states,
- Existing seabed structures and obstacles such as cables, pipelines, etc., and
- Fishing and other marine activities.

A desk study alone is not sufficient for detailed engineering purposes, but will lead to a sensible operational plan that considers the environmental factors that may affect the work. It will identify an appropriate level of technical specification to meet the objectives while allowing for the unforeseen eventuality. A desk study can also address the peripheral issues of regulations, standards and permitting.

16.2.3 Specifications

Assuming that an operation will be intrinsically safe, and that all the statutory and legal issues are correctly addressed, specifications tend to fall into one of the following four classes:

Same as last time. Where it can be shown that the parameters for a new work program are essentially the same as a previous job, then using the last specification is a reasonable choice. However, few jobs fall into this category even though, at the first glance, the conditions appear similar. The end product must always be the first consideration;

an earlier work program for, say, a template emplacement will be substantially different to that for an anchoring operation in the same vicinity. Apart from the very different geotechnical requirements, reliance on a previous specification will lead to erroneous design assumptions, technical failure and financial risk.

Best technique. The best technical solution for a particular engineering problem may still not be the correct choice. For many reasons, it may not be feasible because of time constraints, or the remoteness of location, or over cost grounds. The choice of the best technical solution should always be based upon a cost-benefit analysis.

Lowest cost. Here the question must always be, "Does the solution offered meet the requirements of the objective?" Apart from the obviously inappropriate, the solution provided by the lowest bid is frequently technically marginal. The risks are considerable when the results from an investigation, depending on the marginal techniques, do not provide adequate design information or, worse, do not identify any potential hazards or weaknesses. The risks of damage and/or failure of the end product structure are very high; remedial action or intervention costs will escalate as also the hazards posed to the environment.

Most reliable on timing. A properly conducted desk study will inevitably lead to a reasonable estimate of time required. An appropriate proposal that meets the technical requirement and offers a reliable timing (assuming this is sensible) can be evaluated simply on cost-benefit terms.

16.2.4 Applications

Pipelines for Oil and Gas Product Transport

Pipelines by their very nature demand protection from their environment and vice versa. In areas of seabed engineering, or other activity, or where the soils offer maximum cost-efficient protection, pipelines are invariably trenched and either left to back-fill naturally or are back-filled mechanically with the excavated soil or are covered with a rock berm. Where pipelines are laid on the seafloor or partly trenched, rock dumping or a layer of concrete "mattress" affords protection.

Geophysical surveys, using the side-scan sonar for imagery and multibeam echo sounders for bathymetry, provide information on the topographic and seabed surface texture, while the sub-bottom profilers provide information on the structure of the soils and the rocks beneath. Geotechnical investigations, using coring and Cone Penetration Tests (CPTs), provide the "ground truth" for the remotely sensed data and information on the soil and the rock types to determine seabed-loading characteristics.

Pipelines are also prone to seabed sediment movements, seawater currents and fish action that result in scouring and suspended sections. An environmental assessment and seabed stability studies identify these risks and can suggest suitable remedies and precautions.

Submarine Telecommunications and Power Cables

Submarine telecommunication cable systems are especially vulnerable to damage between their landfalls and the edge of the continental shelf. Damage to these systems is costly to repair and the loss of revenue from a single day's downtime can easily exceed $1 million. Fish bites, scouring and chafing are all sources of potential damage. In regions of mobile sand, a buried cable can quickly become exposed and, in areas of fishing activity, cables are at a great risk from trawls.

Vessel-anchoring is another source of danger, especially in softer sediments where the anchors tend to drag before finding a holding ground. To protect cables from deep-water fishing activities, in vulnerable areas down to 1500 m water-depth, cables are now frequently buried. Cable burial is normally performed simultaneously with the lay using a special plough or, in softer sediments, a high-pressure water jet. The burial depths vary up to 3 m, occasionally even deeper, although the current norm is 1–2 m.

The nature of the seabed soils dictates the method of burial; to ascertain these parameters, geophysical and geotechnical investigations are mandatory. The seabed morphology is imaged using multibeam echo sounders and side-scan sonar, while the sub-bottom profilers determine the sediment layers and may identify zones of buried boulders and surface cobbles. Once a potential cable route is settled on, geophysical tools such as refraction seismic and resistivity systems, and geotechnical tools such as cone penetration tests (CPT), soil cores and grabs samples, provide the all-important cable burial assessment study (BAS) data. These data are used to select burial methods and optimise the ploughing system configurations.

Seabed Founded Structures and Platforms

Foundation engineering studies are critical for all structures placed on, founded in, or anchored to the seabed. The impact of the proposed structures has also to be assessed for their effect on other structures and their influence on the local (and regional) environment (e.g., scour).

Surveyors and geophysicists use high-resolution geophysical systems to image the proposed work location(s), to assist engineers with their preliminary studies, and to generate data on the surrounding area for environmental impact assessments. After site selection, the same tools provide a detailed topographic and morphological information of the sites and information on the sub-surface conditions.

The soil types, strengths and characteristics are assessed from the soil and rock samples recovered by drilling and coring, augmented by grids of CPTs and other in situ tests.

The oilfield subsea structures are connected with a network of control "bundles", umbilical and communication and power cables. This infrastructure is crucial to the safe and economic operation of a field and the demands for protection are great. The trenching, backfill and rock dump protection methods are all employed and all require detailed geotechnical, geophysical and environmental impact assessment studies to determine the safest and most appropriate method of risk reduction.

Seabed Stability Studies

Very few areas of the world's seas and oceans are benign; seawater currents, temperature gradients, unstable soils, tides and wave action directly or indirectly affect the shallow soils of the seabed. In the higher latitudes, glacial and post-glacial activity has left complex and often unstable seabed conditions. Gas leaking through the sands can produce very hard concretions or, in soft clays and silts, potentially volatile "pock-marks" or gas-charged sediments. In some areas, the mobile sands traverse the seabed resulting in sand bedforms that range from small ripples to the larger "mega-ripples" up to dune-size masses.

The movement of the mobile sands and the thinner sediments alternately cover and uncover structures placed in their path while the current eddies cause scouring in loose sands and softer sediments; pipelines are particularly vulnerable to these effects.

In the deep oceans, the extreme pressure and the low temperatures can result in potentially hazardous frozen gas hydrates. Even on the gentlest of slopes, mudslides can develop that travel for many kilometers, added to which the swift currents and the near-freezing conditions make the deepwater a particularly challenging environment.

Seabed stability studies depend upon high quality data; geophysical surveys using side-scan sonar that provide clear images of seabed morphology, easily identifying mobile sands and boulder fields, while the multibeam echosounders provide the accurate topographic detail for slope determination and the exposed size of geological features. Sub-bottom profilers image the seabed identifying the complexity of the soils, the possible presence of zones of buried boulders, faulting and fissures, gas leaks and signs of trapped gas pockets. The geotechnical samples and in situ tests provide the ground truth data for the geophysical interpretation.

Seabed Protection "Glory Holes"

In active areas where seabed damage can be extreme, such as from iceberg scouring, seabed structures and their infrastructure can be protected within large, man-made, "glory holes". Typically, these holes can be up to 100 m across and 10 m to 15 m deep. A geophysical survey will provide information on the penetrating depth of scouring and hence the minimum depth of the hole. The successful excavation of glory holes depends on a detailed geotechnical study to determine the soils' characteristics, strengths and friction angles in order to design the program and select the most appropriate excavation method.

Anchoring Studies

Increasingly, engineers are recognising that temporary heavy mooring anchors (for example of semi-submersible drilling units) require as much geotechnical consideration as permanent anchoring systems. A geotechnical study will allow calculation of the most appropriate anchor size and best fluke angles for maximum penetration and holding strength. As exploration and production moved into progressively deeper waters, floating platforms have become a design of choice. These "floaters" need to be anchored to the seafloor by means of driven or drilled and grouted anchor piles, suction caissons, dragged plate anchors and suction-embedded plate anchors.

In problematic grounds, the traditional method of anchor tensioning of a semi-sub can take five or more days. In extreme cases, this can lead to a complete re-appraisal of the drilling location. The cost associated with a five-day overrun in mooring-up, including lost production time and increased weather downtime risks, can easily exceed several million dollars. A geotechnical study of an anchor pattern will lead not only to correct anchor choice and set-up parameters, but will also quickly identify any weak or unsuitable grounds at an early enough stage to avoid costly re-design or re-appraisal.

Environmental Impact Studies

Protection of the environment from engineering or other human intervention, and to preserve the natural balance, begins with a careful appraisal. Geophysical surveys can map the terrain, identify its boundaries and provide the framework of topology, but do not necessarily provide any qualitative information on the eco-system. On the other hand, geotechnical sampling, especially box corers, will preserve undisturbed seabed samples of benthic colonies and worm populations upon which other life forms, such as fish stocks, depend.

Geotechnical methods, geophysics and other remote sensing methods, can all be employed to identify and examine the habitats of the endangered corals, chemosynthetic life forms and other oceanic populations.

16.3 Geophysical Techniques

16.3.1 General

High resolution geophysical information is used by the geotechnical engineer for the purpose of siting structures. In siting studies, we identify the geological and the geotechnical conditions that may influence design, placement, construction and safe operation of drilling and production platforms, submarine pipelines and other engineered structures. Significant geologic features are faults, submarine landslides, irregular seafloor topography, areas of seafloor scour, unusually soft sediments, shallow hard bedrock and sediments containing gas or gas hydrates.

Geophysical data acquired for a drilling hazards survey often can be used for a structural siting study. However, the data must be re-evaluated since conditions which are *not* drilling hazards may be significant to structural siting and design. Siting studies always include at least one geotechnical soil-boring.

In most structural siting surveys, geophysical data are obtained along a series of closely spaced survey lines. These survey lines usually are arranged in an orthogonal grid pattern centered on the site of interest.

Various tools are used to acquire graphic records of the seafloor and the subseafloor geologic conditions. Tools include:

- Echo sounder (water depth)
- Subbottom profiler (shallow penetration)
- Medium penetration profiler
- Deep penetration system
- Side-scan sonar
- Marine magnetometer

These tools are run simultaneously and are attached to the hull of the vessel or towed astern. Figure 16.1a shows the layout of the high-resolution acoustic profiling systems aboard a typical survey vessel. These tools will be discussed in greater detail in the

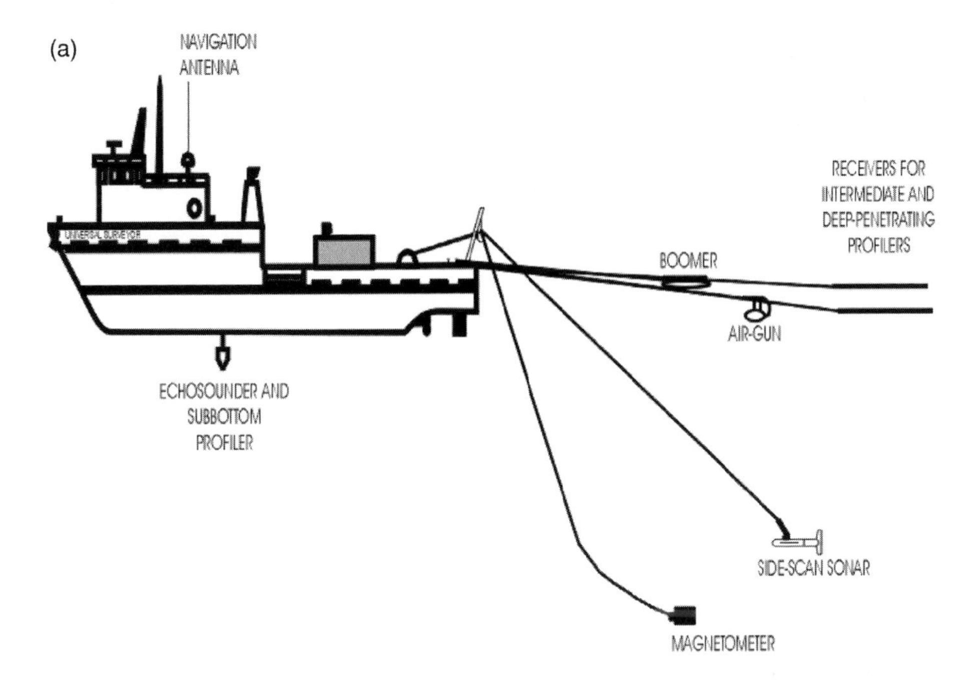

Figure 16.1 (a) Multi-sensor geophysical survey vessel arrangement; (b) example of geophysical survey ship – the Geo Surveyor

following subsections, starting with the highest frequency equipment and generally moving toward the lowest frequency systems.

16.3.2 High-Resolution Reflection Systems

Function and Applications

Geophysical surveys make measurements of the seabed and the sub-seabed using sound or, at close quarters, laser light. The sensors tend to fall into three categories:

- Seabed measuring sensors, e.g. echo sounders, multibeam sounders
- Imaging sensors, e.g. side-scan sonar, laser-scan, acoustic scanning systems
- Sub-bottom profilers, e.g. pingers, boomers, etc.

The most common combinations of system sensors for engineering applications are:

- Echo sounder – for measuring the water depth directly beneath the vessel. This also acts as a calibration device to the multibeam sounder.
- Swathe bathymetry – for measuring a wide swathe of seabed soundings either side of the survey vessel.
- Side-scan sonar – for generating a scaled image of the seabed morphology and features.
- Sub-bottom profiler – for determining the stratification of soils to a depth of, perhaps, 50 m beneath the seabed, depending on frequencies and energy levels.

Support Vessels and Deployment Systems

Major geophysical surveys tend to be conducted from specialised survey vessels (fig. 16.1b) specifically fitted for deploying and handling both the geophysical and the geotechnical systems. Onboard, the surveyors, the geophysicists and others specialist personnel are provided with laboratories, workshops and computer processing and plotting facilities. These vessels can remain at sea for many weeks.

Near the seabed, geotechnical surveys can be performed from most vessels equipped with an A-frame or other suitable crane handling systems. Where office or cabin space is at a premium, special containerised workshops and laboratories can be installed, e.g. on back decks of workboats. Inshore and coastal surveys are normally conducted from launches or from small vessels such as fishing boats.

The smaller geotechnical apparatus, such as grabs, gravity corers, vibrocorers and light-weight CPT systems, can be deployed from survey ships or other of the larger sort of vessels. Heavy or specialist geotechnical systems require dedicated specialist geotechnical vessels fitted with heavy duty A-frames, cranes and winches.

Calculating the size of the cranes and the A-frames is the work of the marine engineer; the safe working loads (SWL) are calculated based on the mass of the tool and its tow or lifting cable together with the maximum dynamic stresses likely to be encountered retrieving the tool from the seabed. Some tools, like the corers, have to be pulled out of the seabed and the mass of grab samplers increases threefold as they collect their large samples. Other tools, such as the deep-towed bodies, or refraction seismic systems which are towed across the seabed impose considerable strains on their tow cables and systems.

System Technology and Science

Acoustic energy (sound) is the most common source for underwater measuring and sensing systems. Over very short distances, in higher quality water, a new generation of scanning systems use laser light but these systems are beyond the scope of this textbook.

In operation, an acoustic energy source generates a pulse of sound that travels through the water column and, when powerful enough, penetrates into the seabed. The sound energy is reflected back as an echo to a receiver system. The lapse in travel time from transmission to reception is converted into ranges.

The media through which the sound passes affects the acoustic signal in various ways. The denser a medium, the faster is the speed of sound; hence, as the wave front passes through different water densities, its rate of progress varies. At the interface between the media, a change in the properties will cause some energy to be reflected; this is most prominent at the water/soil interface and at the boundaries between soil strata.

The two fundamental characteristics of the acoustic wave used in geophysical survey are amplitude and frequency. Different acoustic and seismic tools operate within different amplitude and frequency ranges, and provide information on the different aspects of the physical environment. In the simplest term, high frequency, low amplitude signals provide high-resolution information in the water layer and shallowest depths sub-seabed, and have a shorter range. A low frequency, high amplitude signal will travel further into the earth, but has a lower resolution. This concept is illustrated in fig. 16.2.

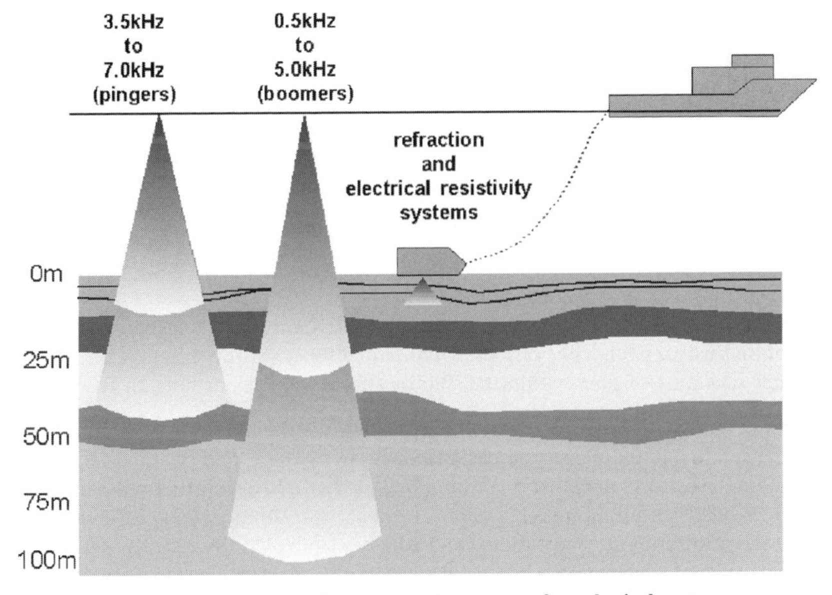

Figure 16.2 Typical seafloor penetration ranges of geophysical systems

To generate different frequencies and amplitudes of acoustic energy, transducers of many types are used. Electromechanical transducers generate acoustic pulses in echosounders, side-scan sonar, pingers, boomers and chirp sonar. Electrical discharges generate acoustic energy in sparker systems. The air gun systems convert compressed air pressure into high-energy acoustic pressure waves in seismic sources. Returning signals are detected using pressure sensitive transducers and hydrophones. The pressure pulses are converted to electrical energy for measurement and storage.

16.3.3 Sounders

Echo Sounders

The echo sounder measures the water depth by measuring a two-way travel time of a high frequency pulse emitted by a transducer (fig. 16.3a). The system must be calibrated to allow for errors introduced by temperature and salinity and other factors that affect sound velocity in the water column. The choice of echo sounder depends on many factors including accuracy requirements, depth of water and resolution. Typical frequencies range from 10 to 200 kHz.

Water-depth records generally employ a single piezoelectric transducer that both transmits and receives the acoustic pulse. The systems operate at a very high frequency of 12–200 kHz so that little or no energy penetrates the seafloor. This information is recorded graphically to produce a seafloor profile that may be used later for computer-aided bathymetric mapping. A typical echo sounder record is shown on fig. 16.3b.

Until the introduction of multibeam instruments, echo sounders were single beam devices, operating vertically below the survey vessel to gather a single line of sounding.

Swath(e) Echo Sounders

Multi-beam or interferometric swath (or swathe) echosounders have become increasingly common and provide the geophysicist with a powerful seabed-modelling tool. Each transducer produces a fan of acoustic beams to provide sounding information on either side of the vessel's track. The high-performance systems have wide-angle swaths that cover an area up to 10 times water depth; more typically, the swath width is twice the water depth. As water depth increases, range increases, but maximum range becomes limited due to acoustic energy depletion of the outer beams.

The accuracy of the swathe systems is critically dependent on the correction applied for vessel motion (heave, pitch, roll, yaw, etc.); consequently, a swathe system is integrated into many other specialist sensors within the ship or subsea vehicle, such as an ROV or AUV.

The chief advantage of the swathe bathymetry systems is the high rate of productivity and the excellent data sample density, especially in deeper water. The swathe systems can be hull mounted in the ship, installed in a towed body (tow-fish) or in other remotely operated platforms. While the hull mounted systems are easier to calibrate than the towed systems, a towed system offers more portability and can be deployed closer to the seabed. Many swathe bathymetry systems also record backscatter (reflected energy) from the seabed, similar to side-scan sonar images (as further discussed below).

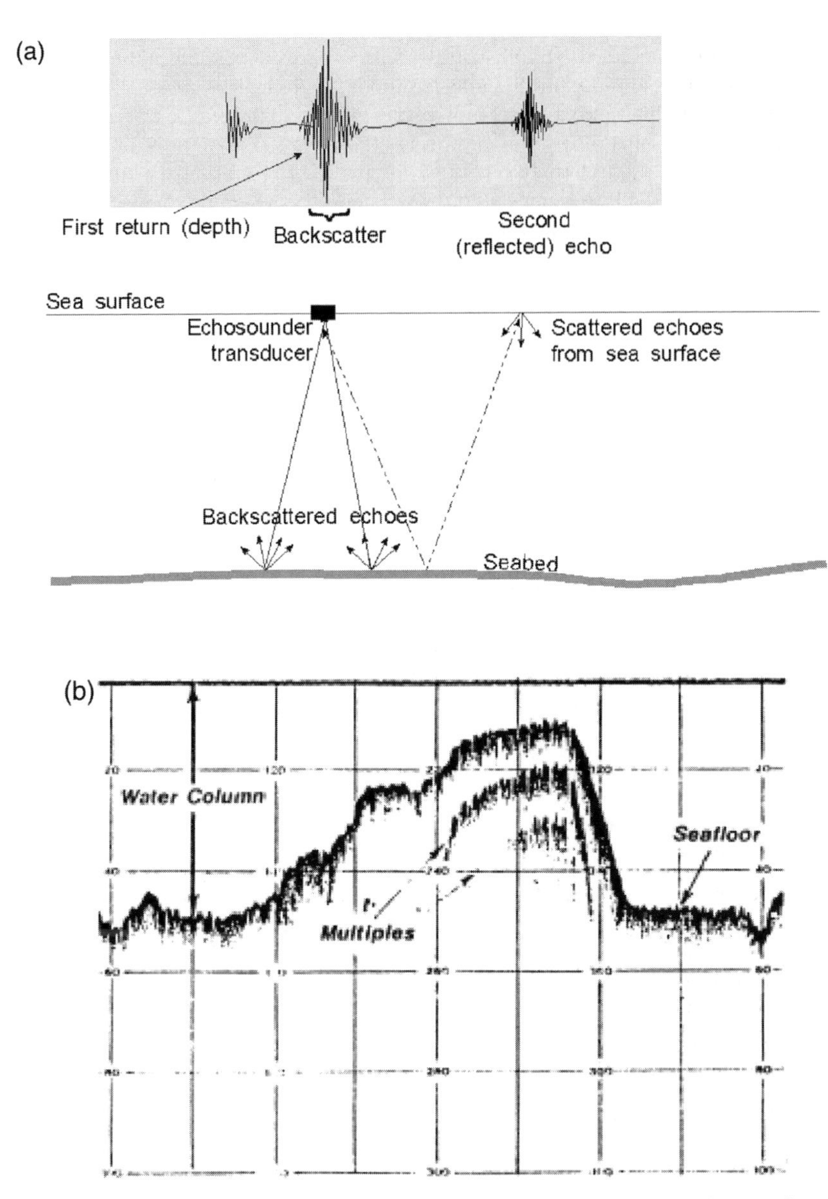

Figure 16.3 (a) Operating principles of an echo sounder; (b) typical echo sounder record

Advantages and Limitations

Excluding the more sophisticated deepwater systems, echosounders can be fitted to most vessels either by an over-the-side mount or through a special opening in the ship's hull.

Multibeam echosounders come in a wide variety of sizes depending upon their function. The large deepwater and oceanic systems require large transducer arrays (4–7 m long) that have to be purpose built into a ship's hull (a very expensive procedure), hence are restricted to specialist survey ships. For water depths less than say 500 m, multibeam systems can be installed on over-the-side mounts but function at their optimum when fitted as purpose-built installations. Shallow water (< 100 m) systems, being more compact, are normally fitted as temporary installations.

All echosounders require careful installation to avoid sources of interference, such as cavitations or acoustic noise. They require calibration that, in the case of the multibeam, is a complex procedure that can take six or more hours to complete; time must be allowed for this critical procedure. Frequent measurements of seawater density and salinity are also needed to determine the ever-changing speed of sound; these can be performed underway using disposable SV (sound velocity) probes, or by stopping the vessel at intervals to take a "SV cast".

16.3.4 Side-Scan Sonar

Side scan sonar systems provide graphic records that show two-dimensional (map) views of seafloor topography and of objects on the seafloor. They are the equivalent to the aerial photos on land. The side scan towfish (fig. 16.4) is deployed so that it remains about 30–120 ft above the seafloor. The beams are perpendicular to the direction of vessel travel and are broad enough in the vertical plane to extend from beneath the towfish to a maximum of 1600 ft on either side of the vessel travel line.

Side-scan sonars provide an acoustic "oblique photograph" of the seafloor. By ensonifying a swath of seabed and measuring the amplitude of the back-scattered return signals, an image is built up of objects on the seabed (fig. 16.5) and information on the morphology (the different material and features comprising the seabed surface) is obtained.

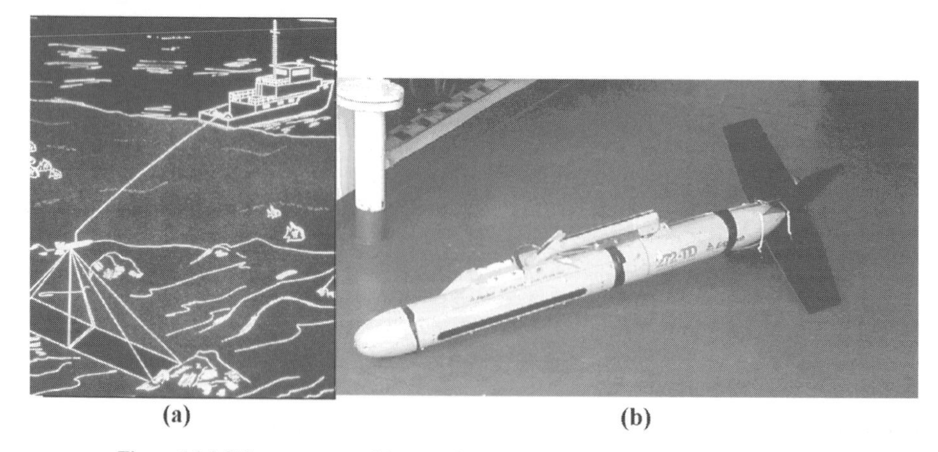

Figure 16.4 Side-scan sonar: (a) operation principles, (b) example – Edgetech fish

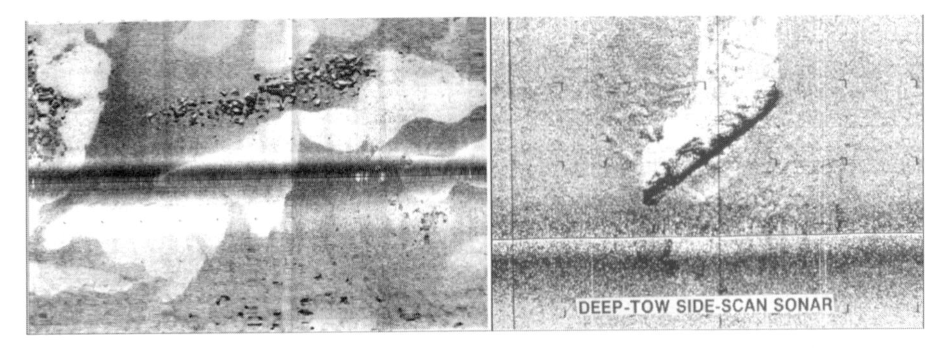

Figure 16.5 Example of side-scan sonar image of seabed

High frequency sonar (e.g. 500 kHz) provide high-resolution images, but with short (100 m) ranges. Lower frequency systems (e.g. 60 kHz) provide long ranges (500 m), but with lower resolution. Side-scan sonar tow-fish can be towed deep or shallow depending on requirements. Alternatively, the systems can be mounted in steerable ROTVs (remotely operated towed vehicles), ROVs (remotely operated vehicles) and AUVs (autonomous underwater vehicles).

In deeper water, tracking a towed side-scan fish is problematic since the acoustic tracking systems are typically limited to a range of approximately three to four kilometers; in 1500 m of water, at least 5 km of cable is required to position the fish at the required depth. Developments to overcome this problem include using a second vessel (chase boat) to track the fish directly from above (costly), or deploying the side-scans on remote platforms, as discussed later.

Advantages and Limitations

Side-scan sonar is probably one of the most useful tools developed for imaging the seabed. The clarity of the image, especially from the latest systems, is extraordinary. Developments in sonar imaging continue to move forward rapidly. Its use in the seabed classification systems is discussed below.

Side-scan sonars in towed fish require a powered winch and a suitable system for running out the cable; normally an A-frame. The smaller, shallow water systems can be deployed from most vessels but the deeper towed systems operating at, say 1000 m depth, require a cable some 5000 m long and, therefore, a large winch. The so-called "deep-tow" systems are very large towfish, 4 or 5 m long, and are heavy. They require a large powered winch and special launch and recovery systems and, therefore, are restricted to specialist survey vessels. The normal tow speed for a side-scan survey is about 4 knots; however, as operating depth increases, so the drag and strain on cables increase. A deep-tow system operating at 2000 m will reduce tow speed to 1 or 2 knots, greatly adding to the time (and cost) required for a survey.

Owing to the long length of the tow cable, surveyors have to allow for a "run-in" and "run-out" equivalent to the length of the tow to ensure that the required area is covered.

Likewise, the turning time with long cables increases such that a deep-tow can take several hours to complete a line turn. These factors must be taken into consideration when planning and costing an operation.

Typical seafloor records are shown on fig. 16.5. Reflected signals normally appear as dark areas on the record, whereas shadows behind the objects appear as light areas. Features less than 1 ft in height can be detected. These data are used to map boulders, sandwaves, reefs, seafloor instability features, pipelines, wellheads and ship wrecks.

16.3.5 Sub-Bottom Profilers

Sub-bottom profilers, sometimes also referred to as single channel systems, are used throughout the industry for the shallowest seabed profiling.

Pingers

The subbottom profiler (sometimes called a pinger profiler, or tuned transducer) is a shallow-penetration, seismic-profiling system designed to provide extremely high-resolution records. This system only penetrates about 100 ft in soft soils. The data are very useful in interpreting slump and creep features, shallow buried channels, gas seeps and erosional unconformities.

Most commonly used systems consist of a transducer element either fixed to the survey vessel or mounted in a towfish (fig. 16.6), a transceiver power package and a suitable recorder. Transmitted and reflected signals are received by the same transducer.

Pingers, so-called because of their high frequency acoustic "pings", operate on a range of single frequencies between 3.5 and 7 kHz, with an energy output of about 10 J. They can achieve seabed penetration from just a few metres to more than 50 m, and are capable of resolving soil layers to approximately 0.3 m. The high frequency profilers are particularly useful for delineating shallow lithology features such as faults, gas accumulations and relict channels.

Figure 16.6 Examples of pinger sub-bottom profilers

Boomers

These instruments have a broader band acoustic source between 500 Hz and 5 kHz and typically can penetrate to between 30 and 100 m with resolution of 0.3–1.0 m and are excellent general-purpose tools.

CHIRP

The CHIRP sub-bottom profiler is a recent introduction to a geophysical survey. Designed to replace the pingers and boomers, the CHIRP systems operate around a central frequency that is swept electronically across a range of frequencies (i.e. a "chirp") between 3 and 40 kHz. This method can improve resolution in suitable near-seabed sediments.

Medium Penetration Profiler (Sparkers)

Typical intermediate penetration systems are the Boomer and the Minisparker systems. The Boomer system (fig. 16.7) uses an electromechanical source and provides typical subseafloor penetrations of about 200–300 ft. The seismic pulse from a Minisparker is generated by a spark discharge in the water, which creates a steam bubble. This bubble expands rapidly, creating an initial pulse. The bubble then collapses, creating a cavitation pulse, which can be greater in amplitude than that produced by the initial pulse. The Minisparker provides records of the upper 200–500 ft of sediments. The greater depth of penetration of the Minisparker occasionally makes this the preferred intermediate penetration source. Little penetration may occur in hard rock or in very dense sediments.

Figure 16.7 Boomer sled

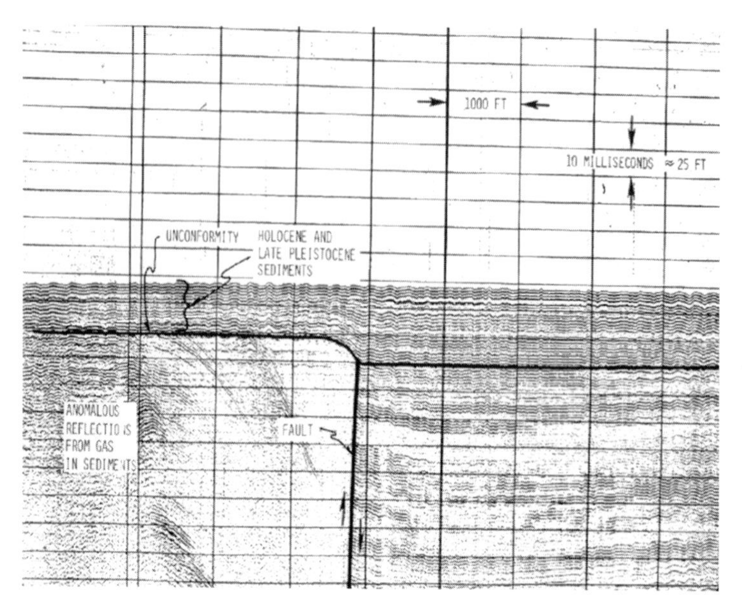

Figure 16.8 Typical boomer record

The intermediate range systems provide an output pulse of 200–4000 Hz. The reflected signals are received by a hydrophone array towed at or near the water surface. Signals outside of the desired frequency range are removed by analog filtering. A typical profile is shown on (fig. 16.8).

These very powerful instruments can penetrate soils and rocks to 1000+ m but, because of their unstable pulse waveform, they are not in such common use as in the past.

Deep Penetration Systems (Air Guns)

Several high-resolution systems can produce continuous seismic profiles to depths as great as 6000 ft below the seafloor. These systems use a variety of energy sources, but typically employ an air gun. All of these systems emit relatively low-frequency pulses, mostly in the range of 50–750 Hz.

The acoustic signals from an air gun are produced by discharging high-pressure air into the water. The resulting air bubble expands rapidly, then collapses, producing the seismic pulse. The seismic waves are reflected off the seafloor and subsurface layers. These data are typically received by a multi-channel hydrophone, or string of receivers. The deep seismic data are recorded digitally on tape. Post-survey processing of this data migrates geologic features to their spatially correct locations and removes water bottom multiples and other anomalous noise.

Deep penetration data are useful to recognise faults, gas zones, buried or deep seated landslides and other geologic features. Their usefulness is enhanced when correlated with shallow profiler and borehole data. An example of deep-penetration data is presented on fig. 16.9.

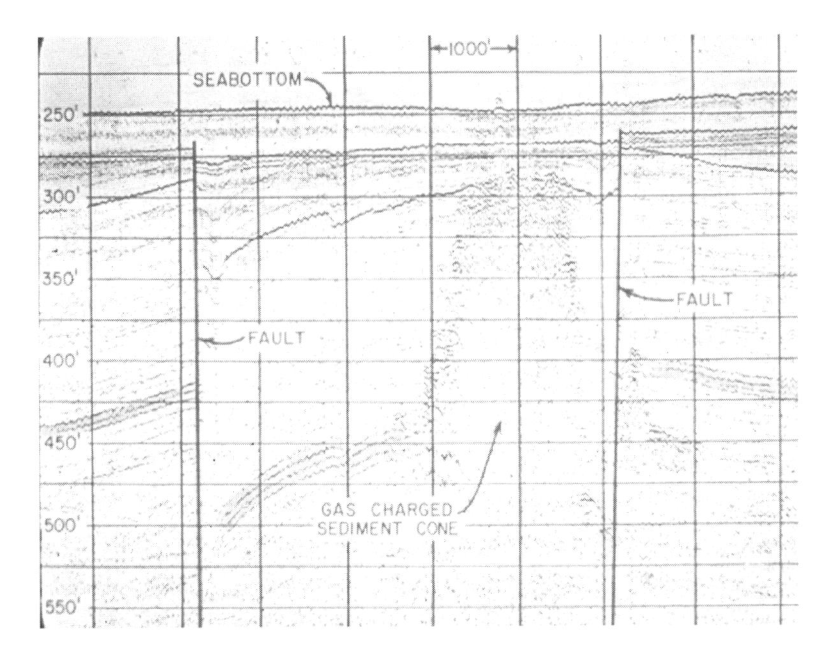

Figure 16.9 Example of deep penetration record

Comments on Single Channel Systems

The single channel acoustic systems provide an excellent range of tools for remotely imaging near-surface soils and rocks. Care is needed not to overreach their capabilities; for example, as the depth of soil penetration increases, the single channel systems begin to suffer from decreasing signal-to-noise ratios and from multiple reflections. These multiple reflections are the result of acoustic energy being reflected between pairs of horizons before returning to the receiver. The so-called "ghost" echoes become superimposed on real data causing masking and interpretation difficulties. The problem of "multiples" is particularly acute within the water column as the sea surface and seabed interfaces are strong acoustic reflectors. These strong reflectors give rise to "seabed multiples" of real reflections confusing the record. The same factors affecting side-scan cables apply, although the length/depth ratio is somewhat less. A limitation with the higher frequency profilers is that, in the presence of gas or hard soils or biologic colonies, acoustic penetration can be severely reduced or even arrested.

16.3.6 Marine Magnetometer

A marine magnetometer system is used to detect anomalous magnetic intensities in the earth's magnetic field. This tool is useful for locating pipelines, ship wrecks, and other ferrous metal objects on or just below the seafloor. The total intensity of the magnetic field is recorded as a single line on a chart. Ferrous-metal objects are represented on the charts as sharp peaks and/or depressions (fig. 16.10). The amplitude and shape of the peak principally depend on the size of the object and its distance from the sensor.

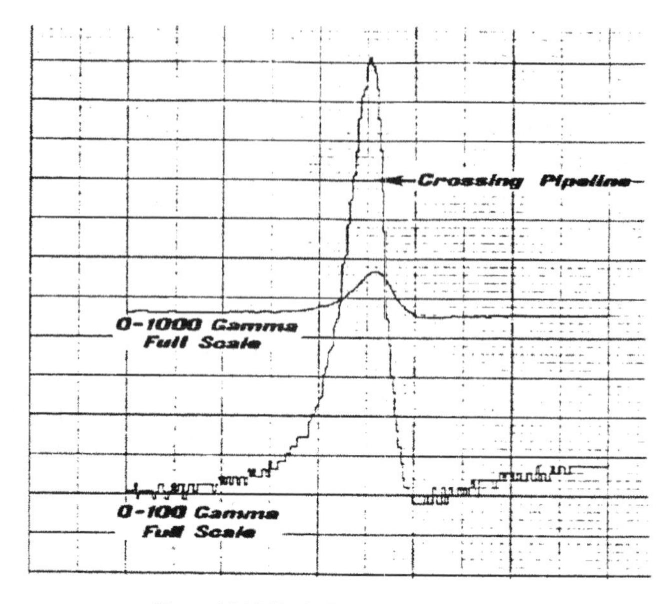

Figure 16.10 Typical magnetometer record

16.3.7 Use of Data

The data interpretation and site assessment phase of a site study should provide the user with as much information as possible about seafloor topography and geologic conditions of engineering importance. This information must be presented in a useful format for the engineer.

The engineering geologist compiles information obtained from each seismic profile into a series of interpretive maps and cross-sections. Maps and cross-sections are the most effective methods of portraying the location and area extent of three-dimensional geologic features. Bathymetric, shallow structure, isopach (thickness) and geologic feature maps (fig 16.11) commonly are prepared for most types of site surveys. Cross-sections are typically prepared in conjunction with maps to portray the seafloor profile, material units, shallow structure and other important features.

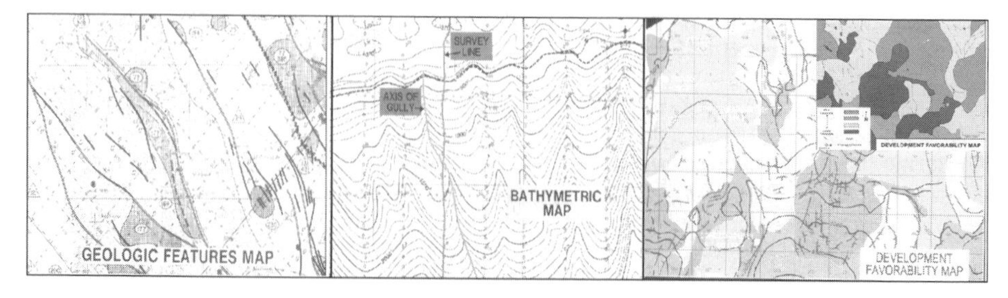

Figure 16.11 Examples of geologic features, bathymetry and development favourability maps

16.4 Remote Geophysical Platforms

16.4.1 Remotely Operated Vehicles (ROVs)

The Remotely Operated Vehicles (ROVs) have, for many years, been used as platforms for geophysical sensors. Linked to the mother vessel via an optical and electric umbilical, survey and inspection ROVs are frequently fitted with side-scan sonar and multibeam echosounders (fig. 16.12). These vehicles have the advantage of great maneuverability, under direct human control, and a constant source of power.

Typically, in shallower water, an ROV can fly at 2–3 knots but, in deeper waters, the drag of the long umbilical reduces its velocity considerably. The ROVs are ideal for inspection but can offer some disadvantages for geophysical survey, such as noise generated by their propulsion systems and other acoustic interference sources. Because they require substantial handling systems, ROVs capable of carrying geophysical sensors are limited to specialist ROV vessels.

16.4.2 Autonomous Underwater Vehicles (AUVs)

The advent of the AUVs offers a new concept in geophysical surveying. These vehicles (fig. 16.13) can be equipped with a multibeam echosounder, side-scan sonar and high frequency sub-bottom profiler. Some AUVs can also carry a magnetometer or other sensors making them extremely flexible and powerful tools.

Although the AUVs have been used for ocean research and in military operations for many years, they made their first appearance in commercial survey operations only in early 2001. Powered by special battery technology or energy fuel cells, AUVs have mission endurance ranging from 12 to 48^{+} hours and some can reach depths of 6000 m.

Figure 16.12 Example of an ROV (Sea Demon)

Figure 16.13 Example of an AUV (The 5.6 m long Boeing-Oceaneering-Fugro AUV)

Typically, they operate at 3–4 knots (independent of depth) and eliminate the time required for line turns or deviations.

The smaller ones (<2.5 m LOA) can be deployed from any vessel that has a suitable handling system. For the larger vehicles, which can reach lengths of 6 m, special launch and recovery systems are used and, hence, these vehicles are generally restricted to larger vessels.

The AUVs produce, and store very high-quality data because, unlike the towed platforms, they are capable of operating continuously at optimum sensor heights above the seabed and can adjust their aspects to meet changing environmental factors.

16.5 Seabed Classification Systems

Function and Applications

A capability to classify a seabed material without the need for costly sampling devices has its obvious advantages. Seabed classification systems do exist and their effectiveness is improving; however, they are not yet a panacea.

System Description

Seabed classification is a processed solution depending on a proprietary software and electronics package. The measures of combined roughness and hardness can provide quantitative information on seabed types but will not be reliable enough to determine detailed soil characteristics.

A side-scan sonar can identify seabed morphological boundaries very well. By combining the bounding attributes of a side-scan with the roughness/hardness ratios of a seabed classification system, areas with similar properties can be identified with high reliability. The final step is to use seabed sampling, for example with a box corer, grab sampler or drop corer, to recover examples of the topmost soils and correlate these to the roughness/hardness ratios. In this way, a reliable model of the seabed topsoil is possible.

Advantages and Limitations

Seabed classification using remote sensing is a rapid method that does not require additional in-sea equipment. However, a side-scan is necessary to detect seabed objects, and determine the morphological boundaries. Additionally, if reliable seabed interpretation is required, then the seabed samples are required.

16.6 Seismic Refraction Systems

Seismic refraction is a method of speedily acquiring high-resolution information of soil sedimentary structures. These systems are used typically where fine detail is required of the first 3 m of the seabed, and especially the topmost 1 m. The most common application is as a burial assessment tool for submarine cable installation; they are also used for pipeline route investigations. Other applications include site investigations for harbours and coastal developments and pre-dredge areas.

Until recently, these seabed refraction systems were limited to shallow water depths but recent developments have increased operating depths to over 1500 m. Results obtained are independent of the water depth.

System Technology and Science

The seismic refraction methods have been used for many years as an exploration reconnaissance tool and for civil engineering applications on land. In recent years, the technique has been applied with great success to shallow marine soil investigations.

A seismic source at the seabed is used to induce an acoustic pressure wave into the soil. Typically, in shallow water, an air gun is used but for deepwater operation, a mechanical percussion device provides a better option. As the pressure wave passes through the soil layers, some of its energy is refracted along sedimentary boundaries before returning to the soil surface where it is picked up by a hydrophone streamer. The length of the streamer and the number of hydrophones determines the depth of recorded penetration and the resolution of the information – the longer the streamer the greater the depth of penetration recorded but the lower is the resolution. For detailed imaging of the topmost 3–5 m, a typical streamer is 24–30 m in length containing some 48 hydrophones.

Time–distance curves are produced by plotting the first time of arrival (first break) of the refracted waves versus distance from the seismic source. The analysis of the slope of these curves provides a direct determination of the depth of the various soil layers. The compression wave's velocity (Vp) provides the geoscientist with information that can be used to characterise each soil layer.

The spacing between "shots" is of the order of 15–25 m and each observation requires 2–4 s. During this period, the seismic refraction system needs to remain quiescent to keep extraneous noise to the minimum. The refraction method can measure seismic velocities to better than 50 m/s with soil penetration accuracy of about 10% of depth (i.e. a soil layer at 2 m depth could be resolved to ±0.2 m). The main weakness of the method is that it falls short in resolving inversion velocity problems (i.e. situations where a softer layer underlies a stronger one).

Compressive wave velocities are linked to the mechanical properties of soils and provide quantitative information on soil stiffness. Soil classification of marine sediments based on their seismic velocity is also under development. However, at present, geotechnical information is usually obtained using CPT and/or coring samples taken at, say, 1 km intervals is used to discriminate between soils of similar velocity and to obtain shear strength properties indispensable for estimating burial conditions (i.e. achievable burial depth and magnitude of towing forces).

System Description

A typical deep-water seabed refraction system (fig. 16.14) comprises a steel reinforced instrument sled that is dragged across the seabed. Within the sled (fig. 16.15) are housed the attitude sensors, the pressure/depth and temperature sensors, the tension meters for the tow cable and the multiplexing electronics for passing the data to the support vessel. The sled is positioned using acoustic positioning such as an ultra-short baseline (USBL) system. Also, within the sled is the air-powered sleeve gun or mechanical percussion device for generating the seismic pulse. Trailing behind the sled is the hydrophone streamer for receiving the refracted signals. Depending on depth configuration, the sled system can weigh between 1 and 2.5 tonnes.

To tow the system across the seabed, a composite tow and power/communications cable connects the sled to the winch system installed on the surface support vessel. Each refraction-measuring cycle requires the sled to be stationary while the vessel continues to steam ahead at 3–4 knots. This is achieved by using a stop-go, or "yo-yo", device that pays out cable while the sled is stopped and pulls in cable (faster than the ships motion) to bring the sled to its next observing location. For water depths less than 300 m,

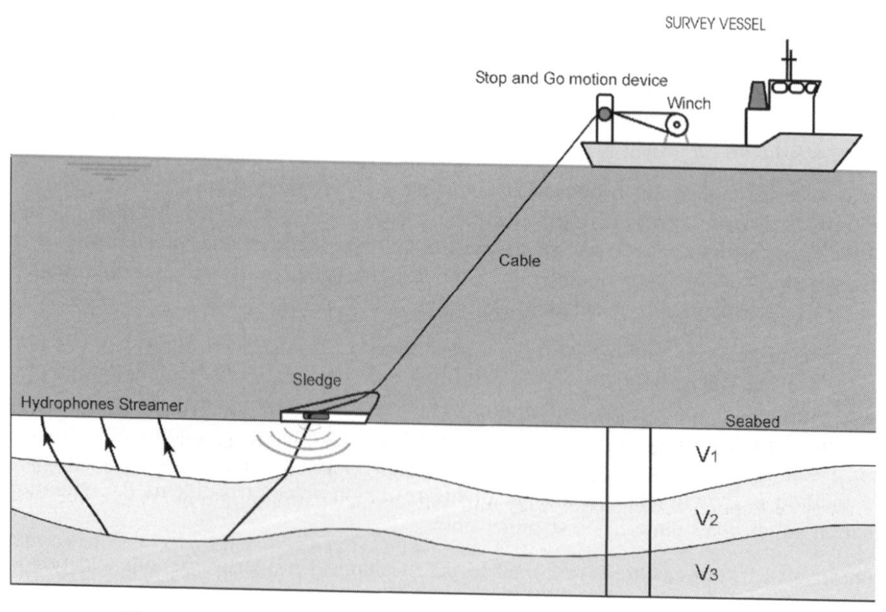

Figure 16.14 Refraction system operating principles – GAMBAS® system

Figure 16.15 GAMBAS® seismic refraction system – seabed tow sled

the yo-yo is normally mounted on the ship while in greater depth it is better to have the yo-yo included within the sled.

Seismic refraction systems require ships equipped with 1.5–5 tonnes A-frames for sled deployment and tow. A deck space of about 100 m^2 is required for handling the system, for the heavy cable winch and, possibly, an air-line winch, and for some storage. Usually, refraction systems are deployed from specialist survey vessels, geotechnical vessels or larger workboats. For inshore surveys, smaller equipment and boats can be used.

In operation, it is best practice to first perform a geophysical survey to ascertain suitable (cable/pipeline) routes before employing a refraction system. This practice is the most cost-beneficial method and will identify rough or hazardous seabed across which a refraction survey would not be feasible.

Advantages and Limitations

High-resolution seismic refraction is an efficient technique for ascertaining detailed information in the top meters of the sub-surface. The technique provides an accurate quasi-continuous profile of sub-seabed sediments, giving simultaneously a high-resolution definition of the soil layering and a quantitative characterisation of their materials.

Information is acquired in real time and can, firstly, be used to define the subsequent geotechnical programme and optimise the number and location of samples (CPT or coring). A detailed analysis is performed during office interpretation where integrated alignment charts are compiled showing lithology and soil characteristics all along the profile.

A high-resolution seismic refraction is an ideal tool for any kind of burial assessment purposes. The continuous profile aids in eliminating geotechnical uncertainties that, in turn, reduce the risk of ploughing downtime. Due to the variable tow speed, dimensions of the tow and noise created by reflection surveys, coincident geophysical (sonar) surveys cannot be performed simultaneously. Specialist geotechnical vessels are preferred, although the larger sort of workboats can also be used.

16.7 Electrical Resistivity Systems

Function and Applications

Seabed electrical resistivity profiling is a semi-continuous method of measuring the bulk resistivity of a volume of soil near the seabed. The technique is performed using a towed sled from which, in turn, is towed a multi-electrode streamer cable.

For surveys requiring penetration depths of 3–5 m (e.g. for a cable burial assessment), streamer lengths are typically 20 m. For deeper penetration and other applications, such as drilling site surveys, pre-dredge surveys or harbor/coastal investigations, a longer streamer is used.

System Technology and Science

By injecting an electrical square wave current into the seabed through a pair of electrodes (A and B in fig. 16.16), an electrical potential is created that can be measured between the reference electrode (N) and, typically, 13 potential electrodes (M1, . . . , M13).

To compensate for the self-potential effects of the soil, the injected current's polarity is alternated at 1 Hz. The resistivity of the ambient seawater is measured by using a short, low-intensity, square wave injected into the sea by a short quadripole antenna. The ratio of seabed resistivity to that of the seawater is called the Formation Factor. The potential difference is measured at each of the 13 electrodes at a sampling rate of 1–10 Hz. The depth of investigation is a function of the electrodes separation; short spacing produces values associated with the upper part of the soil mass while increasing separation provides information on progressively deeper sediments.

The Formation Factor in saturated marine sediments is directly linked to the material's porosity. Its value provides qualitative information on soil type and the state of consolidation.

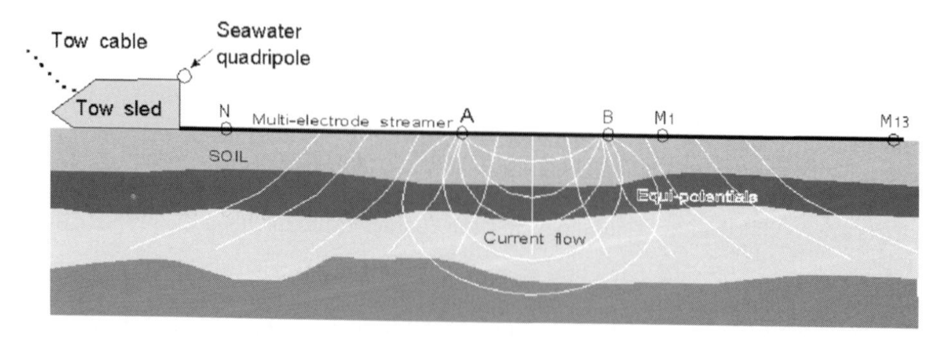

Figure 16.16 A typical seabed resistivity system

Obtaining layered resistivity versus depth is theoretically possible by implementing inversion-modelling techniques. However, currently, this approach has yet to provide convincing results.

An interpretation of resistivity measurements should always be supported by geotechnical information obtained from CPTs, vibrocoring, drop coring or other sampling methods.

System Description

A seabed resistivity system (fig. 16.16) comprises a steel reinforced sled in which are housed the electronics, acquisition unit, power unit, attitude sensors, temperature and pressure sensors and the cable tension meters. Behind the sled is towed the 24–30 m long multi-electrode streamer in which is housed the two 24 V/10 A current injection electrodes (A and B). The sled is hauled across the seabed from a tow/power/communications cable attached to a surface support vessel. Some systems are fitted with a yo-yo device that permits the sled to halt during measurements, thus, improving the signal to noise ratio.

Typically, resistivity systems can operate down to 2000 m water depth and can be towed at upto 2–3 knots. Soil penetration depths are in the order of 5 m, although it is possible to get greater depth (ca. 30 m) using wider spacing for the electrodes and sacrificing resolution and accuracy. The sled, which is similar to but lighter than the refraction sled, is deployed and towed from an A-frame fed by a 2000–6000 m capacity cable winch.

Advantages and Limitations

As results are dependent on the water depth and the salinity, great care is needed when calibrating the system and attention to details required in the operating procedures and interpretation methods.

Resistivity surveys provide continuous profiles and fill gaps where normal acoustic systems are unreliable (e.g. in gas-charged sediments, and between CPTs in the more homogenous soils). They can be employed in conjunction with refraction seismic surveys as an augmentation/ bulk sampling system.

The technique is a bulk sample of a volume of soil rather than discrete elements and its dynamic range is very short. Like reflection, seismic resistivity requires ground truth data in order to provide meaningful soil type information. Marine resistivity techniques are also limited in that they cannot reliably differentiate between discrete soil layers.

16.8 Underwater Cameras

Function and Applications

The visualising systems used for structural inspection can often assist in solving remotely sensed ambiguities. In situ examination of uncontaminated soil color, condition and context provide valuable information for the geologist or benthic scientist for environmental assessment and impact studies.

System Technology and Science

For operation in shallow depths during daylight, there is a range of off-the-shelf cameras. However, daylight tends to become totally absorbed in seawater below 300–500 m;

even at 100 m the amount of light available is often barely perceptible. Two options are available (a) camera lighting systems, (b) low-light cameras.

Lighting systems are housed in-pressure resistant housings and a variety of light emission types are available depending on the need and receptive media employed. Low-light cameras depend on light-enhancement systems (like night-vision glasses) while in extreme dark, solid-state photon detectors are used to collect any available light.

System Description

The common sorts of cameras for sub-sea visualisation are:

- Television (real-time) color or black and white
- Video cameras (self-recording)
- Movie film (now uncommon, but special sensitive films are still occasionally used)
- Still cameras (film), normally 35 mm format
- Digital stills cameras (rapidly becoming the preferred choice)
- Low light /SIT cameras

The most common form of deployment for lightweight cameras (stills, video) is by diver. For prolonged excursions, real-time visualisation and in hazardous or remote locations, cameras and lighting systems are normally installed on an ROV. Either a small observation class vehicle or full size survey vehicle can be used. Cameras can also be lowered to the seafloor from a reinforced power and control cable.

Advantages and Limitations

Seabed visualisation is a valuable tool providing high-resolution and discriminatory information. Color, texture and benthic life forms can all be studied in great detail. Diver deployment in shallow water is relatively inexpensive but deeper water requires saturation diving and costs become extremely high. The alternative is to use an ROV; the small observation class can be operated relatively inexpensively from most vessels but the larger survey class ROVs are limited to specialist survey vessels or the larger sort of workboats.

Visual sampling with an ROV in deep water (e.g., 2000 m) is a time consuming process; a dive to the seabed can take over four hours and a similar time to return to the surface. Once at the seafloor, an ROV can operate for many hours, or even days, and therefore it is more cost-efficient to combine visualisation with other remote sensing operations.

16.9 Geotechnical Techniques

16.9.1 General

A soil investigation serves three main purposes:

- Determine water depth,
- Delineate the soil stratigraphy, and
- Determine the soil properties.

At least one soil boring is essential for any offshore platform investigation. Several borings are sometimes required, depending on the type and size of the structure, the soil conditions and the potential for lateral soil variability. Soil borings can be supplemented with in situ testing such as cone soundings, remote vane, pressuremeter tests, piezoprobe tests and hydraulic fracture tests. These in situ tests can be performed downhole or from the seafloor.

16.9.2 Vessels and Rigs

The drilling rig used offshore is essentially the same as used onshore, but has some modifications to adapt it to marine work. A convenient rig is one that has drawworks, rotary, pump and power plant unitised on a common skid. A principal requirement is that it has sufficient depth capacity to cope with the combined water depth and boring penetration when drilling is done from a floating vessel. The rated capacity of a drilling rig should be reduced to allow for surge loads that may occur. Power swivel equipment lends itself well to soil boring operations when supplemented by hoisting equipment and a pump. Efforts have been made to develop and use submerged remotely operated (fig. 16.17) and diver-operated drilling equipment (fig. 16.18). These rigs are slow to operate and do not provide good quality samples. Nothing has the capability and versatility of equipment operated above the water surface.

The most commonly used method of performing offshore borings is a wet rotary drilling rig on a platform or floating vessel. The rig can be portable and can be operated from an oil field class supply boat (fig. 16.19) or can be a fixed derrick on a specialised geotechnical drillship such as the R/V *Seaprobe* (fig. 16.20a), the M/S *Mariner* (fig. 16.20b), and the M/S *Bucentaur* (fig. 16.20c).

Each type of operation serves a specific purpose and has its advantages and disadvantages. The supply boat is the most common and, generally, the most economical method of drilling soil borings on the continental shelves of the world. Supply boats have worked in water depths up to 1100 ft. Drillships such as the M/S *Bucentaur* (fig. 16.20c), the M/V

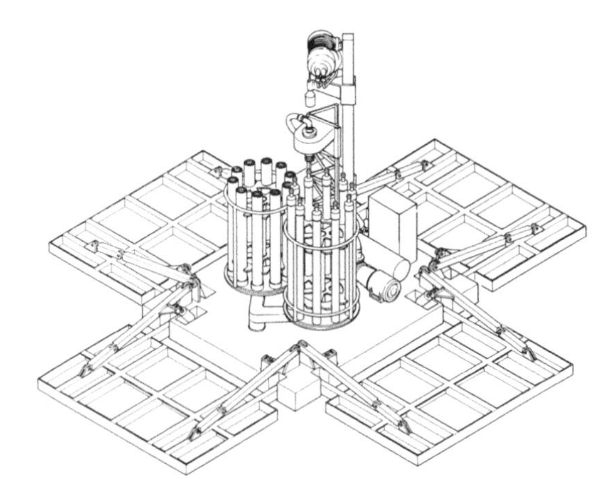

Figure 16.17 NCEL remotely controlled seafloor corer

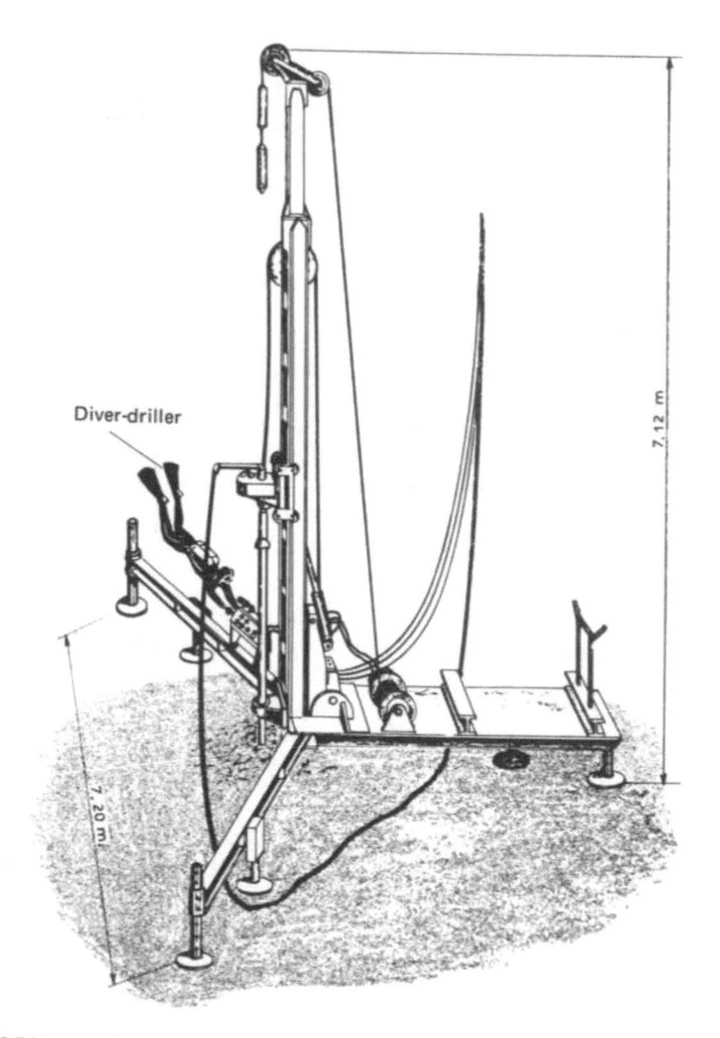

Figure 16.18 Diver operated rig (Reproduced with permission from Le Tirant, P., Seabed Reconnaissance and Offshore Soil Mechanics for the Installation of Petroleum Structures, English translation by John Chilton Ward; Technip, 1979, p. 155)

Bavenit (fig. 16.20d) or the M/S *Norskold* (fig. 16.20e) can work in water depths up to 5000 ft and are less subject to environmental conditions. Larger vessels, such as the DP vessel *Fugro Explorer* (fig. 16.21a) and the semi-submersible the MSV *Uncle John* (fig. 16.21b), can work in water depths as great as 10,000 ft and drill down to over 2000 ft below mudline.

Soil investigations are also conducted from small lift barges (fig. 16.22a), jack-up exploration barges (fig. 16.22b) or large oil well drilling semi-submersibles (fig. 16.20f). The drilling operations must be located to avoid the jack-up rigs' spud cans or mat foundation, and not interfere with the main drilling operations.

Figure 16.19 Typical supply vessel with skid mounted drill rig – M/V Perkins

Figure 16.20 Examples of geotechnical drillships: (a) R/V Seaprobe, (b) M/S Mariner, (c) M/S Bucentaur, (d) M/S Bavenit, (e) M/S Norskold, (f) drilling rig on oil-well semi-submersible

Figure 16.20 Examples of geotechnical drillings (Continued)

Figure 16.21 Examples of deepwater geotechnical drillships: (a) DP Fugro Explorer, (b) MSV Uncle John

In remote areas of the world, where rotary drilling rigs may not be available, the main derrick on an exploration rig can be used. The drilling contractor's crews operate the equipment, and a geotechnical crew is present to supervise the proper sampling intervals and maintain good quality samples. Close coordination between the drill crews and the geotechnical crews is mandatory for a successful and meaningful soil boring.

Ships conducting coring or in situ testing operations have to maintain station vertically above the core/test location during the operation. This is best achieved if the vessel has a dynamic positioning system or joystick–controlled thrusters. Alternatively, a multi-point anchoring system may suffice, although this can increase operational times and is usually impracticable in deepwater.

Attention to details is required when selecting a vessel; a low-cost vessel can easily turn into a financial liability and seriously jeopardise a project. The vessel's weather keeping attributes are vital in harsher environments where a sea-state can easily terminate an operation with an ill-considered ship. A vessel's capacity for deploying and recovering systems requires closest attention, especially if it is new to the work. If cranes or A-frames have to be fitted, then the ships structure needs to be surveyed to ensure its integral strength is sufficient.

Any survey or sampling vessel must meet modern health and safety requirements and have fully up-to-date certification for her life-saving aids, communications and navigation, as well as for work systems such as cranes, winches, etc.

For most geotechnical investigations requiring seabed penetration greater than 10 m, drilling methods will be required. The exceptions to this are the use of long piston corers, of 20–30 m in length, which can be used in soft deepwater clay deposits, and the bigger seabed CPT systems, both of which require large vessels with specialist deployment equipment and sufficient deck space and facilities for a safe operation.

A detailed description of the geotechnical drilling systems and operations is outside the scope of this textbook, but a brief summary of the main methods employed is given below.

Geotechnical Drillships

As discussed earlier, most deep geotechnical investigations are performed from dedicated, purpose built or converted vessels. Since drilling operations can take several days per

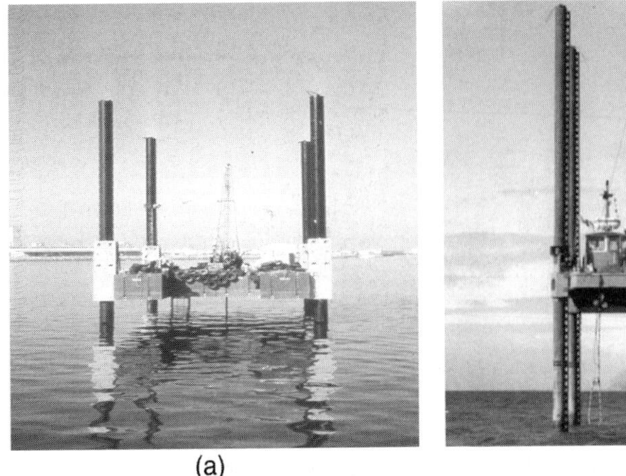

(a) (b)

Figure 16.22 Examples of geotechnical jack-up drill rigs

borehole and verticality of the drill string is critical, the use of dynamic positioning or a four-point (minimum) anchoring system is essential. A heave-compensated rotary drilling technique is used, typically utilising 5-in.-O.D. steel drill pipe and an open-faced dragbit. In ultra deep waters, aluminum drill string is usually required. Sampling and in situ testing is performed via wireline operated down-hole tools. The highly controlled nature of the sampling and testing operations means that, for a majority of ground conditions, this will provide the highest achievable quality of samples and test data.

The size and favourable weather-keeping characteristics of such vessels can, in many situations, also make them cost-effective for shallow penetration investigations.

Geotechnical Jack-up Drilling Rigs

Drill ships can, in favourable circumstances, operate in water depths as shallow as 20 m. In extreme circumstances, shallow-penetration investigations may be feasible in water depths as shallow as 10 m. However, the primary method for drilling boreholes in water depths from around 20 m to shore – including the inter-tidal zone – is with a jack-up drilling platform (fig. 16.22a and b). Such platforms are typically capable of both rotary and percussive drilling techniques, high quality sampling and in situ testing.

16.9.3 Methods of Drilling and Sampling

Early exploration techniques in the late 1940s were an extrapolation of those used on land. Drilling and sampling were done from a coring platform (fig. 16.23) or large anchored barges (fig. 16.24), using casing to permit hole reentry and re-circulation of drilling fluid (fig. 16.25). Samples were obtained by hydraulically pushing a thin-walled tube or a split barrel sampler. This method is still used for shallow water and shallow borehole applications.

In about 1962, it became necessary to drill and sample in water depths up to about 500 ft deep. Use of slender casing and conventional drilling and sampling was not practicable

Figure 16.23 Example of early coring platform (ca. 1947)

Figure 16.24 Example of coring from an anchored barge (ca. 1955)

in these water depths, and new techniques had to be employed. The open-hole drilling (fig. 16.26) and wireline sampling (fig. 16.27) techniques, which are still used today, were developed at that time. In this method, the drill pipe is the only connection to the seafloor.

The soils are sampled through the drill pipe and open centre bit (fig. 16.27). The drill pipe is internally flush at the joints, so that the centre opening is the same size as the inside diameter of the drill pipe. When the boring has been advanced to the desired sampling depth, the kelly or swivel is removed from the drill pipe. The pipe is supported by slips or elevators, with the bit raised off the bottom of the hole. A wireline percussion sampler is then passed through the drillpipe. Since no casing is used, all pumped drilling fluids emerge from the annulus at the seafloor. A constant supply of new drilling fluid is thus required. Seawater is readily available as the basic fluid. Saltwater gel or chemical polymer is added to the seawater to produce suitable gel properties and viscosity. Once a cohesionless

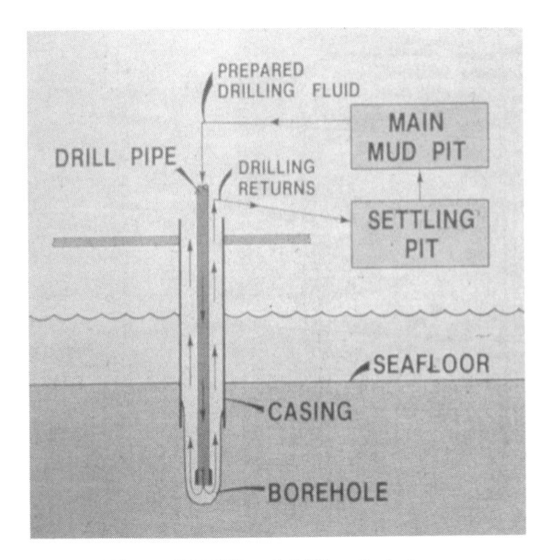

Figure 16.25 Land drilling technique

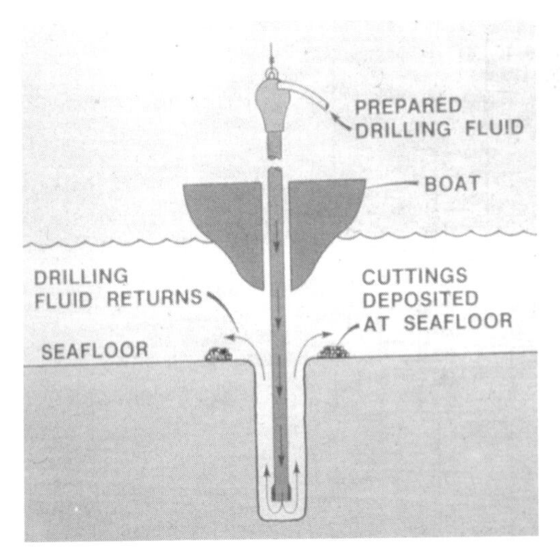

Figure 16.26 Open hole drilling technique

(granular) soil has been encountered, use of one of these mud materials must be continued to the bottom of the hole to avoid caving of the formation and getting the drill pipe stuck. The open-hole method leads to higher mud consumption, but is quite fast and is the most feasible and the most economical method in moderate to deep water.

The wireline percussion sampler (fig. 16.28) consists of a lower tubular section to which the soil sampler is attached, a rod with telescopes within the tubular section, and an upper hammer section which is connected to the wireline by a swivel. Sampler penetration is

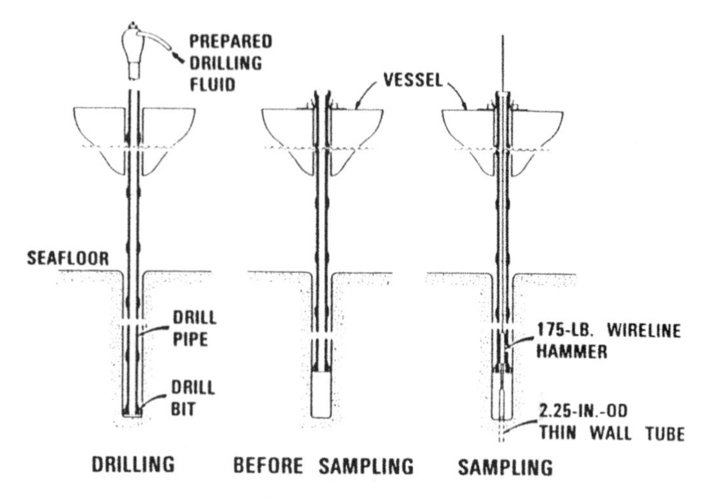

Figure 16.27 Wire-line sampler operation

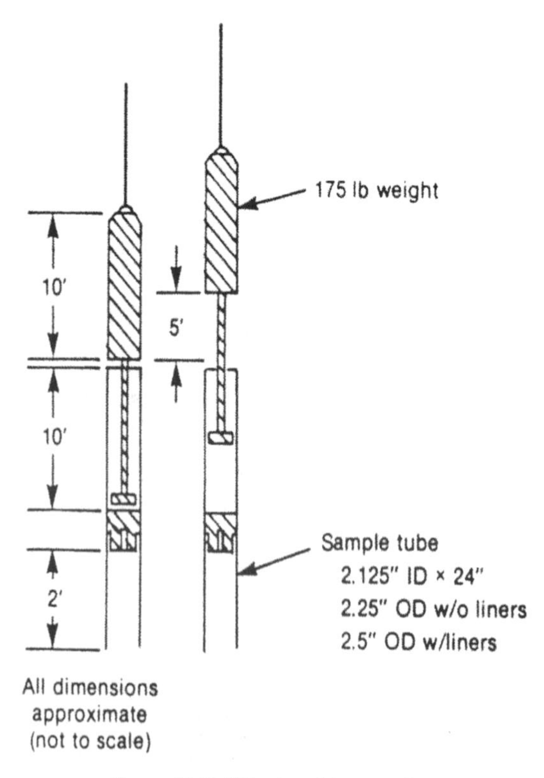

Figure 16.28 Wire-line drive sampler

obtained by alternately raising and dropping the hammer section. The telescoping portion has a stroke of 10 ft. Only 5 ft of this stroke is used for driving; the remaining 5 ft provides tolerance to vertical vessel motion during the sampling operation. The sampler is driven a maximum of 2 ft; in hard or dense soils, the penetration may be limited to that required to produce about 30 blows per foot. The sampler is retrieved by simply pulling on the wire line.

Soil Sampling Tools

The sampling tools used offshore are the same types that have a long history of successful use on land. For soils, these are the liner samplers, the thin-walled tubes (often called the Shelby tubes) and the split-barrel (or the split spoon) samplers (fig. 16.29), and the core barrels for rocks.

Liner samplers are used to sample extremely soft soils that cannot stand under their own weight. The sampler is pushed into the soil, and samples are forced into plastic liners. The samples are capped and retained in the liners for future testing.

The thin-walled tube is the most common device used to sample fine-grained soils. The object is to obtain the least disturbed sample so as to best represent the in situ

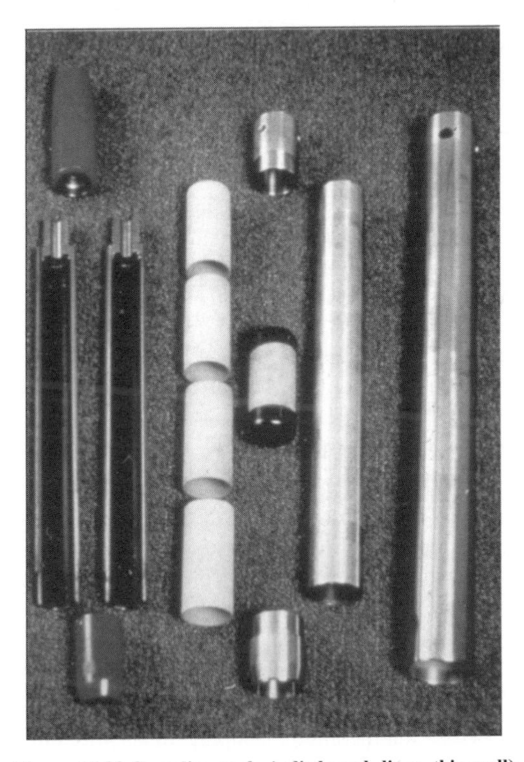

Figure 16.29 Sampling tools (split-barrel, liner, thin wall)

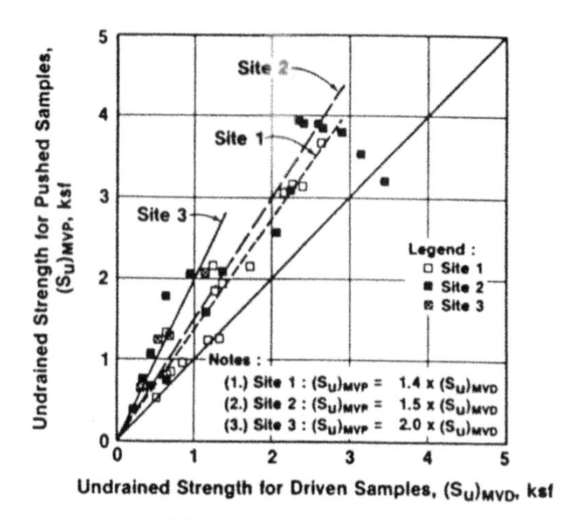

Figure 16.30 Comparison of miniature vane test results for pushed and driven samples

shear strength and stress-strain properties of the soil. The tubes should be cleaned and smoothened to reduce friction, and be sharpened by having an outside bevel.

The best quality samples are obtained by pushing in a rapid and continuous motion. Hammering samplers into the soil has a detrimental effect. Figure 16.30 shows that shear strengths obtained from the driven samples can be 20–100% smaller than those obtained from the push samples. Unless corrections are made for the sample disturbance, the resulting design will likely be too conservative.

A sample of undisturbed sand is virtually impossible to obtain because of its cohesionless nature. Onshore, the most common procedure used in sampling sand is one which gives an indication of the sand density, while recovering a specimen for examination, and testing in the disturbed state. A procedure called the "Standard Penetration Test" (SPT) (fig. 16.31) has been used for many years onshore. It involves driving a 2 in. split-barrel sampler using a controlled driving effort. The resultant number of blows to drive the sampler one foot is called penetration resistance (N), and this value is used to judge the sand relative density (fig. 16.32). Offshore, the split-barrel sampler is used principally to obtain specimens of the hard-to-sample soil such as gravel, coral and any other material that damages the thin-walled tubes.

The actual SPT test *cannot* be performed offshore, as there is no practical way to meet the ASTM specifications, which involve driving at the top of a drill string of rods. Offshore, the driving procedure in sand departs from the SPT method, as constant height of drop and friction in the wireline system cannot be controlled. Therefore, blow count information is *not* equivalent to the SPT "N" value and it should *not* be used to correlate density of granular soils. Only *crude* correlations can be made using the blow counts obtained with the wireline downhole hammer.

Similarly, the wireline percussion method of sampling in clay does not follow the preferred procedure which calls for penetration of the tube to be achieved in one fast, continuous

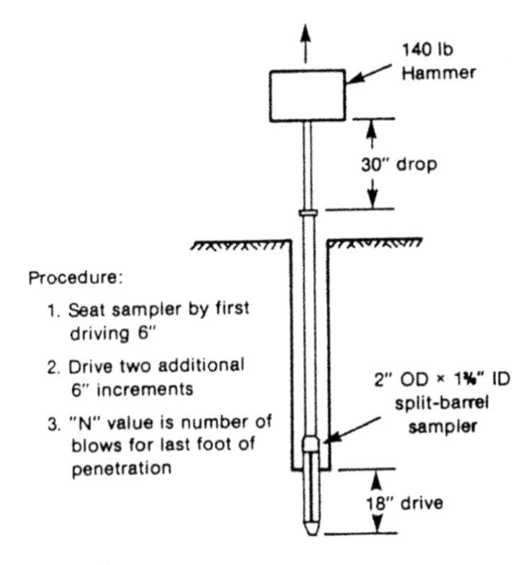

Figure 16.31 Standard penetration test (ASTM D-1556)

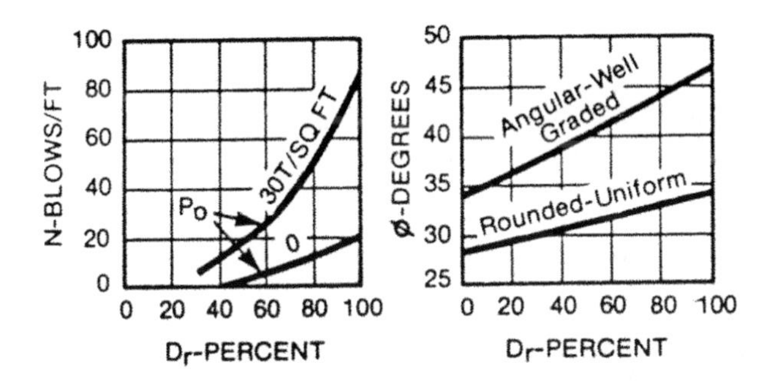

N = Number of blows from SPT

P_o = Overburden pressure

D_r = Relative Density

ϕ = Angle of internal friction

Figure 16.32 Typical application of results of standard penetration test (SPT)

motion. The percussion sampler provides a simple and economical means of obtaining soil samples from a floating vessel, but results in a loss of the sample quality.

More recent (mid to late 1970s) technology has led to improvements in the sampling techniques. The push samples can also be taken offshore using the wireline technique. The push sampler is lowered by the wireline, latched into the bit and pushed into the soil formation in one smooth, continuous motion. The force required to push the sampler can be provided either by a seafloor-based jacking unit (fig. 16.33) or by the weight of the

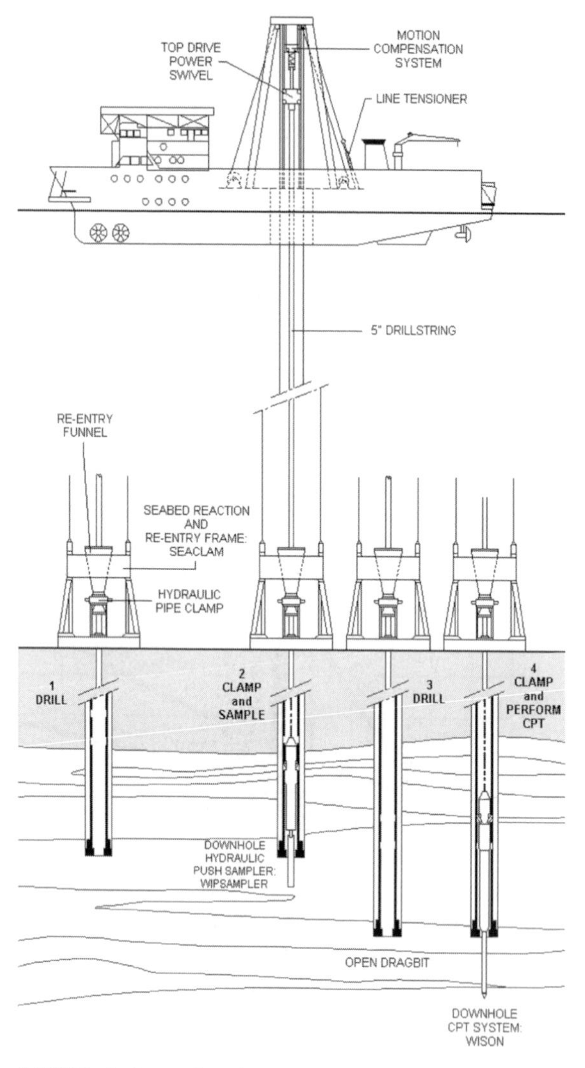

Figure 16.33 Typical procedure for drilling push sampling and in situ testing

drill string. The pushed sample is retrieved by first pulling with the drill string to free the sampler from the formation and then, retrieving the sampler to the deck by pulling on the wireline.

An even more recent development is the adaptation of a fixed piston sampler (fig. 16.34) for offshore use. The fixed piston sampler (fig. 16.35) is considered to obtain the highest quality samples. The sampler is pushed into the soil, while a piston inside the tube remains stationary. One of its major advantages is that the fixed piston prevents the entrance of

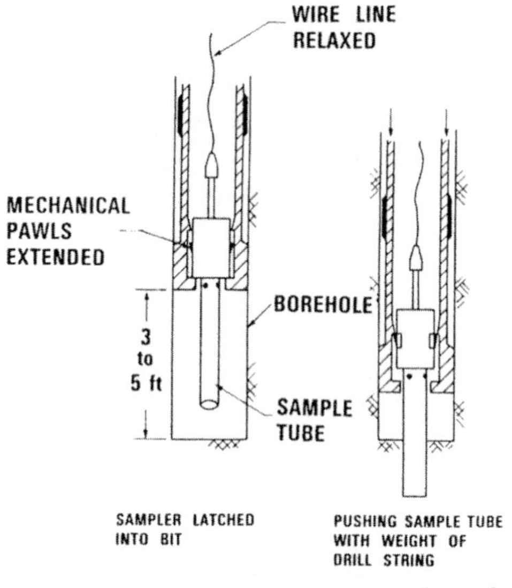

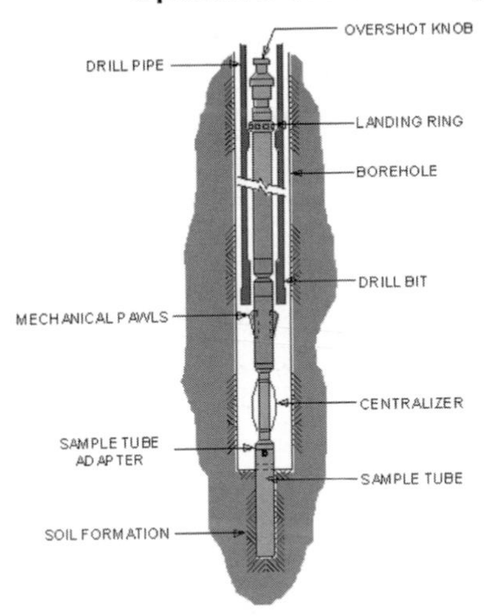

Figure 16.34 Dolphin push sampler

excess soil at the beginning of sampling, thus precluding recovery ratios greater than 100%. Another advantage is that the piston provides suction at the sample's top and acts more positively to retain the sample.

Rock Coring

The primary purpose of a core drilling is to obtain an undisturbed, intact sample representative of the in situ material. Coring is the primary method of obtaining

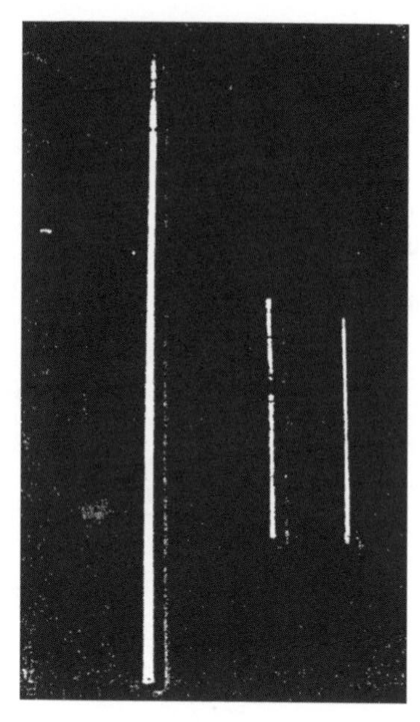

Figure 16.35 Piston sampler

samples of the soft rock and the cemented soil that are encountered in many areas of the world.

Wireline core barrels are the primary type of coring tools used in geotechnical investigations. Figure 16.36 shows the basic components of a rock core barrel. The tool's primary components are a bit, an outer barrel, and a latch-in inner barrel. The bit and the outer barrel turn with the drill pipe and cut a core. As the bit advances, the core is forced into the stationary inner barrel. After the coring run is complete, the inner barrel and the core are retrieved with an overshot.

The proper bit selection is critical to successful rock coring. Rock type, hardness and integrity will dictate the type of bit required. The two main types of bits are diamond chip and carbide. Many variations of the tooth shape, spacing and the diamond chip placement exist. The best bit for a certain rock formation is usually determined by trial and error.

16.9.4 Shallow Soil Sampling and Rock Coring Systems

16.9.4.1 Rock Coring Systems

Function and Applications

Underwater rotary rock corers are used to recover the undisturbed core samples of harder soils and rock, usually in shallow water. They are particularly well-suited for:

- Pre-dredging investigations
- Engineering developments in ports and harbours

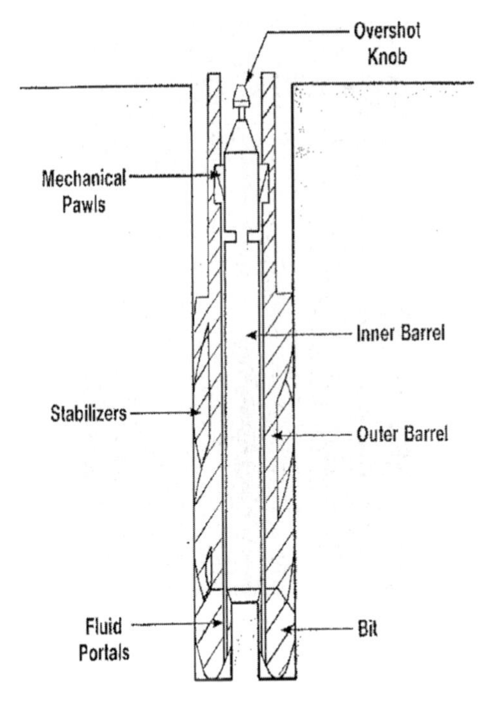

Figure 16.36 Wireline core barrel

- Long sea-outfall and pipeline and cable landfall investigations
- Mineral prospecting

System Technology and Science

To recover a high-quality and undisturbed core sample, the core tube has to be static. Rotary rock corers are designed as double or triple tube devices where the innermost tube acts as a core liner, the middle tube, if it is present, acts as a "holder" and the rotating outer tube carries the hollow drill bit. As the hollow center bit cuts down through the soils and rock, the core created passes into the liner in a relatively undisturbed state.

System Description

Seabed rock corers come in a variety of shapes and sizes, capable typically of recovering cores from 25 to 150 mm in diameter and 2 to 6 m in length.

Most rock corers comprise some form of coring tower (see fig. 16.37) mounted on a base plate or tripod footing, with dimensions and weights usually within the following ranges:

Height	:	4–8 m
Maximum base width	:	2–6 m
Weight, in air	:	1–8 tonnes

Figure 16.37 Example of seabed mounted rock corer – the Sorotel rock corer

The rotary drilling mechanism can be electrically or hydraulically driven via umbilicals to the surface. Some systems incorporate a video camera on the seabed frame to improve operational monitoring and control. The drilling fluid used to lubricate and cool the drilling process can be either water or a "mud" flush.

Since coring can take 1 or 2 hours, the deployment vessel needs to have good station-keeping capabilities. Dynamic positioning, joystick controlled thrusters or a multi-point anchoring system are normally required.

Advantages and Limitations

The primary advantage of these systems is their ability to core harder soils than other seabed sampling devices and the ability to core rock that would otherwise require a surface operated drilling rig.

Most systems are designed for operation in water depths of 200 m or less, since it is close to the coastline where one is most likely to encounter seabed rock outcrops. As mentioned, coring depths are usually limited to a few metres. The percentage recovery and core quality may also be lower than for surface operated systems.

16.9.5 Basic Gravity Corer

Function and Applications

Gravity corers provide a rapid means of obtaining a continuous core sample in water depths down to several thousand meters. Depending upon their deployment and operating systems, gravity corers can be deployed from a wide range of vessels.

Gravity coring applications cover nearly all facets of seabed soils investigation including:

- Dredging and inshore engineering
- Offshore oil and gas engineering
- Route surveys for pipelines and cables

Gravity core samples are useful for providing soil type control for geophysical surveys.

System Technology and Science

One of the simplest geotechnical devices, the impetus of gravity acting on the heavy, free-falling device is the motive force that drives the corer into the soil.

System Description

A gravity corer (fig. 16.38) consists of a steel tube in which a plastic liner is inserted to retain the core sample. The penetrating end of the tube is fitted with a cutter and a concave spring-steel core-catcher to retain the sample when the corer is retracted from the soil and recovered to the ship.

A set of heavy weights, up to 750 kg, is attached at the top end of the tube above which is a fin arrangement to keep the corer stable and vertical during its fall to the seabed.

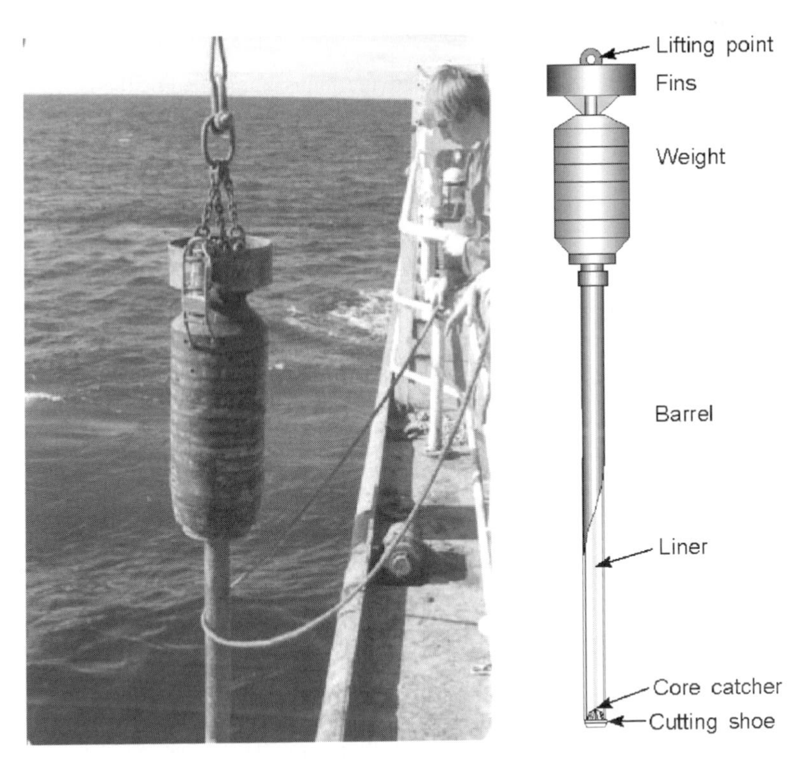

Figure 16.38 Example of a gravity corer

A deployment and recovery line is attached to the top of the corer. Normal practice is to lower the device to within 10 m of the seabed before releasing it. The gravity core tubes range in length from 1 to 6 m. The standard tube has a 102 mm external diameter with a 90 mm external diameter plastic core liner.

The deployment is normally from a deck crane, up to 2 tonnes SWL (depending on size of the corer) with a free-fall winch capability.

Advantages and Limitations

Gravity corers are only really appropriate for collecting very soft to firm clays, as penetration in stiffer clays or sands is usually limited. Furthermore, the samples thus obtained are generally of average to mediocre quality.

16.9.6 Kullenberg Device

Function and Applications

As mentioned earlier, gravity corers are typically released at a height of 10 m above the sea floor. To ensure optimum free-fall, the Kullenberg release device is used.

The device (fig. 16.39) is most appropriate when handling long piston corers and in deep water, or other circumstances, where a controlled free-fall distance is required.

System Technology and Science

Simply stated, the Kullenberg device is a release mechanism activated by the weight coming off a trigger line.

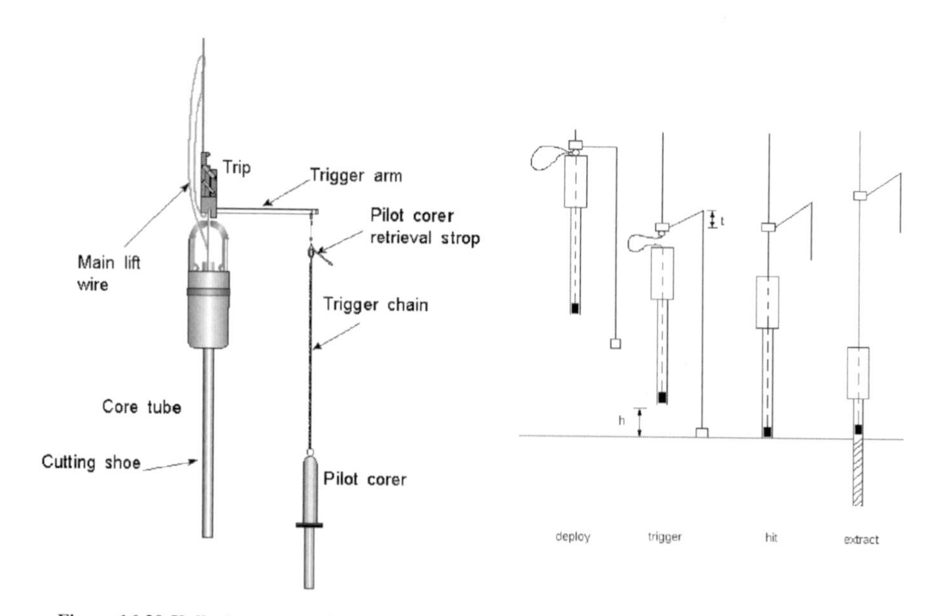

Figure 16.39 Kullenberg-type trigger mechanism and deployment-triggering-extraction sequence

System Description

The device comprises a latch that retains the gravity corer, a boom attached to the lift line and from which is suspended a weight (sometimes another sampling tool) that holds the latch closed. The weight line is made to equal the length of the corer plus the desired free-fall distance. When the weight touches the seabed, the tension comes off, the latch is released and the gravity corer falls to the seabed.

Advantages and Limitations

The Kullenberg-type device eliminates the need for a deployment winch with free-fall capability. It also allows for a controlled free-fall distance that can be varied to suit the prevailing soil conditions.

The release mechanism, however, is a temperamental arrangement and requires skilled operatives to function satisfactorily and safely. It is chiefly used for deepwater sampling.

16.9.7 Piston Corer

Function and Applications

For higher quality gravity core samples in soft soils, the liners of the core barrels are fitted with a "static" piston. The samples from the piston corers allow for more detailed soil sequencing and more accurate strength analysis.

Piston corers are suited to investigations where high-quality information is needed and applications include:

- Anchor holding assessment
- Suction caisson design
- Seabed structure foundation and installation studies
- Slope stability analysis

System Technology and Science

The suction caused when withdrawing a core barrel from a soft soil such as clay, can pull the sample from the barrel or in other ways disturb its homogeneity. By inserting a piston above the sample, when the barrel is withdrawn the suction caused above the piston keeps the sample from being pulled out of the tube.

System Description

Piston corers (fig. 16.40a) can have barrels up to 30 m in length (fig. 16.40b). Handling such a long device requires (optionally) a Kullenberg-type trigger mechanism and a purpose-built deployment and recovery system. The barrel of the device is recovered to the horizontal and must be supported at points along its length to prevent buckling. The operating water depth of the system is solely a function of its deployment winch and cable capacity.

Advantages and Limitations

The piston corer, when correctly designed and operated, can produce good quality samples in soft soils. The long, deepwater piston corers can, in some cases, eliminate the

Figure 16.40 (a) Preparing a long piston corer for deployment and retrieving it to deck; (b) STACOR mega piston corer being deployed along side of vessel

requirement for a drilling vessel and, in theory, be deployed from a wide range of vessels. However, the realisation of a safe and efficient operation requires the use of large well-equipped vessels and, usually, the mobilisation of a high capacity deployment winch and handling system together with structural modifications to the vessel.

16.9.8 Abrams Corer

Function and Applications

The Abrams gravity corer (fig. 16.41) is a self-contained coring system designed to improve the standard gravity coring techniques and increase production rates. In other respects, the application of the Abrams device is the same as for standard gravity corers. It is particularly suited to deepwater sampling where a controlled free-fall is optimal.

System Technology and Science

In most respects similar to the gravity corer, the Abrams' has a larger internal diameter behind the sample as it is forced into the liner, which reduces hydraulic resistance effects, hence increasing soil penetration.

System Description

The standard Abrams system consists of the corer described above with a maximum barrel length of up to 18 m. The barrel and the liner have the same diameters as the standard

Figure 16.41 Abrams coring system

gravity corer. The weights used tend to be greater, one or more tonnes. The corer comes complete with its own special handling/deployment and recovery system. The recovery system incorporates an A-frame and hydraulic swivel that allows the corer to be recovered inboard thus making the operation safer and more efficient.

The operating depth of the system is a function of its winch and cable capacity.

Advantages and Limitations

The Abrams corer provides scope for better soil penetration than the standard piston gravity corer. A practical benefit is that being a self-contained system, it can be deployed safely from vessels not equipped with cranes or A-frames or where these are in use by other systems. The handling system provides flexibility to use barrel lengths up to 18 m.

16.9.9 Vibrocorer

Function and Applications

Vibrocorers are used wherever soil conditions are unsuited to gravity corers or where greater penetration of the seabed is necessary. The Aimers McLean type (fig. 16.42) is one of the standard industry designs for use in sands and denser/stronger soils; it is the

Figure 16.42 Aimers McLean type vibrocorer

next step down from the rotary rock corer. Vibrocorers are used widely throughout the geotechnical investigation industry and can be deployed in water depths up to 1000 m. They are chiefly employed to recover samples for:

- Pre-dredge soil investigations
- Offshore oil and gas pipeline investigations
- Mineral and aggregates prospecting
- Environmental impact studies
- Civil engineering for ports and harbours
- Inshore geotechnical investigations

Vibrocorers are occasionally used for cable route investigations in particularly difficult soils.

System Technology and Science

To penetrate soils such as dense sands and gravels, or to reach deeper into stiff clays, rather than depending on a gravity free-fall, the corer's barrel is vibrated thus facilitating its penetration. In other respects, the barrel and sample retention systems are similar to those of the gravity corers.

System Description

The typical vibrocorer consists of a tall steel frame and tripod support (fig. 16.42). Within the frame is a standard 102 mm mild steel coring barrel within which is inserted a 90 mm PVC liner to contain the sample. A spring steel core-catcher is fitted to the cutting shoe as for the gravity corer.

Two linear electric motors enclosed in a pressure housing provide the vibratory movements. The core barrel is attached directly to the motor housing. Power ($415V_{DC}$) is fed to the motors via an electrical control line from the surface support vessel. Once in motion, the heavy motor housing provides the mass to drive the core barrel into the seabed. Depending on the water depth and other operating parameters, lifting lines are used to deploy and recover the system from the ship.

A typical 6 m vibrocorer will weigh nearly two tonnes and requires a crane for deployment and recovery. A separate generator installed on deck normally provides power. Vibrocorers come with barrel lengths of 3, 6 and 8 m; a normal coring operation in North Sea depths will take about one hour.

Advantages and Limitations

Vibrocorers provide valuable information for shore-based laboratory testing for soil classification purposes. The shear strength measurements must be interpreted with caution due to sample disturbance caused by the shaking motion. Owing to their size and power demands, substantial sized ships are required. Further, because coring is a protracted process, the ship must be capable of remaining on station and will preferably either have a DP or a good joystick control, otherwise excessive position excursions may cause the core barrels to bend, which could lead to a total loss of the system and other financial loss through downtime.

16.9.10 High Performance Corer^{TM}

Function and Applications

For coring in very strong soils, in gravelly or dense sands, the High Performance Corer^{TM} (HPC^{TM}), shown in fig. 16.43, offers better penetrating powers than the Aimers McLean type. This type of vibrocorer is usually found working in the aggregates and dredging industry, although they do have applications in offshore oil and gas and for civil engineering investigations. Their typical maximum operating water depth is around 300 m.

System Technology and Science

In common with the Aimers McLean vibrocorer, the HPC^{TM} relies on vibration to agitate the barrel into the soils, and the mass of the motors to provide the downward force.

System Description

The chief differences between the HPC^{TM} and the Aimers McLean type is that its motor can be controlled to optimise its excitation frequency and vibration amplitude to suit various soil conditions. At its most powerful, the HPC^{TM} will apply twice the power and up to 5 times the vibration amplitude of the standard vibrocorer. The steel frame is also of a heavier build.

The core barrels are standard and come in lengths of 3, 6 and 8 m. A complete system weighs in the order of 3 ¼ tonnes and requires a deck crane for its deployment and recovery.

Figure 16.43 High performance corer^{TM}

Advantages and Limitations

As for the standard vibrocorer, the HPC^{TM} provides samples for shore-based laboratory testing. Their great advantage is that they can penetrate faster and deeper in sands, gravels and stiff clays than the conventional vibrocorers.

16.9.11 Box Corers

Function and Applications

The seabed box corer is used to recover relatively undisturbed block samples of mudline material in soft, cohesive sediments.

System Technology and Science

The box corer is a very simple device that envelops an area of seabed then seals the base of its box to retain the sample from further disturbance during recovery.

System Description

The standard box corer (fig. 16.44) consists of a steel frame incorporating the sample box surmounted by a 200–300 kg weight. When activated by a self-release trigger system,

Figure 16.44 Box corer

the box is closed at the bottom by a swiveling base. The total weight is in the order of 1.5 tonnes and the sample volume of the box is about 25–30 litres (7–8 gallons).

During operation, the box corer is lowered to the seabed and, on contact, the self-release trigger is primed. The sample box is then pushed 40–50 cm into the seabed by the action of the weight. The trigger mechanism releases a latch that allows the swivel base to close off the captured sample, before the whole unit is recovered to the surface where the sample is removed for examination.

Advantages and Limitations

The great advantage of these devices is that they can recover a large, relatively undisturbed and high quality sample for study and for laboratory testing. Their value in environmental assessments is that they preserve well any benthic life forms and habitants and facilitate visual examination of a portion of the seabed surface.

16.9.12 Push-In Samplers

Function and Applications

Push-in samplers were originally developed as downhole tools for geotechnical drilling operations to reduce the high level of sample disturbance common to the older type of percussion driven samplers.

Seabed versions of this technique (fig. 16.45) have evolved to allow higher quality samples to be recovered in softer soils. Typical applications include:

- Offshore oil and gas pipeline and small structure investigations
- Civil engineering studies for ports and harbours
- Inshore geotechnical investigations

System Technology and Science

The standard downhole push-sampling technique involves latching the sampler behind the drill bit and inserting it into the soil in a controlled manner through the application of a hydraulic pressure or using the drill string's self-weight.

Most standard push-sampling tubes are fabricated from stainless steel and are 1 m long with an internal diameter of about 75 mm. The wall thickness is around 1.5 mm and the tube has a sharpened cutting edge. The tube is attached to a "sample head" incorporating a one-way valve system. As the soil enters the tube, water trapped above it is efficiently expelled. When the tube is being extracted, the valve closes and creates suction above the sample to ensure its retention.

System Description

The seabed push-samplers are usually incorporated in a seabed frame, either as a stand-alone tool, or in tandem with a cone penetration test (CPT) device. Most of them are capable of pushing in a sample tube between 1.0 and 1.5 m length with a diameter of between 75 and 100 mm. The seabed frame can weigh from 1 or 2 tonnes up to 3–5 tonnes if a CPT unit is incorporated.

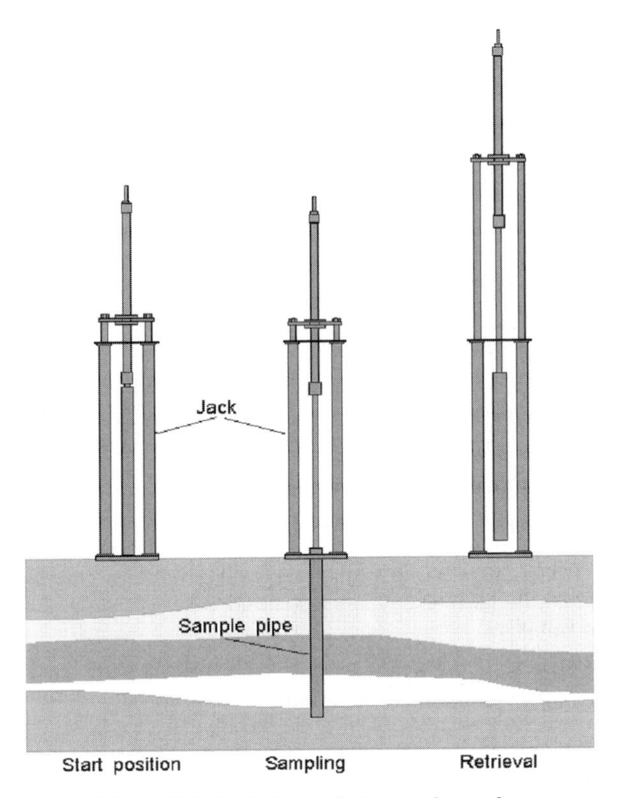

Figure 16.45 A simple stand-alone push sampler

Advantages and Limitations

The advantages of these devices are (a) they provide higher quality samples in soft soils than other seabed coring devices, and (b) if incorporated with a CPT system, they reduce the deployment time and provide the samples directly adjacent to CPT tests for correlation.

16.9.13 Grab Samplers

Function and Applications

Grab samplers are one of the most common methods of retrieving soil samples from the seabed surface. The information they provide, although coarse, can be applied in a number of applications such as:

- Bulk sampling for seabed minerals
- Marine aggregate prospecting
- Environmental sampling
- Pre-dredge investigations
- Ground-truth for morphological mapping and geophysical survey

Grabs can be used on any seabed to recover samples although care is needed in selecting the right size unit for the task.

System Technology and Science

The grab sampler is a device that simply grabs a sample of the topmost layers of the seabed by bringing two steel clamshells together and cutting a bite from the soil.

System Description

The grab sampler (fig. 16.46) comprises two steel clamshells acting on a single or double pivot. The shells are brought together either by a powerful spring (the Shipek type) or powered hydraulic rams operated from the support vessel.

In operation, the grab is lowered to the seabed and activated, either automatically or by remote control. The shells swivel together in a cutting action and by doing so by remove a section of seabed. The sample is simply recovered to the surface for examination.

Geotechnical investigations normally require large samples and favour the bigger hydraulic clamshell grab. These systems can retrieve samples of 0.35 m^3 or 700 kg mass. A typical hydraulic grab will weigh around half a tonne and can operate in water depths down to 200 m. Typical performance rates are between three and four samples per hour.

Advantages and Limitations

The smaller Shipek type grab sampler is only useful for ground truthing geophysical surveys and in basic hydrography. The more massive hydraulic grabs are capable of recovering relatively intact samples of consolidated soils.

In areas of large cobbles or boulders, the grabs can become inadvertently jammed open and their contents washed away during recovery to the surface. However, hydraulic grabs are

Figure 16.46 Hydraulic grab sampler

more likely to recover cobbles and small boulders than any other system and in this respect are invaluable.

The small grabs can be operated from virtually any sort of vessel. The large hydraulic grabs require at minimum a 2½ tonne SWL crane and 10 kVA, $415V_{DC}$ generator and a 210 bar hydraulic power pack and, therefore, demand the use of a larger sort of vessel.

The geotechnical value of soil samples obtained by the grab sampler may be limited due to the washing out of finer cohesionless materials during recovery and due to the level of disturbance imparted to cohesionless soils.

16.10 In situ Testing Systems

The greatest value of in situ testing is that it permits the evaluation of important physical characteristics of soil, and sometimes rock, in its natural state. In cohesionless soils, it is often the only means of determining certain engineering parameters such as the relative density of sands.

A secondary, but extremely valuable, benefit is the immediate availability of results, with most types of test, allowing decisions to be made on site without having to wait for the results of laboratory testing.

In situ testing is a means of avoiding sample disturbance, obtaining a better delineation of soil stratigraphy and measuring soil properties that cannot be determined from soil samples. In situ testing provides important supplemental and complementary data but it should not be considered to replace a soil boring. The following paragraphs introduce various types of in situ tools, present typical data and discuss the advantages of obtaining these data.

16.10.1 Cone Penetration Testing (CPT) Systems

Function and Applications

The cone penetration test (CPT) is the most widely used in situ test for marine engineering applications. Its prime use is providing information on the soil type and stratification, as well as the shear strength in clays and the relative density and friction angles in sand.

CPTs have a wide range of applications that include:

- Offshore oil and gas pipeline route investigations, trenching and stability studies
- Geotechnical investigations for seabed structures and anchors
- Submarine cable route surveys and burial assessment studies
- Inshore civil engineering studies
- Pre-dredge investigations
- Ground truth for geophysical survey and morphological mapping

System Technology and Science

The cone penetrometer provides an accurate and continuous profile of soil stratification. This tool provides valuable information about granular and cohesive soils. In cohesive soils, cone data provides a continuous profile of shear strength and eliminates strength anomalies caused by the sampling process. The cone is more important in granular soils because it is the only tool available that can provide reliable density information.

An original "Dutch" cone was used to measure point resistance as the cone was jacked into the ground. A later version incorporated a friction sleeve above the point. The point and the sleeve were alternately jacked ahead to obtain point resistance and side friction. Modern electrical friction cone equipment (fig. 16.47a) simultaneously monitors both the point resistance and the friction as the cone device is steadily advanced into the soil.

The cone penetration test provides an empirical assessment of seabed soils based on the resistance of the soil to a cone-tipped probe, or a penetrometer, as it is pushed into the seabed at a constant rate of penetration (2 cm/s). The standard cones have a tip angle of 60° and a cross-sectional area between 5 and 20 cm^2, with 10 and 15 cm^2 cones being the most common. Mini Cones with cross-sections of 1 or 2 cm^2 are also available and are discussed later.

The piezocone (fig. 16.47b) has been used in offshore investigations since the mid 1970s. It measures dynamic pore pressure, as well as point resistance and sleeve friction. As the cone is pushed into the soil, pore pressure is developed ahead of the tip. Coarse materials are fast-draining and will exhibit low pore pressure response. As soils become finer, they drain slower, and pore pressure response will increase. This cone is beneficial to define soil stratigraphy since it is more sensitive to soil type than the tip and sleeve components (fig. 16.47c).

Electrical strain gauges within the cone assembly measure the resistance on the cone tip and friction on a "sleeve" behind the tip. In a "piezocone" penetration test (PCPT), an additional parameter, the soil pore water pressure, is measured via a porous element in the cone face or at the shoulder between the cone tip and the friction sleeve. Note that the PCPT is sometimes referred to as the CPTU, the *u* being geotechnical shorthand for pore pressure. Data are transmitted in real-time to the surface support vessel for recording and analysis, or can be stored in a Remote Memory Unit (RMU) for downloading into a shipboard computer upon retrieval of the tool on deck. The latter method is the preferred one for deepwater site investigations, where the umbilicals would become too costly, too slow to handle, and present power transmission problems.

Soil types are determined by reference to a graph of Cone Resistance (q_c) against the Friction Ratio. Friction Ratio is the sleeve friction (f_s) divided by Cone Resistance (q_c). Other empirical relationships are used to estimate shear strength in clays and the relative density and internal angle of friction in sands.

Measuring the pore pressure provides valuable additional information on a soil's stratification, permeability and stress history (i.e. whether it is "under", "normally" or "over" consolidated).

The Wheel-Drive CPT

One type of remotely operated seafloor cone system (the SEACALF) has seen extensive offshore use (fig. 16.48a). The system is lowered to the bottom with a fixed length of cone rods in place and then operated remotely to get the penetration data to a limited depth.

(a)

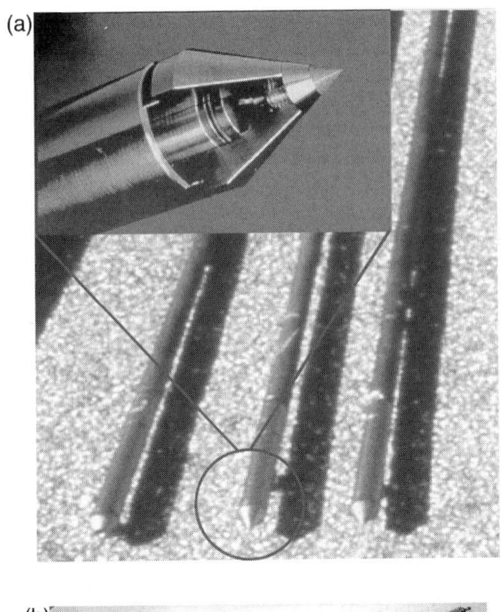

(b)

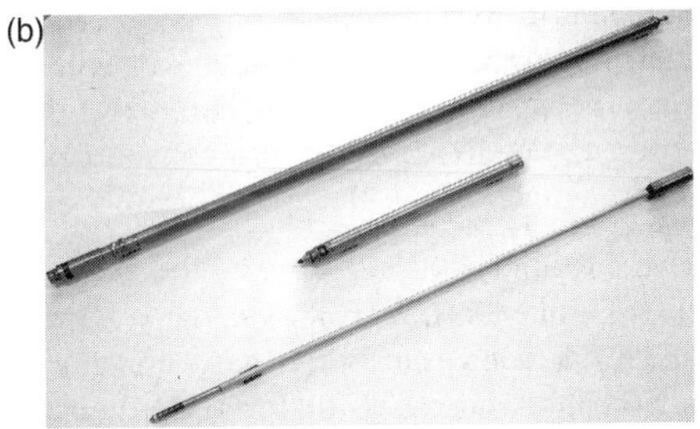

(c)

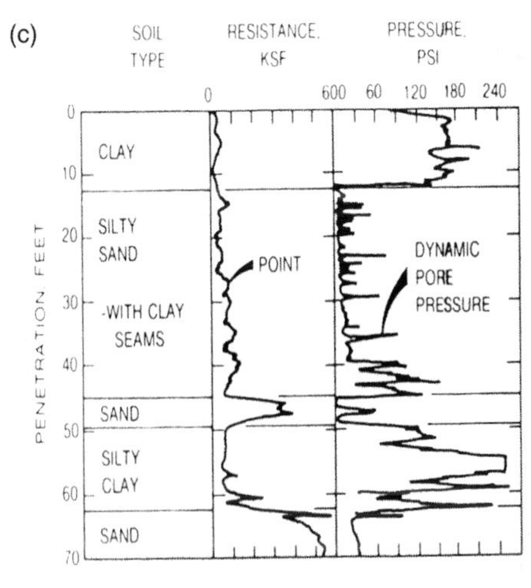

Figure 16.47 (a) Actual cone penetrometers; (b) typical piezocone components; (c) typical piezocone data

Another remotely operated cone device is the SEACLAM. This is a seafloor-resting template which acts as a reaction system. The drillpipe is first lowered through the jaws of the SEACLAM and subsequently clamped using hydraulic power. The pipe is motion-compensated from the deck. A cone penetrometer system is then lowered through the pipe, latched in at the base and pushed into the soil using the SEACLAM for reaction.

In another system, the cone is lowered inside a string of drill pipe, and mud pressure is used to effect cone penetration.

One CPT system used with the SEACLAM is Fugro's downhole "Dolphin" system (fig. 16.48b). This cone has no electrical or hydraulic umbilical and uses mud pressure in the drill string to force the cone into the soil. Data is stored in a remote memory unit (RMU) in the tool and is downloaded to the onboard computer after the tool is retrieved to the deck with a wireline. A reaction mass provides the dead weight for loading. This tool tends to be faster and more reliable than systems that use umbilicals.

The wheel-drive CPT system (fig. 16.48a) comprises a seabed reaction frame containing the wheel-drive mechanism and electronic control and data acquisition systems. Each wheel-drive unit consists of four steel wheels clamped tightly against the CPT cone thrust rod. A 440 V_{AC} electrical power for the hydraulic or electric wheel-drive motor is supplied via an umbilical cable from the support vessel. A guide attached to the lift lines supports the thrust rod, which can be up to 65 m long, in the water column. As the wheels are rotated against the rods, the cone is pushed at a constant rate of 2 cm/s into the soil. The outputs from the strain gauges within the cone assembly are passed up the umbilical to the ship.

The largest wheel-drive CPT systems (fig. 16.48a) weigh some 25 tonnes in air, and have a thrust capacity of 200 kN. In dense sands and hard gravelly clays, typical penetration ranges are about 20 m and between 30–60 m for softer, normally consolidated, clays. Wheel-drives that are more typical weigh 6–13 tonnes and deliver a thrust of up to 100 kN.

These large wheel-drive devices require sizeable handling systems capable of upto 40 tonnes SWL, plus adequate deck space, and hence can only be operated from relatively large vessels. They can be used in water depths down to 1800 m at the present time, but systems capable of operations in 3000 m are under development.

Lightweight Wheel-Drive CPT

The lightweight CPTs (figs. 16.49a and b) are the most popular models for submarine cable route investigations, ploughing assessment and trenching studies. They typically consist of a 4-m-tall frame mounted on a 4-m-diameter seabed base-frame and weigh about 2.3 tonnes. The drive motors, wheel-drive and sensor systems are mounted on the base-frame.

The wheel-drive operates the same as the large version but uses only two steel wheels and can apply only a thrust of up to 15 kN to the 10 cm^2 cone. The CPT cone rod can penetrate up to 2 m into the seabed. Electrical power for the motors is supplied via an armored umbilical cable that is also used to deploy and recover the device from the surface support vessel. Lightweight wheel-drive CPTs normally include an array of

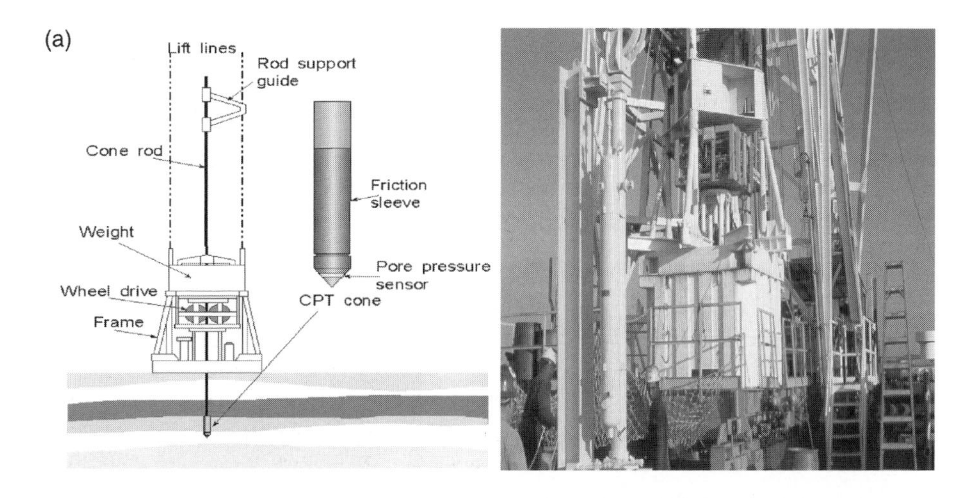

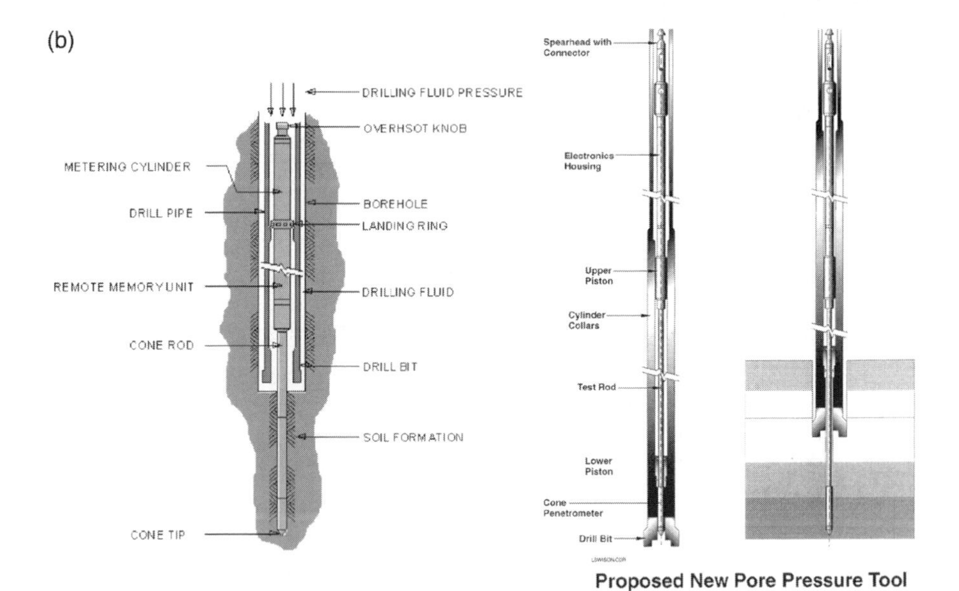

Figure 16.48 (a) Wheel-drive CPT – SEACALF; (b) downhole dolphin CPT shematic

ancillary sensors and samplers and can be deployed in water depths down to 1500 m. Some battery-powered versions can be operated in over 2000 m water depth.

The lightweight wheel-drive systems can be operated from most vessels fitted with 5 tonne SWL cranes or A-frames and having sufficient reach. Adequate deck space is needed

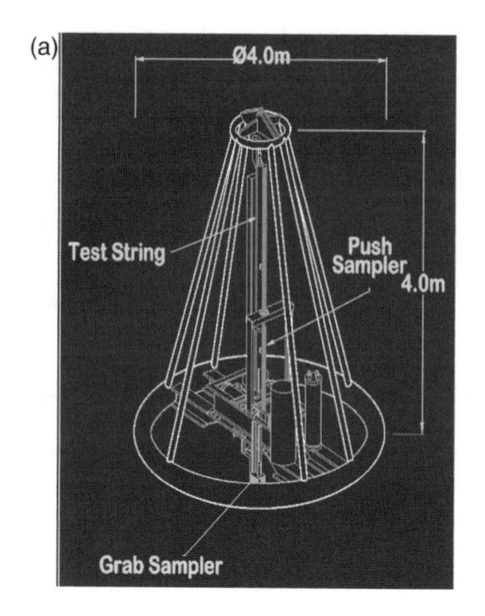

Figure 16.49 (a) SEAROBIN CPT and schematic; (b) Starfish CPT

to accommodate the smaller 20 kVA generators, 1500 m winch and for storage/handling. The design is particularly robust to enable high productivity rates to be maintained.

Advantages and Limitations

The wheel-drive and the lightweight wheel-drive CPT are probably the most versatile and well-proven of the in situ seabed testing systems available. The high productivity and reliability of most systems means they provide a very cost-effective means of geotechnical data acquisition. The ability to evaluate results in real time provides for greater program flexibility and minimises the probability of leaving site with insufficient data.

The lightweight systems can be deployed from most geophysical survey vessels and a wide range of other, suitably equipped, vessels of opportunity. The largest wheel-drive systems require vessels with heavy deployment capabilities, such as drill ships, DSVs and other construction support vessels, but their penetration capability is unsurpassed and can reduce the requirement for borehole data.

16.10.2 Minicones

Function and Applications

Minicones are not CPTs in the strictest sense, falling as they do, outside of the currently defined size range acknowledged by international standards (i.e., 5–20 cm^2). They were developed in order to facilitate improved stratigraphic profiling and soil parameter definition from vessels that could previously deploy only coring and sampling devices, because of limited handling capability. The primary application for this technology was on pipeline and cable route surveys, but this subsequently expanded to include investigations for small subsea structures, anchors and dredging assessments.

System Technology and Science

The cone tip of a minicone system is typically 1 cm^2 or 2 cm^2 (i.e., one-tenth that of a standard cone) and its push rod, instead of being one length of rigid steel, is a coil that must first pass through a straightening device. The cone tip resistance and the sleeve friction are measured as in the conventional CPT, but pore pressure measurement is less common.

The rate of penetration is usually higher than the CPT, typically twice as fast, at around 4 cm/s. Taking into account the scale and rate effects this gives rise to resistances comparable to a full size cone. Minicones capable of measuring soil temperature and/or thermal conductivity are also available.

System Description

The minicone and its peripheral systems are mounted on a seabed framework that comprises a thrust machine, electronic data acquisition unit, hydraulic power pack, coiled push rod and straightener (fig. 16.50). Power for the hydraulic power pack is provided by an umbilical to the support vessel or batteries mounted on the seabed frame. The minicone test (MCT) can penetrate 5–6 m into the seabed with a thrust of 1 tonne. The water depth capability of the minicone systems is usually in the range 1500–2500 m.

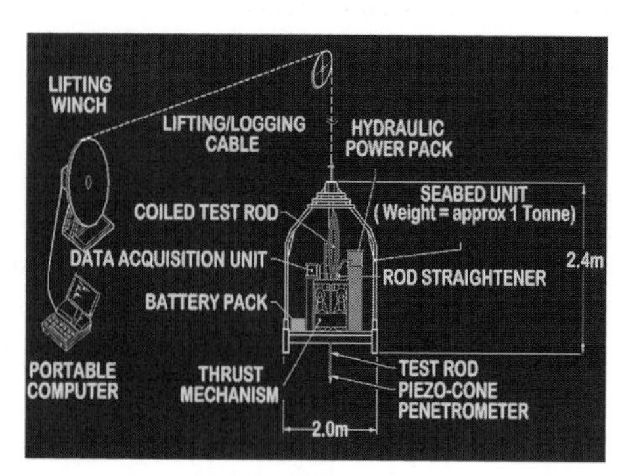

Figure 16.50 SEASCOUT minicone CPT and schematic

Advantages and Limitations

The minicone system can be operated from a wide range of vessel types that are equipped with cranes or A-frames capable of 2.5–5 tonnes SWL and have minimal free deck space. The test is very rapid, requiring as little as 10–15 min on the seabed and results are immediately available for evaluation. They are also more sensitive to the presence of thin lamina within soil layers. This enables, for example, the detection of thin sand layers within a soft clay formation that may dramatically affect the drainage and settlement characteristics of that formation. Greater care needs to be exercised when using the MCT to derive engineering design parameters, because there is not the global database of correlation between MCT results and laboratory test data, or engineering back-analysis, that exists for the CPT. This is particularly pertinent for areas where there is little existing geotechnical knowledge. In such areas, site and regional, specific correlation may be required.

16.10.3 The ROV CPT

Function and Applications

Where a CPT is required at a precise location or where a series of continuous tests are needed along, for example, a cable route or pipeline trench, the ROV CPT provides an excellent tool.

The information available from these units meets a number of applications including:

- Offshore oil and gas pipeline and control cable route investigations
- Soil temperature gradients for pipeline heave and buckling assessments
- Trench backfill investigations
- Submarine cable route studies, real-time plough assessment studies
- Ground truth for morphological mapping and geophysical survey

- Environmental and geotechnical assessments, of drill cutting mounds for example

System Technology and Science

Instead of deploying a CPT from a crane, the ROV CPT can be precisely placed for optimum results. The advantage of mobility allows the CPT operator to select areas based upon visual inspection and to conduct a rapid series of tests.

System Description

The CPT device and its peripheral systems are mounted onto a standard ROV tooling skid (fig. 16.51). A Work-class ROV is then attached to the skid and the CPT and its sub-systems take their power from the vehicle's supply.

The 1 m long CPT push rod is thrust into the soil by a hydraulic ram. Due to the limited reactive force provided by an ROV, the CPT's cone area is reduced to 5 cm^2.

The water depth capabilities of ROV CPTs are often only limited by the capacity of the vehicle. Hence, deployment down to 3000 m is quite feasible, although 1500 m is more the norm.

Advantages and Limitations

The ROV CPT can be operated from any vessel equipped for ROV operations. For deepwater pipeline route assessments, the ROV CPT is a useful quantitative instrument. Its great advantage is the precision with which it can be placed, for example right alongside a pipeline or structure. Operated in advance of a configurable cable plough, the ROV CPT can provide real-time ploughing assessment information.

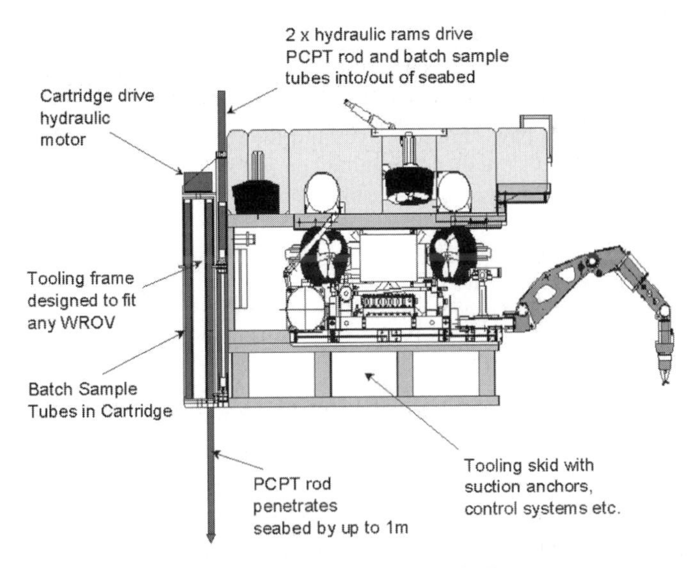

Figure 16.51 ROV mounted CPT

16.10.4 Vane Test

Function and Applications

One of the many tools that can be deployed downhole, or from the seafloor (e.g., wheel-drive machines) or using the stand-alone test rigs, the vane test is a rapid and an accurate means of assessing the in situ undrained shear strength of cohesive soils (e.g., soft clays).

System Technology and Science

The vane test comprises of pushing a cruciform steel vane into a clay soil and applying a torque. The torque resistance is measured until the soil fails at its natural shear strength.

System Description

The vane test consists of a steel vane typically between 38 and 65 mm in diameter and 75 and 130 mm high. The vane is attached to a shaft that turns at a constant rate between 6° and 12° per minute. After the soil fails, the rate of rotation can be increased to around 60° per minute and a measure of remolded strength obtained. The measure of torque vs. rotation is transmitted to the surface support vessel through an umbilical link or it can be stored in a solid-state remote memory unit (RMU) on the seabed frame, or in the downhole tool.

Vane tests can be attached to wheel-drive CPT rods, as wire line deployed sensors within a borehole or deployed to the seabed in a purpose built frame. In downhole operation, the vane test is pushed into the soil, typically 0.5 m below the bottom of the borehole, before being activated. At the end of each test, the vane can be pushed in further and the procedure repeated at a different elevation.

Downhole In Situ Vane: The remote vane shear device is principally used for in situ testing of clay soils. The tool is designed to be compatible with the open-hole and wireline methods. The device is run to the bottom of the hole through the bore of the drill pipe (fig. 16.52a). Electrically or mechanically operated pawls are extended after the device passes the bit, and the weight of the drill pipe is used to insert the vane and its reaction system into the soil. The bit is then raised, the pawls are retracted, and the vane motor is actuated to perform the test. Different size vanes are used, depending on soil strength (fig. 16.52b). Results are recorded on a chart or displayed digitally on a surface readout. The newest tools use a remote memory unit (RMU) to record the vane data downhole, without the use of an umbilical. The remote vane tool is then retrieved by a wireline and overshot. The recorded data is then downloaded from the RMU into a computer on deck. The direction of vane rotation can be reversed, and cyclic tests can be performed.

Seabed In Situ Vane System. The in situ seabed vane systems are commonly used in very soft to soft, normally consolidated clay soils. A system used over the entire world, but most often used in the Gulf of Mexico, is the Halibut vane system (fig. 16.53). The Halibut is a self-contained, remote seabed system that uses no umbilical and can work in up to 10,000 ft of water.

The tool is deployed from a seafloor template (basket) and weighs about 1000 lb. The system is deployed on a single lift line using a relatively small winch over the side of a vessel or rig. The Halibut performs two tests at the same depth (in different locations) and the

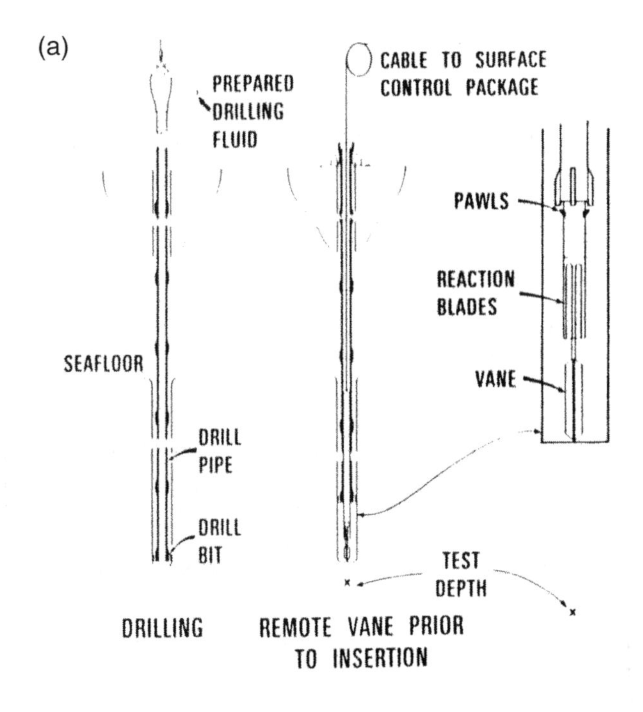

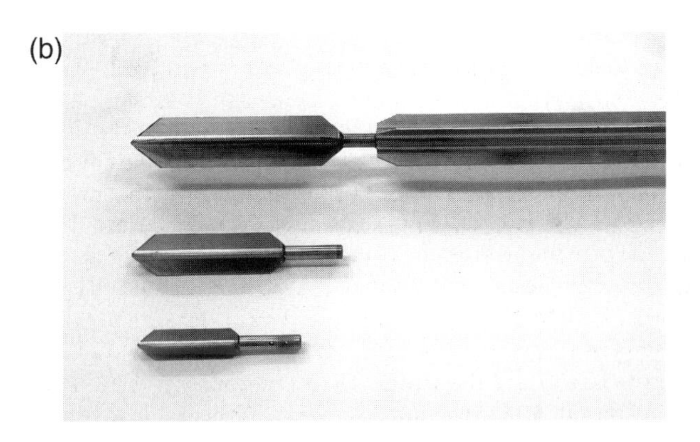

Figure 16.52 (a) Downhole in situ vane; (b) different sizes of vanes

system is retrieved to the deck and test results are downloaded from the onboard memory unit. After resetting the tool's memory and extending the vane rods below the seabed frame to a different depth, the tool is again deployed. This system may also operate as a wireline tool to obtain "real time" data. The disadvantage of this tool is the time required to obtain the data. Multiple deployments are required because only one test depth is obtained for

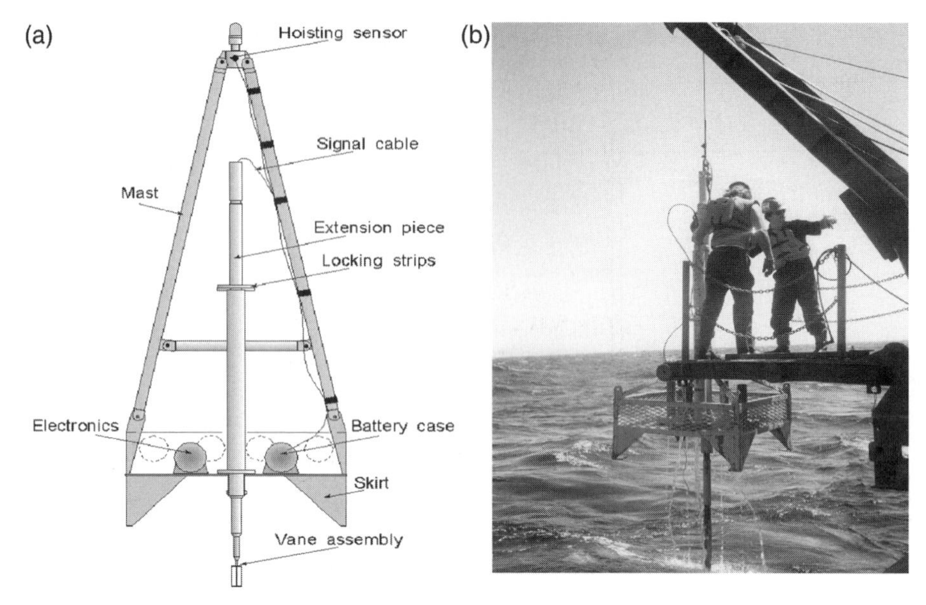

Figure 16.53 Halibut vane basket: (a) schematic of stand-alone vane test, (b) Halibut vane test rig during recovery

each deployment. In 5000 ft of water, a series of tests from the mudline down to 28 ft require about 12–15 hours to perform.

It is difficult, if not impossible, to get accurate strength measurements on recovered "gassy" and expansive soil samples. In gassy sediments, the pore fluid contains gas. This gas comes out of solution during retrieval of the sample to the surface, due to the pressure relief, much like what happens to a carbonated beverage when the bottle is opened. The Remote Vane provides consistent data that is not affected by gassy soils. Figure 16.54 shows the differences in shear strength that can occur in gassy soils.

Relying on the shear strengths measured on samples can lead to an improper interpretation and a very conservative design.

16.10.5 T-Bar Test

Function and Applications

The T-bar test is similar to the CPT but was designed to hopefully provide a more accurate assessment of shear strength in very soft soils.

System Technology and Science

As for the CPT, the test provides an empirical assessment of seabed soils based on the resistance of the soil to a constant thrust as the bar is pushed into the seabed.

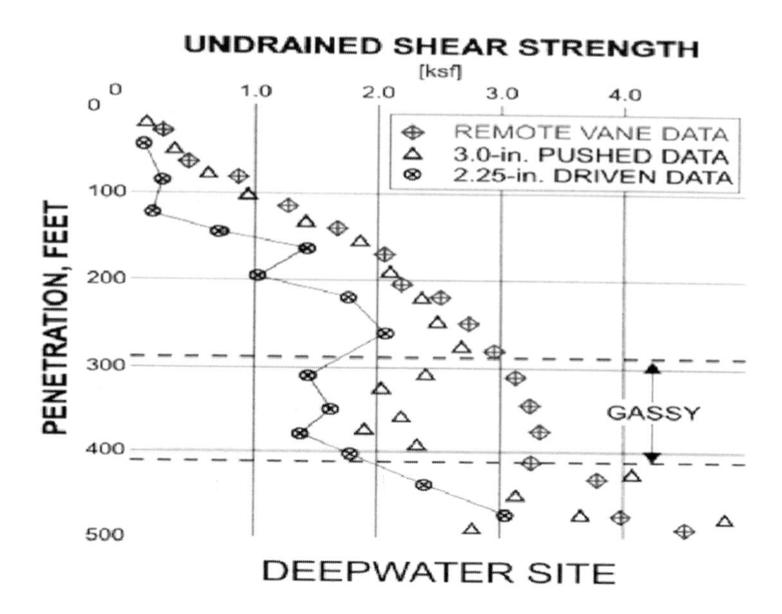

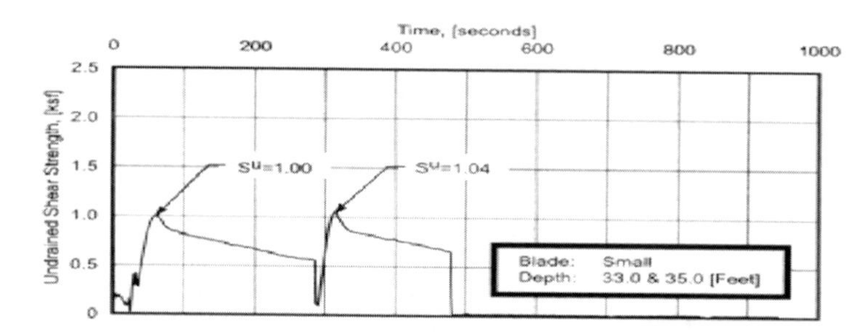

Figure 16.54 Comparison of remote vane and laboratory data in clay

The T-bar comprises a short cylindrical bar that is attached perpendicularly to the penetrometer rods. As it is pushed into the soil, a load cell situated immediately behind the bar measures the resistance. The higher resistances generated in very soft soil, combined with a more rigorous analytical solution of the soil failure mechanism, enables shear strengths to be determined with greater confidence.

System Description

The standard T bar, used with wheel-drive seabed CPT systems, is 250 mm long and has a diameter of 40 mm (fig. 16.55). In addition to the resistance load cell, two pore pressure transducers are incorporated in the bar and an inclinometer in the shaft. Its penetration rate is the same as for the CPT, at 2 cm/s. A smaller version has been developed for downhole deployment.

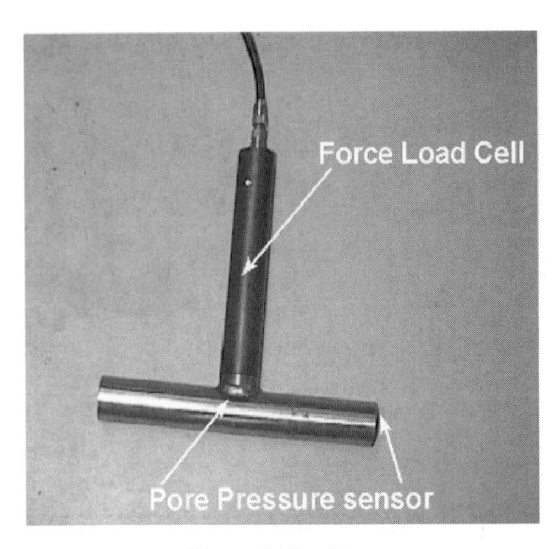

Figure 16.55 T-bar

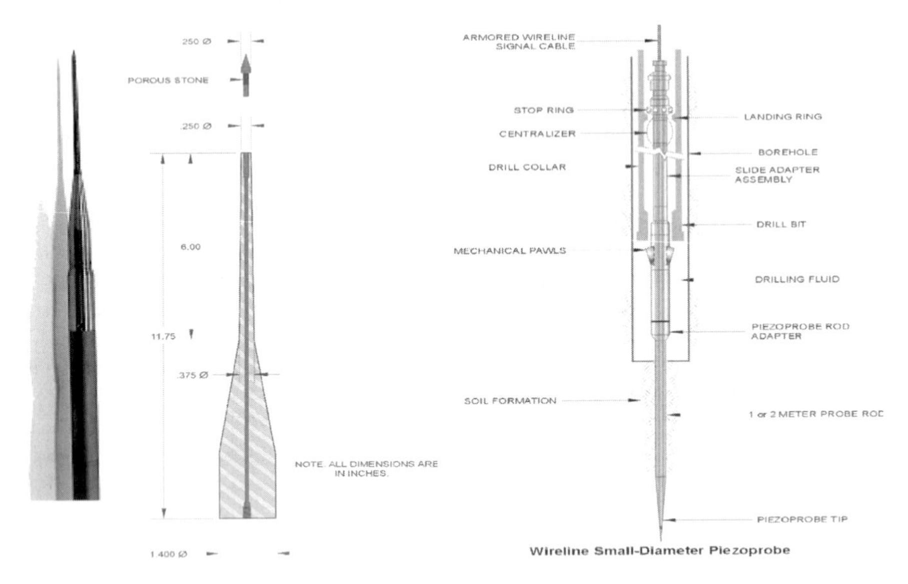

Figure 16.56 Piezoprobe and schematic

16.10.6 Piezoprobe Test

Piezoprobe tests are performed using a wireline operated small diameter probe (fig. 16.56). Pore pressures in the soil formation are measured by a pressure transducer which is in contact with the sediment through a porous stone. A temperature transducer is also incorporated in the tool and the data is used to correct the measured pressures.

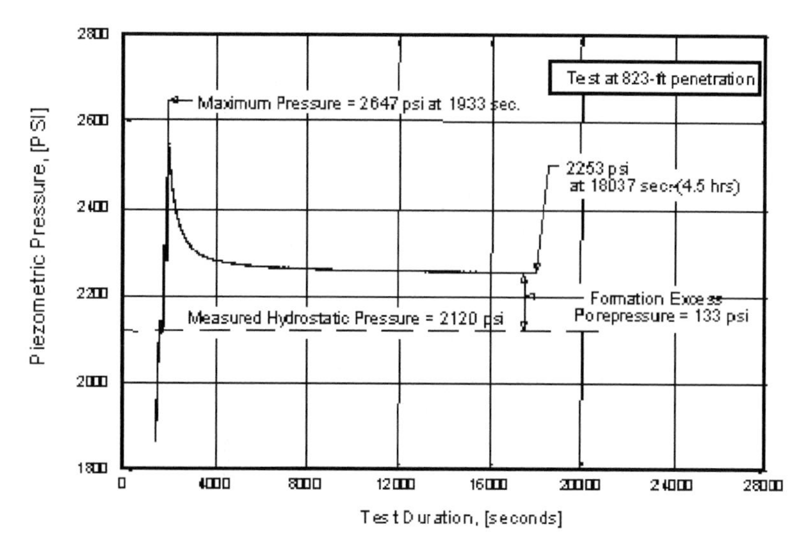

Figure 16.57 Typical piezoprobe test results

After the borehole is drilled to the desired test depth, the drill bit is raised about 3 ft from the borehole bottom. The piezoprobe is then lowered through the drill pipe until it rests on a catch ring in the drill bit. The piezoprobe is allowed to sit for about 30 s to measure the hydrostatic pressure. To insert the piezoprobe tip into the virgin soil below the bottom of the borehole, the drill bit is raised about 12 ft above the bottom of the borehole to clear the mechanical pawls and subsequently lowered to engage the pawls. The weight of the drill string pushes the piezoprobe tip into the sediment. The length of the push is usually about 1–3 ft beyond the bottom of the borehole, depending on the stiffness of the soils. After the tool is pushed, the drill string is raised about 5 ft to prevent contact with the pawls and the drill bit while the piezoprobe acquires data. The pressure transducer transmits electronic signals through an armored cable to a computer on deck, where the data are continuously displayed in real time and stored digitally. The pore pressures are sampled at a relatively fast rate (e.g. one reading per second) during the test. When sufficient data have been obtained to define the pore pressure dissipation curve adequately, the tool is then retrieved.

Typical data plots (fig. 16.57) show peak pressure after insertion followed by pore pressure dissipation stabilising at an ambient condition. The excess pore pressure in the soil formation is the difference between the equilibrium piezometric pressure (ambient pressure) and the measured hydrostatic pressure.

16.10.7 Other In Situ Tests

A range of additional sensors can be deployed by means of the conventional seabed CPT technology, including:

Thermal Conductivity Probe – for measuring a soil's heat dissipation characteristics. Especially important in deepwater geotechnical investigations where the insulating

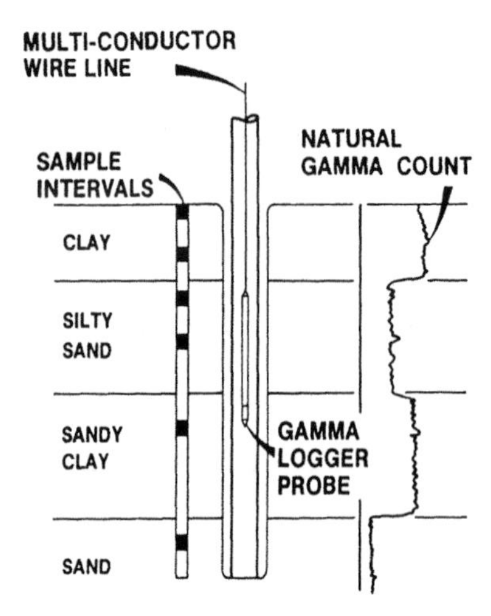

Figure 16.58 Gamma logging operations

characteristics of soils are important considerations for designing anti-waxing solutions in pipelines.

Electrical Conductivity Cone – the electrical conductivity of a soil depends on the soil type, porosity, water content and pore water composition. The primary applications for this measurement are in determining corrosion potential for pipelines and structures, detecting pore water pollutants and assessing changes in porosity.

Seismic Cone – incorporates a triaxial geophone in the cone shaft. Used with a shear wave generator on the seabed reaction frame, it provides information on the low strain stiffness of a soil; particularly useful in large foundations subject to dynamic loads. The shear wave velocity profile it provides is also useful in deep seismic processing.

Natural Gamma – a sensor for detecting the natural gamma radiation of a soil as a means of ascertaining soil type. The gamma logger is the fastest and the least expensive method of obtaining a continuous profile of soil stratigraphy. Since it is not practical to sample a borehole continuously, the gamma log is extremely useful in detecting strata breaks between sample intervals. The unit works on the principle that various soil types emit different levels of natural gamma radiation. Clays emit higher levels of radiation than sands due to the greater percentage of minerals containing radioactive particles (fig. 16.58). The gamma logger is normally run through the drill pipe at the completion of the drilling and sampling program.

16.11 Operational Considerations

16.11.1 Horizontal Control or Positioning

Close horizontal control is needed so that an investigation is performed at the eventual platform location. Surveying may be performed using line of sight methods under certain conditions or may employ one of the hyperbolic systems utilising fixed onshore stations and a mobile offshore station. In most cases, satellite positioning is now used to locate vessels on site. The surveying is handled by installing the surveying equipment on the drilling vessel. A separate vessel is sometimes needed to assist in setting anchors for the drilling vessel and to provide a standby in the area in the event of an emergency.

Work performed in deepwater and ultra deepwater usually requires a vessel with dynamic positioning (DP) capabilities. The vessel's DP system uses a series of thrusters controlled by a computer which continually monitors the position of the vessel and directs the vessel to apply thrust in a given direction to maintain the vessel's position (i.e. "to maintain station"). All movements are controlled automatically to keep the vessel within an acceptable watch circle. The positioning information is typically provided by continual communication through satellite systems such as Starfix-MN8 System, which covers all active oil and gas exploration areas worldwide, or other Global Positioning Systems (DGPS). These systems usually have a fully redundant backup system.

16.11.2 Water Depth Measuring Procedures

Measuring water depth sounds easy; however, it is extremely difficult, particularly for the combinations of deep water, currents and the soft bottom conditions. An approximate water depth can be obtained by use of the ship's fathometer. However, an accurate water depth is needed so that sampling can begin exactly at the mudline; it is also needed for jacket design. The water depth measurements are usually taken at the boring location by three methods: a sounding weight, an electronic bottom sensor and the first sample recovery during the coring operation.

An initial water depth is estimated using a sounding weight and wireline. A Cavins wireline counter is attached to the weighted wireline, and the sounding weight is lowered until the seafloor is encountered.

After this approximate water depth has been obtained, the drill pipe can be run to a depth somewhat above that indicated. A pressure-sensitive electrical device or bottom sensor (fig. 16.59) is used to measure the water depth more accurately. The bottom sensor is an electronic seafloor sensor that is operated on an electric cable and latches into the drill bit. The drill string is slowly lowered to the seafloor until the electronic sensor signals that its tip has contacted the soil, as indicated on a meter at the surface. Water depth can be computed by simply tallying the length of pipe below water level at the time of contact.

The water depths obtained from these first two methods are then confirmed by recovery of the first soil sample. The Cavins wireline counter is used in conjunction with the sampling equipment to obtain the mudline sample. The water depth measurement thought to be most representative is indicated on the boring log.

After completing the soil boring, a confirmatory water depth measurement is sometimes taken. The coring vessel is repositioned, and the water depth is confirmed. Some variations in these procedures may occur due to available equipment and conditions at the time of the boring.

In areas of significant tidal variations at the time of sounding and of the first sampling attempt should be recorded (fig. 16.60). Subsequently, the tide change can be observed by

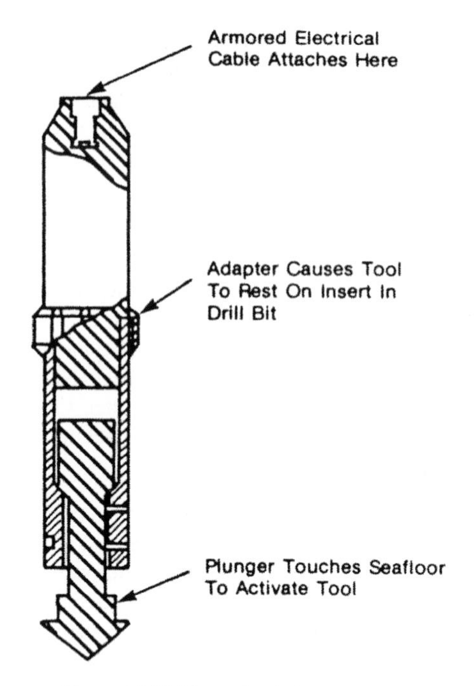

Figure 16.59 Water depth bottom sensor

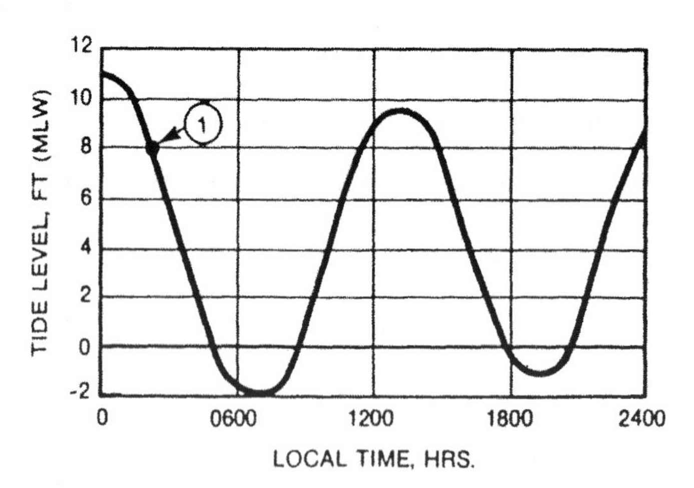

Figure 16.60 Tide change vs. time

intermittent operation of a suitable fathometer or a pressure measuring device sitting on the seafloor so that corrections can be made to drilling and sampling depths. Tidal records also provide a means of converting water depth to the desired reference datum.

16.11.3 Borehole Stability

In the open-hole drilling method, it is essential that mud having suitable viscosity and gel properties be used once a granular formation has been encountered. The mud stabilises the walls of the hole and prevents caving of the cohesionless soils. Very soft underconsolidated cohesive soils can also create a drilling problem unless weight material is added to the drilling fluid. Unless the weight of soil removed is counterbalanced by mud, these soils have a tendency to squeeze or flow into the drilled hole. Drilling mud that is too heavy can fracture very soft clay formations. The detrimental effects on sample quality are obvious.

16.11.4 Blowout Prevention

Shallow gas has been adopted many times in soil borings and has been responsible for blowouts and fires. Another gas hazard to a floating vessel is loss of support due to decreased buoyancy of the water. Artesian flow of water can also be experienced from offshore sediments. Since the conventional use of a blowout preventer is impossible with the open-hole drilling method, use of weighted drilling mud affords the only protection against blowout. In a typical soil boring operation, the mud weight may gradually be increased to about 11 lbs per gallon by a penetration of 400 ft. As a minimum, the completed boring should be left full of heavy mud; allowances should be made for pipe displacement as it is withdrawn. If obvious gas problems exist, or if regulations require, a boring should be cemented upon completion.

16.12 Industry Legislation, Regulations and Guidelines

The following are examples only and are by no means complete or updated:

- *Application for Consent to Drill Exploration, Appraisal and Development Wells*. DTI PON 4, May 1996. Requirement to investigate for shallow hazards and hydrogen sulphide.
- *Coast Protection Act, 1949 (UK) – Section 34(1)*. The consent of the Secretary of State for Transport) is required for the construction, alteration, or improvement of works on the seashore that may cause an obstruction or danger to navigation.
- *Conduct of mobile drilling rig site surveys, Vol 1*. UK Offshore Operators Association Ltd (Survey & Positioning Committee), London and Aberdeen, 1997.
- *Conduct of mobile drilling rig site surveys, Vol 2*. UK Offshore Operators Association Ltd (Survey & Positioning Committee), London and Aberdeen, 1997.
- *Continental Shelf Act, 1964 (UK) – Section 4(1)*. Extend the provisions of Part 2 of the Coast Protection Act to "any part of the seabed in the designated area".
- *Design, construction, operation and maintenance of offshore hydrocarbon pipelines*, 2nd edition. American Petroleum Institute, 1993.

- *Draft international standard for design and operation of subsea production systems, Part 1 – General requirements and recommendations.* International Standards Organization/DIS 13628-1, 1997.
- *Environmental Guidelines for Worldwide Geophysical Operations.* IAGC, January 1994.
- *Guidance notes on geotechnical investigations for marine pipelines*, Rev 3, Offshore Soil Investigation Forum, 1999.
- *Guidance notes on geotechnical investigations for subsea structures*, Rev 2, Offshore Soil Investigation Forum, 2000.
- *Guide to the Offshore Installations (Safety Case) Regulations 1992.* Guidance on Regulations, HSE, L30, 1992.
- *Guidelines and Recommended Practice for the Site Specific Assessment of Mobile Jack-up Rigs.* American Society of Naval Architects and Marine Engineers, May 1994.
- *Liaison with other bodies.* HSE, Offshore Operations Division. Operations Notice 3, February 1995.
- *Minerals Workings (Offshore Installations) Act, 1971* (UK). Provides for the safety, health and welfare of persons on installations.
- *New Guidance on the Coast Protection Act – Consent to Locate and the Marking of Offshore Installations.* HSE Offshore Safety Division (OSD), Operations Notice No. 14, February 1995.
- *NORSOK – Norsok Standard; Common Requirements, Marine Soil Investigations –* G-GR-001 – Rev. 1. May 1996.
- *Notification of Geophysical Surveys.* DTI PON 14, May 1996.
- *Offshore installations: Guidance on design, construction and certification.* Section 14, Site investigations. HSE (formally Department of Energy (DoE)). Fourth Edition (June 1990).
- *Offshore installations: Guidance on design, construction and certification.* Section 20, Foundations. HSE (formally DoE). Fourth Edition (June 1990).
- *Offshore Safety Act, 1992* (UK). Brought existing offshore legislation within the scope of the HSWA.
- *Petroleum Act (Production) (Seaward Areas) Regulations, 1988, No. 1213* (UK).
- *Petroleum and Submarine Pipelines Act, 1975 (UK)* – Schedule 2, Clause 17 provides for the consent to drill requirements as set out in the Department of Trade and Industry (DII) PON 4.
- *Pipeline subsea: design construction and installation*, British Standards Institute, BS8010 Part 3, 1993.
- *Recommended practice for design and operation of subsea production systems*, 2nd edition, American Petroleum Institute, RP17A, 1996.
- *Recommended practice for planning, designing and constructing fixed offshore platforms.* Recommended practice note 2A-WSD (RP 2A), 20th edition, American Petroleum Institute, Washington 1995.
- *Record and Sample Requirements for Surveys and Wells.* DTI PON 9, May 1996.
- *Rules for submarine pipeline systems*, Det Norske Veritas, 1996.
- *Rules for the design, construction and inspection of offshore structures.* Det Norske Veritas, Hovik, Norway 1988.

- *Site Specific Assessment of Mobile Jack-up Units,*. SNAME Technical and research Bulletin 5-5A, 1994.
- *The Health and Safety at Work Act*, 1974 (HSWA) (UK).
- *The Offshore Installations (Safety Case) Regulations, SI 1992, No. 2885*. Pertain to safe and proper operations including seabed soil and subsoil investigations, the meteorological and oceanographic conditions, the depth of water, properties of the seabed and subsoil.
- *The Offshore Installations and Wells (Design and Construction, etc.)* (UK) Regulations, SI 1996, No 913.
- *Technical Notes for the Conduct of Mobile Drilling Rig Site Surveys (Geophysical and Hydrographic)* – UK00A, 1990.

16.13 Laboratory Testing

This section is not meant to represent an exhaustive treatise on the complex and wide subject of soil testing and foundation design. There are many excellent textbooks on the subject, and the reader is encouraged to refer to such textbooks for additional and more detailed information. Rather, this section is meant to provide enough of an overview to the non-geotechnical person for him/her to understand and appreciate the subject, and be better equipped to interact with geotechnical engineers and consultants, ask the right questions and better understand their answers. It is recommended that a non-geotechnical engineer seek the advice and help of a geotechnical engineer when it comes to selecting laboratory tests, interpreting their results and designing offshore foundations.

16.13.1 General

Selective laboratory testing is performed offshore concurrently with the soil boring. The purpose of testing in the field is two fold: (1) provide the necessary information to decide at what depth to stop the boring (generally done by making sure the selected pile size(s) will provide the required capacity) and (2) test the samples at the earliest after retrieval, in order to minimise degradation of soil shear strength due to further stress relief.

An estimate of the undrained shear strength of cohesive samples may be obtained from simple devices such as the Torvane and a pocket penetrometer. Selected samples are tested in the field using the miniature vane and unconsolidated-undrained triaxial compression device. The moisture content and bulk unit weight are also determined offshore. Classification tests such as the grain size, the Atterberg limits and the carbonate content could be conducted offshore, but would require additional equipment and personnel and generally would not meet ASTM specifications. As such, they are almost always performed onshore.

Representative portions of each recovered sample are appropriately packaged in the field for shipment to the onshore laboratory. Onshore laboratory testing can be separated into two phases: additional conventional testing to supplement field testing and an advanced testing especially designed to evaluate soil parameters that are pertinent to the particular foundation elements under consideration.

The onshore *conventional* testing generally consists of further classification tests such as the Atterberg Limits, the grain size analyses (sieve, percent soil passing a number 200 sieve and hydrometer), the quantitative carbonate content (if applicable), and the additional moisture content tests. The primary goal of these tests is to further classify the soils encountered based on their physical properties. If necessary, additional undisturbed and remolded strength tests can be performed. These generally consist of the undisturbed and the remolded miniature vane and unconsolidated undrained (UU) triaxial tests. The primary aim of these tests is to better define the shear strength profile developed in the field or to replace or supplement suspicious field test results. The controlled environment of an onshore laboratory is more conducive to laboratory testing than its field counterpart.

Advanced laboratory testing is also carried out onshore when special considerations must be evaluated for a particular type of foundation element or environmental loading scenario. Advanced testing may include consolidation tests and cyclic or dynamic soil tests.

16.13.2 Conventional Laboratory Testing

Salinity Content: This test is performed to investigate the possible presence of high salt concentration within the soil samples. The amount of soluble salt is measured by extracting the pore fluids on selected sample recovered from the borings and using the procedure recommended in ASTM D-4542. This parameter helps ascertain if unusual geologic conditions (e.g. salt dome uplift) exist. This test is performed in an onshore laboratory.

Organic Matter Content Tests: Determination of the organic content of soils can be performed on selected samples using loss on ignition procedures recommended in ASTM D-2974. A crucible containing about 20 g of oven-dried soil is heated to 800°C. The organic matter content lost through ignition is then determined by the percentage loss in weight. The results of these tests are used as an aid to classify of the samples. This test is performed in an onshore laboratory.

Specific Gravity: Specific gravity (G_s) is defined as the unit weight of soil solids divided by the unit weight of water and using the procedures recommended in ASTM D-0854. This parameter is used in various calculations. This test is performed in an onshore laboratory.

Grain-Size Distribution: The grain-size distribution of the granular and cohesive samples is determined by sieve and hydrometer analyses, respectively. The grain-size distribution test results are used as an aid to classification of the samples. This test has rarely been performed offshore, and is routinely performed in an onshore laboratory using the procedures recommended in ASTM D-0422.

Classification Tests: Plastic and liquid limits, collectively termed the Atterberg limits, can be determined for cohesive samples to provide classification information. These tests are performed onshore and using the procedures recommended in ASTM D-4318. Natural water content determinations can be made for triaxial compression and miniature vane test specimens. The total unit weight for cohesive samples, including each triaxial compression test specimen, can be measured in the field by weighing a sample of known volume immediately after extrusion (fig. 16.61). These last two tests are routinely performed both offshore and onshore.

Figure 16.61 Total unit weight

Degree of Saturation: During any deepwater geotechnical investigation, a major source of sample disturbance is due to stress relief associated with removing a soil sample from a great depth below the sea level and bringing it to the surface. Due to this stress relief (i.e. reduction in hydrostatic pressure), there is a tendency for the sample to swell or expand because of gas coming out of solution from the pore fluids. The magnitude of negative pressure or suction that develops in the pore water, due to stress relief of the sample, determines the degree of sample swelling or expansion. An indication or measure of the degree of disturbance due to stress relief is provided by the degree of saturation of the soil sample [Whelan, 1979]. The sample expansion results in a reduction in the degree of saturation in the samples and a decrease in the measured unit weight and soil shear strength. The degree of saturation (S_r) can be calculated by the following equation:

$$S_r = \frac{\gamma_t - [\gamma_t/(1+w)]}{\gamma_{w\,\text{fresh}}[1 - (\gamma_t/(\gamma_{w\,\text{fresh}} \times G_s \times (1+w)))]}$$

where:

γ_t = measured total unit weight;
$\gamma_{w\,\text{fresh}}$ = unit weight of fresh water;
w = natural water content, decimal; and
G_s = specific gravity adopted from laboratory test results.

For simplicity, the above equation considers the salt particles in seawater as solids within the total volume of soil sample. This assumption is conservative, since correcting the computed degree of saturation for salt content in the pore fluid (seawater) would lead to slightly higher values. This determination can be performed offshore using an assumed value for G_s.

Submerged Unit Weight: During the field phase of the investigation, the total unit weights are measured on the soil samples recovered from the borings. The density of seawater is subtracted from the measured total unit weight to obtain an estimate of the submerged unit weight of the sample. As discussed earlier, sample expansion results in a reduction in the degree of saturation and a decrease in the measured total unit weight of the samples. To further investigate the effect of sample expansion on the measured unit weights, the submerged unit weight is computed from natural water content and specific gravity data, with the assumption that the soils are 100% saturated in situ. The submerged unit weights are computed using the following equation:

$$\gamma' = \gamma_{w\ \mathrm{fresh}}\left[\frac{G_s(1+w)}{1+wG_s}\right] - \gamma_{w\ \mathrm{sea}}$$

where:

γ' = theoretical submerged unit weight;
$\gamma_{w\ \mathrm{fresh}}$ = unit weight of fresh water;
$\gamma_{w\ \mathrm{sea}}$ = unit weight of sea water;
w = natural water content, decimal; and
G_s = specific gravity.

Additional submerged unit weight determinations are performed in the onshore laboratory.

Calcium Carbonate Content Test: Selected soil specimens can be tested for solubility in diluted hydrochloric acid solution (10% concentration) using the gasometric method, which approximates the quantity of carbonate material, by weight, in the test specimen. In this method, 3 g of dried soil is treated with 25 g of diluted hydrochloric acid in an enclosed reactor vessel. Carbon dioxide gas is emitted during the reaction between the acid and carbonate fraction of the test specimen. A pressure gauge, attached to the reactor, is pre-calibrated with reagent grade calcium carbonate to provide a direct measurement of the carbonate content. This test is performed in an onshore laboratory using the procedures recommended in ASTM D-4373.

Torvane: The Torvane (fig. 16.62) is a small, hand-operated device consisting of a metal disc with thin, radial vanes projecting from one face. The disc is pressed against a flat surface of the soil until the vanes are fully embedded and is rotated through a torsion spring until the soil is sheared. The device is calibrated to indicate the shear strength of the soil directly from the rotation of the torsion spring. Such tests are performed routinely in offshore and onshore laboratories.

Pocket Penetrometer: The pocket penetrometer (fig. 16.63) is a small, hand-held device consisting of a flat-faced cylindrical plunger and a spring encased in a cylindrical housing. The plunger is pressed against a flat soil surface, compressing the spring until the soil experiences a punching type bearing failure. The penetrometer is calibrated to indicate the shear strength of the soil directly from compression of the spring. Such tests are performed routinely both offshore and onshore.

Miniature Vane: The miniature vane test (fig. 16.64) is used to measure the undrained shear strength of cohesive soils. In this test, a small, 4 bladed vane is inserted into either an undisturbed or a remolded cohesive specimen. Torque is applied to the vane through a calibrated spring until soil shear failure occurs. The undrained shear strength is determined by multiplying the rotation, in degrees, by the spring constant. Such tests are performed routinely both offshore and onshore using the procedures recommended in ASTM D-4648-94.

Figure 16.62 Torvane test

Figure 16.63 Pocket penetrometer

Unconsolidated-Undrained Triaxial Compression: In the unconsolidated-undrained (UU) triaxial compression tests (fig. 16.65), either an undisturbed or a remolded soil specimen is enclosed in a thin rubber membrane and subjected to a confining pressure at least equal to the computed effective overburden pressure. The specimen is then

Figure 16.64 Miniature vane

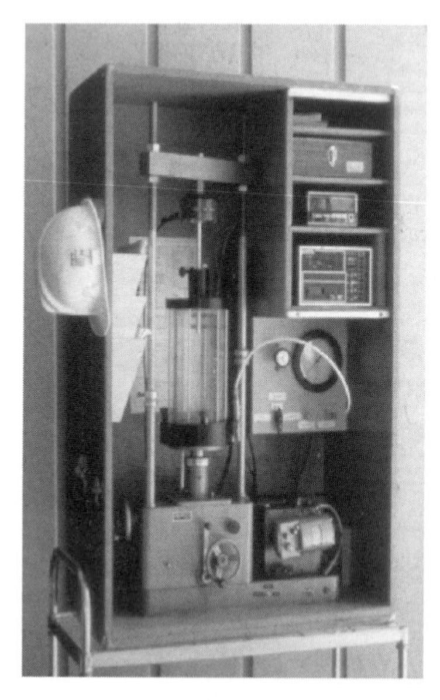

Figure 16.65 UU triaxial testing package

axially loaded to failure at a nearly constant-rate-of-strain without allowing drainage. The undrained shear strength of cohesive soils is computed as one-half the maximum observed deviator stress. The undrained shear strengths from the UU tests are the main basis for estimating a shear strength profile, provided advanced testing is not performed.

Additionally, the strain at 50% of ultimate load (deviator stress) is used as an input parameter for lateral load (*p-y*) pile response. Such tests are performed routinely both offshore and onshore using the procedures recommended in ASTM D-2850.

16.13.3 Advanced Laboratory Testing

Advanced laboratory testing is done exclusively onshore. Preferably advanced laboratory testing is performed on the test specimens primarily selected from the "saved tube" samples. A "saved tube" is a thin-walled tube sample that was *not* processed onboard the geotechnical survey vessel and from which soil was not extruded. The soil remains in the tube until extruded for testing in the laboratory, hence the name "saved tube". The "saved tube" samples or portions thereof remains refrigerated during storage. The "saved tube" samples are better preserved and less disturbed compared to other types of soil samples.

It is recommended that all of the "saved tubes" be X-rayed to facilitate and enhance the sample/specimen selection and processing. The procedure is recommended to follow the recommendations in the ASTM Test Methods D 4452 – 85 (1995). An X-ray radiography provides a qualitative measure of the internal structure of the samples by showing varying shades of gray resulting from variations in the ability of X-rays to penetrate matter. These varying shades of gray enable the evaluation or determination of the following:

- sample quality as noted by signs of voids, drilling wash, separations in the soil caused by gas expansion, unusual changes in bedding planes or layering, etc.;
- presence of inclusions in the sample, such as shells and/or calcareous nodules; and
- presence of naturally occurring fissures, shear planes, bedding planes, voids, layering, gravel and silts seams.

The X-ray radiographs are used to identify anomalies that might affect the test results, quantify the amount of testable soil and select specific sections of the samples for testing.

The selected portions of the saved tubes are then cut into segments with a mechanical hacksaw. A wire saw is later used to separate the soil from the surrounding tube in an effort to reduce potential disturbance upon extrusion. An advanced laboratory testing is then carried out on the selected soil samples.

The purpose of the advanced laboratory testing program is to determine selected index, drained and undrained engineering properties of the samples, and stress history of the soil deposits. These tests are further described below.

Incremental Load (INC) Consolidation Test: A consolidation test specimen is trimmed into a 2.5 in. i.d. by 0.75 in. high stainless steel ring. The testing is performed in general accordance with ASTM D2435-90. Deformation data are recorded and plotted using an automated data-acquisition system, and are corrected for the deformation of apparatus, filter stones, filter paper, etc. (where applicable).

The total load is doubled during each load increment, with the maximum applied stress ranging between 0.02 and 8.0 ksf. The duration of each loading increment is equal to the time needed to reach 90% primary consolidation, as defined by Taylor's square root of time fitting method plus one to two hours (i.e. t_{90} + 1–2 h), or overnight. For increments left overnight, the stress, strain, etc. data are presented at t_{90} + 1–2 h and upon completion of

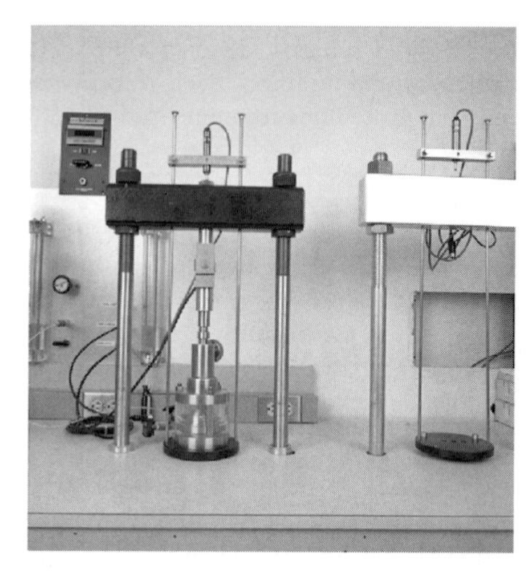

Figure 16.66 CRS consolidation test

the increment (two points). The laboratory measured maximum past stress (σ'_{vm}) applied to the test specimens can be interpreted from the consolidation curves using Casagrande's (1936) procedures and the Work Per Unit Volume method [Becker, et al 1987].

1-D Controlled-Strain Loading (CRS): To determine the soil deposit's stress history (pre-consolidation stress) and the compressibility characteristics versus depth, one-dimensional (1-D) consolidation (CRS) tests (fig. 16.66) are performed on soil samples at different depth intervals. The one-dimensional (1-D) consolidation tests with loading at a constant rate-of-strain (CRS) is performed in general accordance with ASTM Test Method D 4186–89 (1998), using an updated consolidometer and testing methodology. The consolidation test specimens are trimmed into a stainless steel ring and placed in a specially made cell. The set-up of the test specimen into the cell is performed with the entire cell under water so that there is no air trapped in the system that would affect the pore pressure response during loading. After the cell has been fully assembled, it is placed in a loading frame where the test specimen is loaded in increments that usually double the previous load. The data readings are used to compute the vertical strain response of the specimen under vertical pressure.

Static Strength Tests: The key assumption in the implementation of this portion of the testing program is that the concepts of SHANSEP (*S*tress *H*istory *A*nd *N*ormalised *S*oil *E*ngineering *P*roperties), as presented by Ladd, C.C. (1991) could be used to mitigate the effects of sampling and specimen preparation disturbance and enhance the presentation and the application of the test results.

Typically, the first part of the SHANSEP methodology calls for the consolidation of each test specimen to an induced overconsolidation ratio (OCR) of either one or greater and a uniform amount of secondary consolidation. An induced OCR of 1 is obtained by consolidating the specimen well into the virgin consolidation region (i.e. stress level greater than the preconsolidation stress (σ'_p) by a factor of about 1.5–3.0 or an axial strain of at least about 12% for plastic clays). Consolidation is usually accomplished under K_o or anisotropic conditions through the application of staged stress increments. In the second part, the SHANSEP methodology calls for the normalisation of the undrained stress-strain and shear strength parameters (e.g. $c_u/\sigma'_{v,c}$).

Static Direct Simple Shear Tests: To determine the direct simple shear (CK_oU'-DSS) strength characteristics versus depth; static K_o consolidated-undrained (constant volume) DSS with strain controlled loading tests can be performed using the procedures recommended in ASTM D-6528-00 and an apparatus as described by Doroudian and Vucetic (1995).

Each test specimen has a diameter of 2.60 in. (66 mm) and height of about 0.75 in. (19 mm) and top and bottom drainage boundaries during consolidation and shearing. The volume of the test specimens are kept constant during shearing by keeping the specimen's height constant. As a result, undrained conditions (no volume change) are maintained during shearing. Therefore, it can be assumed that the change in vertical stress is equivalent to the change in pore water pressure (PWP).

Each specimen is incrementally consolidated to an induced OCR ≥ 1, with this final effective-vertical stress ($\sigma'_{v,c}$) maintained constant for about 24 h (curing or simulated aging) or one log cycle of time past the time to reach 90% consolidation (t_{90}). The samples are then sheared at a strain rate of about 5%/h.

Static Direct Simple Shear Creep Tests: To investigate the effects of the magnitude and duration of shear stress application on the resulting shear strain behaviour, a series of static direct simple shear creep (CkoU′-DSS-CR) tests can be performed on trimmed specimens 2.62 in. (66.55 mm) in diameter and about 0.71 in. (18 mm) in height at different depth below the mudline. They are laterally confined by a wire-reinforced rubber membrane, without a water bath.

The undrained-creep shear stress is applied in 10–15 increments using a dead weight hanger system (cable and pulley type). The normalised creep stress ($\tau_{h,cr}/\sigma'_{v,c}$) applied in each test is a percentage of the normalised undrained shear strength of the soil ($c_u/\sigma'_{v,c}$).

The applied creep stress ($\tau_{h,cr}$) in the first test of each test series is typically 95% of the estimated undrained shear strength (c_u). If rupture occurs, then the $\tau_{h,cr}$ in the next test will be reduced to about 90–93% of the c_u. The percentage of c_u in the successive tests will be adjusted to obtain appropriate test results. However, if rupture does not occur, then the $\tau_{h,cr}$ in the next test will be increased to about 97–100% of c_u.

The stress-controlled increments of the undrained shear stress are applied every 10 minutes until the final creep stress is reached ($\tau_{h,cr}$). This creep stress is maintained constant until one of the following: (a) a rupture occurs, (b) a certain time period has passed (about one to three days), or (c) it is evident that rupture will not occur.

TruePath™ K_o Triaxial Tests: TruePath™ K_o consolidated-undrained triaxial test with shearing in either compression or extension can be performed, either as a single test or as part of a series of tests from a sample(s) at an assigned depth(s). If a single test is performed, then it is a compression test with an induced OCR = 1. If two tests are performed in a series, then the series consist of a compression test and an extension test, with both tests having an OCR value equal to unity.

Each test is performed using an automated system (TruePath™) developed by the Fugro Group, the Trautwein and Germaine of MIT. The test procedures follow the technical requirements of the ASTM Test Method D 4767-95 except for: (a) the TruePath™ K_o consolidation, (b) some minor calculation methodologies (volume of specimen before shearing, membrane correction and area correction during shearing) and (c) shearing in extension.

Each specimen (2 × 4 in. or 51 × 100 mm) has a top, bottom and radial drainage boundaries during consolidation. The radial drainage is provided by spirally oriented 0.25 in. (6 mm) wide, Whatman No. 1 filter strips placed at about 0.25 in. (6 mm) spacing.

Each specimen is prepared and mounted in the triaxial testing apparatus. Specimen saturation is achieved through back pressuring at, either an effective isotropic confining stress of 3–7 psi, or a stress which prevents swelling, whichever is smaller. Using the SHANSEP methodology, the specimen is K_o consolidated in a drained state at a controlled rate of strain of about 0.1% /h. Upon reaching the assigned effective-vertical stress ($\sigma'_{v,c}$) or an axial strain of at least 15% the applied stress is maintained constant for a curing period of about 24 h (simulated ageing).

During a shear phase, the chamber pressure is kept constant and specimen drainage is not permitted. An axial loading piston is advanced into (shearing in compression), or retracted from (shearing in extension) the cell at a specific rate-of-strain. The applied rate-of-strain is slow enough (about 0.5%/h) to produce approximate equalisation of excess-pore-water pressures (PWP) throughout the specimen at failure. The static stresses and the PWPs are used to express the measured stress parameters in terms of effective stresses.

Cyclic Strength Tests: The behaviour of clay under a cyclic loading is an important consideration for determining the shear strength degradation and soil damping characteristics for offshore structure foundation design. Wave action on the structure causes the vertical and the horizontal cyclic forces and moments, which degrade soil strength and reduce foundation resistance. To investigate soil response due to cyclic loading, a test program comprises a series of different types of cyclic simple shear and cyclic triaxial tests are performed to determine the shear strength, cyclic degradation, cyclic displacement (causing possible settlement induced from cyclic loading) and the soil damping characteristics. Depending upon the magnitude of applied loads, large strain direct simple shear and small strain resonant column tests can be performed. A combination of cyclic DSS tests and cyclic triaxial tests can be performed to obtain cyclic strain characteristics for "Failure Interaction Diagrams (FID)". Details of the relevant test procedures are explained here.

Cyclic Direct Simple Shear Tests: To determine the stress–strain characteristics of soils under cyclic shear loading; i.e. the threshold cyclic strength and the shape of the $N_{f\text{-isolines}}$ in a "Failure Interaction Diagram", a series of cyclic DSS stress-controlled strength tests (fig. 16.67) can be performed on trimmed specimens 2.62 in. (66.55 mm) in diameter and about 0.71 in. (18 mm) in height at different depths below mudline. The samples are laterally confined by a wire-reinforced rubber membrane, without a water bath.

Upon completion of consolidation, each test specimen is loaded cyclically using an electro-hydraulic closed-loop loading system. The specimens are maintained in an undrained (no volume change) state during cyclic loading. A data acquisition system will be used to collect the data during cyclic loading. The data acquisition system can be programmed to collect, process, store data files and display selected data during a cyclic loading.

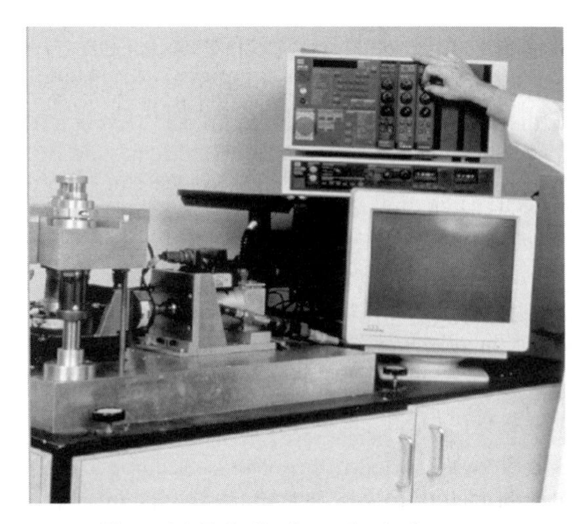

Figure 16.67 Cyclic direct simple shear test

Two types of cyclic strength tests are usually conducted, those with or without an undrained bias shear stress ($\tau_{hu,avg}$) applied during cyclic loading. During cyclic loading, the bias stress becomes the average cyclic shear stress. In the test series without a $\tau_{hu,avg}$, the electro-hydraulic closed-loop loading system is programmed to apply a sinusoidally varying shear stress ($\pm\tau$) at a certain frequency. This method of cyclic loading is typically referred to as "stress-controlled with no bias shear stress". In the test series with a $\tau_{hu,avg}$, the loading system is programmed in the same manner, except the sinusoidal-wave form had an offset of $\tau_{hu,avg}$ from zero. This method of cyclic loading is typically referred to as "stress-controlled with undrained bias shear stress". The $\tau_{hu,avg}$ is applied just prior to cyclic loading by gradually increasing the applied horizontal shear stress using the loading system.

Cyclic loading continues until failure occurs or after approximately 10,000 cycles have been applied. Failure can be readily defined if it occurs within 10,000 cycles; otherwise, it is defined by data extrapolation. The failure criterion established by Malek et al (1987) is used in these tests. In this criterion, failure occurs when the peak positive cyclic shear stress divided by the minimum effective-vertical stress within a given cycle corresponds to the peak effective stress ratio (Ψ'_{DSS}), as determined by the companion static test or an interpolated value.

Anisotropically Consolidated Cyclic Triaxial Tests: To determine the cyclic stress-strain characteristics under vertical loading, a series of cyclic triaxial tests can be performed on samples at different depths below mudline. The specimens are anisotropically consolidated (OCR = 1) and cyclically sheared in compression. Alternatively, the specimen are anisotropically consolidated (OCR = 1) and cyclically sheared in extension. In the anisotropically consolidated-undrained stress-controlled cyclic triaxial tests, each specimen (2×4.5 in., or 51×114 mm) has a bottom, top and radial drainage boundaries during consolidation. The radial drainage is provided by spirally oriented 0.25 in. (6 mm) wide Whatman No. 1 filter strips placed at about 0.25 in. (6 mm) spacing.

Specimen saturation is achieved through backpressuring. Specimens are then anisotropically consolidated (using a *K*-value obtained from static triaxial tests) by applying increments in the vertical and the radial stresses. The specimens are allowed to cure at the prescribed consolidation stresses for about 24 h, prior to cyclic loading.

The loading system that is used in the undrained triaxial tests can be used in the cyclic triaxial tests too. The chamber pressure is kept constant and specimen drainage is not permitted during cyclic loading. The loading system can be programmed to apply a sinusoidally varying cyclic shear stress above the undrained ambient shear stress at a certain frequency. Cyclic loading continues until the average or cyclic axial strain reaches a value of about 15%, or a certain time period has passes.

Thixotropic Characteristics Test: Thixotropic characteristics of the soils can be obtained by conducting a series of Miniature Vane Shear tests with six tests per series. The Miniature Vane Shear tests are performed in general accordance with ASTM designation D 4648-94. Portions of "saved tube" samples, sufficient to fill six cylindrical containers 2 in. (50.8 mm) in diameter and 4 in. (101.6 mm) in height, are selected using the X-ray radiographs. Then, the whole sample is thoroughly mixed to break the soil fabric and remolded into the six cylindrical containers. Miniature Vane Shear test on the first specimen is conducted immediately after remolding to obtain remolded shear strength (S_R). The remaining five specimens are sealed immediately to prevent escape of moisture and stored in the laboratory walk-in cooler room. Samples are marked with preparation time. Remolded shear strengths (S_R) of the specimens are calculated according to the ASTM standard designation 4648-94. The miniature Vane Shear test on the remaining specimens are performed at curing times of 2, 8, 24, 168 and 720 h. The aged shear strengths (S_A) of specimens and thixotropic strength ratios (S_A/S_R) are calculated using the same method used to calculate S_R.

Drained Triaxial Tests:

Resonant Column Tests: Resonant column tests (see fig. 16.68) can be performed to characterise the shear modulus and material damping ratio of the soils at small shear

Figure 16.68 Resonant column test

strains (10^{-5}–10^{-1}%). Resonant column tests are performed on solid, cylindrical-shaped, soil specimen approximately 2.1 in. length. Each test specimen is back-pressure saturated to 20 psi (138 kPa) and then isotropically consolidated to three successive effective confining pressures (three-stage testing) equal to about 1.0, 2.0 and 4.0 times the mean effective in situ overburden pressure. During consolidation at each effective confining pressure, the variation of the shear modulus (G) and the material damping ratio (ξ) with time at low-amplitude shear strains (shear strains less than 10^{-4}%) are measured in a sequence similar to a consolidation test; i.e. 0.5, 1, 2, 4, 8, 15 and 30 min, etc., after the effective confining pressure is applied. The readings continue until approximately one log cycle of time or 24 h, whichever is less, has expired since the end of primary consolidation (T_{100}) to measure a value of maximum shear modulus (G_{max}), which includes some time effects or increase in shear modulus with time after primary consolidation [Anderson and Stokoe, 1978]. These low-amplitude shear moduli are calculated from the data obtained when the frequency of the soil-oscillator system has been adjusted to vibrate at the undamped first mode frequency. A sinusoidal waveform is used to apply torque to the top of the specimen.

Following the low-amplitude testing at each stage, the shear modulus (G) and the material damping ratio (ξ) at high-amplitude shear strains (shear strains greater than 10^{-4}%) are measured with drainage lines closed. At the first and the second stages in a three-stage test, the strain level is increased in steps until the measured G value decreases to between 80 and 90% of G_{max}. At the last stage, the strain level is increased in steps to the limits of the device or oscillator instability, whichever occurs first.

Between each high-amplitude step, a low-amplitude strain is applied, and low-amplitude shear modulus and material damping ratio are measured. After the last high-amplitude step, the specimen is allowed to drain until the measured low-amplitude shear modulus values approach the value measured just before the high-amplitude sequence starts. A failure in regaining this modulus may be indicative of specimen degradation.

Static and Rapid Direct Simple Shear Tests: The strain-rate effect on the static undrained shear strength of cohesive soils can be investigated by performing the "static" DSS tests (at a shear strain rate of 5% per hour) and the "rapid" DSS tests (at shear strain rates of 100 and 2000% per hour). Static and rapid direct simple shear (DSS) tests are performed on clay specimens of 0.7 in. (18.60 mm) in height, trimmed to 2.5 in. (63.5 mm) in diameter. Each specimen has its top and bottom drainage boundaries during consolidation and is confined laterally by a series of polished steel rings. Each specimen is consolidated in increments similar to an incremental consolidation test (i.e. the total load is doubled during each load increment) to the final effective vertical consolidation pressure (σ'_{vc}), which is maintained constant for about one log cycle of time or 24 h, whichever is less, past the end of primary consolidation (T_{100}). For a normally consolidated clay specimen, the final effective vertical consolidation stress (σ'_{vc}) is about 2.0 times the estimated in situ effective vertical stress (σ'_{vo}) to ensure that a normally consolidated state of stress (OCR = 1) is induced in the sample. For an overconsolidated clay sample, the specimen is consolidated to at least a vertical strain of 10% to ensure that an OCR = 1.0 is induced in the sample (consolidation pressure greater than the past consolidation pressure), and then unloaded to an effective vertical consolidation stress so as to induce a specified overconsolidated state of stress. The specimen is then loaded to failure at a constant shear strain rate of about 5% per hour. During loading, the specimen is maintained in an undrained (no volume change) state by keeping the height of the specimen constant. In addition to the 5% per hour strain rate, the DSS tests at shear strain rates of about 100 and 2000% per hour are performed to study the effect of the strain rate on the shear strength.

The vertical and horizontal loads on the sample are measured with load cells and horizontal deformations are measured by an LVDT. During shear, the sample volume is held constant by locking the vertical loading piston in place to prevent further change in height. The change in the vertical stress during shear is used to estimate pore pressure changes in the sample. The maximum shear stress recorded during the test is used to evaluate the in situ undrained shear strength of the soil.

Hydraulic Conductivity (Permeability) Tests: Permeability tests can be performed in accordance with the technical requirements of ASTM Test method D 5084-00 and Method F. Each test series comprises three tests on the same soil/material in a given sequence depending on how the specimens are prepared.

The extruded specimen is trimmed into 6.4 cm (2.5 in.) diameter with a height of about 5–6.4 cm (2–2.5 in.) and placed in a flexible-wall permeameter (triaxial cell). The orientation of the specimen and permeation at the first and the second stages determine the vertical and horizontal permeabilities, respectively. For the third stage the specimen is thoroughly remolded. In each test, permeability determinations are done at two effective isotropic consolidation stress levels representing approximately one-third and two-thirds of the approximate in situ effective-vertical stress ($\sigma'_{v,o}$). Specimen saturation is achieved by back pressuring at either an effective isotropic-confining stress of 34 kPa (5 psi), or a stress which prevents swelling, whichever is smaller.

The test specimen is then isotropically consolidated in increments to the first level of effective isotropic-consolidation stress. Permeation is accomplished using a falling-head constant-volume hydraulic system (e.g. the Trautwein Permeameter and the Permometer). This constant-volume system ensures the continuity of inflow and outflow of the permeant during each permeability measurement.

Upon completion of the vertical permeability measurements, with the drainage lines closed and the cell pressure and backpressure removed, the triaxial cell is disassembled and the test specimen removed. A 3.8–5.0 cm (1.5–2.0 in.) diameter by 5.0 cm (2 in.) high specimen is then trimmed from the original specimen so that the central axis of the new specimen is perpendicular to that of the old specimen.

The newly trimmed specimen is then tested, as discussed above, to determine the horizontal permeability at the same two levels of isotropic consolidation stress. Upon completion of the horizontal permeability measurements, the specimen and all of the earlier saved trimmings are thoroughly remolded in a rubber membrane with a diameter slightly greater than 5.0 cm (2 in.) The remolded specimen is then removed from the membrane and placed in an expanded split-cylinder mold measuring 5.0 cm i.d. (2 in. i.d.) × 5.0 cm (2 in.) in height, clamped and trimmed flush at the ends. The specimen is then tested in a manner consistent with the earlier vertical and horizontal permeability tests.

16.14 Offshore Foundation Design

16.14.1 Pile Design

In shallow water, the steel jacket structure is the most common offshore platform (fig. 16.69). Steel pipe piles are the typical foundation for offshore platforms. The design of offshore piles is different from that of onshore piles because of the loading types, magnitudes and soil conditions. For example, axial loads on offshore piles are generally an order of magnitude greater that those on onshore piles, and offshore piles are subjected to substantial lateral loads (fig. 16.70). A significant proportion of the axial and lateral loads on offshore piles are cyclic in nature. The nature of some offshore soils (e.g. carbonate sands and silts) may give rise to unusual and unexpected behaviours. Therefore, the criteria for design of offshore piles and pile groups include adequate axial and lateral capacities, acceptable load-deformation response and feasibility of installation of the piles (e.g. ensuring that available equipment can drive the piles to the design penetration).

Two major types of offshore piles are encountered frequently: driven piles and bored piles. The driven piles include the open-ended driven steel pipe piles (most common), and the precast concrete piles. The bored piles include grouted piles (fig. 16.71), belled piles (fig. 16.71) and driven primary and straight bored piles.

In determining the number, diameter, length and arrangement of the piles, various analyses are usually required to address the following design issues: axial load capacity, axial deformation, lateral load capacity, lateral deformation, drivability of piles (for driven piles) and dynamic response.

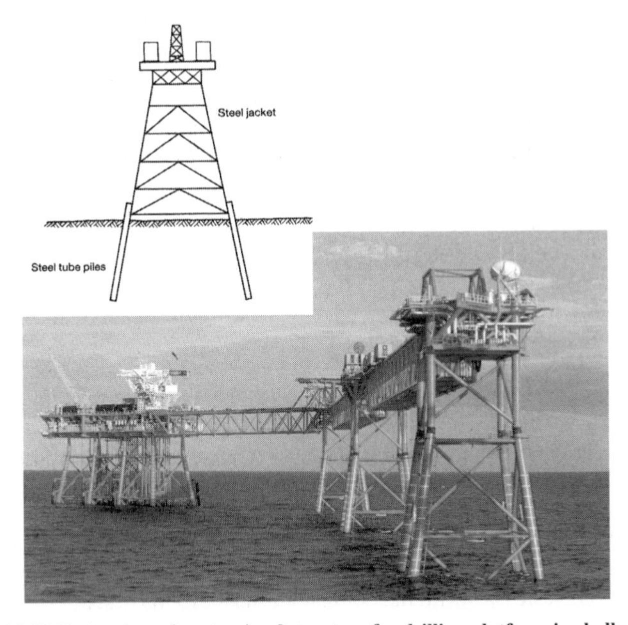

Figure 16.69 Jacket (template-type) substructure for drilling platform in shallow water

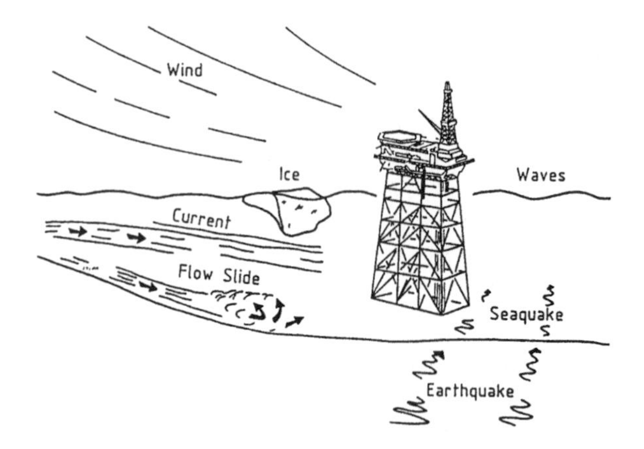

Figure 16.70 Environmental forces on offshore structures

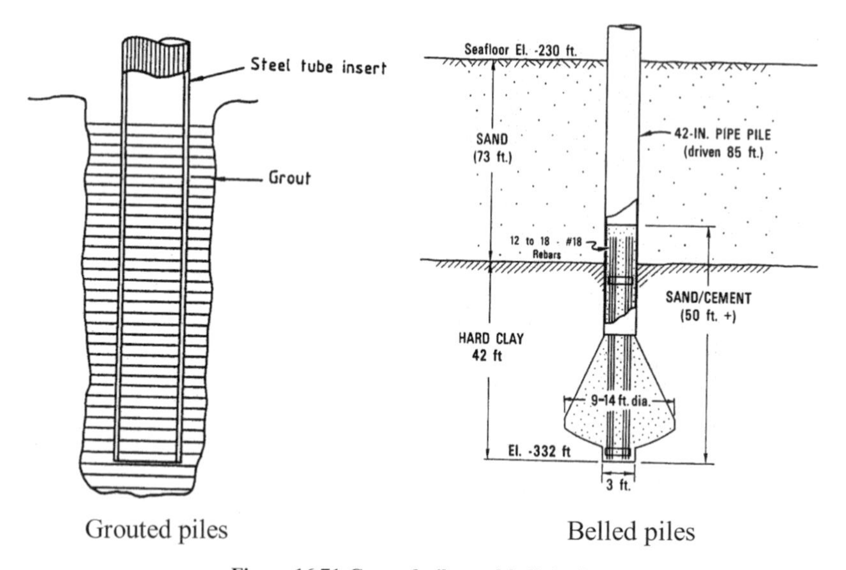

Figure 16.71 Grouted piles and belled piles

16.14.2 Axial Pile Capacity

The following sections will discuss different important aspects of axial pile capacity design for different types of pile foundations, such as, soil properties, procedures to compute axial pile capacity, factor of safety, pile-soil setup, pile group effects and axial pile capacity in carbonate soils. Since the design of offshore foundations generally follows recommendations from various recommended Practice documents published by the American Petroleum Institute (API), those recommendations will be mostly referenced in the following sections.

16.14.2.1 Subsurface Soil Profile and Engineering Characteristics

Usually, a geotechnical investigation boring is drilled at the proposed platform site to gather soil samples and perform in situ tests. If the subsoil profile is very complicated and varies significantly within the spread of the proposed platform site, geophysical and geotechnical surveys may be required to provide a more accurate site characterisation for the proposed platform site. Based on the site characterisation, important cohesionless soil properties are obtained for selecting pile design parameters (fig. 16.72). For cohesive soils, the important soil parameters include the undrained shear strength, unit weight, plasticity and liquidity indices, etc. For granular soils, the important soil parameters include relative density, angle of internal friction, unit weight, gradation and carbonate content. For example, different design parameters for cohesionless siliceous soil (fig. 16.73) are recommended in the API Recommended Practice [API RP 2A, 2000].

Based on the recommended design strength and deformation parameters, the ultimate axial pile capacity curves for different types of piles can be derived accordingly to determine the required penetration of the piles, so as to ensure an adequate safety factor against axial failure.

16.14.2.2 Axial Pile Capacity of Driven Pipe Piles

Ultimate Axial Compressive and Tensile Capacity: The static method of analysis for determining the ultimate compressive capacity of driven pipe piles (open-ended or unplugged

Density	Soil Description	Soil-Pile Friction Angle, δ Degrees	Limiting Skin Friction Values kips/ft^2 (kPa)	N_q	Limiting Unit End Bearing Values kips/ft^2 (MPa)
Very Loose Loose Medium	Sand Sand-Silt* Silt	15	1.0 (47.8)	8	40 (1.9)
Loose Medium Dense	Sand Sand-Silt* Silt	20	1.4 (67.0)	12	60 (2.9)
Medium Dense	Sand Sand-Silt*	25	1.7 (81.3)	20	100 (4.8)
Dense Very Dense	Sand Sand-Silt*	30	2.0 (95.7)	40	200 (9.6)
Dense Very Dense	Gravel Sand	35	2.4 (114.8)	50	250 (12.0)

* Sand-Silt includes those soils with significant fractions of both sand and silt. Strength values generally increase with increasing sand fractions and decrease with increasing silt fractions.

Note: API RP 2A notes that the parameters listed above are intended as guidelines only. Where detailed information, such as in situ cone tests, strength tests on high quality samples, model tests, or pile driving performance is available, other values may be justified.

Figure 16.72 Recommended design parameters for cohesionless siliceous soil [API RP 2A, 2000, reproduced with permission from API]

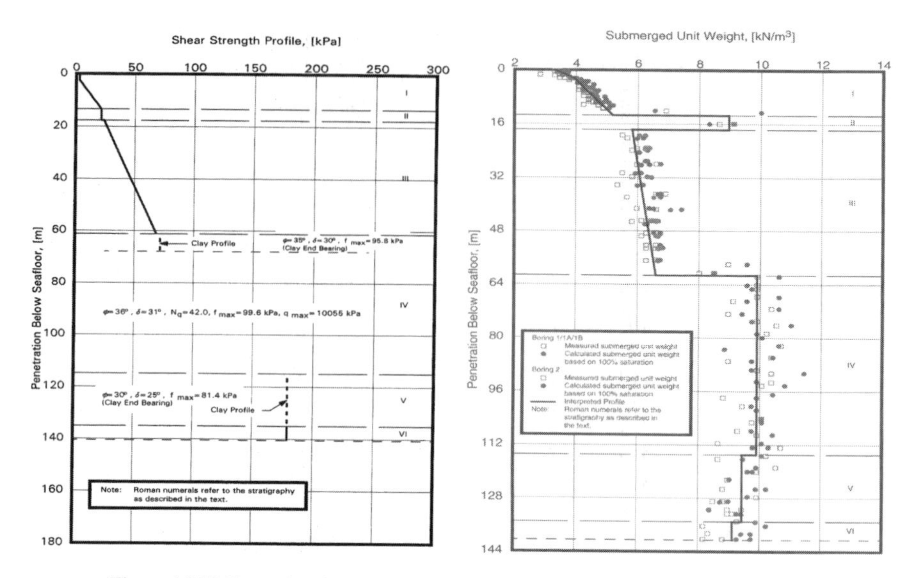

Figure 16.73 Example of design strength and submerged unit weight parameters

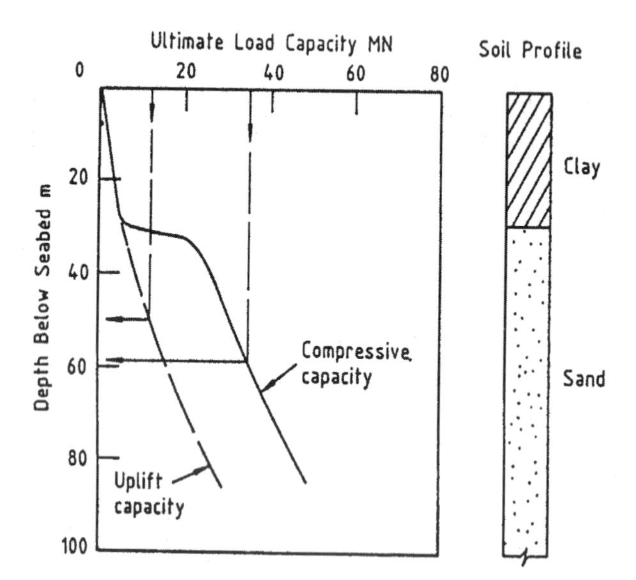

Figure 16.74 Determine the pile penetration with an appropriate FOS

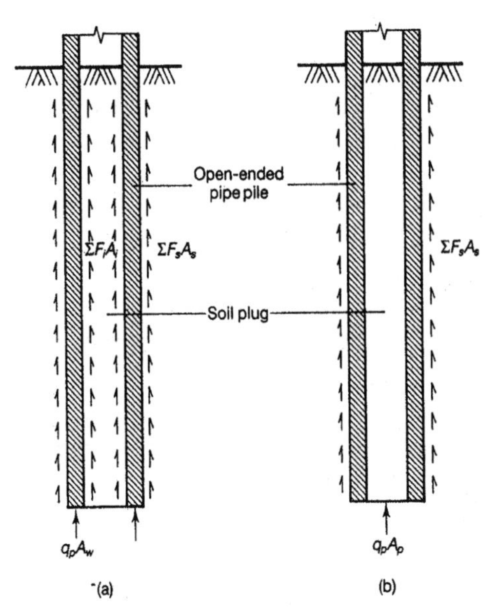

Figure 16.75 Ultimate compressive capacity model of driven pipe piles: (a) open ended, (b) closed ended

and closed-ended or plugged, fig. 16.75) through the following equations:

$$Q_u = \sum f_i A_i + \sum f_s A_s + q_p A_w - W$$

(For an open-ended or unplugged pile)

$$Q_u = \sum f_s A_s + q_p A_p - W$$

(For an closed-ended or plugged pile)

where,

Q_u = ultimate compressive axial capacity
Q_s = frictional resistance between the soil and the surface of the pile
Q_p = end bearing on the pile tip
W = weight of the pile and soil plug
f_i, f_s = inner and outer unit skin friction
A_i, A_s = inner and outer pile shaft area
q_p = unit pile point end bearing
A_w = cross-sectional area of pile wall $= (\pi\, d_o^2)/4 - \left(\pi(d_o - 2t)^2\right)/4$
A_p = pile gross end-bearing area $= (\pi\, d_o^2)/4$

d_o = outer pile diameter
t = pile wall thickness

For open-ended pipe piles, the following conditions are recommended to determine whether the pile is unplugged or plugged:

$\sum f_i A_i > q_p A'_p$ Pile will be plugged (occurs in relatively low end-bearing soils, such as clays and silts)

$\sum f_i A_i < q_p A'_p$ Pile will be unplugged (occurs in sands).

where,

f_i = inner unit skin friction
A_i = inner pile shaft area
q_p = unit pile point end bearing
A'_p = end-bearing area of pile plug = $(\pi\, d_i^2)/4$
d_i = inner pile diameter

For the driven pipe piles subjected to tensile load, the ultimate tensile capacity is derived only from the outer skin friction due to the uncertainty of the soil plug behaviour:

$$Q_t = \sum f_s A_s + W$$

where,

Q_t = pull-out pile capacity
f_s = outer unit skin friction
A_s = outer pile shaft area
W = weight of the pile and soil plug

Unit Skin Friction in Cohesive Soils (Siliceous and Carbonate): Procedures to compute the unit skin friction, f, in cohesive soils are recommended by API RP 2A (2000) and Kolk and Van der Velde (1996) as follows:

$$f = \alpha S_u \leq S_u$$

where,

α = a dimensionless factor, derived as outlined below;
S_u = undrained shear strength of the soil at the point in question.

According to Section 6.4.2, API RP 2A-WSD (2000):

$$\alpha = 0.5(S_u/\sigma'_v)^{-0.5} \leq 1.0 \qquad \text{for } (S_u/\sigma'_v) \leq 1.0$$

$$\alpha = 0.5(S_u/\sigma'_v)^{-0.25} \leq 1.0 \qquad \text{for } (S_u/\sigma'_v) > 1.0$$

where,

σ'_v = effective overburden pressure at the point in question.

Additionally, API RP 2A (2000) recommends the following criteria in the commentary section:

$$\alpha = 1.0 \quad \text{for } S_u \leq 24\ \text{kPa}$$
$$\alpha = 1.0 - [(0.5(S_u - 24))/48] \quad \text{for } 24\ \text{kPa} < S_u < 72\ \text{kPa}$$
$$\alpha = 0.5 \quad \text{for } S_u \geq 72\ \text{kPa}$$

Kolk and Van der Velde (1996) recommended the following equation to calculate α to account for the pile flexibility (which increases as the slenderness (L/D ratio) of the pile increases):

$$\alpha = [0.5(L/D)]^{0.2}\left(S_u/\sigma'_v\right)^{0.3}$$

where,

L/D = pile flexibility

Unit Skin Friction in Granular Soils (Siliceous): The procedure to compute the unit skin friction, f, in granular soils (siliceous) recommended by API RP 2A (2000) is as follows:

$$f = K\sigma'_v \tan\delta \leq f_{\max}$$

where,

σ'_v = effective overburden pressure
δ = angle of friction between soil and pile wall
$f_{\max}$ = limiting (maximum) unit skin friction (recommended by API)
K = coefficient of lateral earth pressure
= 0.8 (compressive loads)
= 0.8 (tensile load as recommended by API)
= 0.5 (modified value for tensile load based on pile load test data [Toolan and Ims, 1988].

Unit Skin Friction in Granular Soils (Carbonate): The state of the art was recently summarised by Alba and Audibert (1999), who reviewed the literature on carbonate soils published during the last 25 yr. Based on this review and Fugro's in-house database, Johnson et al (1999) documented the methodology used by Fugro-McClelland's engineers in Houston to upgrade the parameters used to calculate the axial capacity of driven piles, when cone penetration test data are available.

Piles driven in carbonate ($CaCO_3$ > 50%) sands and silts have been found to develop significantly lower load capacity than would be predicted from the conventional static theory and soil-pile parameters for siliceous material described above. Pile load tests in sands and silts of carbonate origin from offshore Australia [Angemeer, et al 1973; Ripley, et al 1988], offshore Philippines [Puyuelo, et al 1983; Dutt, et al 1985], the Red Sea [Hagenaar, 1982; Hagenaar and Van Seters, 1985] and the Gulf of Suez [Dutt and Cheng, 1984] illustrate this unique feature. It is apparent that the soil-pile parameters summarised for siliceous sands and silts are not appropriate for predicting pile capacity in siliceous carbonate and carbonate sands and silts.

According to some researchers, the young carbonate sediments become semi-lithified almost immediately after deposition, resulting in some degree of cementation without any significant change in density [McClelland, 1974]. A pile driven in such a deposit results in a structural collapse of the material and displaces some of the material into the adjacent porous formation. Very low lateral stresses are mobilised along the pile wall due perhaps to partial cementation of the formation. Consequently, the axial capacity that is developed in carbonate sediments is very low. Another contributing factor is the high compressibility of these sediments. It has been shown that high material compressibility leads to the mobilisation of low skin friction and end bearing [Nauroy and LeTirant, 1983].

Others [Angemeer, et al 1975; Datta, et al 1980] contend that the soft carbonate sediments (Mohs hardness of 3 compared to 7 for quartz sand) undergo significant grain crushing at high stress levels such as those experienced during pile driving. This difference is believed to result in a drastic reduction in the soil-pile friction angle. In addition, cementation between the particles prevents development of lateral pressures on the pile wall.

It should be apparent from the above short discussions that our understanding of the soil-pile interaction problems in carbonate sands and silts is still not clear. Consequently, the current practice uses the conventional method for computing axial pile capacity of a driven pile in a siliceous environment with some modification to the skin friction and end bearing to account for various engineering aspects of carbonate soils. For carbonate sands and silts, the limiting unit skin friction value is selected on the basis of the pile load test data in similar carbonate materials. To account for the reduced limiting unit skin friction value, the $K\tan\delta$ value is limited to 0.14 to account for both the low lateral earth pressure and the low pile-soil friction angle associated with carbonate sands and silts.

Unit End Bearing (Siliceous and Carbonate): The procedure to compute the unit end bearing, q, in cohesive and granular soils, as recommended by API RP 2A (2000), is as follows:

$$q = 9S_u \qquad \text{for cohesive soils}$$

$$q = N_q\sigma'_v \leq q_{\max} \qquad \text{for granular soils}$$

where,

N_q = bearing capacity factor for deep foundation

$q_{\max}$ = limiting (maximum) unit end bearing for granular soils (as recommended by API RP 2A, 2000).

Equivalent Unit End Bearing: For the open-ended driven pipe piles, the end bearing is limited to the frictional resistance of the soil plug developed inside the pile, which defines an equivalent end bearing. In general, the total skin friction on the inside of the pile is assumed equal to the total skin friction on the outside of the pile. Any influence of the driving shoe on the internal skin friction is neglected. The end bearing on the steel end area of the pile is also neglected. The assumptions made in the analyses make no difference in the unit end bearing below the point where the pile plugs (i.e. the equivalent unit end bearing becomes equal to the unit end bearing). Above this point, the unit end bearing is limited by the frictional resistance of the soil plug.

16.14.2.3 Axial Pile Capacity of Drilled and Grouted Piles

Drilled-and-grouted pile design is primarily based on API RP 2A (2000) codes. The following paragraphs discuss some key factors affecting the axial pile capacity of drilled and grouted piles, such as, the drilled hole size, the shear keys, the construction and installation techniques, etc.

Drilled Holes and Shear Keys: According to the API, the diameter of the drilled hole should be at least 6 in. larger than the pile diameter. API RP 2A (2000), Para. 6.4.2 states that the selection of skin friction values should take into account soil disturbance resulting from installation. The API also recommends a check be made of the allowable bond stress between the pile steel and grout (API RP 2A (2000), Para. 7.4.4). The presence of shear keys will increase the strength of the pile-grout interface and move the failure plane to the soil-grout interface. For drilled-and-grouted piles, the limiting value of unit skin friction (f_{max}) at the soil-grout interface is equal to the limiting bond stress between the grout and the steel multiplied by the ratio of the insert pile diameter to the diameter of the drilled hole.

Construction and Installation Technique: Although drilled-and-grouted piles have been used extensively offshore in hard soil, rock, and calcareous and carbonate soils, there has been little application of drilled-and-grouted piles in normally consolidated clays. The construction and installation techniques have a much more pronounced effect on skin friction in drilled-and-grouted-piles than in driven piles. Kraft and Lyons (1974) believed that the most critical results of the construction processes are: (1) the increase in water content and a resulting decrease in shear strength at the soil-grout interface and (2) the influence of drilling mud on the drilled-and-grouted piles. The migration of water from the soil mass towards the edge of the borehole and the use of drilling mud accounts for most of the decrease in skin friction as compared to driven piles.

Computational Method: The ultimate axial capacity of piles is computed using the static method of analysis in general accordance with the API RP 2A (2000). In this method, the total ultimate capacity of a pile is taken as the sum of the skin friction on the pile wall. If the piles will be subjected to sustained tension, the end bearing is not included. For drilled-and-grouted piles in compression, the end bearing component is usually neglected due to the possible presence of cuttings and/or drilling mud in the bottom of the hole. Unit skin friction data for drilled-and-grouted piles are typically related to the shear strength of clays through an alpha (α) factor, viz:

$$f = \alpha\, S_u$$

where:

f = unit skin friction;
α = factor dependent upon installation considerations; and
S_u = undrained shear strength of the soil at the point in question.

Cox and Reese (1976) conducted pullout test on drilled-and-grouted piles in stiff clays. They found that the full shear strength is not developed as a result of the reduction of stress near the borehole wall and increase in water content and decrease in shear strength as a result of the drilling operations. They reported that values of 0.45 to 0.60 could be expected for stiff clays. Kraft and Lyons (1974) reported that for drilled-and-grouted piles, α-values typically range from 0.5 to 1.0 for the normally consolidated clays. The higher α-values tend to be associated with low plasticity clays.

16.14.2.4 Axial Pile Capacity – Factor of Safety

API RP 2A recommends that pile penetrations be selected using appropriate factors of safety or pile resistance factors. For working stress design (WSD), API RP 2A recommends that pile penetrations be selected to provide factors of safety of at least 2.0 with respect to normal operating loads and at least 1.5 with respect to maximum design storm loads. These factors of safety should be applied to the design compressive and tensile loads. For load and resistance factor design (LRFD), the API RP 2A recommends pile resistance factors of 0.7 and 0.8 for operating and maximum storm loads, respectively. Also, appropriate load factors should be used to determine the operating and maximum storm loads for LRFD design.

16.14.2.5 Axial Pile Capacity – Pile-Soil Set-up

The ultimate pile capacity represents the maximum pile capacity during undrained axial loading to failure after dissipation of all excess pore pressures caused by the installation procedures. However, immediately after pile driving, pile capacity in the normally consolidated cohesive deposits can be significantly lower than the ultimate achievable value. Field measurements in the normally consolidated clays [Soderberg, 1962; Azzouz and Baligh, 1984; Whittle and Baligh, 1988; Bogard and Matlock, 1990] have shown that the time required for driven piles to regain ultimate capacity in a cohesive deposit can be relatively long (fig. 16.76). On the other hand, there is hardly any set-up in highly overconsolidated clays [Lehane and Jardine, 1994; Bond and Jardine, 1995].

During continuous driving in a pile installation operation, the normally consolidated clay surrounding the pile is significantly disturbed and large excess pore pressures are generated. This results in reduced adhesion at the soil-pile interface and hence the reduced capacity. After installation, the excess pore pressure begins to dissipate out of the disturbed zone and the surrounding soil mass begins to consolidate until, with time, the pile reaches its ultimate capacity. The rate of consolidation is a function of the coefficient of the radial (horizontal) consolidation, pile radius and plug characteristics (plugged versus unplugged pile). In the case of a driven pile foundation, the capacity of the pile immediately after driving and the increase in capacity with time are important considerations that are needed to evaluate the factor of safety in the foundation during the early stages of the consolidation process.

Bogard and Matlock (1990) studied the behaviour of axially loaded piles in highly plastic, normally consolidated clay from a large number of experiments with an instrumented pile segment model. From the experimental data, they obtained the following empirical correlations between the degree of consolidation (U), time for 50% consolidation (t_{50}), time dependent shear transfer (f) and plug characteristics:

$$U = (t/t_{50})/(1.1 + t/t_{50})$$

$$t_{50} = D^2(85 - 1.7\ D/t_w)$$

$$f/f_{\text{ult}} = 0.2 + 0.8\ U$$

where:

t = time, min;

D = outside pile diameter, in.;

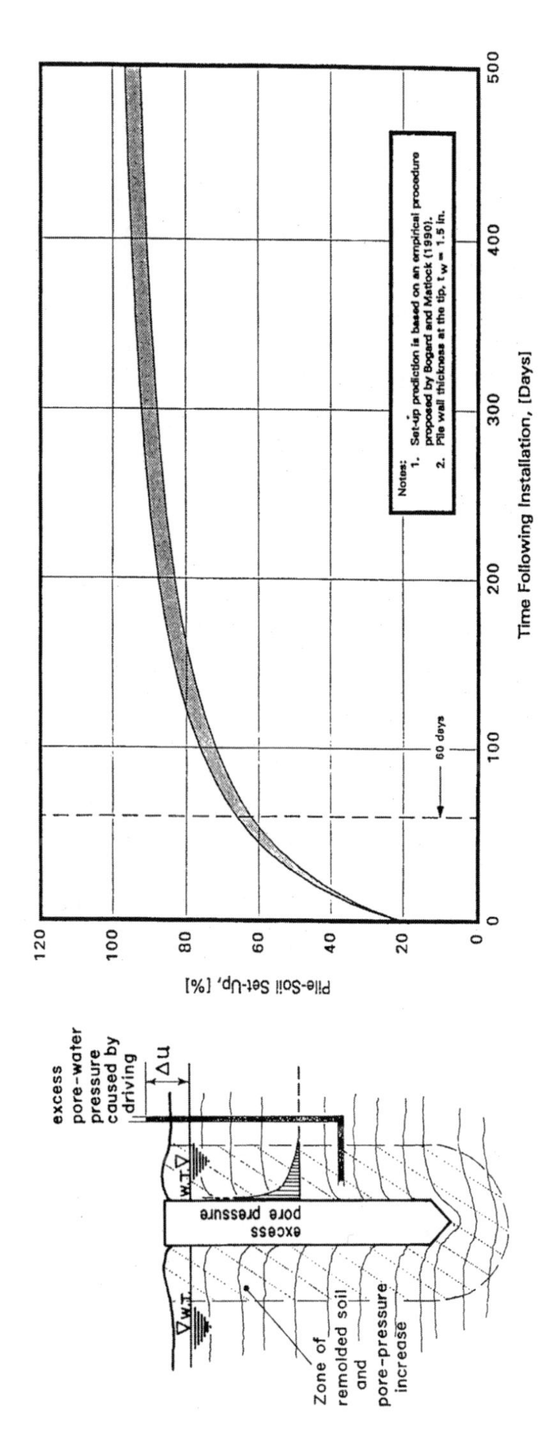

Figure 16.76 Pile-soil set-up (increase in axial capacity with time after driving)

t_w = pile wall thickness at pile tip, in.; and

f_{ult} = shear transfer after dissipation of all excess pore pressure, ksf.

The procedure indirectly accounts for pile plugging by the ratio of pile diameter to pile wall thickness at the pile tip. The higher the ratio, the lower the chances of the pile plugging. The above relationships were verified by comparing the actual pile load test data with load tests performed at different times after driving in the normally consolidated clays. This relationship is only valid for D/t between 2 and 40, which is the range of experimental data.

16.14.2.6 Pile and Spudcan Interaction

When a spudcan penetrates into the seafloor, a cylindrical zone of remolded and lower (degraded) shear strength soil is created. This zone of lower shear strength is called a spudcan depression. Piles located near the existing, or future, spudcan depressions may have degraded axial and lateral capacities. This degradation is a function of the spudcan and pile diameter, depth of spudcan penetration, distance between the spudcan depression and the pile, and the soil type. Consideration should also be given to the effects on pile performance associated with the potential use of jack-up rigs and the formation of future spudcan depressions based on the geometry and layout of the piles and spudcan depressions.

16.14.3 Axial Pile Response

Axial pile performance, specifically the pile head movement during applied loading, may be estimated using a subgrade reaction model. Using this method, the pile is modelled as a series of discrete elements connected by linear springs. The nonlinear local shear load transfer characteristics between each pile element and the surrounding soil are represented by t–z (unit side load vs. side movement along the side of the pile) curves and the end bearing on the pile tip (Q–z data), as shown on fig. 16.77). Recommended procedures to develop the t–z and Q–z relationships are given in API RP 2A (2000).

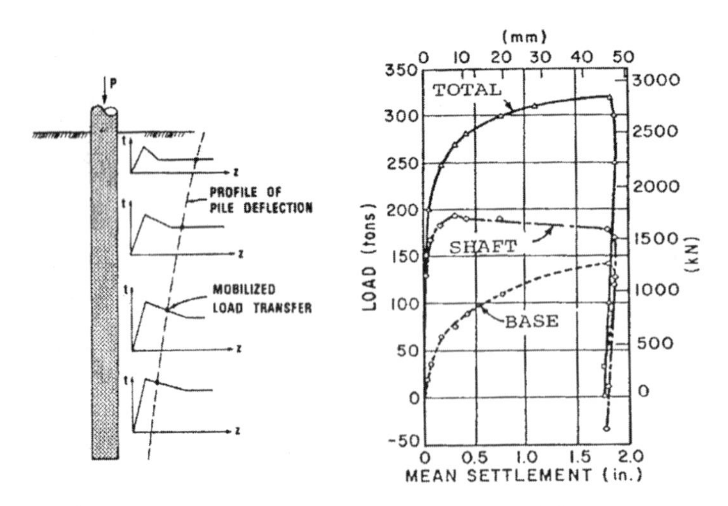

Figure 16.77 Axial load-pile movement analysis

16.14.3.1 Side Friction Versus Pile Movement Data (*t–z* Data)

Axial side load transfer curves are different for cohesive soils (clay) and granular soils (sand). Typical axial side load transfer–displacement (*t–z*) curves for both the material types are discussed below.

Cohesive Soils: The side friction versus pile movement (*t–z*) curve (fig. 16.78) for cohesive soils is given in the API RP 2A (2000) and is the same for compressive and tensile loading for driven pipe piles. As suggested by API RP 2A, without more definitive criteria, the maximum side friction, t_{max}, at the pile-soil interface is taken as the ultimate skin friction, *f*. The post peak adhesion ratio for clays can range from 0.90 to 0.70 for highly plastic, normally consolidated clays, to as low as 0.50 for low plasticity, highly overconsolidated clays.

Granular Soils: The side friction versus pile movement (*t–z*) curve (fig. 16.78) for granular soils is also presented in the API RP 2A (2000). The maximum side friction, t_{max}, at the pile-soil interface is the ultimate unit skin friction, *f*. For sands and silts which are

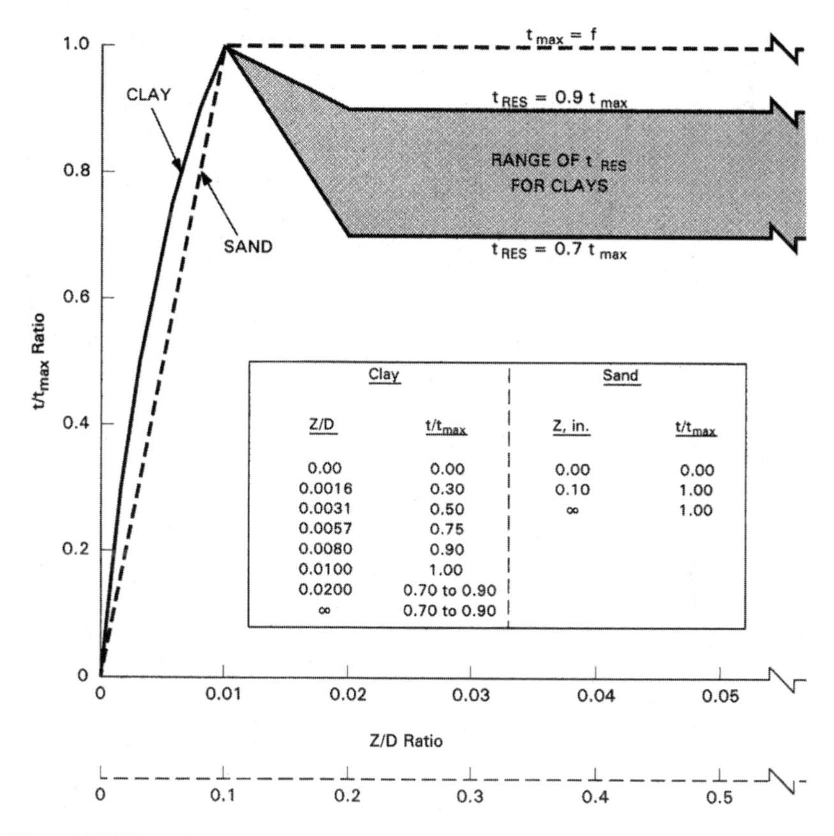

Z/D	t/tmax	Z, in.	t/tmax
0.00	0.00	0.00	0.00
0.0016	0.30	0.10	1.00
0.0031	0.50	∞	1.00
0.0057	0.75		
0.0080	0.90		
0.0100	1.00		
0.0200	0.70 to 0.90		
∞	0.70 to 0.90		

Figure 16.78 Typical *t–z* curve (per API RP 2A, 2000, reproduced with permission from API)

predominantly of carbonate origin ($CaCO_3$ > 50%), there is evidence of significant brittle softening, with residual unit skin friction values between 45 and 73% [Wiltsie, et al 1988] of the peak values.

Based on the results of various researchers, the recommended *t–z* curves for *drilled-and-grouted piles* under static loading conditions have a maximum shear transfer (f_{max}) value on the order of 60 to 75% of the undisturbed shear strength of the soil, with a post-peak minimum shear transfer (f_{min}) value being equal to the remolded shear strength.

16.14.3.2 Tip Load Versus Tip Movement Data

Relatively large axial movements may be required to mobilise full end bearing resistance. End bearing or tip load increases with displacement of the pile tip. The development of full end bearing occurs at a displacement on the order of 10% of the pile diameter according to API RP 2A (2000). In the case of predominantly carbonate sands and silts, three times as much pile tip movement (30%) may be required to mobilise the full end bearing. A typical pile tip load vs. tip movement (*Q–z*) curve is presented in fig. 16.79 from API RP 2A (2000). The end bearing component should not be considered when tensile loads are applied to a pile.

Also, as previously mentioned, for drilled and grouted piles, no end bearing resistance is considered for design due to the possible presence of cuttings and/or drilling mud in the bottom of the hole.

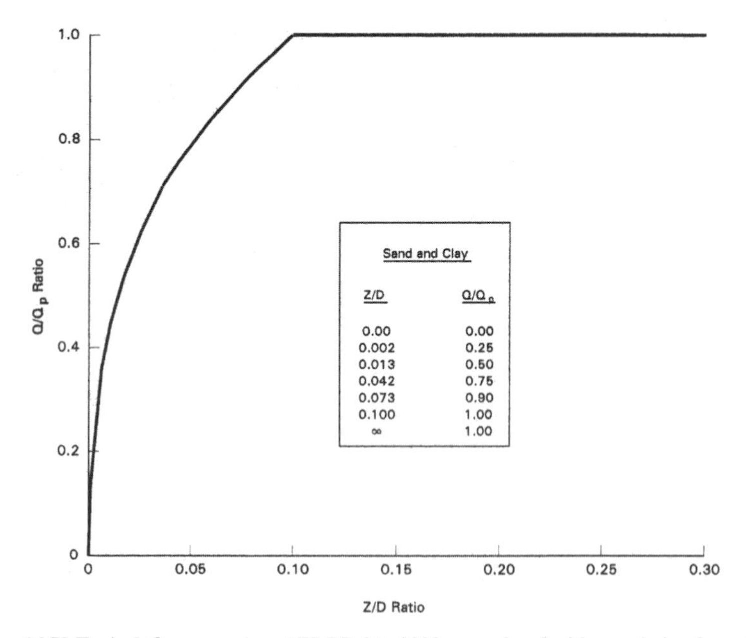

Figure 16.79 Typical *Q–z* curve (per API RP 2A, 2000, reproduced with permission from API)

16.14.4 Lateral Pile Capacity

API RP 2A recommends that pile foundations be designed for lateral loading conditions. The lateral soil structure interaction is complex and the soil response to lateral loading is generally nonlinear. Typically, the response of pilings to lateral loading is approximated by employing a beam-column model. The nonlinear characteristics of soil reaction to lateral pile movement along the length of a pile are represented by *p–y* (soil resistance per unit of pile length vs. pile lateral deflection) curves (fig. 16.80).

16.14.4.1 Cohesive Soils

Soil resistance–pile deflection (*p–y*) data for cohesive soils are developed using the procedure outlined by Matlock (1970) for soft clays subjected to cyclic loads. Interpreted shear strengths, submerged unit weights and strain values at one-half the maximum deviator stress (ε_{50}) are important soil parameters for lateral pile resistance analyses. These strain values are selected based on data from unconsolidated–undrained triaxial compression tests.

According to API RP 2A (2000), the ultimate lateral soil resistance (p_{us}) increases from $3S_uD$ to $9S_uD$ as X increases from 0 to X_R according to the following equation:

$$f = \alpha \, S_u$$

$$p_{us} = 3S_uD + \gamma XD + JS_uX$$

$$p_{ud} = 9S_uD \qquad \text{for } X \geq X_R$$

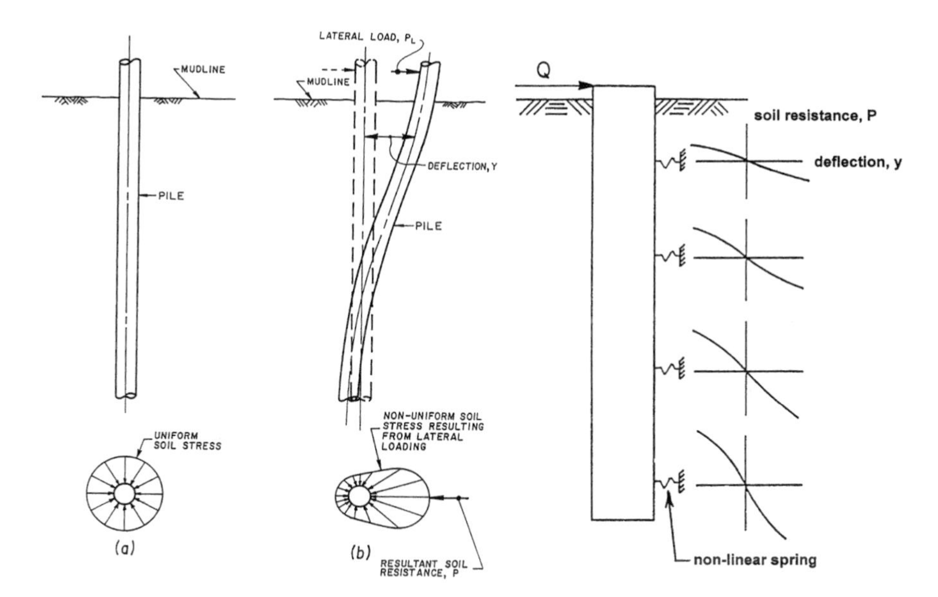

Figure 16.80 Lateral soil resistance–pile deflection (*P–y*) analysis (per API RP 2A, 2000, reproduced with permission from API)

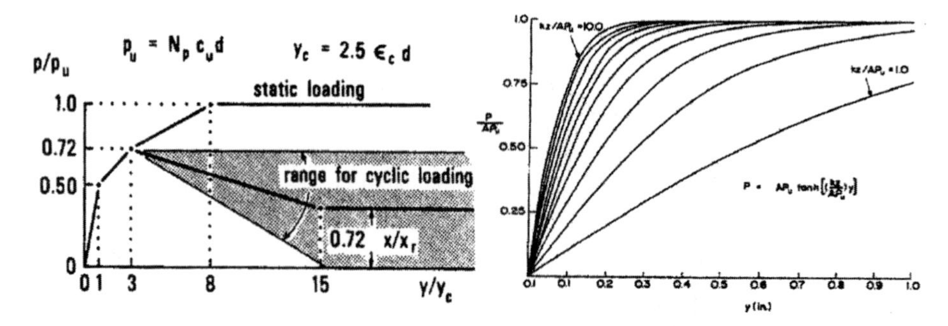

Figure 16.81 Typical lateral soil resistance-pile deflection (p–y) curves (per API RP 2A, 2000)

where:

p_u = ultimate resistance (s = shallow, d = deep),
S_u = undrained shear strength for undisturbed clay soil samples,
D = pile diameter,
γ = effective unit weight of soil,
J = dimensionless empirical constant with values ranging from 0.25 to 0.5 having been determined by field testing. A value of 0.5 is appropriate for Gulf of Mexico clays,
X = depth below soil surface, and
X_R = depth below soil surface to bottom of reduced resistance zone.

The deflection values (y) are a function of the pile diameter and $\varepsilon_{50.}$ Typical curve shapes for cohesive soils are shown on fig. 16.81.

16.14.4.2. Granular Soils (Siliceous)

Soil resistance-pile deflection (p–y) data for granular soils are developed using the procedure outlined by O'Neill and Murchison (1983) for sands subjected to cyclic loading. Input parameters include submerged unit weight, angle of internal friction and the initial modulus of horizontal subgrade reaction. Values of initial modulus of subgrade reaction are selected from the recommendations in API RP 2A based on the interpretation of the soil relative density from cone penetrometer data, sampler driving resistance records and grain size analyses.

At a given depth, the following equation giving the smallest value of p_u should be used as the ultimate lateral bearing capacity in granular soils (per API RP 2A).

$$p_{\text{us}} = (C_1 H + C_2 D)\,\gamma\,H$$

and

$$p_{\text{ud}} = C_3 D\,\gamma\,H$$

where:

p_u = ultimate resistance (s = shallow, d = deep),
γ = effective soil unit weight,
H = depth,
C_1, C_2, C_3 = coefficients, and
D = average pile diameter from surface to depth.

The shape of the p–y curve in granular soils is defined by the following equation:

$$P = A\,p_u\,\tanh\left[\frac{k\,H}{A\,p_u}y\right]$$

where:

A = factor to account for cyclic or static loading condition,
p_u = ultimate bearing capacity at depth H,
k = initial modulus of subgrade reaction,
y = lateral deflection, and
H = depth.

The shape of typical granular p–y curves is illustrated on fig. 16.81.

16.14.4.3 Granular Soils (Carbonate)

Lateral soil resistance-pile deflection (p–y) data for carbonate granular soils ($CaCO_3 > 50\%$) are developed using the procedures outlined by Wesselink et al (1988) and Williams et al (1988). The p–y relationship for carbonate granular material is expressed as:

$$p = R\left[\frac{X}{X_o}\right]^{0.70}\left[\frac{y}{D}\right]^{0.65}D$$

where:

p = soil resistance per unit length along the pile, lb/in.;
R = material constant, psi;
x = penetration below seafloor, ft;
x_o = 3.2808 ft;
y = lateral pile deflection, in.; and
D = pile diameter, in.

The soil resistance (p) determined above is limited to a value determined by the following equation:

$$p_{\mathrm{ult}} = K_p^2\,\sigma'_v\,D$$

where:

p_{ult} = ultimate soil resistance per unit length along the pile, lb/in.;
K_p = coefficient of passive earth pressure;
σ'_v = effective vertical stress, psi; and
D = pile diameter, in.

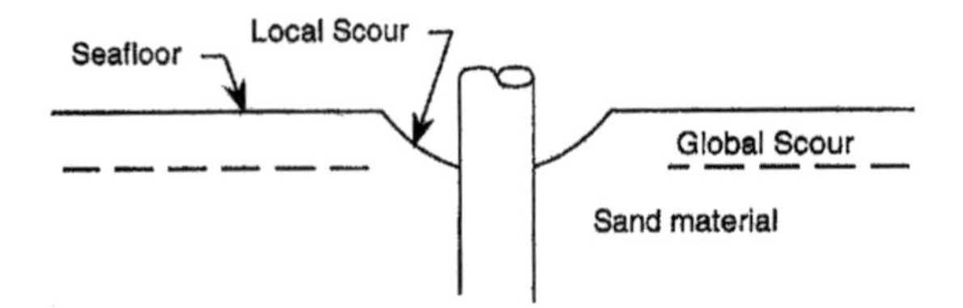

Figure 16.82 Seabed scour

From a strict interpretation of the *p–y* criterion from Wesselink et al (1988) procedures, the *p–y* curves represent soil–pile interaction under monotonic loading. However, on the basis of findings and observations from prototype lateral load tests, the *p–y* data developed using the Wesselink et al (1988) approach can be conservatively used to represent cyclic lateral soil-pile interaction of the carbonate granular soils.

16.14.5 Other Considerations

Other geotechnical considerations should be taken into account for both the lateral and the axial pile performance, such as seabed scour, pile group effects and seafloor instability.

Seabed Scour: Whenever the near-surface soils comprise of granular material, they may be susceptible to scour (fig. 16.82). Scour effects are considered insignificant to axial capacity but can have a large influence on lateral capacity. When scour is considered likely, the *p–y* data are reduced to reflect the potential loss of lateral support from the material scoured away near the seafloor around the pile. Global scour indicates that installation of the structure may cause a layer of material to be removed throughout the area of the platform. Local scour indicates that scour is likely to occur only in the near vicinity of the piles.

Pile Group Action: Consideration should be given to the effects of the closely spaced adjacent piles on the load and deflection characteristics of pile groups. According to the API RP 2A, group effects may have to be evaluated for pile spacing less than eight (8) diameters. Generally, for piles in clay, the group effects will cause the group axial capacity to be less than the individual pile capacity multiplied by the pile numbers in the group. For piles in sand, the opposite applies. In addition, regardless of soil types, the group effects will generally increase the pile head lateral displacement compared to the individual pile subjected to the average load of the corresponding group.

Seafloor Instability: Seafloor instability (fig. 16.83) includes mudflow overrun (fig. 16.84), seafloor failure (fig. 16.85) and wave-actuated soil motions (fig. 16.86). All of the above can significantly increase bending stress along the piles and pile lateral displacements (fig. 16.87).

16.14.6 Pile Drivability Analyses and Monitoring

16.14.6.1 Soil Resistance To Driving

Computation of the soil resistance to pile driving is analogous to the computation of the ultimate axial pile capacity by the static method. The resistance to driving is the

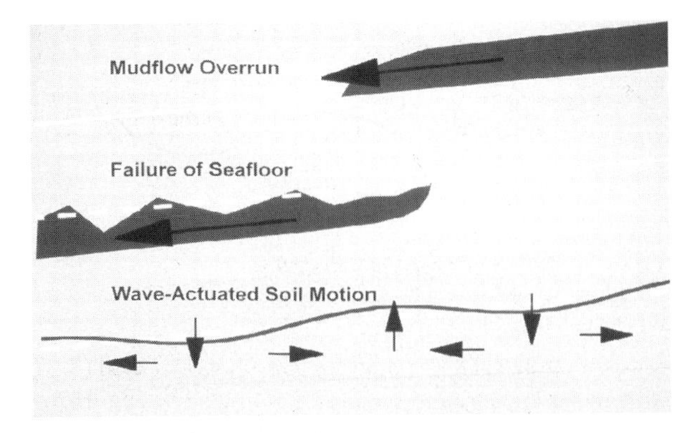

Figure 16.83 Seafloor instability

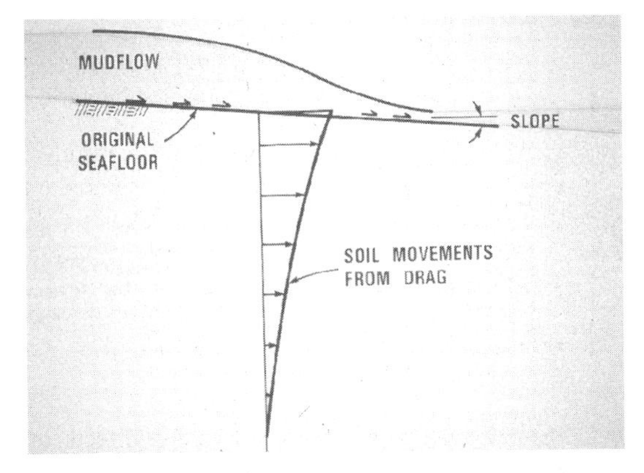

Figure 16.84 Mudflow overrun

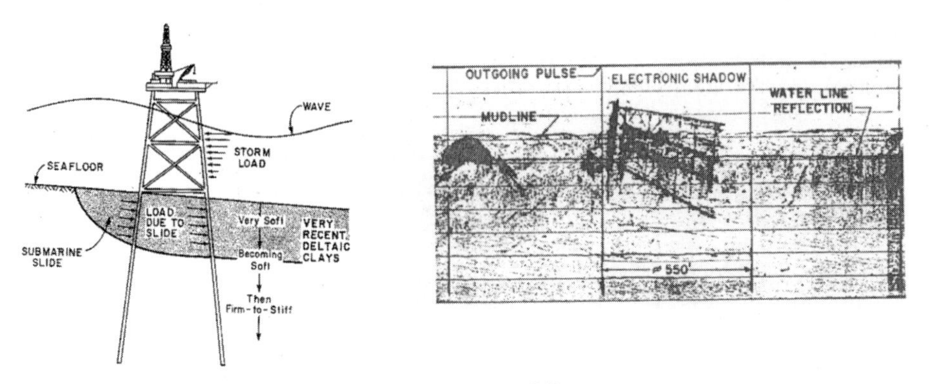

Figure 16.85 Seafloor failure

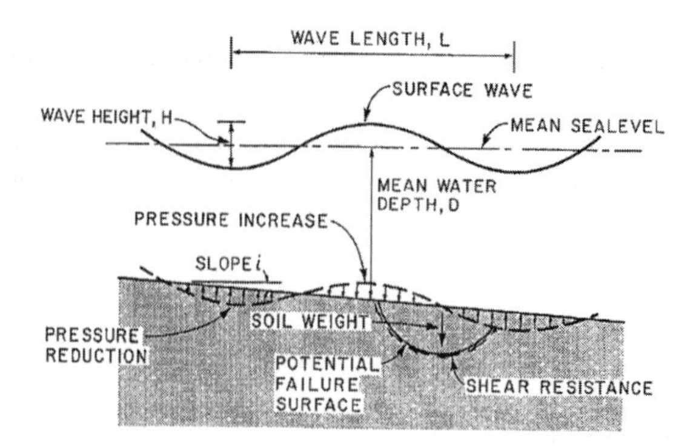

Figure 16.86 Wave-actuated soil motion

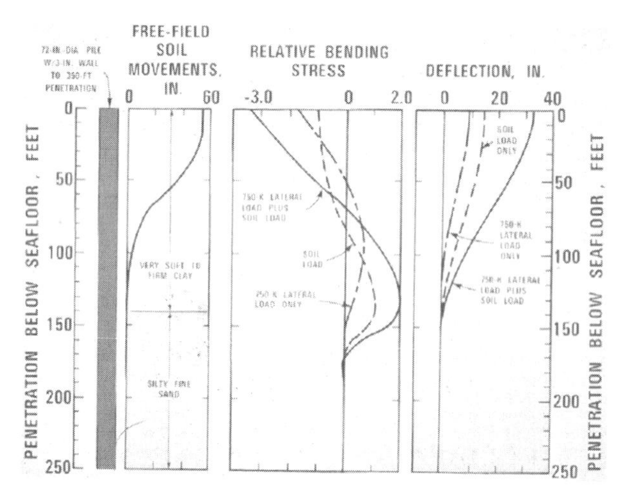

Figure 16.87 Example of additional bending stress and deflection due to soil movements

sum of the shaft resistance and the toe resistance. The shaft resistance is computed by multiplying the average unit skin friction during driving and the embedded surface area of the pile. The toe resistance is computed by multiplying the unit end bearing and the end bearing area.

Stevens, Wiltsie and Turton (1982) proposed computing the lower and the upper bound values of soil resistance to driving for both the coring and the plugged pile conditions. When a pile cores, relative movement between the pile and the soil occurs both on the outside and inside of the pile wall. Skin friction is, therefore, developed on both the outside and the inside pile wall. The end bearing area is equal to the cross-sectional area of steel at the pile toe. When a pile plugs, the soil plug moves with the pile during driving. Skin friction is mobilised only on the outer wall. The end bearing area is the gross area of the pile. The determination if a pile is coring, partially plugged or plugged is based on the soil conditions and pile acceleration. Plugging during continuous driving in cohesive soils is unlikely, as discussed by Stevens (1988).

For piles driven in cohesive soils, Semple and Gemeinhardt (1981) recommended multiplying the unit skin friction after full set-up by a pile capacity factor, such that:

$$f_{dr} = F_p f$$

where:

f_{dr} = unit skin friction during pile driving;
F_p = pile capacity factor; and
f = unit skin friction after full set-up.

A pile capacity factor (F_p) empirically determined from the wave equation analyses performed for six sites is given by:

$$F_p = 0.5(\mathrm{OCR})^{0.3}$$

The overconsolidation ratio (OCR) is estimated using the equation:

$$S_u/S_{unc} = (\mathrm{OCR})^{0.85}$$

where:

S_u = actual undrained shear strength of clay having a given PI; and
S_{unc} = undrained shear strength of the same clay, if normally consolidated at that depth.

Also, according to a relationship developed by Skempton (1957), the undrained shear strength of the same clay, if normally consolidated at that depth, can be estimated as follows:

$$S_{unc} = \sigma'_{vo}(0.11 + 0.0037\ \mathrm{PI})$$

where:

σ'_{vo} = effective overburden pressure; and
PI = plasticity index.

For piles driven in granular soils, the upper bound plugged case is computed by increasing the unit skin friction 30%, and the unit end bearing 50%. A corresponding increase in limiting values for unit skin friction and unit end bearing is assumed.

16.14.6.2 Wave Equation Analyses

The GRLWEAP computer program, originally coded by Goble and Rausche (1986), is recommended to calculate the predicted blow counts. Wave equation analysis of pile driving is based on the discrete element idealisation of the hammer–pile–soil system formulated by Smith (1962). The parameters used in the wave equation analyses can be divided into three groups: (1) hammer parameters, (2) pile parameters and (3) soil parameters. These parameters are discussed in the following paragraphs.

Hammer Parameters: The air/steam hammers are modelled by three segments: (1) the ram as a weight with infinite stiffness, (2) the cushion as a weightless spring with finite stiffness, and (3) the pile cap as a weight with infinite stiffness. The pile driving hammer is described by (1) the rated hammer energy, (2) the efficiency of the hammer, (3) the weight of the ram, (4) the weight of the pile cap, (5) the cushion stiffness, and (6) the coefficients of restitution for the ram hitting the cushion and for the pile cap-pile contact. The rated energy and the weight of the ram and pile cap are obtained from the manufacturer. The hammer efficiency and cushion properties are either the measured driving system performance data (e.g. from in-house databases) or published values. For hydraulic hammers, a cushion is not used. The ram impacts directly on the pile cap. For diesel hammers, a thermodynamic analysis determines the gas pressure in the combustion chamber during compression, combustion delay, ignition and expansion, rather than assuming a constant pressure, and the hammer stroke is calculated rather than using a prescribed value.

Pile Parameters: The pile is divided into an appropriate number of segments of approximately equal length. Each pile segment is modelled as a weight and a spring. The pile parameters consist of the diameter, the wall thickness schedule, modulus of elasticity of the pile material, unit weight of the pile material, free-standing length of pile and penetration below the seafloor.

Soil Parameters: The soil resistance is distributed along the side of each embedded element and at the pile tip. During driving, the static component of resistance on each element is represented by an elastic spring with a friction block used to represent the ultimate static resistance. The dynamic component of resistance is modelled by a dashpot. There are essentially three soil parameters used in the wave equation analyses. These parameters are (1) the quake (also referred to as the elastic ground compression) for the side and tip of the pile, (2) the damping coefficient for the side and tip of the pile and (3) the percentage of the total resistance to driving at the pile tip.

The soil quake and damping parameters recommended by Roussel (1979) can be used in the wave equation analyses. These parameters were determined from a comprehensive correlation study performed for large-diameter offshore piles in which the driving records of 58 piles at 15 offshore sites in the Gulf of Mexico were analysed. The side and the tip quake are assumed equal, with a magnitude of 0.10 in. for stiff to hard clay, silt and sand. Side damping in clay decreases with increasing shear strength, which is in agreement with the laboratory test results of Coyle and Gibson (1970) and Heerema (1979). Tip damping of 0.15 s/ft is recommended for firm to hard clay, silt and sand.

Difference Equations: The equations of motion for the idealised system are written in finite difference form. The pile–soil system is assumed to be at rest with no residual effects from previous blows. The response of the pile under one hammer blow is desired. The computations proceed as follows:

(1) the impact velocity of the ram is calculated and other time-dependent quantities are initialised so as to satisfy static equilibrium;
(2) displacements are calculated for each mass;
(3) compressions of internal springs, forces in internal springs, forces exerted by external soil springs, accelerations and velocities are calculated for each mass; and
(4) the cycle is repeated for successive time intervals.

16.14.6.3 Pile Installation Considerations

During driving, it will be necessary to interrupt driving operations in order to make pile add-ons or change hammers. Interruptions of driving operations may last six to eight hours. Delays on the order of several days may result from bad weather or equipment breakdown. During this time, many clays will gain strength as excess pore pressures dissipate and soil particles reorient themselves. This phenomenon is commonly referred to as set-up (see discussion in subsection 6.2.5). A similar phenomenon may also occur in fine-grained granular deposits. Upon redriving piles after some set-up has occurred, increased blow counts may be experienced. Due to set-up, soil resistance to driving at the beginning of driving may increase to the point of refusal. It is suggested that the driving programme be planned so as to reduce the number and duration of delays in order to reduce the set-up of the soil around the piles.

16.14.7 Supplementary Pile Installation Procedures

The most economical pile installation procedure is to drive alone, without resorting to supplementary procedures. However, in many cases, especially those involving piles driven into hard clays and dense to very dense sands, the piles cannot be installed to the required penetration by driving alone. When techniques other than driving are used to aid pile installation, conditions assumed in computations based on driving alone may not be met. In these cases, computed capacities must frequently be adjusted to fit the actual installation conditions. Sullivan and Ehlers (1972) presented supplementary pile installation procedures that may be used under various circumstances, including the possible effects that these procedures may have on pile capacity. Supplementary procedures should be selected considering not only construction expediency, but also the effects of the procedures on pile capacity.

16.14.7.1 Pile Monitoring

Pile installation monitoring should provide the appropriate information for accepting piles if they encounter refusal, generally in dense to very dense granular soils encountered at penetrations shallower than design grade. To monitor pile driving, strain and acceleration transducers are attached near the pile top, and the following information is determined onsite during or immediately following driving:

(1) the impact stress and velocity,
(2) the hammer efficiency and cushion properties (cushion stiffness and coefficient of restitution),
(3) the maximum energy transmitted to the pile and, therefore, the energy transfer ratio or system efficiency,
(4) the maximum compressive driving stress in the pile (not at the transducer location),
(5) an *estimate* of the soil resistance to driving, and
(6) an *estimate* of the ultimate compressive and tensile pile capacity.

To accept piles refusing in granular strata, the pile wall thickness schedule may have to be modified to permit an underdrive allowance. This alternative should be less expensive than removing the soil plug and redriving in the event of premature pile refusal.

16.15 Shallow Foundation Design

While the platform jacket is temporarily resting on the seafloor prior to pile installation, seafloor support is provided by the soil resistance (skin friction and end bearing) on the jacket leg extensions and the bearing capacity of the soil supporting the lowest horizontal (mudline) bracing members and mudmats. Mudmats with skirted plates are design to sustain significant horizontal loading from the upper structures. Therefore, for a reliable offshore shallow foundation design, both the bearing capacity and horizontal loading resistance need to be considered. Mudmat settlements under loading should be included in the design analyses to prevent possible excessive short-term or long-term settlements.

16.15.1 Bearing Capacity for Mudmats and Skirted Plates

Bearing capacity equations for the near-surface soils are taken from design methods developed by different researchers for different soil profiles.

Clay Profiles: Skempton (1951) recommended the following equations to determine the ultimate bearing capacity for horizontal tubular members and mudmats resting on clay seafloors:

$$Q = S_u\, N_c\, A$$

where,

$S_u = S_{\text{uavg}}$ to $B/2$ below foundation depth
$N_c = 5\,[1+\,0.2(D/B)]\,[1+\,0.2(B/L)] \le 9$ or,
$N_c = (2+\pi)\,(1.2+0.4\tan^{-1} D/B) \le 9$ (Hansen values)
D = Foundation depth
B = Foundation width
L = Foundation length
A = Bearing area

Alternatively, Davis and Booker (1973) also recommended bearing capacity equations to determine the ultimate bearing capacity for horizontal tubular members and mudmats resting on a clay seafloor:

$$q_u = F(6\, S_{\text{uo}} + \rho\, B/4) \qquad \text{for cicular footings}$$

$$q_u = F(5.14\, S_{\text{uo}} + \rho\, B/4) \qquad \text{for strip footings}$$

where,

$S_{\text{uo}} = S_u$ at the mudline
F = theoretical correction factor

Sand Profiles: Terzaghi and Peck (1967) recommended the following equations to determine the ultimate bearing capacity for horizontal tubular members and mudmats resting on a cohesionless (sand) seafloor:

$$q_u = \left[S_u N_c + 0.5\, S_\gamma\, \gamma_1'\, B\, N_\gamma + \gamma_2'\, D\, (N_q - 1)\right]$$

where,

N_c, N_q, N_γ = Bearing capacity factors [Vesic, 1975]
S_γ = Shape factor = $[1 - 0.4B/L]$
D = Foundation depth
B = Foundation width
L = Foundation length

Weaker Clay Over Stronger Clay Profile: Brown and Meyerhof (1969) recommended the following equations to determine the ultimate bearing capacity for horizontal tubular members and mudmats resting on a seafloor made of a weaker clay resting over a stronger clay layer:

$$Q = S_{ut} N_{md}[1 + 0.2D/B]\, A$$

where,

S_{ut} = Average S_u for the upper soil layer
N_{md} = Modified bearing capacity factors
A = Bearing area

For horizontal tubular members penetrating less than one radius, the projected area at the mudline should be used to calculate the ultimate bearing capacity of the members. For members penetrating one radius or more, the diameter should be used. For triangular shaped mudmats, B should be taken as 75% of the least altitude and L should be taken as the longest side. The ultimate bearing capacity of the near seafloor soils is a function of the size and configuration of the mudmats and jacket structure.

For WSD design, the API RP 2A recommends that a safety factor of at least 2.0 be used with the ultimate bearing capacity determined from the above equations. For an LRFD based design, a resistance factor of 0.67 is recommended. Also, an appropriate load factor should be used to determine the jacket load in the LRFD design procedure. The ultimate bearing (load-carrying) capacity of a horizontal tubular member or mudmat may be calculated as the ultimate bearing capacity of the soil multiplied by the base area of the mat or member. The equations for ultimate bearing capacity presented above are based on static bearing capacity conditions. Significant vertical platform velocities at the time of jacket placement could cause large or uneven jacket settlements.

16.15.2 Horizontal Sliding Resistance

Mudmat Design: The horizontal sliding resistance for mudmats comes from the shear strength of the soil at the base of the mudmat. Therefore, the following equation is recommended for horizontal sliding resistance for mudmat resting on the seafloor of clay profile:

$$Q_1 = S_u\, A_r$$

where:

Q_l = the ultimate lateral resistance,
S_u = the intact shear strength at the base of the mudmat, and
A_r = the area of the mudmat.

Skirted Rectangular Footings: The horizontal sliding resistance for skirted rectangular footings comes from the base shear resistance, skirt side shear resistance and passive soil pressure from the soil in front of the footings. Therefore, the following equation is recommended for skirted rectangular footings in clay profile:

$$Q_l = c_t\, A_{\text{base}} + 2\,\tau_s\, H\, B + 2\,\tau_s\, H\, L + (\gamma H^2/2)\, L$$

where:

Q_l = the resisting force,
c_t = the shear strength below the top of the skirt plates,
$A_{\text{base}} = B\, L$,
B = the length of the skirt parallel with the direction of movement,
L = the length of the skirt normal to the direction of movement,
H = the depth of penetration of the skirt plates,
τ_s = the average shear strength over the depth of the skirts, and
γ = the average submerged unit weight of the soil.

16.15.3 Shallow Foundation Settlement Analyses

Usually, mudmats resting on clay soil have more settlement than that on sand soil. Therefore, the following paragraphs briefly discuss the mudmat short-term and long-term settlements in clay.

Short-Term Settlement (Clay): The short-term settlement is treated as elastic deformation of clay under loading. Three different analysis methods have been recommended as follows:

(a) *API Load-Settlement*

$$u = \left[(1-\nu)/(4 * G_{\text{avg}} * R)\right] Q$$

where,

ν = Poisson's ratio
G_{avg} = average shear modulus
Q = vertical load applied to footing
u = elastic settlement

(b) *DNV Load-Settlement*

$$u = \left[\left((Q * K * (1-\nu)^2 * B)/E_{avg}\right) * /(A - (Q/\alpha * F_d))\right] Q$$

where,

K = load shape factor ($\pi/4$)
E_{avg} = average elastic modulus
α = failure ratio
F_d = ultimate design load

(c) *Skempton Load-Settlement*

$$q/q_u = (\sigma 1 - \sigma 3)/\,(\sigma 1 - \sigma 3)_f$$

$$u = 2B\varepsilon$$

where,

$q_u = 5\ S_u\ (1+0.2\ B/L)$

E_{avg} = average elastic modulus

$(\sigma 1-\sigma 3)/\ (\sigma 1-\sigma 3)_f$ = normalised deviator stress

ε = strain corresponding to $(\sigma 1-\sigma 3)/\ (\sigma 1-\sigma 3)_f$

Long-Term Settlement (Clay): As excess pore water pressures generated during initial loading dissipates, the clay mass consolidates with time under load and, thus, additional long-term settlement occurs. The API RP 2A recommends that an estimate of the vertical settlement of a clay layer under an imposed vertical load can be made using the following equation:

$$u_v = hC/(1 + e_o)\log_{10}((q_o + \Delta q)/q_o)$$

where:

u_v = the vertical settlement,
h = the layer thickness,
e_o = the initial void ratio of the soil,
C = the compression index of the soil over the load range considered,
q_o = the initial effective vertical stress, and
Δq = the added effective vertical stress.

16.16 Spudcan Penetration Predictions

Multiple-legged self elevating barges are widely used in offshore exploration and construction work. The static load from the barge is generally assumed to be distributed equally among the legs. During preload, the legs will penetrate into the soil and stop at some depth below the mudline, depending on the encountered soil resistance. Determination of leg penetration is important as sometimes the soil conditions are such that the barge may run out of leg before the legs stop penetrating into the seafloor. This is especially true offshore the Mississippi delta, where extremely soft soil conditions are encountered, and where leg penetrations well in excess of 100 ft have been experienced.

Using the bearing capacity methods recommended by the API RP 2A, the bearing capacity of the soil can be computed based on the leg pad (spudcan) geometry amd dimensions, the shear strength of the cohesive soil (average over some distance below the pad), the friction angle of granular soil strata, and the volume of the pad.

Some simplifying assumptions are made when performing the bearing capacity analysis:

(1) Static loading is assumed, with a uniform stress distribution beneath the entire pad (no eccentricity is considered),
(2) The pads are assumed to be rigid plates (without flexibility),
(3) Short term loading is considered (long term settlement due to consolidation is not considered), and
(4) The soil above the pad is assumed to flow back on to top of the pads during penetration.

It is important to note that many catastrophic failures have been associated with the siting of jack-up rigs. The main two causes for these failures were:

- The sudden punch-through of the spudcans, when spudcans stopped penetrating when they encountered a (thin) layer of sand (underlain by clay) which subsequently gave way, resulting in the sudden plunging of one of the legs, thus leading to failure of the leg(s) and capsizing of the rig.
- When the seafloor consists of granular material (sands, silts), scour may develop around the spudcans or under the foundation mat (pads), leading to a progressive loss of bearing capacity and a sudden settlement of one of the legs of the rig when the bearing capacity is exceed. As before, this may lead to buckling of the leg(s), and the eventual capsizing of the rig.

Therefore, it is highly recommended that a rig operator seek the advice of a geotechnical engineer before siting a jack-up rig.

16.17 ASTM Standards

D 0422 – 63 (1998), Test Method for Particle-Size Analysis of Soils

D 0854 – 00, Test Methods for Specific Gravity of Soil Solids by Water Pycnometer

D 1140 – 00, Test Method for Amount of Material in Soils Finer than the No. 200 (75-μm) Sieve

D 2166 – 98a, Test Method for Unconfined Compressive Strength of Cohesive Soil

D 2216 – 98, Test Method for Laboratory Determination of Water (Moisture) Content of Soil and Rock

D 2435 – 96, Test Method for One-Dimensional Consolidation Properties of Soils

D 2487 – 98, Practice for Classification of Soils for Engineering Purposes (Unified Soil Classification System)

D 2488 – 93, Practice for Description and Identification of Soils (Visual–Manual Procedure)

D 2850 – 95, Test Method for Unconsolidated–Undrained Compressive Strength of Cohesive Soils in Triaxial Compression

D 3999 – 91 (1996), Test Method for the Determination of the Modulus and Damping Properties of Soils Using the Cyclic Triaxial Apparatus

D 4015 – 92 (1995), Test Methods for Modulus and Damping of Soils by the Resonant-Column Method

D 4186 – 89 (1998), Test Method for One-Dimensional Consolidation Properties of Soils Using Controlled-Strain Loading

D 4318 – 00, Test Methods for Liquid Limit, Plastic Limit, and Plasticity Index of Soils

D 4373 – 96, Test Method for Calcium Carbonate Content of Soils

D 4452 – 85 (1995), Test Methods for X-Ray Radiography of Soil Samples

D 5084 – 00, Test Methods for Measurement of Hydraulic Conductivity of Saturated Porous Materials Using a Flexible Wall Permeameter

References

Alba, J. L. and Audibert, J. M. E. (1999). "Pile design in calcareous and carbonaceous granular materials: a historical overview", *Second International Conference Engineering on for Calcareous Sediments*, Bahrain.

American Petroleum Institute (1993). "Recommended practice for planning, designing, and constructing fixed offshore platform-load and resistance factor design", API Recommended Practice 2A-LRFD (RP 2A-LRFD), (1st ed.). API, Washington, D.C.

American Petroleum Institute (2000). "Recommended practice for planning, designing, and constructing fixed offshore platforms-working stress design", API Recommended Practice 2A-WSD (RP 2A-WSD), (21st ed.). API, Washington, D.C.

Anderson, D. G. and Stokoe, K. H. (1978). "Shear modulus: a time-dependent soil property", *Dynamic Geotechnical Testing*, ASTM STP 654, American Society for Testing and Materials, pp. 66–90.

Angemeer, J., Carlson, E. D., and Klick, J. H. (1973). "Techniques and results of offshore pile load testing in calcareous soils", *Proceedings, Fifth Offshore Technology Conference*, Houston, Vol. 2, pp. 677–692.

Angemeer, J., Carlson, E. D., Stroud, S., and Kurzeme, M. (1975). "Pile load tests in calcareous soils conducted in 400 feet of water from a semi-submersible exploratory rig", *Proceedings, Seventh Offshore Technology Conference*, Vol. 2, pp. 657–670.

Azzouz, A. S. and Baligh, M. M. (1984), "*Behavior of Friction Pile in Plastic Empire Clays*", Vol. 2, Report No. R84-14, Constructed Facilities Division, Civil Engineering Department, Massachusetts Institute of Technology.

Becker, D. E., Crooks, J. H. A., Been, K., and Jefferies, M. G. (1987). "Work as a criterion for determining in situ and yield stresses in clays" *Canadian Geotechnical Journal*, Vol. 24.

Bogard, J. D. and Matlock, H. (1990). "Application of model pile tests to axial pile design", *Proceedings, Twenty-Second Annual Offshore Technology Conference*, Houston, Paper No. 6376.

Bond, A. J. and Jardine, R. J. (1995). "Shaft capacity of displacement piles in a high OCR clay", *Geotechnique 45*, No. 1, pp. 3–23.

Brown, J. D. and Meyerhof, G. G. (1969). "Experimental study of bearing capacity in layered clays", *Proceedings, Seventh International Conference on Soil Mechanics and Foundation Engineering*, Mexico, Vol. 2, pp. 45–51.

Cox, W. R. and Reese, L. C. (1976). "Pullout tests of grouted piles in stiff", *Proceedings, Eighth Offshore Technology Conference*, OTC 2473, Houston, Vol. 2, pp. 539–551.

Coyle, H. M. and Gibson, G. C. (1970). "Empirical damping constants for sands and clays". *Journal, Soil Mechanics and Foundations Division*, ASCE, Vol. 96, No. SM3, pp. 949–965.

Datta, M., Gulhati, S. K., and Rao, G. V. (1980). "An appraisal of the existing practice of determining the axial load capacity of deep penetration piles in calcareous sands", *Proceedings, Twelfth Offshore Technology Conference*, Houston, Vol. 4, pp. 119–130.

Davis, E. H. and Booker, J. R. (1973). "The effect of increasing strength with depth on the bearing capacity of clays". *Geotechnique 23*, No. 4, pp. 551–563.

Doroudian, M. and Vucetic, M. (1995). "A direct simple shear device for measuring small-strain behavior". *Geotechnical Testing Journal*, ASTM, GTJOJ, Vol. 18, No. 1, March, pp. 69–85.

Dutt, R. N. and Cheng, A. P. (1984). "Frictional response of piles in calcareous deposits", *Proceedings, Sixteenth Offshore Technology Conference*, Houston, Vol. 3, pp. 527–534.

Dutt, R. N., Moore, J. E., Mudd, R. W., and Rees, T. E. (1985). "Behavior of piles in granular carbonate sediments from offshore Philippines", *Proceedings, Seventeenth Offshore Technology Conference*, Houston, Vol. 1, pp. 73–82.

Goble, G. G. and Rausche, F. (1986). "*Wave Equation Analysis of Pile Foundations WEAP86 Program*", U.S. Department of Transportation, Federal Highway Administration.

Hagenaar, J. (1982). "The use and interpretation of SPT results for the determination of axial bearing capacities of piles driven into carbonate soils and coral", *Proceedings, Second European Symposium on Penetration Testing*, ESOPT II, Vol. I, Amsterdam, pp. 51–55.

Hagenaar, J. and Van Seters, A. (1985). "Ultimate axial bearing capacity of piles driven into coral rock and carbonate soils", *Proceedings, Eleventh International Conference on Soil Mechanics and Foundation Engineering*, San Francisco, Vol. 3, pp. 1599–1602.

Heerema, E. P. (1979). "Relationships between wall friction, displacement velocity and horizontal stress in clays and in sand, for pile drivability analysis". *Ground Engineering*, Vol. 12, No. 1, pp. 55–61,65.

Johnson, S., Audibert, J. M. E., and Stevens, R. F. (1999). "CPT-based method for selecting improved pile foundation design parameters in carbonate soils", *Second International Conference on Engineering for Calcareous Sediments*, Bahrain.

Kolk, H. J. and Van der Velde, E. (1996). "A reliable method to determine friction capacity of piles driven into clays", *Proceedings, 28th Offshore Technology Conference*, OTC 7993, Houston, May 6–9.

Kraft, L. M. Jr. and Lyons, C. G. (1974). "State of the art: ultimate axial capacity of grouted piles", *Proceedings, Sixth Offshore Technology Conference*, OTC 2081, Houston, Vol. 2, pp. 485–503.

Ladd, C. C. (1991). "Stability evaluation during staged construction". *The Twenty-Second Karl Terzaghi Lecture, Journal of Geotechnical Engineering*, ASCE, Vol. 117, No. 4, April, pp. 537–615.

Lehane, B. M. and Jardine, R. J. (1994). "Displacement pile behavior in glacial clay". *Canadian Geotechnical Journal*, Vol. 31, pp. 79–90.

Malek, A. M., Azzouz, A. S., Baligh, M. M., and Germaine, J. T. (1987). "Undrained cyclic simple shear behavior of clay with application to piles supporting tension leg platform", *MIT Sea Grant Report 87—20*, Department of Civil Engineering, Massachusetts Institute of Technology, Cambridge, Mass.

Matlock, H. (1970). "Correlations for design of laterally loaded piles in soft clay", *Proceedings, Second Offshore Technology Conference*, Houston, Vol. 1, pp. 577–594.

McClelland, B. (1974). "Design of deep penetration piles for ocean structures", *Journal of the Soil Mechanics and Foundation Engineering Division*, ASCE, Vol. 100, No. GT7, pp. 705–748.

Nauroy, J. F. and LeTirant, P. (1983). "Model tests on piles in calcareous sands", *Geotechnical Practice in Offshore Engineering*, ASCE Conference, Austin, pp. 356–369.

O'Neill, M. W. and Murchison, J. M. (1983). "An evaluation of p–y relationships in sands", *Report PRAC 82-41-1, Prepared for the American Petroleum Institute*, Houston, May.

Puyuelo, G. J., Sastre, J., and Soriano, A. (1983). "Driven piles in granular carbonate deposit", *Geotechnical Practice in Offshore Engineering*, ASCE Conference, University of Texas, Austin, pp. 440–456.

Ripley, I., Kewlers, A. J. C., and Creed, S. G. (1988). "Conductor load tests", *Engineering for Calcareous Sediments*, Perth, Vol. 2, pp. 429–438.

Roussel, H. J. (1979), *Pile Driving Analysis of Large Diameter High Capacity Offshore Pipe Piles*, Ph.D. Thesis, Department of Civil Engineering, Tulane University, New Orleans.

Semple and Gemeinhardt (1981). "Stress history approach to analysis of soil resistance to pile driving", *Proceedings, 13th Offshore Technology Conference*, OTC 3969, Houston, May 4–7.

Skempton, A. W. (1951). "The bearing capacity of clays", *Proceedings, Building Research Congress*, Institute of Civil Engineers, London, pp. 180–183.

Skempton, A. W. (1957). Discussion on "Planning and design of the New Hong Kong airport", *Proceedings, Institution of Civil Engineers*, London, Vol. 7, pp. 305–307.

Smith, E. A. L. (1962). "Pile driving analysis by the wave equation", Transactions, ASCE, Vol. 127, Part 1, pp. 1145–1193.

Soderberg, L. O. (1962). "Consolidation theory applied to foundation pile time effects". *Geotechnique*, Vol. 12, No. 3, March, pp. 217–225.

Stevens, R. F., Wiltsie, E. A., and Turton, T. H. (1982). "Evaluating pile drivability for hard clay, very dense sand, and rock", *Proceedings, 14th Offshore Technology Conference*, Houston, Vol. 1, pp. 465–481.

Stevens (1988). "The effect of a soil plug on pile drivability in clay", *Proceedings, 3rd International Conference on Application of Stress-Wave Theory to Piles*, Ottawa, Canada, May, pp. 861–868.

Sullivan, R. A. and Ehlers, C. J. (1972). "Practical planning for driving offshore pipe piles", *Proceedings, Fourth Offshore Technology Conference*, Houston, Vol. 1, pp. 805–822.

Terzaghi, K. and Peck, R. B. (1967), "Soil Mechanics in Engineering Practice", (2nd ed.). John Wiley & Sons, p. 729.

Toolan, F. E. and Ims, B. W. (1988). "Impact of recent changes in the API recommended practice for offshore piles in sands and clays". *Underwater Technology*, Vol. 14, No. 1 (Spring 1988), pp. 9–13.29.

Wesselink, B. D., Murff, J. D., Randolph, M. F., Nunez, I. L., and Hyden, A. M. (1988). "Analysis of centrifuge model test data from laterally loaded piles in calcareous sand" *Engineering for Calcareous Sediments*, Perth, Vol. 1, pp. 261–270.

Whelan, T. (1979). "Methane in marine sediments", Lecture Notes for Technical Session Presented to McClelland Engineers, Inc., Houston.

Whittle, A. J. and Baligh, M. M. (1988). *The Behavior of Piles Supporting Tension Leg Platforms*, Final Report Phase III, Constructed Facilities Division, Department of Civil Engineering, Massachusetts Institute of Technology.

Williams, A. F., Dunnavant, T. W., Anderson, S., Equid, D. W., and Hyden, A. M. (1988). "The performance and analysis of lateral load tests on 356 mm dia. piles in reconstituted calcareous sand", *Engineering for Calcareous Sediments*, Perth, Vol. 1, pp. 271–280.

Wiltsie, E. A., Hulett, J. M., Murff, J. D., Hyden, A. M., and Abbs, A. F. (1988). "Foundation design for external strut strengthening system for bass strait first generation platforms", *Proceedings, International Conference of Calcareous Sediments*, Perth, Australia, Vol. 1, pp. 321–330.

Index